Identification of Continuous-Time Systems

International Series on

MICROPROCESSOR-BASED SYSTEMS ENGINEERING

VOLUME 7

Editor

Professor S. G. Tzafestas, *National Technical University, Athens, Greece*

Editorial Advisory Board

Professor C. S. Chen, *University of Akron, Ohio, U.S.A.*
Professor F. Harashima, *University of Tokyo, Tokyo, Japan*
Professor G. Messina, *University of Catania, Catania, Italy*
Professor N. K. Sinha, *McMaster University, Hamilton, Ontario, Canada*
Professor D. Tabak, *Ben Gurion University of the Negev, Beer Sheva, Israel*

The titles published in this series are listed at the end of this volume.

Identification of Continuous-Time Systems

Methodology and Computer Implementation

edited by

N. K. SINHA

McMaster University, Hamilton, Ontario, Canada

and

G. P. RAO

Indian Institute of Technology, Kharagpur, India

KLUWER ACADEMIC PUBLISHERS

DORDRECHT / BOSTON / LONDON

Library of Congress Cataloging-in-Publication Data

Identification of continuous-time systems : methodology and computer implementation / edited by N.K. Sinha, G.P. Rao.
p. cm. -- (International series on microprocessor-based systems engineering ; v. 7)
Includes bibliographical references and index.
ISBN 0-7923-1336-4 (HB : alk. paper)
1. Automatic control--Congresses. 2. System identification--Congresses. 3. Microprocessors--Congresses. I. Sinha, N. K. (Naresh Kumar), 1927- . II. Prasada Rao, Ganti, 1942- . III. Series.
TJ212.2.I3245 1991
629.8--dc20 91-20058
CIP

ISBN 0-7923-1336-4

Published by Kluwer Academic Publishers,
P.O. Box 17, 3300 AA Dordrecht, The Netherlands.

Kluwer Academic Publishers incorporates
the publishing programmes of
D. Reidel, Martinus Nijhoff, Dr W. Junk and MTP Press.

Sold and distributed in the U.S.A. and Canada
by Kluwer Academic Publishers,
101 Philip Drive, Norwell, MA 02061, U.S.A.

In all other countries, sold and distributed
by Kluwer Academic Publishers Group,
P.O. Box 322, 3300 AH Dordrecht, The Netherlands.

Printed on acid-free paper

Printed in the Netherlands

Preface

In view of the importance of system identification, the International Federation of Automatic Control (IFAC) and the International Federation of Operational Research Societies (IFORS) hold symposia on this topic every three years. Interest in continuous time approaches to system identification has been growing in recent years. This is evident from the fact that the number of invited sessions on continuous time systems has increased from one in the 8th Symposium that was held in Beijing in 1988 to three in the 9th Symposium in Budapest in 1991.

It was during the 8th Symposium in August 1988 that the idea of bringing together important results on the topic of Identification of continuous time systems was conceived. Several distinguished colleagues, who were with us in Beijing at that time, encouraged us by promising on the spot to contribute to a comprehensive volume of collective work. Subsequently, we contacted colleagues all over the world, known for their work in this area, with a formal request to contribute to the proposed volume. The response was prompt and overwhelmingly encouraging. We sincerely thank all the authors for their valuable contributions covering various aspects of identification of continuous time systems.

In view of the present scarcity of such publications, it is hoped that this volume would serve as a useful source of information to researchers and practising engineers in the areas of automatic control and operations research. An attempt has been made to make the style as uniform as possible in spite of the diversity of the authorship, and all chapters contain sufficient background material for independent reading.

We thank the publishers for their cooperation and encouragement. Special thanks are due to Professor Spyros Tzafestas, who is the Editor of the Kluwer International Series on Microprocessor-based Systems Engineering and to Dr. Nigel Hollingworth, Acquisitions Editor for Kluwer Academic Publishees for their help.

July 1991

Naresh K. Sinha
Ganti Prasada Rao

CONTRIBUTORS

Bapat, V.N.	Walchand College of Engineering, Sangli, India
Bingulac, S	Virginia Polytechnic Institute and State University, Blacksburg, VA,
Bohn, E.V.	University of British Columbia, Vancouver, Canada
Boje, E.	University of Durban, South Africa
Chotai, A.	Lancaster University, Lancaster, U.K.
Cooper, D.C.	Virginia Polytechnic Institute and State University, Blacksburg, VA,
Dai, H.	McMaster University, Hamilton, Canada
Datta, K.B.	Indian Institute of Technology, Kharagpur, India
de Moor, B.	ESAT Katholieke Universiteit Leuven, Belgium
Gerencsér, L.	McGill University, Montréal, Canada
Goodwin, G.C.	University of Newcastle, Australia
Jiang, Z.H.	E.T.H., Zurich, Switzerland
Lastman, G.J.	University of Waterloo, Waterloo, Canada
Matko, D.	University of Ljubljana, Yugoslavia
Moonen, M.	ESAT Katholieke Universiteit Leuven, Belgium
Mukhopadhyay, S.	Indian Institute of Technology, Kharagpur, India
Ninness, B.M.	University of Newcastle, Australia
Palanisamy, K.R.	Government Engineering College, Salem, India
Patra, A	Indian Institute of Technology, Kharagpur, India
Pintelon, R.	Vrije Universiteit Brussel, Belgium
Rao, G.P.	Indian Institute of Technology, Kharagpur, India
Roy, B.K.	Regional Engineering College, Silchar, India
Sagara, S.	Kyushu University, Japan
Saha, D.C.	Indian Institute of Technology, Kharagpur, India
Schaufelberger, W.	E.T.H., Zurich, Switzerland
Schoukens, J.	Vrije Universiteit Brussel, Belgium
Sinha, N.K.	McMaster University, Hamilton, Canada
Tych, W.	Lancaster University, Lancaster, U.K.
Tzafestas, S.	National Technical University, Athens, Greece
Unbehauen, H.	Ruhr-University, Bochum, Germany
Vágó, Zs.	Technical University of Budapest, Hungary
Van den Bos, A.	Delft University of Technology, The Netherlands
Van hamme, H.	Vrije Universiteit Brussel, Belgium
Vandewalle, J.	ESAT Katholieke Universiteit Leuven, Belgium
Wang, S-Y.	East China Institute of Technology, Shanghai, China
Young, P.C.	Lancaster University, Lancaster, U.K.
Zhao, Z.Y.	Kyushu University, Japan

CONTENTS

Continuous-time models and approaches

Ganti Prasada Rao
Department of Electrical Engineering
Indian Insititute of Technology
Kharagpur (W.B.) 721302, India

and

Naresh K. Sinha
Department of Electrical & Computer Engineering
McMaster University
Hamilton, Canada L8S 4L7

Abstract

In this introductory chapter we distinguish between the discrete time (discrete-time) and continuous time approaches. We point out situations where the latter are preferable by drawing attention to certain problems associated with the conventional shift operator, z. Some viable alternatives for a continuous-time treatment are given. This is followed by an outline of the contents and organization of the various contributions to this book.

1. Introduction

The development of methods and tools to handle scientific and technological problems is naturally guided by happenings in fields of parallel and simultaneous growth. In order to exploit the highly developed facilities in those allied fields, the characterization of problems in a given field is, for obvious reasons, tailored to match the nature of those allied fields. An example of such tailoring action may be clearly seen in the case of systems, signals and control. Despite the continuous-time character of most real world physical processes, totally discrete-time descriptions of such processes are usually chosen as the sole basis of operations. Discrete time descriptions of systems and signals offer the following conveniences:

- Reduction of the calculus of continuous-time processes into algebra, that is, mathematics devoid of derivatives and integrals.

- Ease in implementation of dynamic strategies in real time in digital data processing systems. The basic block in simulation of dynamic strategies is the delay element, together with those performing simple arithmetic.

N. K. Sinha and G. P. Rao (eds.), Identification of Continuous-Time Systems, 1–15.

- The support of a well established theory in the discrete-time domain in deterministic and stochastic situations for systems, signals and control.

Taking advantage of these conveniences and the great advances in technology, the attention received by the discrete-time methods is so enormous that the continuous-time counterpart was completely overshadowed until a few years ago. In fact, people who talked of continuous-time treatment were considered either as "old timers of the analog age" or "those having academic interest only". However, the situation is gradually changing nowadays as the relevance of continuous-time treatment has been established (Harris and Billings 1981, Gawthrop 1982, Unbehauen and Rao 1987, Rao 1983, Saha and Rao 1983, Sagara and Zhao 1987) in a number of practical cases. The few early advocates of continuous-time treatment gave the following points in its favour:

- Modelling of physical systems is based on laws established in continuous-time, for example, Newton's laws, Faraday's laws etc. In the resulting differential equations, the coefficients are closely related to the corresponding physical parameters in the system. This makes a sound "physical" understanding of the related problems possible. Since there are no equivalent physical laws in discrete-time, models of physical systems are not *born* digital.

- Most of those concerned with systems, signals and control had their early education and understanding of the subject through the medium of the continuous-time domain.

- Early developments in the field itself were continuous-time-based. For example, Laplace transforms and frequency response methods were used in systems and control, while classical filtering techniques were used in signal processing.

Attempts by some researchers in the 1970's to advocate the use of continuous-time models for identification appear to have made little impact on the large community of researchers and practising engineers deeply involved with the extensive use of discrete-time treatment only. The question of compatibility (on the surface) of the continuous-time treatment within the environment of digital technology also appears to have acted as a deterrent to the case. This is because of the following presumptions:

- Signals and systems have to be inevitably discretized in the time domain.

- Certain time-domain operations (e.g. differentiation, integration, etc.) which are essential in the continuous-time description of dynamical systems cannot be truly realized in the digital environment. Such operations naturally suited analog devices and systems, which are, of course, the best in real time.

- The environment of analog devices and systems, despite its natural suitability to continuous-time treatment and real-time operation, is subjected to uncertainties in its own setup due to physical parameter drifts, noise etc. The digital counterpart is, however, free from these. It has, on the other hand, numerical effects due to certain factors such as word length, truncation and roundoff. Such factors are usually modelled by stochastic methods.

- Analog devices and systems are not as versatile as their digital counterparts in "executing a sequence of operations". The mathematics of decision making, being inherently discrete in nature, is not easily realizable on analog hardware.

- Incorporation of features of artificial intelligence (AI) on purely analog hardware is not easy.

Consequently, at this stage it may appear as if there is hardly anything left in favour of a continuous-time treatment. This is, however, not true as will be seen in the following pages. In order to make the setting of our discussions clear, we recall the following:

- In the language of mathematics, systems and signals are not distinguishable from each other and a description in continuous-time (of a signal or a system) transformed into discrete-time is acceptable only under certain conditions. The discrete-time form results in a loss of information during the sampling interval.

- The discrete-time form is uniquely obtained for a continuous-time version. The reverse is not true. To get back to the continuous-time form requires certain assumptions on the information during the sampling interval.

- The discrete-time forms (without hold devices) of continuous-time descriptions are not usually treated as function approximations of the latter. They are “transforms”. In place of L_2 function space theories, spectral theories are used.

- Continuous-time and discrete-time information may be transformed naturally into s and z (complex) domains, respectively. These domains are connected by the transcendental relation

$$z = e^{sT}$$

  where T is the sampling interval.

It is generally agreed that a problem has to be "discretized" for handling by a digital computer. With the advent of advanced programming languages (e.g., LISP), analytical expressions can also be derived as the output of a program. This removes the commonly felt limitation that problems can be handled by a digital computer only in their discrete-time form. For the purpose of real time processing, however, the continuous-time variable has to be either lumped or discretized for sequential release of information over the lumps (or instants of time). If discretization is not limited to the concerned domain alone, there are situations in which a problem can be discretized leaving the domain of the problem in the continuous form. To clarify this, consider a linear time-invariant causal system having *y(t)* and *u(t)* as the output and input, respectively. The input-output relation may be modelled by the nondiscrete form (continuous-time)

$$\left.\begin{aligned} y(t) &= \int_0^t g(\tau)u(t-\tau)\,d\tau \\ \text{or,}\quad y(t) &= \int_0^t g(t-\tau)u(\tau)\,d\tau \end{aligned}\right\} \qquad (1)$$

Alternatively, if the model is parameterized as, say

$$\frac{dy(t)}{dt} + ay(t) = bu(t) \qquad (2)$$

we still have a continuous-time form. However, it may be seen that equation (2) is in terms of two parameters a and b. Notice that for actual processing, we have to reduce a problem to the finite-dimensional form. Thus equation (1), the non-parametric or infinite dimensional problem has been reduced to finite dimensional form and equation (2) is such a reduction. This is a model of two parameters (but order one). It is also possible to resort to time discretization of equation (1). The resulting number (dimension) of "parameters" in the (discrete-time) form depends on the sampling interval itself. On the other hand, the finite dimensional model (2) of (1) has a fixed dimension and number of parameters. In spite of being finite dimensional in this sense, these continuous-time descriptions still pose *infinite dimensionality* to digital processors due to the presence of continuous-time functions and their derivatives. Some kind of discretization of the time variable in the parameterized continuous-time form is required to suit the digital environment. The widely used shift operator (z-transform) form is the result of one standard process of discretization. The term discrete-time is associated with such a form. There are other methods in which time discretization takes place, but the parameters in the resulting form either remain the same or retain close links with them. For this reason, despite the inevitably introduced discretization for digital processing, such a treatment is viewed as one on continuous-time basis. The shift operator (z-transform) form alone is referred to as the discrete-time version and the related methods in the z-plane are called discrete-time approaches.

2. Time discretization and the *z*-plane

As stated above, our main concern is the effect of discretizing a system model using the *z*-transform. Although the concept of mapping between the *s*-plane and the *z*-plane is intuitively appealing, it causes some problems, as described in this section.

Mapping from the *s*-plane to the *z*-plane:

The *z*-plane is a transformation of the *s*-plane through the transcendental relation: $z = e^{sT}$, where T is the sampling interval. The left and the right halves of the *s*-plane are mapped respectively into the regions inside and outside the unit circle centred at the origin of the *z*-plane. The $j\omega$-axis of the *s*-plane maps into the unit circle of the *z*-plane. In practice, the sampling frequency must be much higher than the bandwidth of the system. One rule of thumb is to select T such that

$$\lambda_m T \leq 0.5,$$

where λ_m is the magnitude of the largest eigenvalue of the system. In practice, it is often desirable to make the sampling interval much smaller than the value suggested by the above equality. The result of such a choice is to force all the poles to lie in a small lens-shaped region in the *z*-plane, even though they may be far apart in the *s*-plane. This is illustrated in the following figure.

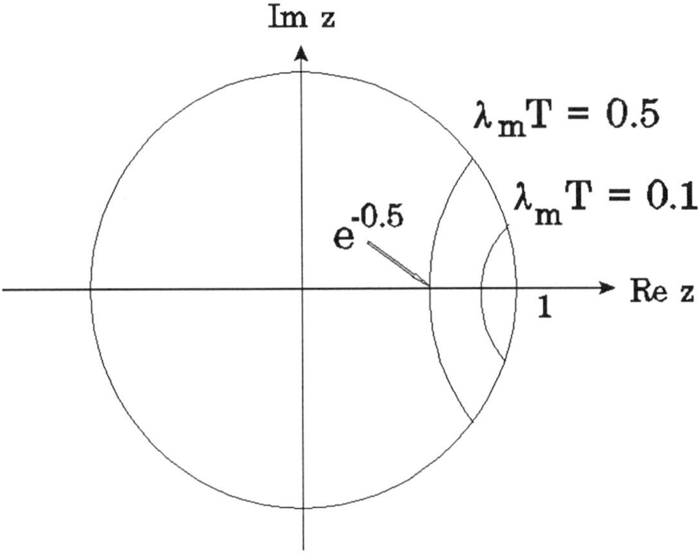

Figure 1 The region of normal operation in the *z*-plane

Due to the very nature of the mapping, the entire left half of the s-plane is squeezed into the unit circle of the z-plane. As $T \to 0$, the point (1,0) in the z-plane becomes the target of convergence for the working area of the s-plane, very much like a "black hole". The working area consequently reduces to the close neighbourhood of the black hole giving rise to numerical ill-conditioning in the z-plane. The aim of some continuous-time-like approaches is mainly to circumvent this problem.

Mapping of transfer functions of linear time-invariant systems from the s-plane to the z-plane :

Poles and zeros are important parameters in the transfer functions of linear time-invariant systems. The poles p_i of a continuous-time system are transformed to $e^{p_i t}$ in the z-plane. Unfortunately, the zeros are not related through such a simple transformation, and the relationship is further complicated by the type of hold circuit used in the sampling process. In other words, with reference to poles and zeros as points in the s- and z-planes, the point-to-point mapping between the two planes cannot be applied in general, without considering the properties of the hold circuit and the "type" of the point.

Åström et al. (1984) have discussed the placement of the s-plane zeros into the z-plane when the sampling process contains a zero-order hold. A strictly proper continuous-time rational function $G(s)$ transformed into $H(z)$ in the z-plane remains rational and possesses generically $n-1$ zeros, where n is the number of poles. These zeros cannot be expressed in a closed form in terms of the s-plane parameters and T. If $G(s)$ is an m-zero n-pole transfer function with zeros at z_i and poles at p_i, $m < n$, then as $T \to 0$, the $n-1$ zeros of $H(z)$ in two groups approach two limits as given below:

Group 1: m zeros approach 1 as $e^{z_i T}$.
Group 2: n-m-1 zeros approach the zeros of the polynomial

$$B_{n-m}(z) = \sum_{i=1}^{n-m} b_i^{n-m} z^{n-m-i}$$

where, $$b_k^{n-m} = \sum_{\ell=1}^{k} (-1)^{k-\ell}\, \ell^{n-m} \begin{pmatrix} n-m+1 \\ k-\ell \end{pmatrix}, \quad k = 1, 2, \cdots, (n-m).$$

For n-$m \geq 2$, the zeros of group 2 lie on or outside the unit circle. Thus, there are continuous-time systems with left half plane zeros whose pulse transfer functions have zeros outside the unit circle.

The influence of fractional delay on the zeros in the z-plane:

Since a discussion based on a very general expression is likely to be complex, let us consider the problem of finding $H(z)$ corresponding to $e^{-s\tau}/(s+a)$, $0 < \tau \leq T$, incuding a zero-order hold. Here, we have

$$G_{ZOH}(s)\,G(s) = \frac{1-e^{-Ts}}{s} \times \frac{e^{-s\tau}}{s+a}$$

$$= (1-e^{-Ts})e^{-Ts} \times \frac{e^{s(T-\tau)}}{s(s+a)}$$

Partial fraction expansion leads to

$$\frac{e^{s(T-\tau)}}{s(s+a)} = \frac{1}{as} - \frac{e^{a(\tau-T)}}{a(s+a)} \Rightarrow \frac{1}{a(1-z^{-1})} - \frac{e^{a(\tau-T)}}{a(1-qz^{-1})}$$

where $q = e^{-aT}$. Recombining and multiplying by $(1 - z^{-1})z^{-1}$, we obtain

$$H(z) = z^{-1}\frac{b_0 + b_1 z^{-1}}{a(1-qz^{-1})}$$

where $b_0 = 1 - e^{a(\tau-T)}$

and $b_1 = -q + e^{a(\tau-T)} = e^{-aT}(e^{a\tau}-1)$ [Note that $b_1 = 0$ if $\tau = 0$].

The zero of $H(z)$ is seen to be

$$z_1 = -\frac{b_1}{b_0} = \frac{\tau}{\tau - T}, \qquad \text{for small } T$$

For $\tau < T/2$, this lies within the unit circle disc, but moves outside for the case when $\tau > T/2$.

Thus, due to fractional delay arising out of the sampling process, zeros of $H(z)$ occur in the region outside the unit circle in the z-plane.

Sensitivity of poles in the z-plane:

The i-th pole in the s-plane is mapped into the z-plane to be placed at $q_i = e^{p_i T}$. As $T \rightarrow 0$, $q_i \rightarrow 1$. The poles in the s-plane are, thus, mapped into a close cluster near $z = 1$ in the z-plane. Due to the effect of finite precision in computation, this clustering leads to numerical ill-conditioning in the calculations for the design of a controller. In order to examine this, consider monic polynomials $A(s)$ and $A^*(z)$. Let a_i and a_i^* be the coefficients of s^i and z^i in A and A^*, respectively. Let the poles of A and A^* be p_i and q_i, $i = 1, 2, \ldots, n$, respectively.

Let $$A_i(s) = \frac{A(s)}{s-p_i}, \quad \boldsymbol{p}^T = [p_1\ p_2\ \cdots\ p_n] \quad \text{and} \quad \boldsymbol{a} = [a_1\ a_2\ \cdots\ a_n]^T$$

Similarly, let

$$A_i^*(z) = \frac{A^*(z)}{z-q_i}, \quad \boldsymbol{q}^T = [q_1\ q_2\ \cdots\ q_n] \quad \text{and} \quad \boldsymbol{a}^* = [a_1^*\ a_2^*\ \cdots\ a_n^*]^T$$

Then,

$$J = \frac{\partial \boldsymbol{a}}{\partial \boldsymbol{p}} = -[a_{ki}] = [J_{ik}]$$

{Note that a_{ki} is the coefficient of s^{k-i} in $A_i(s)$}

$$J^{-1} = -\left[p_i^{k-i} \prod_{l \neq i}^{n} (p_i - p_l)^{-1}\right] = \left[(J^{-1})_{ik}\right] \tag{3}$$

similarly,

$$J^* = \frac{\partial \boldsymbol{a}^*}{\partial \boldsymbol{q}} = -\left[a_{ki}^*\right] = \left[J_{ik}^*\right]$$

{Note that a_{ki}^* is the coefficient of z^{k-i} in $A_i^*(z)$}

$$J^{*-1} = -\left[q_i^{k-i} \prod_{\ell \neq i}^{n} (q_i - q_\ell)^{-1}\right] \tag{4}$$

For proof see Mantey (1968).

Recalling that $q_i = e^{p_i T}$, we may write this as

$$J^{*-1} = -\left[e^{p_i (k-i) T} \prod_{l \neq i}^{n} (q_i - q_l)^{-1}\right]$$

If $T \to 0$, we may write

$$J^{*-1} = -\left[(1 + p_i(k-i)T) \prod_{\ell \neq i}^{n} (1 + p_i T - 1 - p_\ell T)^{-1}\right]$$

$$= -\frac{1}{T^{n-1}}\left[(1 + p_i(k-i)\,T) \prod_{\ell \neq i}^{n} (p_i - p_\ell)^{-1}\right] = \left[(J^*)_{ik}^{-1}\right] \tag{5}$$

For small vector changes $\tilde{a}^*$ and $\tilde{q}^*$

$$\tilde{q} = \tilde{J}^{*^{-1}} \tilde{a}^* \tag{6}$$

Notice that as $T \rightarrow 0$, $J^{*^{-1}}$, tends to develop identical columns, thereby tending to be singular. This phenomenon is numerically undesirable. In terms of the s-plane poles, as $T \rightarrow 0$,

$$\left[J^{*^{-1}}\right] = -\frac{1}{T^{n-1}} \prod_{l \neq i}^{n} (p_i - p_l) \tag{7}$$

and since $$q_i = e^{p_i T},$$

$$\frac{dq_i}{dp_i} = Te^{p_i T} \rightarrow 0 \tag{8}$$

This shows that q_i become increasingly insensitive to p_i forming a cluster at the point (1,0) in the z-plane. This phenomenon is also numerically undesirable.

3. Other viable alternatives

In view of the undesirable sensitivity and numerical problems associated with the manipulation of numbers in the z-plane, other possible alternatives have been considered. Some of these are aimed at improved numerical conditioning and maintenance of harmony of the discrete-time description with the continuous-time counterpart as $T \rightarrow 0$. Others are intended solely to leave the system parameters undisturbed in their s-domain values. These are briefly outlined below.

The delta operator :

The delta operator is expressed in terms of z (Gupta 1966) as

$$\delta = \frac{z-1}{T} \tag{9}$$

This may be seen to be the Euler approximation for the derivative. When the coefficients of the system are handled in the δ-plane, the numerical conditioning improves. At first the cluster problem is resolved as $z = 1$ at high rate of sampling (for $T \rightarrow 0$) maps itself into $\delta = 0$ and division by T provides appropriate zooming in the area. The hardware or software necessary to implement δ as the building block is only marginally more complex than the block implementing the shift operation z^{-1}. Middleton and Goodwin (1986) have demonstrated improvement in finite word length performance of the δ-operator. They have also shown how, as $T \to 0$, the results approach those in continuous-time. The so called δ-transform, defined as

$$F_T(\delta) = TF_z(1+T\delta) = T\sum_{i=0}^{\infty} f(kT)(1+T\delta)^{-k} \tag{10}$$

converges to the usual Laplace transform as $T \to 0$.

The notion of the δ operator may be seen in numerical analysis as the first divided difference operator. In the control literature, it is used to relate continuous-time and discrete-time transfer functions. The numerical properties of similar operators have been studied in the context of digital filtering (Agarwal and Burrus 1975, Orlandi and Martinelli 1984) and parameter estimation (Edmumds 1985).

Signal characterization in terms of orthogonal functions:

When the signals in continuous-time dynamical systems are expressed in terms of orthogonal functions, the time-domain operators such as differentiation, integration, delay, etc., are approximated by operational matrices in the sense of least squares (Unbehauen and Rao 1987, Rao 1983). The system parameters are retained in their original continuous-time form (s-domain). For instance if

$$y(\mathrm{t}) = \mathbf{y}^{\mathrm{T}}\,\boldsymbol{\Theta}(\mathrm{t}),$$

where

$\mathbf{y} = [y_1 \;\; y_2 \cdots \; y_{\mathrm{m}}]^{\mathrm{T}}$, the vector of spectral components

$\boldsymbol{\Theta}(\mathrm{t}) = [\theta_1(\mathrm{t}) \;\; \theta_2(\mathrm{t}) \;\; \cdots \; \theta_{\mathrm{m}}(\mathrm{t})]^{\mathrm{T}}$, the vector of orthogonal basis functions,

and $y_i = \langle y(\mathrm{t}),\;\; \theta_{\mathrm{i}}(\mathrm{t})\rangle$ (inner product)

then

$$\int_0^t y(t)\,dt \approx \mathbf{y}^T \boldsymbol{E}\,\boldsymbol{\Theta}(t)$$

in which $\boldsymbol{E}$ is the so called operational matrix for integration. Using such operational matrices, the continuous-time domain operations specified in a system description can be realized in terms of the spectral components. The coefficients or parameters related to a continuous-time description are left untouched, while all the operations are performed on the signals themselves. Of the various systems of orthogonal functions, block pulse functions (BPF) have made real-time operations possible (Patra and Rao 1989c). Recently a framework of general hybrid orthogonal functions (GHOF) has been introduced by Patra and Rao (1989a, b) as a flexible toolbox for approximating continuous-time signals and systems. The use of GHOF has been illustrated in analysis and identification of systems in which signals possess mixed features of continuity andjumps. Block-pulse functions, which are shown to arise out of GHOF as a special case, have been used by Patra and Rao (1989c) in a self tuning control algorithm which is completely continuous-time model based.

Numerical integration formulae realized through digital filters in system identification

Sagara and Zhao (1987) convert models of dynamical systems into integral equations and realize the related signal integral in real-time through integrating filters, exploiting the well developed digital filtering techniques. In this process the parameters of the system are again left undisturbed in their original continuous-time form. Incidentally, their work is in identification of continuous-time systems. Their methods are based on linear filtering of process data. The input and output signals are subjected to a linear dynamic operator $\mathbb{R}_{\mathcal{L}\mathrm{D}}$ before use. For instance, state variable filters, Poisson filters (Saha and Rao 1983), chains of integrators, etc., are realizable in digital form but the main work is addressed to continuous-time systems.

Thus, discrete-time characterization in systems and signals clearly shows certain problems which are mainly numerical in nature. The source of these problems is in the very nature of the link between s and z domains. A transcendental transformation which squeezes an entire half space of the s-plane into a unit disc in the z-plane gives rise to numerical ill-conditioning of the related computations. Methods aimed at retaining the parameters either in the original continuous-time domain or close to it (e.g., δ-operator) are discussed in the book. An actual implementation can be in the usual discrete-time domain, although the other manipulations necessary in design, etc., on the system parameters are recommended to be performed in continuous-time domain. Some recent efforts in this direction are also briefly reviewed. The authors are of the opinion that approaches to the solution of problems should be chosen to suit the actual situation rather than by following what is widely in vogue.

4. Identification of continuous-time models from sampled data

The identification of continuous-time multivariable systems is a very important problem in Control Engineering and has been the subject of a great deal of research. One approach, called the "indirect method" is to utilize the samples of the input-output data to first estimate a discrete-time model and then determine an equivalent continuous-time model (Sinha and Lastman 1982). The main attraction of this approach is the decomposition of the original problem to two simpler problems. Several well-known methods can be used to determine a "good" discrete-time model with the help of a digital computer. The second part of the problem, i.e., obtaining an equivalent continuous-time model is deterministic and is solved rather easily. It is important to note that the results obtained by using this approach are highly dependent on the choice of the sampling interval for obtaining the data (Sinha and Puthenpura 1985, Puthenpura and Sinha 1985).

As shown in earlier work, difficulties are caused whenever the sampling interval is either too large or too small. Whereas a large sampling interval leads to loss of information, making it very small creates numerical problems due to the fact that all the poles are constrained to lie in a very small region of the z-plane. The latter difficulty can be largely overcome by using the δ-operator for representing the discrete-time model which can be identified from the samples of the input-output data, followed by a suitable procedure for determining the corresponding continuous-time model. Another difficulty is caused by the fact that the basic assumption, that the input is held constant between the sampling instants, may not be satisfied in practice. In such cases, often a better approximation is obtained by assuming that the input varies linearly between the sampling instants (Bingulac and Sinha 1989).

It follows that there is a great need for an alternative method for identifying continuous time system. Several chapters of this book will be devoted to some of these so-called "direct" methods for identification of continuous time systems for the samples of the data, without the need for identifying an intermediate discrete time model.

5. Scope of the book

This book is aimed towards researchers and practising engineers who are interested in modelling and control of continuous-time systems. Since most physical systems are of the continuous-time nature, it is worthwhile to investigate various methods for obtaining the continuous-time model. Some of these may utilize discrete-time techniques using the samples of input-output data, but eventually these lead to continuous-time models in most practical situations.

The remaining 19 chapters of the book can be divided into five main groups. Chapters 2, 3 and 4 belong to the first group and attempt to present the relationships between continuous-time and discrete-time models. Chapter 2 describes a general framework for discrete-time modeling and identification of continuous-time systems while retaining the system parameters with their physical interpretation. The relationship between discrete-time and continuous-time estimation is discussed in chapter 3, which also describes the use of the delta operator and shows how it is much better than the shift operator for obtaining discrete-time models for continuous-time systems. Chapter 4 is concerned with the transformation of discrete-time models to equivalent continuous-time models. It is assumed that the sampling interval has been selected carefully so that a suitable discrete-time model has been obtained from the samples of the input-output data. The treatment here is general and applies to multivariable linear system. Transformation of discrete-time models obtained using the delta-operator is also described.

Chapters 5, 6, 7, and 8 are concerned with the use of orthogonal functions for identification of continuous-time systems from the samples of input-output data. Chapter

5 describes methods for Walsh coefficient evaluation of integral functions and their application to system parameter estimation. Chapter 6 presents the use of the block-pulse operator for the identification of continuous-time systems from samples of input-output data. Application to the identification of nonlinear and distributed-parameter systems are also given. This is followed by a discussion of the application of a recursive block-pulse function method for direct estimation of the parameters of a single-input single-output continuous-time system from the samples of the input-output data. The method can also be applied to the identification of multi-input multi-output linear systems, linear systems with time delays an Hammerstein model nonlinear systems. Chapter 8 describes identification using different types of orthogonal polynomials. These include Legendre, Laguerre, Hermite and Tchebyscheff polynomials. Both lumped-parameter and distributed-parameter continuous-time systems are considered.

Chapters 9, 10, 11, and 12 are concerned with numerical techniques for direct identification of the parameters of continuous-time systems. Chapter 9 describes the use of numerical methods of integration. The state equations of a linear system are integrated numerically, using the samples of the input-output data. These are then utilized for estimating the parameters of a canonical model of the system. The effect of noise on the data is also investigated. Chapter 10 describes the use of digital pre-filters for obtaining a discrete-time identification model which retains the continuous-time model parameters. It has been shown that if the pre-filters are properly designed, even the least-squares algorithm gives satisfactory results if the noise level is low, otherwise bias-eliminating algorithms, such as instrumental variable methods should be used. Chapter 11 presents some new results on the use of the Poisson moment functional technique for combined parameter and state estimation in a linear time-invariant system based on a recursive least squares algorithm, which can be extended to the case of time-varying systems. A recursive instrumental variable algorithm, based on Poisson moment functionals, is developed to reduce the bias in the estimates if the noise level is high. Chapter 12 describes identification, estimation and control of continuous-time systems described by delta operator models.

Chapters 13, 14, and 15 concentrate on the identification of multivariable continuous-time systems. Chapter 13 presents a parameter estimation algorithm for such systems and shows how to treat systems with non-zero initial conditions and measurement offset or bias, using a minimal order input-output representation. Chapter 14 describes a discrete identification method for identifying linear continuous-time multivariable systems that does not require structural identification. Chapter 15 presents subspace methods for identification of such systems that are based on singular value decomposition, a numerical technique that is known to be very robust and accurate when dealing with noisy data. These state space methods are viewed as the better alternatives to polynomial model identification, owing to the superior numerical conditioning obtained, especially for high-order multivariable systems.

Chapters 16 to 19 describe other approaches to the identification of continuous-time systems. These include the use of multiharmonic test signals (Chapter 16), adaptive model approaches (Chapter 17), nonparametric approaches (Chapter 18), stochastic systems (Chapter 19).

Finally, Chapter 20 discusses the issues in real time identification of continuous-time systems. Starting with some aspects of the form of the plant model, measurement systems and preprocessing schemes, algorithms for model estimation are considered. Further postprocessing techniques for parameter estimates and residual are examined for special applications like adaptive control, fault detection and condition monitoring. Salient features of hardware and software for practical implementation are discussed briefly and a practical example of real-time parameter estimation is presented.

References

Agarwal, R.C. and Burrus,C.S. (1975), "New recursive digital filter structures having very low sensitivity and roundoff noise", *IEEE Trans. Circuits and Systems*, Vol. CAS-22, pp. 921-927.

Åström, K.J., Hagander, P. and Sternby, J. (1984), "Zeros of sampled systems", *Automatica*, Vol.20. pp. 31-38.

Bingulac, S. and Sinha, N.K. (1989), "On the identification of continuous time multivariable systems from samples of input-output data", Proc. Seventh Int. Conf. on Mathematical and Computer Modelling, (Chicago, Ill, August 1989), pp. 203-208.

Edmunds, J.M. (1982), "Identifying sampled data systems using difference operator model", Report UMIST, CSC-601, Manchester, U.K.

Gawthrop, P.J. (1982), "A continuous time approach to self tuning control", *Optimal Control-Applications and Methods*, Vol. 3, pp. 394-414.

Gupta, S.C. (1966), "*Transform and State Variable Analysis in Linear Algebra*", John Wiley and Sons, New York

Harris, C.J. and Billings, S.A. (1981), "*Self Tuning and Adaptive Control: Theory and Applications*", Peter Peregrinus, Stevenage, U.K.

Mantey, P.E. (1968), "Eigenvalue sensitivity and state variable selection", *IEEE Trans. Auto. Control*, Vol. AC-13, pp. 263-269.

Middleton, R.H. and Goodwin, G.C. (1986), "Improved finite word length characteristics in digital control using delta operator", *IEEE Trans. on Auto. Control*, Vol. AC-31, pp. 1015-1021.

Orandi, G. and Martinelli, G. (1984), "Low sensitivity recursive digital filters obtained via delay replacement", *IEEE Trans. Circuits and Systems*, Vol. CAS-31, pp. 654-657.

Patra, A. and Rao, G.P. (1989a), "General hybrid orthogonal functions and some potential applications in systems and control", *IEE Proc., Part D.*, Vol. 136, pp. 157-163.

Patra, A. and Rao, G.P. (1989b), "General hybrid orthogonal functions - A new tool for analysis of power electronic systems", *IEEE Trans. Ind. Elec.*, Vol. IE- 36, pp. 413-424.

Patra, A. and Rao, G.P. (1989c), "Continuous time approach to self tuning control: Algorithm, implementation and assessment", *IEE Proc., Pt-D*, Vol. 136, pp. 333-340.

Puthenpura, S. and Sinha, N.K. (1985), "A procedure for determining the optimal sampling interval for system identification using a digital computer", *Can. Elec. Eng. J.*, Vol. 10, pp. 152-157.

Rao, G.P. (1983), "*Piecewise Constant Orthogonal Functions and Their Application to Systems and Control*", Springer Verlag, Berlin.

Sagara, S. and Zhao, Z.Y. (1987), "On-line identification of continuous systems using linear integral filter", 19th JAACE Symp. on Stochastic systems theory and its applications, (Fukuoka, Japan), pp. 41-46.

Saha, D.C. and Rao, G.P. (1983), "*Identification of Continuous Dynamical Systems: The Poisson Moment Functional (PMF) Approach*", Springer Verlag, Berlin.

Sinha, N.K. and Lastman, G.J. (1982), "Identification of continuous time multivariable systems from sampled data", *Int. J. Control*, Vol. 35, pp. 117-126.

Sinha, N.K. and Puthenpura, S. (1985), "Choice of sampling interval for the identification of continuous-time systems from samples of input output data", *IEE Proceedings, Pt.D.*, Vol. 132, pp. 263-267.

Unbehauen, H. and Rao, G.P. (1987), "*Identification of Continuous Systems*", North Holland, Amsterdam.

Discrete-time modeling and identification of continuous-time systems: a general framework¶

H. Van hamme, R. Pintelon and J. Schoukens
Vrije Universiteit Brussel
Department ELEC
Pleinlaan, 02
1050 Brussels, Belgium

Abstract

A general framework for modeling of a time-varying continuous-time SISO system from its sampled input and output while retaining the system parameters with their physical interpretation is presented. The theory can be specialized to the Poisson moment functional approach, the integrated sampling approach, the instantaneous sampling approach or the use of state variable filters. In all methods, the initial conditions can be removed by applying an appropriate discrete-time operator. Digital filtering is used to explicitly or implicitly reconstruct the time-derivatives of the sampled continuous-time signals involved. A thorough study of the approximations resulting from converting the continuous-time model to a discrete-time version is presented. It is shown how these errors can be controlled and that system parameter estimates can be obtained with an arbitrary accuracy. The digital filtering approach to integrated sampling and instantaneous sampling exhibit optimal properties among all methods that fit in the general framework. Moreover, using digital filter designs instead of numerical integration allows slower sampling. A maximum likelihood estimator is derived for time-invariant systems in an errors-in-variables stochastic framework. Finally, the theory is verified through simulations.

Key words–continuous-time systems, discrete-time models, time-varying systems, Poisson moment functionals, integrated sampling, instantaneous sampling, state variable filters, errors-in-variables, total least squares estimator, maximum likelihood estimator.

1. Introduction.

The basic problem of parametric continuous-time system identification from sampled data is the correct handling of the time-derivatives. Many methods to circumvent the need to reconstruct these derivatives have been devised. A first

---

¶ This work is supported by the Belgian National Fund for Scientific Research (NFWO) and the Flemish Community (concerted action IMMI).

N. K. Sinha and G. P. Rao (eds.), Identification of Continuous-Time Systems, 17–77.

possibility is the use of special excitation signals. The plant is excited with a known input of a special form and the transient response is measured at the output. Special choices for the input signal are steps and impulses. For time-invariant (TI) linear systems one can resort to the literature on the problem of robust analysis of a superposition of complex exponentially damped sinusoids which has become vast since the publication of the work of Kumaresan and Tufts (1982). These methods then provide ways to find the system poles and their residues with a low sensitivity to noise. Some authors have paid special attention to the problem of identifying linear systems using this approach, e.g. Henderson (1981) and Jain *et al.* (1983).

A sum of harmonically related sinusoids (a multisine) is another possible excitation not requiring the explicit reconstruction of the time-derivatives. The transfer function of a TI linear system can be estimated using the method described in Pintelon and Schoukens (1990a), which is a maximum likelihood errors-in-variables method based on frequency-domain relations.

Often it is not possible or too expensive to choose the excitation signal of the device under test (DUT) and the operating conditions impose an arbitrary excitation. Moreover, the observation of the excitation may be contaminated by noise of the same, if not a higher, level than the noise that corrupts the output. In that case, the above methods suffer from modeling errors or become unrealizable. To avoid the reconstruction of the time-derivatives, the differential equations are transformed into algebraic equations via (multiple) integration with some weighting function. These approaches and methods to estimate the system parameters have been surveyed by Young (1979) and Unbehauen and Rao (1987, 1990). The main drawback of these methods is that the sampling rate must be chosen high with respect to the bandwidth of the DUT. Fast sampling results obviously in an increased hardware complexity and cost and can involve prohibitively long data records in off-line estimation. The method presented in this text will be able to work with a reduced sampling rate.

We will concentrate on approaches that generate auxiliary signals and their time-derivatives. These signals are related to the input and to the output and are constructed in such a way that they satisfy a model equation from which it is possible to recover the original system parameters. In particular, we will examine systems where the input u(t) and the output y(t) are related by the model equation

$$\sum_{i=0}^{n} \alpha_i(t) \frac{d^i y(t)}{dt^i} = \sum_{i=0}^{m} \beta_i(t) \frac{d^i u(t)}{dt^i}.$$

It is our intention to estimate the time-varying (TV) system parameters $\alpha_i(t)$ and $\beta_i(t)$ from samples of the input and the output. We stick to this model in which the parameters bear a relation to the physical system properties since we are not pleased with the parameters of some model that can explain the input/output relation but in which the relation with physical quantities is lost. The differential equation will be transformed to the form

$$\sum_{i=0}^{n} a_i(k)\, \mathbf{qy}_i(k) = \sum_{i=0}^{m} b_i(k)\, \mathbf{qu}_i(k) \qquad k \in \mathbb{N}$$

where, under mild assumptions, $a_i(k)$ and $b_i(k)$ are delayed samples of the functions $\alpha_i(t)$ and $\beta_i(t)$. The vector quantities **qu** and **qy** are obtained from the available samples of u(t) and y(t) by digital filtering. This digital filtering approach is a general framework which can be specialized to the digital implementation of the Poisson moment functional approach (e.g. Unbehauen and Rao, 1987), the integrated sampling approach as seen by Sagara and Zhao (1990) or Schoukens (1990) or the instantaneous sampling approach (e.g. Wahlberg, 1987 or Pintelon and Kollár, 1991). Also, we present an elegant solution to eliminate the initial condition problem in all approaches. Methods that use classical numerical integration or differentiation make the transformation of the model equation more accurate by increasing the sampling rate. In our digital filtering method, this transformation can be achieved with any accuracy by increasing the order of the digital filtering involved or by increasing the sampling rate. Formulated in terms of the sampling integration approach, the integral of a signal over a number of samples also takes samples outside the integration interval into account. For the correctness of this approach, it is important that the signal to be integrated is band-limited. This is achieved by preprocessing it by an analog and a digital filter.

Extensions of the method to models of systems that are nonlinear in the input-output signals and their time-derivatives, but linear in the system parameters are possible.

This text is organized as follows: section 2 contains a discussion of the building blocks required for the implementation of a measurement setup that allows slow sampling using the concept of digital filtering. Section 3 presents an analysis of the approximation errors involved in the process of converting the differential equation to its algebraic counterpart. In section 4, a Gaussian maximum likelihood estimator (MLE) for time-invariant linear systems is derived. Time-varying estimators do not introduce fundamental problems, but are not considered in this text since the implications of the approximation errors will be easier to track and present for TI systems. The TI-MLE is used in section 5 to compare the performance of the proposed modeling technique using digital filters with existing methods, in particular with the frequency-domain approach and those using numerical integration.

2. Construction of a measurement setup.

Consider a linear possibly time-varying single-input-single-output (SISO) system described by the following differential equation:

$$\sum_{i=0}^{n} \alpha_i(t)\, y^{(i)}(t) = \sum_{i=0}^{m} \beta_i(t)\, u^{(i)}(t) \tag{2.0.1}$$

where $u^{(i)}(t)$ is the i-th time-derivative of the input signal and $y^{(i)}(t)$ is the i-th time-derivative of the output signal (the zero-th derivative is the signal itself). We will address the problem of estimating $\alpha_i(t)$ and $\beta_i(t)$ from N equally spaced samples of the (filtered) input and output. For real-life systems, we have $n \geq m$, but the method presented in this text doesn't require this assumption. Having $m > n$ can be useful if the true system model is not identifiable from data confined to the limited frequency band in which the measurements are situated. For example, an active differentiator whose poles lie well out of the measurement frequency band is best modeled by $y(t) = \beta_1\, u^{(1)}(t)$.

We will not use (2.0.1) directly, but instead we will transform it by a linear operator G[.]. Let g(t) be the impulse response of this operator and let G(s) denote its Laplace transform. Then define

$$G[x(t)] = \int_{-\infty}^{t} g(t-r)\, x(r)\, dr \tag{2.0.2}$$

Applying a linear operator G[.] to equation (2.0.1) will not affect its validity:

$$\sum_{i=0}^{n} G[\alpha_i(t)\, y^{(i)}(t)] = \sum_{i=0}^{m} G[\beta_i(t)\, u^{(i)}(t)] \tag{2.0.3}$$

Instead of using (2.0.3), the estimators of the unknown system parameters will be based on the model equation:

$$\sum_{i=0}^{n} a_i(t)\, G[y^{(i)}(t)] = \sum_{i=0}^{m} b_i(t)\, G[u^{(i)}(t)] \tag{2.0.4}$$

where $a_i(t)$ and $b_i(t)$ are time functions from which it is possible to recover the original system parameters $\alpha_i(t)$ and $\beta_i(t)$. The approximation involved in using (2.0.4) instead of (2.0.3) is discussed in the next section.

The linear operator G[.] can be interpreted as an analog filter applied to $u^{(i)}(t)$ and $y^{(i)}(t)$. In this respect, the issue of the initial conditions of these filters will be addressed later.

2.1 Conditions on G[.] for time-varying system parameter estimation.

Assume that the system parameters $\alpha_i(t)$ and $\beta_i(t)$ are band-limited with upper frequency ω_a and that the linear operator G[.] satisfies

$$G(j\omega) = G_0\, e^{-j\omega\tau}\, (1 + \Delta(j\omega)) \tag{2.1.1}$$

where $\Delta(j\omega)$ is small for $|\omega| < \omega_b$ ($\omega_a < \omega_b$) and $G(j\omega)$ is the Fourier transform of the impulse response of G[.]. In this text we use $G(j\omega)$ for the Fourier transform of the function g(t), while the Fourier transform of the sequence $\{g_k\}$ is denoted by $G(e^{j\omega})$. Applying lemma 2.1 and corollary 2.1 to the model equation (2.0.3) shows that using (2.0.4) instead of (2.0.3) introduces a systematic error which is small for $|\omega| < (\omega_b - \omega_a)$. Hence, the systematic error can be controlled in a limited frequency band by a proper design of G[.], i.e. $G(j\omega)$ should have a linear phase characteristic and a flat amplitude response ($\Delta(j\omega) = 0$). In this context, we simply have:

$$\alpha_i(t-\tau) = a_i(t) \qquad i = 0 \ldots n \tag{2.1.2}$$

and $$\beta_i(t-\tau) = b_i(t) \qquad i = 0 \ldots m$$

The linear operator G[.] will be composed of several linear operators upon which we will elaborate in the subsequent sections of § 2. These linear operators will be chosen with the purpose to reconstruct samples of the signals $G[u^{(i)}(t)]$ and $G[y^{(i)}(t)]$ with an arbitrary precision. The flow of the signals through the measurement channels thus obtained is depicted in Fig. 2.1.

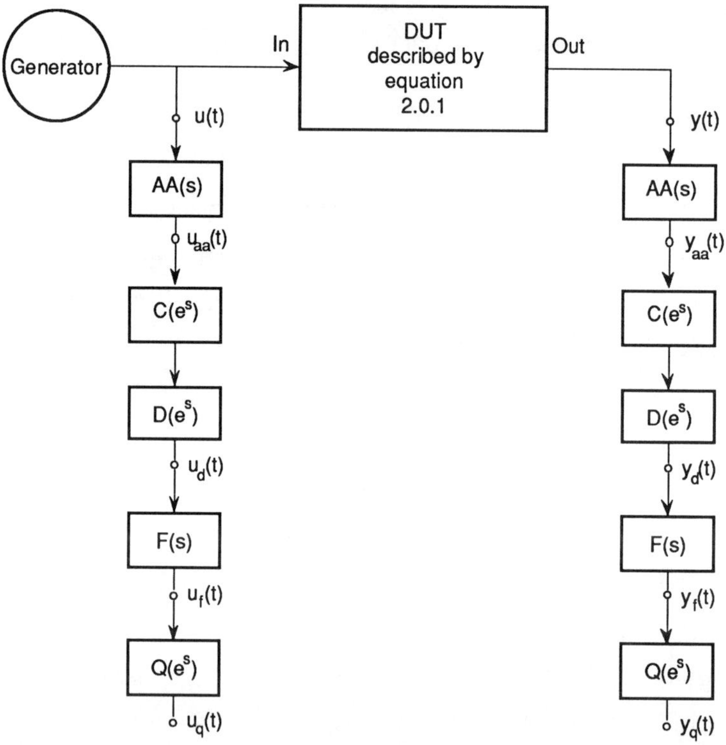

Figure 2.1: the plant's input signal u(t) and its output signal y(t) are processed by a linear operator G[.] which is composed of an anti-alias filter (AA), a compensation filter (C), a de-emphasis filter (D), an observation filter (F) and an initial condition removing filter (Q).

The signals u(t) and y(t) are fed into the measurement filter chain containing:

1. an anti-alias filter, characterized by AA(s) and producing $u_{aa}(t)$ and $y_{aa}(t)$,
2. an anti-alias compensation filter, characterized by $C(e^s)$,
3. a de-emphasis filter, characterized by $D(e^s)$ and producing $u_d(t)$ and $y_d(t)$,
4. an observation filter, characterized by F(s) and producing $u_f(t)$ and $y_f(t)$,
5. an initial condition removing filter, characterized by $Q(e^s)$ and producing $u_q(t)$ and $y_q(t)$.

The observation filter and the initial condition removing filter will later be compressed into a single approximating filter named $H(e^s)$.

2.2 The observation filter.

In general, the derivatives of u(t) and y(t) cannot be measured directly on the plant. Through implementation of the linear operator F[.] as an analog state variable filter (e.g. Young, 1979), it is possible to construct signals $u_f^{(i)}(t)$ and $y_f^{(i)}(t)$ that satisfy a model equation containing the original system parameters (Fig. 2.2). This model equation will then be used to estimate the system parameters. The order q of the state variable filter (or observation filter) is chosen greater than or equal to the maximum of n and m. Assuming

$$F(s) = \frac{K_f}{\sum_{i=0}^{q} f_i s^i} = \frac{K_f}{P(s)} \qquad f_q = 1 \text{ and } K_f \in \mathbb{R},$$

equation (2.0.4) is rewritten as

$$\sum_{i=0}^{n} a_i(t)\, Q[y_f^{(i)}(t)] = \sum_{i=0}^{m} b_i(t)\, Q[u_f^{(i)}(t)] \tag{2.2.1}$$

where $u_f^{(i)}(t)$ and $y_f^{(i)}(t)$ are now derived from the states of the observation filters which are driven by $u_d(t)$ and $y_d(t)$ and which ought to satisfy $u_f^{(i)}(t) = F[u_d^{(i)}(t)]$ and $y_f^{(i)}(t) = F[y_d^{(i)}(t)]$.

In reality, the observation filter is activated at some finite time, which we will assume to be zero without loss of generality. Hence, the convolution in (2.0.2) will not start at $t = -\infty$ and a problem of initial conditions arises. For now, we will assume that the anti-alias filter, the compensation filter and the de-emphasis filter were "hooked" to u(t) and y(t) at $t = -\infty$, i.e. the initial condition problem arises for F[.] only. The assumption will be relaxed in § 2.5. Hence, we can interchange the differentiation operator and the "D[.] after C[.] after AA[.]" operator:

$$u_d^{(i)}(t) = D\big[C[AA[u^{(i)}]]\big]$$

$$\text{and } y_d^{(i)}(t) = D\big[C[AA[y^{(i)}]]\big] \tag{2.2.2}$$

Assume that two identical controllable, observable, strictly stable polynomial observation filters are implemented in their completely controllable canonical state-space form:

$$\frac{d\mathbf{x}(t)}{dt} = \mathbf{A}\,\mathbf{x}(t) + \mathbf{b}\,v(t) \qquad (2.2.3)$$

$$\mathbf{z}(t) = \mathbf{C}\,\mathbf{x}(t) + \mathbf{d}\,v(t)$$

with $\mathbf{f} = (f_0, f_1, \ldots, f_{q-1})^t$ (superscript t denotes transpose)

$\mathbf{A} = (\mathbf{e_2}, \mathbf{e_3}, \ldots, \mathbf{e_q}, -\mathbf{f})^t \in \mathbb{R}^{qxq}$

$\mathbf{b} = K_f\,\mathbf{e_q} \in \mathbb{R}^{qx1}$

$\mathbf{C} = (\mathbf{I}_{qxq}, -\mathbf{f})^t \in \mathbb{R}^{(q+1)xq}$

$\mathbf{d} = K_f\,\mathbf{e_{q+1}} \in \mathbb{R}^{(q+1)x1}$

where $\mathbf{x}(t)$ is the q-dimensional state vector $\mathbf{x}(t)$, $v(t)$ is the scalar input, $\mathbf{z}(t)$ is a q+1 dimensional output and $\mathbf{e_k} = (0, \ldots, 0, 1, 0, \ldots, 0)^t$ is a vector of appropriate dimension with the "1" in the k-th position.

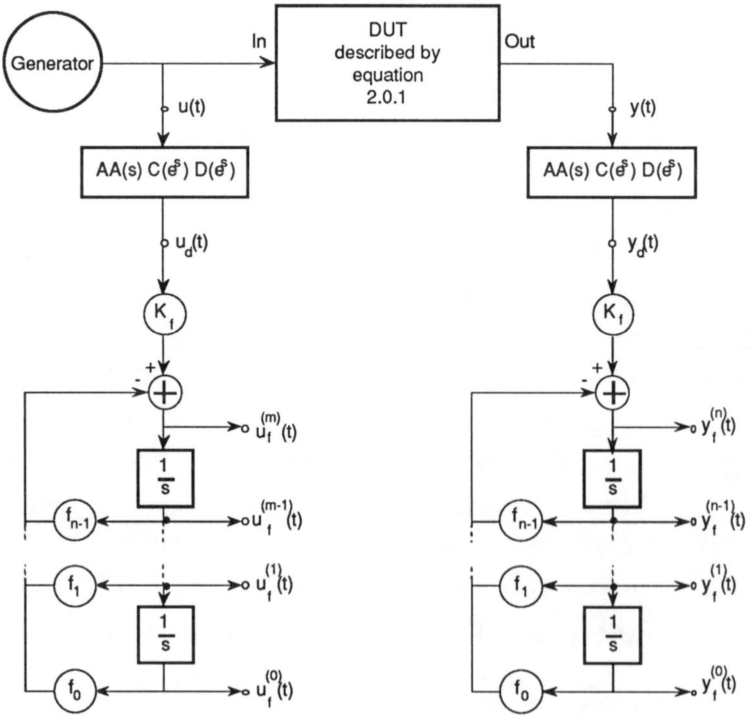

Figure 2.2 : observation filters yielding $u_f^{(i)}(t)$ and $y_f^{(i)}(t)$ that satisfy a model equation containing the original system parameters.

This special structure of observation filter is used because it allows to access $\mathbf{z}_1(t)$ and its time-derivatives up to the q-th order. Denote the state vector of a filter (2.2.3) when the input is $u_d^{(i)}(t)$ by $\mathbf{xu}(i,t)$ and the output vector when the input is $u_d^{(0)}(t)$ by $\mathbf{zu}(t)$. Similar notations are introduced for $\mathbf{xy}(i,t)$ and $\mathbf{zy}(t)$. Now, applying (2.2.2) and lemma 2.2 to the following expression yields:

$$G\left[\sum_{i=0}^{m} \beta_i(t)\, u^{(i)}(t) - \sum_{i=0}^{n} \alpha_i(t)\, y^{(i)}(t)\right]$$

$$\approx \sum_{i=0}^{m} b_i(t)\, Q[F[u_d^{(i)}(t)]] - \sum_{i=0}^{n} a_i(t)\, Q[F[y_d^{(i)}(t)]]$$

$$= \sum_{i=0}^{m} b_i(t)\, Q[\mathbf{e}_1^t\, \mathbf{xu}(i,t)] - \sum_{i=0}^{n} a_i(t)\, Q[\mathbf{e}_1^t\, \mathbf{xy}(i,t)]$$

$$= \sum_{i=0}^{m} b_i(t)\, Q[\mathbf{e}_{i+1}^t\, \mathbf{zu}(t)] - \sum_{i=0}^{n} a_i(t)\, Q[\mathbf{e}_{i+1}^t\, \mathbf{zy}(t)]$$

$$- K_f \sum_{i=0}^{m} b_i(t) \sum_{k=1}^{i} Q[\mathbf{e}_k^t e^{\mathbf{A}t}\, \mathbf{e}_n]\, u_d^{(i-k)}(0) + K_f \sum_{i=0}^{n} a_i(t) \sum_{k=1}^{i} Q[\mathbf{e}_k^t e^{\mathbf{A}t}\, \mathbf{e}_n]\, y_d^{(i-k)}(0)$$

(2.2.4)

Without the Q-filter, the model equation based on (2.2.4) is a generalization of Whitfield and Messali (1987). They derive their equation by applying an n-fold integration ($\mathbf{f} = \mathbf{0}$) starting at $t = 0$ to the time-invariant version of (2.0.1).

Assume that two identical filters (2.2.3) are initially at rest and are fed with u(t) and y(t). The transformed model equation based on (2.2.4) can then be used to estimate the system parameters and $a_i(t)$ and $b_i(t)$. Several strategies can be devised to tackle the presence of the initial condition:

1. Expression (2.2.4) is only evaluated for t sufficiently large such that the initial condition terms make an unmeasurable contribution. The roots of P(s) are chosen to have a real part that is as negative as possible such that $\| e^{\mathbf{A}t} \|$ vanishes as quickly as possible. Obviously precious measurement time is lost by waiting.

2. The initial condition $u_d^{(i)}(0)$ (i = 0 ... m) and $y_d^{(i)}(0)$ (i = 0 ... n) is estimated along with the system parameters. This constitutes an extra computational burden.
3. The model equation, or equivalently the observation filters outputs, are sampled and filtered by a discrete filter that removes the initial condition terms. This approach is discussed in § 2.3.

2.3 A discrete-time filter to remove the initial condition.

The estimator will be implemented on a digital computer, and therefore we will assume, purely for the theoretical development of the method, that equally spaced simultaneous samples of all components of the observation filter outputs **zu**(kT) and **zy**(kT) are available where T is the sampling period and k = 0 ... N. Without loss of generality, we will assume that the sampling period T equals 1 s. The results for T ≠ 1 s can be derived by rescaling the analog time and frequency appropriately. Hence, the observation filter outputs **zu**(t) and **zy**(t) are evaluated at t = k and the causal discrete operator Q[.] is applied to the resulting vector sequence. Here,

$$Q(z) = \prod_{i=1}^{M}(1-e^{s_i l} z^{-l})^{m_i} = \sum_{i=0}^{nl} q_i z^{-i} \qquad (2.3.1)$$

where s_i is an m_i - fold pole of the continuous-time observation filter F(s). Each factor $1-e^{s_i l} z^{-l}$ has the purpose of removing an exponential response $e^{s_i t}$ by taking a weighted difference over any positive integer number of samples l. Similar to (2.0.2), the discrete operator Q[.] applied to a vector sequence $\{\mathbf{x}_k\}$ is defined as:

$$Q[\mathbf{x}_k] = \sum_{i=0}^{nl} q_i \mathbf{x}_{k-i}.$$

Lemma 2.3 applied to (2.2.4) then allows to write:

$$\sum_{i=0}^{m} \beta_i(k-\tau)\, G[u^{(i)}(t)] - \sum_{i=0}^{n} \alpha_i(k-\tau)\, G[y^{(i)}(t)]$$

$$= \sum_{i=0}^{m} \beta_i(k-\tau)\, Q[\mathbf{e}_{i+1}^t\, \mathbf{zu}(k)] - \sum_{i=0}^{n} \alpha_i(k-\tau)\, Q[\mathbf{e}_{i+1}^t\, \mathbf{zy}(k)]$$

With the definitions $\mathbf{qu}_k$ = Q[**zu**(k)] and $\mathbf{qy}_k$ = Q[**zy**(k)], the expression above yields the following model equation:

$$\sum_{i=0}^{m} \beta_i(t\text{-}\tau)\,\mathbf{e}^{t}_{i+1}\,\mathbf{q}\mathbf{u}_k = \sum_{i=0}^{n} \alpha_i(t\text{-}\tau)\,\mathbf{e}^{t}_{i+1}\,\mathbf{q}\mathbf{y}_k \qquad \text{for } k = nl \ldots N \qquad (2.3.2)$$

The price that has to be paid to eliminate the initial condition term in (2.2.4) is the loss of n l samples to set up the initial conditions of the Q-filter. The Q-filter places zeros at the impulse-invariant transformed pole locations of the F-filter. These poles and zeros match better if the roots of P(s) are closer to the origin. Spectrally, we have multiplied the transfer function of the observation filter by nearly its inverse and hence another implication of the introduction of the Q-filter emerges: we loose the freedom to choose the spectral properties of the observation filter. Equation (2.3.2) is a generalization of Sagara and Zhao (1990) and Schoukens (1990), who set $\mathbf{f} = \mathbf{0}$ and examine TI systems.

2.4 A discrete-time approximation to the continuous-time observation filter.

In practice, one cannot produce two identical observation filters. Even if they would be realized or digitally corrected (Kollár *et al.*, 1991) within an allowable tolerance, it would have to be re-built for each estimation problem: its order would determine the maximal order of the DUT one can tackle and its bandwidth should bear some relation to the frequency range in which we wish to examine the DUT. Moreover, at least n+m+2 outputs ought to be sampled simultaneously. If possible, one would prefer to start from a record of sampled observations of the plant's (SISO-filtered) input and output. Therefore, the hybrid cascade filter "Q[.] after F[.]" will be approximated by a discrete-time filter H[.] which is fed by a scalar record and produces up to q+1 outputs. H(z) is designed such that the approximation error involved is small in the frequency band 0 Hz to f_m (f_m < 0.5 Hz) and increases outside this band. It is only in this frequency band that the estimator will be allowed to make full use of the approximate model equation thus obtained. Frequencies outside this band should make only minor contributions in the estimation procedure. The frequency-dependent behavior of the estimator will be treated in § 4.4 and an example of how the choice of f_m is made will be given in § 5. An extra advantage of introducing the Q-filter in § 2.3 emerges here: due to the removal of the exponential decay it is easily felt that good H-filters can be of the FIR-type with a small number of taps.

Summarizing, each observation filter is replaced by a discrete-time system with one input and q+1 outputs. Denote the transfer function from the input of this SIMO system to its i-th output by $H_i(z)$ and introduce the approximation errors $\Delta_i(j\omega)$ in the interval $-\pi \leq \omega \leq \pi$ as:

$$H_i(e^{j\omega}) = (j\omega)^{i-1} (1 + \Delta_i(j\omega)) H_0(e^{j\omega}) \qquad i = 1 \ldots q+1 \qquad (2.4.1)$$

The function $\Delta_i(j\omega)$ is defined such that the factor $(j\omega)^{i-1} (1 + \Delta_i(j\omega))$ is periodic in ω with period 2π.

2.5 The anti-alias filter, the de-emphasis filter and the correction Filter.

In general, we do not have full control over the excitation signal. Hence, it may contain spectral components that are "folded" by the sampling action and this may introduce serious modeling errors. Indeed, harmonic components with frequency $f+k$ (k is any integer number) cannot be distinguished after sampling and only a "folded" frequency response model can be fitted to the measurements. Even if the excitation signal is a pure sinewave of unknown frequency, one would never know how to "unfold" the observed frequency to obtain the frequency at which the measurement took place. Hence the need of a continuous-time anti-alias filter AA[.] when arbitrary excitation signals are allowed.

As mentioned in § 2.4, the observation filter approximation error is frequency-dependent. If the signal fed to the approximated observation filters is band-limited to f_m, the model error introduced by approximating the filters is small. This ideal situation is approached by introducing the discrete-time de-emphasis filter D[.], which attenuates possible spectral components above f_m. Hence, since only signals with small components at frequencies above f_m will be presented to the observation filters, they will produce small errors.

Finally, since it is desirable that the total measurement channel has a "flat" magnitude response and a linear phase response, and to compensate the production tolerances of the analog anti-alias filters, a discrete-time compensation filter C[.] is introduced.

The frequency-domain design criteria for AA(s), C(z) and D(z) will be derived later, but one time-domain criterion is obvious: the impulse response of these filters must exhibit a fast decay. When activated at a finite time, these filters contribute to the model equation error with a transient that dies away like their state transition matrix, similar to derivation (2.2.4). Because these filters have definite spectral constraints, a trick like § 2.3 cannot be applied here. The initial condition removing digital filter would spoil their deliberate spectral shaping. The demand on AA(s) can be relaxed if the anti-alias filters can be "hooked" to u(t) and y(t) well before the acquisition starts. For the remaining discrete-time filters, the finite-impulse-response (FIR) type is preferred since their "initialization error" becomes exactly zero after a finite time.

The total setup is shown in Fig. 2.3. The samples of $u_{aa}(t)$ and $y_{aa}(t)$ are taken at the sampling instants k (k = 0 ... N). If the initialization of the digital filters requires M+1 samples, with (2.1.2) and (2.3.2), the model equation becomes:

$$\mathbf{h}_k^t \, \mathbf{p}_k = 0 \qquad \text{for } k = M \ldots N \qquad (2.5.1)$$

with

$$\mathbf{p}_k = (a_0(k), \ldots, a_m(k), b_0(k), \ldots, b_n(k))^t$$

$$\mathbf{h}_k = (\mathbf{q}\mathbf{y}_k^t, -\mathbf{q}\mathbf{u}_k^t)^t$$

The system parameters $a_i(t)$ and $b_i(t)$ are related to $\alpha_i(t)$ and $\beta_i(t)$ as in (2.1.2) where the total delay of the measurement channel τ and the measurement channel DC-gain G_0 are chosen to minimize $\Delta(s)$ in some sense in:

$$AA(s)\,C(e^s)\,D(e^s)\,H_0(e^s) = G_0\,e^{-\tau s}\,(1+\Delta(s)) = G(s) \qquad (2.5.2)$$

Model equation (2.5.1) is only approximately valid: it doesn't account for the errors due to the time-varying aspect of the system (§ 2.1), the alias phenomenon due to the sampling introduced in § 2.3 and the observation filter approximation error introduced in § 2.4. Analyzing the consequences on estimates based on model (2.5.1) will be the subject of the next section.

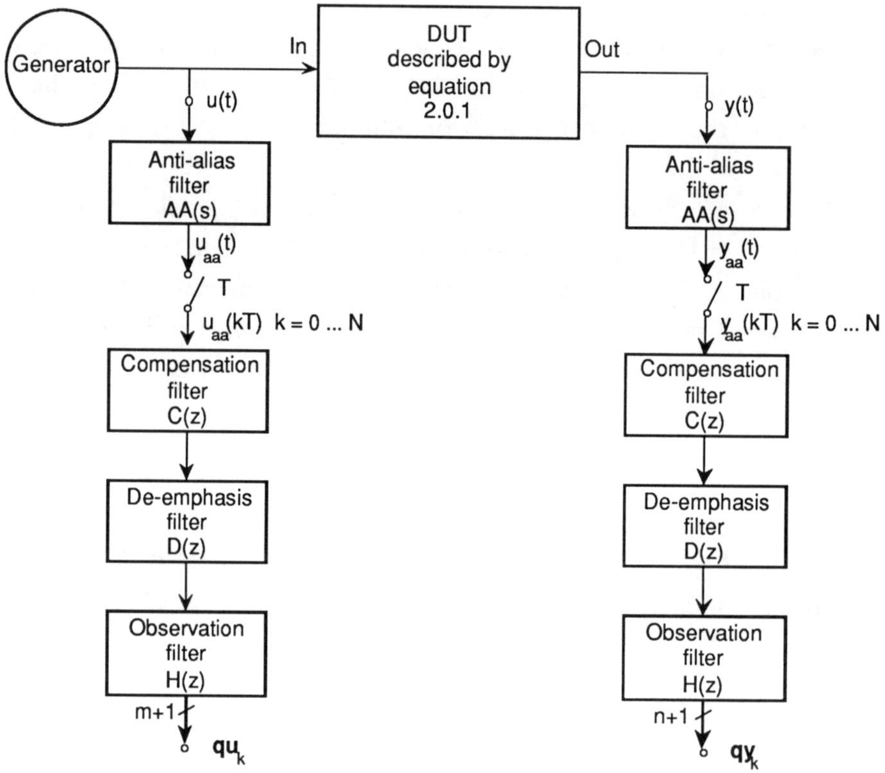

Figure 2.3 : the complete measurement setup as it will be implemented.

Appendix 2.A

Lemma 2.1

Assume that the frequency response of the G[.] is given by:

$$G(j\omega) \quad = G_0\, e^{-j\omega\tau}\,(1 + \Delta(j\omega)) \qquad (2.A.1a)$$

and let f(t) be a band-limited signal with Fourier transform

$$F(j\omega) \quad = 0 \qquad \text{for } |\omega| > \omega_a. \qquad (2.A.1b)$$

Define

$$\Gamma(s) \quad = \frac{d\Delta(s)}{ds}$$

and

$$e(t) \quad = G[f(t)\, x(t)] - f(t-\tau)\, G[x(t)]$$

then the Fourier transform of e(t) is given by

$$E(j\omega) = G_0\, e^{-j\omega\tau} \int_{-\omega_a}^{\omega_a} \big(\Delta(j\omega)-\Delta(j(\omega-\alpha))\big)\, F(j\alpha)\, X(j(\omega-\alpha))\, d\alpha$$

$$\approx G_0\, e^{-j\omega\tau} \int_{-\omega_a}^{\omega_a} \Gamma(j(\omega-\alpha))\, j\alpha\, F(j\alpha)\, X(j(\omega-\alpha))\, d\alpha \qquad (2.A.2)$$

Proof:

$$E(j\omega) = G(j\omega)\,(F(j\omega) * X(j\omega)) - (e^{-j\omega\tau}\, F(j\omega)) * (G(j\omega)\, X(j\omega))$$

$$= G_0\, e^{-j\omega\tau}\,(1 + \Delta(j\omega)) \int_{-\omega_a}^{\omega_a} F(j\alpha) X(j(\omega-\alpha))\, d\alpha$$

$$- G_0 \int_{-\omega_a}^{\omega_a} e^{-j\alpha\tau}\, F(j\alpha)\, e^{-j(\omega-\alpha)\tau} \big(1 + \Delta(j(\omega-\alpha))\big)\, X(j(\omega-\alpha))\, d\alpha$$

$$= G_0\, e^{-j\omega\tau} \int_{-\omega_a}^{\omega_a} \big(\Delta(j\omega)-\Delta(j(\omega-\alpha))\big)\, F(j\alpha)\, X(j(\omega-\alpha))\, d\alpha$$

Provided that the changes of $\Delta(j\omega)$ in the interval $\omega-\omega_a$ to ω are well-approximated by its first order truncated Taylor series, we have

$$E(j\omega) \approx G_0\, e^{-j\omega\tau} \int_{-\omega_a}^{\omega_a} \Gamma(j(\omega-\alpha))\, j\alpha\, F(j\alpha)\, X(j(\omega-\alpha))\, d\alpha$$

Equation (2.A.2) can be expressed in the time domain:

$$e(t) \approx G_0 \frac{df(t-\tau)}{dt}\, \Gamma[x(t-\tau)]$$

where $\Gamma[.]$ is the linear operator corresponding to $\Gamma(j\omega)$. □

Corollary 2.1

If the changes of $\Delta(j\omega)$ are small or $\Gamma(j\omega)$ is small for $|\omega| < \omega_b$ ($\omega_a < \omega_b$), then $|E(j\omega)|$ is small for $|\omega| < (\omega_b-\omega_a)$.

Lemma 2.2

Consider a controllable and observable linear SIMO system with characteristic polynomial

$$P(s) = \sum_{i=0}^{q} f_i s^i \qquad f_i \in \mathbb{R} \text{ and } f_q = 1 \qquad (2.A.3)$$

and which is described by the state space model (2.2.2). The roots of P(s) are assumed to have a strictly negative real part. Let $\mathbf{x}(i,t)$ be the state vector when the input is chosen equal to $u^{(i)}(t)$. Then its components are given by

$$\mathbf{e}_1^t\,\mathbf{x}(i,t) = -K_f \sum_{k=1}^{i} \mathbf{e}_k^t e^{\mathbf{A}t}\,\mathbf{e}_q\,u^{(i-k)}(0) + \mathbf{e}_{i+1}^t\,\mathbf{x}(0,t) \quad \text{for } i = 1 \dots q\text{-}1$$

$$\mathbf{e}_1^t\,\mathbf{x}(q,t) = -K_f \sum_{k=1}^{q} \mathbf{e}_k^t e^{\mathbf{A}t}\,\mathbf{e}_q u^{(i-k)}(0) - \mathbf{f}^t\,\mathbf{x}(0,t) + K_f\; u(t) \qquad (2.A.4)$$

Proof:

The state vector $\mathbf{x}(i,t)$ resulting from an input $u^{(i)}(t)$ is given by:

$$\mathbf{x}(i,t) = e^{\mathbf{A}t}\left\{\mathbf{x}(i,0) + \int_0^t e^{-\mathbf{A}r}\,\mathbf{b}\,u^{(i)}(r)\,dr\right\}$$

Assuming $\mathbf{x}(i,0) = \mathbf{0}$ and using partial integration:

$$\mathbf{x}(i,t) = \mathbf{b}\,u^{(i-1)}(t) - e^{\mathbf{A}t}\,\mathbf{b}\,u^{(i-1)}(0) + e^{\mathbf{A}t}\int_0^t \mathbf{A}\,e^{\mathbf{A}r}\,\mathbf{b}\,u^{(i-1)}(r)\,dr$$

$$= \mathbf{b}\,u^{(i-1)}(t) - e^{\mathbf{A}t}\,\mathbf{b}\,u^{(i-1)}(0) + \mathbf{A}\,\mathbf{x}(i\text{-}1,t)$$

The last step is justified by the fact that $\mathbf{A}$ and $e^{\mathbf{A}t}$ commute.
Using $\mathbf{e}_k^t\mathbf{b} = 0$ and $\mathbf{e}_k^t\,\mathbf{A} = \mathbf{e}_{k+1}^t$ for k = 1 ... q-1 yields:

$$\mathbf{e}_k^t\,\mathbf{x}(i,t) = -K_f\;\mathbf{e}_k^t\,e^{\mathbf{A}t}\;\mathbf{e}_q\,u^{(i-1)}(0) + \mathbf{e}_{k+1}^t\,\mathbf{x}(i\text{-}1,t) \quad k = 1 \dots q\text{-}1 \qquad (2.A.5a)$$

while identities $\mathbf{e}_q^t\mathbf{b} = K_f$ and $\mathbf{e}_q^t\,\mathbf{A} = -\mathbf{f}^t$ result in:

$$\mathbf{e}_q^t\,\mathbf{x}(1,t) = K_f\;u(t) - K_f\;\mathbf{e}_q^t\,e^{\mathbf{A}t}\;\mathbf{e}_q\,u(0) - \mathbf{f}^t\,\mathbf{x}(0,t) \qquad (2.A.5b)$$

Applying recurrence equation (2.A.5a) i times to $\mathbf{e}_1^t\,\mathbf{x}(i,t)$ proves the lemma for i = 1 ... q - 1. For i = q, relation (2.A.5a) is applied q-1 times to $\mathbf{e}_1^t\,\mathbf{x}(q,t)$ with the final recurrence (2.A.5b).

□

Lemma 2.3

Let $\mathbf{A} \in \mathbb{R}^{qxq}$ be a nonderogatory matrix (Barnett and Cameron, 1985) with characteristic polynomial

$$P(s) = \prod_{p=1}^{M} (s-s_p)^{m_p} \qquad \text{with } m_p \geq 1 \text{ and } \sum_{p=1}^{M} m_p = q.$$

The polynomial with roots placed at the impulse-invariant transformation of the roots of P(s) is:

$$Q(z) = \prod_{p=1}^{M} (z - e^{s_p T})^{m_p} = \sum_{i=0}^{q} q_i z^i.$$

where T is any real number. Then

$$Q(e^{\mathbf{A}T}) = \sum_{i=0}^{n} q_i e^{\mathbf{A} iT} = 0.$$

Proof:

First we prove that Q(z) is the characteristic polynomial of $e^{\mathbf{A}T}$. The result then immediately follows from the Cayley-Hamilton theorem.

Write **A** in its Jordan form:

$\mathbf{A} = \mathbf{M}^{-1} \mathbf{J} \mathbf{M}$ with $\mathbf{M} \in \mathbb{R}^{qxq}$ nonsingular

and $\mathbf{J} = \begin{pmatrix} \mathbf{J}_1 & & \mathbf{0} \\ & \ddots & \\ \mathbf{0} & & \mathbf{J}_M \end{pmatrix}$ where $\mathbf{J}_i = \begin{pmatrix} s_i & 1 & & \mathbf{0} \\ & \ddots & \ddots & \\ & & s_i & 1 \\ \mathbf{0} & & & s_i \end{pmatrix} \in \mathbb{R}^{m_i x m_i}$

The characteristic polynomial R(z) of $e^{\mathbf{A}T}$ is:

$$\begin{aligned} R(z) &= \det\left(z\mathbf{I}_{nxn} - e^{\mathbf{A}T}\right) \\ &= \det\left(z\mathbf{I}_{nxn} - e^{\mathbf{J}T}\right) \end{aligned}$$

Now, $e^{\mathbf{J}T} = \begin{pmatrix} e^{\mathbf{J}_1 T} & & \mathbf{0} \\ & \ddots & \\ \mathbf{0} & & e^{\mathbf{J}_M T} \end{pmatrix}$

and $e^{\mathbf{J}_i T} = e^{s_i T}\, \mathbf{U}(T)$ where $\mathbf{U}(T)$ is an upper diagonal Toeplitz matrix with unit main diagonal. Hence, $z_i = e^{s_i T}$ is an m_i-fold root of R(z) and R(z) = Q(z). □

3. Controlling the approximation errors.

In this section we will study the systematic errors in the model equation due to the approximations introduced in § 2. The result will be a set of design considerations or even design criteria for the measurement setup and for the estimator such that these errors can be reduced to an insignificant or acceptable level. Hence, unless explicitly disaffirmed, it is assumed in this section that the input and the output of the plant can be observed free of noise.

The time-varying system parameters are assumed to be band-limited by $\omega_a \ll \pi$. Rewrite equation (2.5.1) symbolically as:

$$w_k \left(\sum_{i=0}^{n} a_i(k)\, \mathbf{qy}_i(k) - \sum_{i=0}^{m} b_i(k)\, \mathbf{qu}_i(k) \right) = 0 \qquad k = -\infty \dots +\infty \qquad (3.0.1)$$

where the observation window sequence $\{w_k\}$ satisfies:

$w_k = 1$ for $k = M \dots N$

and $w_k = 0$ for $k = -\infty \dots M-1$ and $k = N+1 \dots \infty$.

The periodic convolution of two functions $X(\omega)$ and $Y(\omega)$ is defined as:

$$X(\omega) *_p Y(\omega) = \frac{1}{2\pi} \int_{-\pi}^{\pi} X((\theta)_{2\pi})\, Y((\omega-\theta)_{2\pi})\, d\theta$$

where $(\omega)_{2\pi}$ denotes "ω modulo 2π". Taking the Fourier transform of the sequence (3.0.1) implies that:

$$W(e^{j\omega}) *_p E(e^{j\omega}) = 0 \qquad -\pi \le \omega \le \pi \qquad (3.0.2)$$

where

$$E(e^{j\omega}) = \sum_{i=0}^{n} A_i(j\omega) *_p L_i(e^{j\omega}) - \sum_{i=0}^{m} B_i(j\omega) *_p R_i(e^{j\omega})$$

and
$$L_i(e^{j\omega}) = C(e^{j\omega})\, D(e^{j\omega})\, H_{i+1}(e^{j\omega}) \sum_{p=-\infty}^{\infty} Y(j\omega+jp2\pi)\, AA(j\omega+jp2\pi)$$

$$R_i(e^{j\omega}) = C(e^{j\omega})\, D(e^{j\omega})\, H_{i+1}(e^{j\omega}) \sum_{p=-\infty}^{\infty} U(j\omega+jp2\pi)\, AA(j\omega+jp2\pi) \qquad \text{for } -\pi \le \omega \le \pi$$

The bandwidth condition on the system parameters allows the Fourier transform of the sequence $\{a_i(k)\}$ to be obtained directly from the Fourier transform of the continous-time function $a_i(t)$ (similar for b_i). The terms $U(j\omega+jp2\pi)AA(j\omega+jp2\pi)$

are the aliased spectra that are present because the excitation signal and the plant's response are not necessarily band-limited. The left hand side of (3.0.2) will only be identically zero if the observations are free of noise, if the measurement setup is perfect and if the exact system parameters are substituted. An estimator will choose the sequences $\{a_i(k)\}$ and $\{b_i(k)\}$ subject to their prior information constraints such that $W(e^{j\omega}) *_p E(e^{j\omega})$ is minimized in some sense. We wish to obtain design criteria that are independent of the measurement window and hence we will examine $E(e^{j\omega})$. For a short observation window there may be a considerable difference between minimizing a frequency-domain measure of $E(e^{j\omega})$ and of $W(e^{j\omega}) *_p E(e^{j\omega})$.

With (2.4.1) and (2.5.2), we find

$$L_i(e^{j\omega}) = (j\omega)^i (1 + \Delta_{i+1}(j\omega)) \sum_{p=-\infty}^{\infty} Y(j\omega+jp2\pi)\, G(j\omega+jp2\pi)$$

and $$R_i(e^{j\omega}) = (j\omega)^i (1 + \Delta_{i+1}(j\omega)) \sum_{p=-\infty}^{\infty} U(j\omega+jp2\pi)\, G(j\omega+jp2\pi)$$

The error term $E(e^{j\omega})$ can be split into its basic components:

$$E(e^{j\omega}) = E_0(e^{j\omega}) + E_t(e^{j\omega}) + E_a(e^{j\omega}) + E_d(e^{j\omega}) \tag{3.0.3a}$$

where E_0 is the equation error for a perfect setup, E_t results from interchanging G[.] and the time-varying system parameters in the original model equation as discussed in § 2.1, E_a is due to the alias phenomenon and E_d is due to the observation filter approximation.

Removing the error terms due to the observation filter approximations Δ_i and those due to the alias phenomenon yields:

$$E_0(e^{j\omega}) + E_t(e^{j\omega}) = \sum_{i=0}^{n} A_i(j\omega) *_p L_i^0(e^{j\omega}) - \sum_{i=0}^{m} B_i(j\omega) *_p R_i^0(e^{j\omega}) \tag{3.0.3b}$$

with $$L_i^0(e^{j\omega}) = (j\omega)^i G(j\omega)\, Y(j\omega)$$

$$R_i^0(e^{j\omega}) = (j\omega)^i G(j\omega)\, U(j\omega)$$

With identity (2.A.2), expression (3.0.3b) can be written as a function of time-domain quantities

$$E_0(e^{j\omega}) = \mathcal{F}\left\{\sum_{i=0}^{n} \alpha_i(t)\, y^{(i)}(t) - \sum_{i=0}^{m} \beta_i(t)\, u^{(i)}(t)\right\} G(j\omega) \quad \text{for } |\omega| < \pi - \omega_a \qquad (3.0.3c)$$

where $\mathcal{F}\{\ \}$ denotes Fourier transform. When the true system parameters are substituted in (3.0.3c), $E_0(e^{j\omega})$ becomes zero for $|\omega| < \pi - \omega_a$. We will consider all other terms in (3.0.3a) as disturbances whose magnitude must be controlled. Also, from (3.0.3b) and (2.A.2) we find that in the same frequency range:

$$E_t(e^{j\omega}) \approx G_0\ \mathcal{F}\left\{\sum_{i=0}^{n} \frac{d\alpha_i(t-\tau)}{dt} \Gamma[y^{(i)}(t-\tau)] - \sum_{i=0}^{m} \frac{d\beta_i(t-\tau)}{dt} \Gamma[u^{(i)}(t-\tau)]\right\} \qquad (3.0.3d)$$

$E_t(e^{j\omega})$ can be made arbitrarily small for $|\omega| < \omega_b - \omega_a$ by making $\Gamma(j\omega)$ arbitrarily small for $|\omega| < \omega_b$.

The alias error can be expressed as:

$$E_a(e^{j\omega}) = \sum_{i=0}^{n} A_i(j\omega) *_p L_i^a(e^{j\omega}) - \sum_{i=0}^{m} B_i(j\omega) *_p R_i^a(e^{j\omega}) \qquad (3.0.3e)$$

with $\quad L_i^a(e^{j\omega}) = (j\omega)^i (1 + \Delta_{i+1}(j\omega))\, Y_a(j\omega)$

$$R_i^a(e^{j\omega}) = (j\omega)^i (1 + \Delta_{i+1}(j\omega))\, U_a(j\omega)$$

$$Y_a(j\omega) = \sum_{\substack{p=-\infty \\ p\neq 0}}^{\infty} Y(j\omega + jp2\pi)\, G(j\omega + jp2\pi)$$

$$U_a(j\omega) = \sum_{\substack{p=-\infty \\ p\neq 0}}^{\infty} U(j\omega + jp2\pi)\, G(j\omega + jp2\pi) \qquad (3.0.3f)$$

$E_a(e^{j\omega})$ can be controlled for $-\pi \le \omega \le \pi$ by assuring that $G(j\omega)\,Y(j\omega)$ and $G(j\omega)\,U(j\omega)$ are small and tend sufficiently fast to zero for $|\omega| > \pi$.

Finally, the equation error due to the observation filter approximation is given by:

$$E_d(e^{j\omega}) = \sum_{i=0}^{n} A_i(j\omega) *_p \left((j\omega)^i \Delta_{i+1}(j\omega)\, G(j\omega)\, Y(j\omega)\right)$$

$$- \sum_{i=0}^{m} B_i(j\omega) *_p \left((j\omega)^i \Delta_{i+1}(j\omega)\, G(j\omega)\, U(j\omega)\right) \qquad (3.0.3g)$$

$E_d(e^{j\omega})$ can be made arbitrarily small for $|\omega| < \omega_m - \omega_a$ by making all $\Delta_i(j\omega)$ arbitrarily small for $|\omega| < \omega_m$.

If these considerations regarding the measurement channel are respected, the estimator must be designed such that the weight of $E(e^{j\omega})$ is sufficiently small for frequencies above $\max(\omega_m-\omega_a, \omega_b-\omega_a)$. In this respect, it is important to notice that $G(j\omega)$ acts as a frequency-domain weighting in all terms of (3.0.3).

The above considerations show the theoretical ability of the method to track TV systems. The design constraints of the measurement setup will now be made more explicit for TI systems, in which case the periodic convolutions become products. The remainder of § 3 will be devoted to examining the magnitude of the error terms in (3.0.3) for such systems. In the TI case, (3.0.3d) becomes identically zero and with the definitions

$$A_d(j\omega) = \sum_{i=0}^{n} (j\omega)^i a_i \Delta_{i+1}(j\omega) \tag{3.0.4a}$$

$$B_d(j\omega) = \sum_{i=0}^{m} (j\omega)^i b_i \Delta_{i+1}(j\omega) \tag{3.0.4b}$$

we find:

$$E_0(e^{j\omega}) = G(j\omega)\left(A(j\omega)\,Y(j\omega) - B(j\omega)\,U(j\omega)\right)$$

$$E_a(e^{j\omega}) = (A(j\omega) + A_d(j\omega))\,Y_a(j\omega) - (B(j\omega) + B_d(j\omega))\,U_a(j\omega)$$

$$E_d(e^{j\omega}) = G(j\omega)\left(A_d(j\omega)\,Y(j\omega) - B_d(j\omega)\,U(j\omega)\right)$$

After substitution in (3.0.3), the equation error becomes

$$\begin{aligned} E(e^{j\omega}) &= \left(1 + \frac{A_d(j\omega)}{A(j\omega)}\right)(Y(j\omega)\,G(j\omega) + Y_a(j\omega))\,A(j\omega) \\ &\quad - \left(1 + \frac{B_d(j\omega)}{B(j\omega)}\right)(U(j\omega)\,G(j\omega) + U_a(j\omega))\,B(j\omega) \end{aligned} \tag{3.0.5}$$

Due to the approximation errors and because the model has only a limited number of degrees of freedom, it will in general be impossible to satisfy $E(e^{j\omega}) = 0$ exactly at all frequencies. Instead, the estimated transfer function will minimize $E(e^{j\omega})$ in some sense and will approximately be given by:

$$\frac{B(j\omega)}{A(j\omega)} \approx \frac{1+\dfrac{A_d(j\omega)}{A(j\omega)}}{1+\dfrac{B_d(j\omega)}{B(j\omega)}} \frac{G(j\omega)\,Y(j\omega)+Y_a(j\omega)}{G(j\omega)\,U(j\omega)+U_a(j\omega)} \tag{3.0.6}$$

For small relative approximation errors (and if the ratios make sense), the fitted transfer function is (for $|\omega| < \pi$):

$$\frac{B(j\omega)}{A(j\omega)} \approx \frac{Be(j\omega)}{Ae(j\omega)}\left(1-\frac{B_d(j\omega)}{B(j\omega)}+\frac{A_d(j\omega)}{A(j\omega)}-\frac{U_a(j\omega)}{U(j\omega)\,AA(j\omega)}+\frac{Y_a(j\omega)}{Y(j\omega)\,AA(j\omega)}\right) \tag{3.0.7}$$

where the suffix "e" denotes exact system quantities. To assure that the spectral fitting error doesn't exceed a certain maximum, it is now a matter of upper bounding the magnitude of the relative errors on the right hand side of this expression.

3.1 The alias error.

In order to simplify the exposition, the reasoning will be held for u(t) only. Identical considerations will be valid for y(t). To control the spectral error due to alias, a first constraint on $G(j\omega)$ and hence on $AA(j\omega)$ emerges: $G(j\omega)\,U(j\omega)$ must tend sufficiently fast to zero as $|\omega|$ increases beyond π, and this with respect to its behavior for $|\omega| < \pi$. Asymptotically, the condition is automatically satisfied for any realizable anti-alias filter and any signal with finite energy. In practice, high values of p will make an insignificant contribution to (3.0.3f) if the difference in order of the numerator and denominator of AA(s) is increased. In this respect, polynomial anti-alias filters are preferred.

Equation (3.0.5) reveals that $U_a(j\omega)$ acts as a disturbance on $G(j\omega)\,U(j\omega)$ and will be treated as such. For a signal u(t) contaminated by observation noise, one might reason that it is sufficient to make the energy content of the aliased signal smaller than the energy content of the noise. If the measurement setup must be universal and the noise level is not known a priori, the analysis must be taken one step further.

Assume that $G(j\omega)\,U(j\omega)$ tends sufficiently fast to zero as ω increases, so only $p = \pm 1$ makes a significant contribution to U_a

$$U_a(j\omega) = G(j\omega+j2\pi)\,U(j\omega+j2\pi) + G(j\omega-j2\pi)\,U(j\omega-j2\pi).$$

If $U(j\omega)$ and $U(j\omega \pm j2\pi)$ are of the same order of magnitude and it is our goal to achieve a relative spectral error of the order of ε_a, $G(j\omega \pm j2\pi)$ must suppress $U(j\omega \pm j2\pi)$ by a factor of ε_a. This can be concluded from (3.0.7), where the ratio of $U(j\omega \pm j2\pi)$ and $G(j\omega)\,U(j\omega)$ appears as a relative transfer function estimation error.

In case $U(j\omega)$ is zero for some ω, which happens in particular if u(t) is a sum of periodic signals, (3.0.6) and (3.0.7) have to be interpreted with the necessary precaution and it is better to resort to (3.0.5). At some frequencies we might have $U(j\omega \pm j2\pi) \neq 0$ while $U(j\omega) = 0$, and hence their ratio is undefined. If the number of periodic components in u(t) with frequency in the interval 0 to f_m is much greater than the number of degrees of freedom in the parametric model, one might expect that the transfer function estimate follows the quantity on the right hand side of (3.0.6) or (3.0.7) in a smoothed way, i.e. we expect a smoothing operation from the model.

$G(j\omega)$ is chosen approximately constant and equal to G_0 in the interval 0 to ω_b and attenuates $U(j\omega)$ above ω_b to achieve a certain weighting. Considering the way $U_a(j\omega)$ occurs in (3.0.5) relative to $G(j\omega)\,U(j\omega)$, we can treat both $U(j\omega \pm j2\pi)$ for $|\omega| < \pi$ and $G(j\omega)\,U(j\omega)$ for $\omega_b < |\omega| < \pi$ as small perturbations to $G_0\,U(j\omega)$ that drive the estimated frequency response away from its true value, and this proportional to the magnitude of these perturbations. Hence we obtain the design criterion:

$$\boxed{|G(j\omega \pm j2\pi)| < G_0\,\varepsilon_a \text{ for } |\omega| < \pi.}$$

The generalization to unconstrained $U(j\omega)$ or to more terms in (3.0.3f) is straightforward and will imply a stronger demand on G[.].

3.2 The observation filter error.

It is our goal to control the relative spectral error introduced by the observation filters. Expressions (3.0.5), (3.0.6) and (3.0.7) suggest that we focus our attention on the ratio of $B_d(j\omega)$ and $B(j\omega)$ and the ratio of $A_d(j\omega)$ and $A(j\omega)$. We wish to discuss which choices regarding structure and spectral properties of the observation filters exhibit a controlled behavior of these ratios for a certain class of DUTs. Also, properties like the complexity and the number of samples required to initiate the filters are important design considerations. The exposition will be given for u(t) knowing that identical considerations are valid for y(t).

As an example to illustrate the importance of the structure of the observation filter, assume that it is realized as the cascade of identical discrete-time approximations to a continuous-time differentiator:

$$H_0(e^{j\omega}) = 1$$

Each digital differentiator exhibits a spectral approximation error of $\delta(j\omega)$:

$$H_{i+1}(e^{j\omega}) = (j\omega)^i\,(1+\Delta_{i+1}(j\omega)) = (j\omega)^i\,(1+\delta(j\omega))^i \approx (j\omega)^i\,(1+i\,\delta(j\omega)) \qquad i = 0 \dots q$$

thus $\Delta_{i+1}(j\omega) \approx i\,\delta(j\omega)$

Now suppose that $B(s) = (s^2 + 2\alpha s + \alpha^2 + \beta^2)^{\frac{m}{2}}$ (m is even), i.e. all its roots coincide at $-\alpha \pm j\beta$, then with $B_d(s) = s\frac{dB(s)}{ds}\delta(s)$ we find

$$\frac{B_d(s)}{B(s)} = m\,\frac{s\,(s+\alpha)}{s^2 + 2\alpha s + \alpha^2 + \beta^2}\,\delta(s) \tag{3.2.1}$$

On the other hand, suppose that there exists no special relation among the $\Delta_i(s)$. In particular, if all outputs of the discrete observation filter are realized exactly, except for the k+1-th output which suffers from a spectral error $\Delta_{k+1}(s)$, we have:

$$\frac{B_d(s)}{B(s)} = \frac{b_k}{b_m}\,\frac{s^k}{(s^2 + 2\alpha s + \alpha^2 + \beta^2)^{\frac{m}{2}}}\,\Delta_{k+1}(s) \tag{3.2.2}$$

When α is small and for $s = j\omega$ in the neighborhood of $j\beta$, expression (3.2.2) becomes proportional to $\alpha^{-\frac{m}{2}}$, while the extreme over ω of (3.2.1) for $s = j\omega$ is proportional to α^{-1}. This undesirable behavior of (3.2.2) extends to clusters of poles and arbitrary relations among the $\Delta_i(s)$. This example shows that some relations among the approximation errors can be less fortunate for some DUTs. These relations are strongly connected to the structure and the spectral properties of the observation filters. We will now consider a few particular cases.

3.2.1 Chain of differentiators.

To model a chain of differentiators within the framework of § 2, we assume that in the useful frequency interval 0 to f_m, we have $|H_0| = 1$. This can be achieved by making all components of the vector of filter coefficients **f** sufficiently large with the constraint $f_0 = K_f$, such that the poles of the observation filter lie outside the useful frequency band. The filter will then also exhibit a fast exponential decay in its impulse response, so in the limit a Q-filter to remove the transients will not be required.

The actual realization of the observation filter is a chain of digital filter approximations to a differentiator. The approximation will only be acceptable regarding filter order or spectral error if we allow that a pure delay is introduced. The "noise-accentuating" properties of the differentiator are taken into account in the estimator : the higher frequencies where the noise is re-enforced should be weighted inversely proportional to their noise amplification. Notice that a possible high-frequency component in the useful signal undergoes the same boosting by the differentiators and the same out-weighting by the estimator. Hence, for a carefully selected estimator, the "noise-accentuation" is not a problem. Furthermore, the observation filters operate in a limited frequency band 0 to f_m. If the band is well-filled with the useful signal, the numerical problem that the noise component swamps the useful component will not occur. Moreover, intrinsic differentiation due to the Q-filter is present for all choices of the observation filter.

The observation filter is depicted in Fig 3.1. Each digital differentiator exhibits an approximation error of $\delta(j\omega)$ whose magnitude is assumed to be much smaller than 1, and is realized within an integer delay τ_d. Let Di(z) denote the z-transform of an individual differentiator. Then

$$Di(e^{j\omega}) = (j\omega)\, e^{-j\omega\tau_d}\, (1 + \delta(j\omega))$$

$$H_{i+1}(e^{j\omega}) = e^{-j(q-i)\omega\tau_d}\, Di(e^{j\omega})^i \qquad i = 0 \ldots q \qquad (3.2.3)$$

and $H_0(e^{j\omega}) = H_1(e^{j\omega}) = e^{-jq\omega\tau_d}$

where q = max(m, n). The extra delay of $(q-i)\tau_d$ in H_{i+1} is added to make all components of the observation filter output contemporary. Hence

$$\begin{aligned} H_{i+1}(j\omega) &= e^{-j\omega q\tau_d}\, (j\omega)^i\, (1+ \Delta_{i+1}(j\omega)) \\ &= e^{-j\omega q\tau_d}\, (j\omega)^i\, (1 + \delta(j\omega))^i \\ &\approx e^{-j\omega q\tau_d}\, (j\omega)^i\, (1 + i\, \delta(j\omega)) \end{aligned}$$

thus $\Delta_{i+1}(j\omega) \approx i\, \delta(j\omega) \qquad i = 0 \ldots q$

With definition (3.0.4b) we find $B_d(j\omega) \approx (j\omega)\, B'(j\omega)\, \delta(j\omega)$ where $B'(s)$ is the first derivative of $B(s)$ with respect to s. If s_i ($i = 1 \ldots m$) are the roots of $B(s)$, we can write:

$$\frac{B_d(j\omega)}{B(j\omega)} \approx \sum_{i=1}^{m} \frac{j\omega}{j\omega - s_i}\, \delta(j\omega) \tag{3.2.4a}$$

Define the "quality factor" of a root $s_i = -\alpha_i + j\beta_i$ as

$$Q_i = \frac{\sqrt{\alpha_i^2 + \beta_i^2}}{2\,|\alpha_i|}$$

then
$$\left|\frac{B_d(j\omega)}{B(j\omega)}\right| \le 2 \sum_{i=1}^{m} Q_i\, |\delta(j\omega)| \tag{3.2.4b}$$

In the neighborhood of a root s_p with for all $k \neq p, |\alpha_p| \ll |\beta_p - \beta_k|$, expression (3.2.4) is simplified even further. Under those assumptions, the "gain" of $\delta(j\omega)$ is approximately $2\, Q_p$.

Other important design considerations are: (3.2.5)

1. $n_{\varepsilon,d}$ = the order of a single differentiator required to reduce $|\delta(j\omega)|$ below a given level of ε_d for all ω in a given interval 0 to ω_m,
2. τ_d = the delay with which the differentiator is realized,
3. $M_{init,d}$ = the number of samples required to initialize all differentiators, reduced by one. This number quantifies the amount of information that is lost to initialize the observation filters,
4. c_d = a number that is proportional to the total number of arithmetic operations required to produce one evaluation of $\mathbf{h}_k$ in (2.5.1),
5. $\tau_{t,d}$ = the total estimation delay introduced by the observation filters.

Notice that τ_d, $M_{init,d}$, c_d and $\tau_{t,d}$ depend on $n_{\varepsilon,d}$.

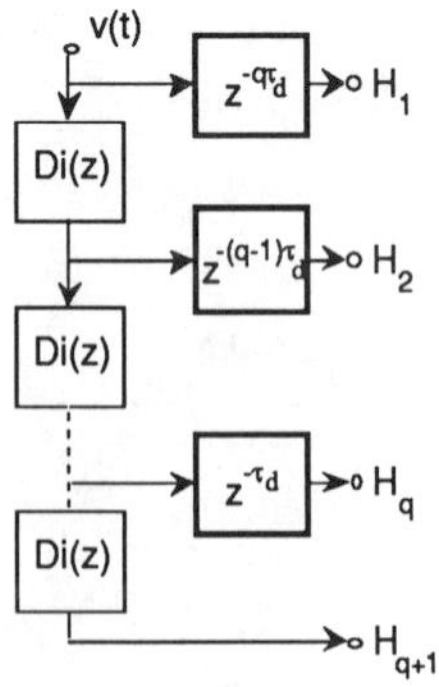

Figure 3.1 : a chain of digital differentiators with appropriate equalizing delays can be used as an observation filter.

These criteria will be compared for several choices of the observation filter and hence we require standardized testing conditions. We idealize $G(j\omega)$ such that

$$G(j\omega) = 1 \qquad \text{for } |\omega| < 2\pi f_b$$
$$= 0 \qquad \text{for } |\omega| \geq 2\pi f_b$$

with $f_b = 0.25$ Hz, i.e. spectrally, we measure up to half the Nyquist frequency. Since the signals that are fed to the observation filters will contain no spectral components above f_b, it is appropriate to choose $f_m = 0.25$ Hz in their design.

The filters are designed as in Pintelon & Schoukens (1990b). Because of the absence of poles and the simplicity of implementation, the FIR filter structure is preferred over the IIR structure. The IIR filter structure may have a degree of the numerator that is greater than the degree of the denominator. See Pintelon and Kollár (1991) for an extensive comparison of these structures. Table 3.1 lists the performance for so-called FIR halfband (f_m is half the Nyquist frequency) differentiators. The measure of goodness of a digital differentiator is defined as:

$$\varepsilon_d = \max_{|\omega| < 2\pi f_m} |\delta(j\omega)|$$

The value for τ_d in the table is the positive integer delay in samples that renders the smallest value of ε_d for a given FIR filter order $n_{\varepsilon,d}$ ($n_{\varepsilon,d} + 1$ coefficients). The designs turn out to be nearly equiripple for $0 \leq \omega \leq 2\pi f_m$. If equiripple designs are used as in Pintelon and Kollár (1991), all approximation errors in Table 3.1 can be reduced by a couple of dB's, but this would not affect our conclusions.

For an m-fold digital differentiator for the input and an n-fold differentiator for the output we have:

$$\boxed{\begin{aligned} M_{init,d} &= q\, n_{\varepsilon,d} \\ c_d &= (m+n)\,(n_{\varepsilon,d}+1) \\ \tau_{t,d} &= q\, \tau_d \end{aligned}} \qquad (3.2.6)$$

where $n_{\varepsilon,d}$ and τ_d are read from Table 3.1 for any given value of ε_d. Notice that the total number of differentiators required is m+n, not 2q.

Table 3.1 : performance of FIR halfband differentiators.

filter order $n_{\varepsilon,d}$	approx. error ε_d in dB	optimal delay τ_d
7	-47	3
8	-56	4
9	-63	4
10	-71	5
11	-79	5
12	-87	6
13	-95	6
14	-103	7
15	-111	7

3.2.2 Chain of first order differenced filters.

In this special case, the characteristic polynomial of the observation filters is chosen as $P(s) = (s + \lambda)^q$, where $q = \max(m, n)$. For notations regarding the observation filters, refer to (2.2.1) and (2.2.2). Notice the close relation with the observation filters used in the Poison moment functional approach (e.g. Saha and Rao, 1983). If λ is chosen to be zero, the integrated sampling approach emerges (Sagara and Zhao, 1990 and Schoukens, 1990). Choosing $K_f = (l\lambda)^q\,(1-e^{-l\lambda})^{-q}$, where l is any strictly positive integer number, this linear system can be realized as a cascade of first order sections with transfer function

$$F_0(s) \quad = \frac{l\lambda}{(1-e^{-l\lambda})\,(s+\lambda)}.$$

If the new realization of the observation filter is to be equivalent with (2.2.2), the transfer function from the scalar input v(t) to the i-th component of the output vector **z**(t) must be

$$F_i(s) \quad = \left(\frac{l\lambda}{1-e^{-l\lambda}}\right)^q \frac{s^{i-1}}{P(s)} \qquad\qquad i = 1 \ldots q+1$$

Denote the state vector of this linear system by **w**(t), while its controller form realization (2.2.2) has state vector **x**(t). The theory of similarity transforms of state-space realizations guarantees the existence of a nonsingular matrix $\mathbf{W} \in \mathbb{R}^{qxq}$ such that

$$\mathbf{x}(t) = \mathbf{W}\,\mathbf{w}(t).$$

Define the nonsingular matrix

$$\mathbf{V} = \begin{pmatrix} \mathbf{W} & \mathbf{0}_{qx1} \\ -\mathbf{f}^t\mathbf{W} & K_f \end{pmatrix}$$

then $\mathbf{z}(t) = \mathbf{V}\begin{pmatrix} \mathbf{w}(t) \\ v(t) \end{pmatrix}$. (3.2.7)

Writing (3.2.7) in terms of transfer functions, we obtain:

$$\left(\frac{l\lambda}{1-e^{-l\lambda}}\right)^q \frac{1}{P(s)} \begin{pmatrix} 1 \\ s \\ s^2 \\ \cdot \\ \cdot \\ \cdot \\ s^{q-2} \\ s^{q-1} \\ s^q \end{pmatrix} = \mathbf{V} \begin{pmatrix} F_0^q(s) \\ F_0^{q-1}(s) \\ \cdot \\ \cdot \\ \cdot \\ F_0^2(s) \\ F_0(s) \\ 1 \end{pmatrix} \tag{3.2.8}$$

Taking the derivative of (3.2.8) with respect to s yields:

$$\left(\frac{l\lambda}{1-e^{-l\lambda}}\right)^q \frac{1}{P(s)} \begin{pmatrix} 0 \\ 1 \\ 2s \\ \cdot \\ \cdot \\ \cdot \\ (q-2)\,s^{q-3} \\ (q-1)\,s^{q-2} \\ q\,s^{q-1} \end{pmatrix} - \left(\frac{l\lambda}{1-e^{-l\lambda}}\right)^q \frac{P'(s)}{P^2(s)} \begin{pmatrix} 1 \\ s \\ s^2 \\ \cdot \\ \cdot \\ \cdot \\ s^{q-2} \\ s^{q-1} \\ s^q \end{pmatrix} = -\frac{1-e^{-l\lambda}}{l\lambda} F_0(s)\,\mathbf{V} \begin{pmatrix} q\,F_0^q(s) \\ (q-1)\,F_0^{q-1}(s) \\ \cdot \\ \cdot \\ \cdot \\ 2\,F_0^2(s) \\ F_0(s) \\ 0 \end{pmatrix} \tag{3.2.9}$$

This identity will be used later.

Fig. 3.2 illustrates the successive steps in the derivation. The observation filter in its controller form and its discrete-time FIR filter to remove the initial conditions (the Q-filter as discussed in § 2.3) are depicted in Fig. 3.2a. The necessity of the extra delay line of $q\,\tau_\lambda$ added after the Q-filter will become clear soon. The transfer function of the Q-filter is written as a function of the Laplace variable:

$$Q(e^s) = Q_0^q(e^s)$$

$$Q_0(e^s) = 1 - e^{-l\lambda} e^{-ls}$$

Fig. 3.2b shows the cascade form of the continuous-time observation filter with its discrete-time companion. When approximating the cascade of $F_0(s)$ and $Q_0(e^s)$ by a discrete filter $\overline{FQ}$, an integer delay of τ_λ will be allowed:

$$\overline{FQ}\,(e^s) = e^{-\tau_\lambda s} F_0(s)\, Q_0(e^s)\,(1 + \delta(s)) \tag{3.2.10}$$

When cascading q of these $\overline{FQ}$-filters, an inevitable delay of $q\tau_\lambda$ results and was therefore already incorporated in Fig 3.2a. After rearranging the Q_0-filters and the equalizing delays, the observation filter, as it would be implemented, is shown in Fig 3.2c. The matrix multiplication would not be executed at each sampling instant. Instead, the system parameters would be transformed by $\mathbf{V}^{-1}$ and (2.5.1) would be written in these transformed variables. We will now derive the expressions of the global approximation errors $\Delta_i(s)$.

The transfer functions from the input of the observation filter to its outputs are given by

$$\begin{pmatrix} H_1(e^s) \\ H_2(e^s) \\ \cdot \\ \cdot \\ \cdot \\ H_q(e^s) \\ H_{q+1}(e^s) \end{pmatrix} = \mathbf{V} \begin{pmatrix} \overline{FQ}^{\,q}(e^s) \\ e^{-\tau_\lambda s} Q_0(e^s)\, \overline{FQ}^{\,q-1}(e^s) \\ \cdot \\ \cdot \\ \cdot \\ (e^{-\tau_\lambda s} Q_0(e^s))^{q-1}\, \overline{FQ}\,(e^s) \\ (e^{-\tau_\lambda s} Q_0(e^s))^q \end{pmatrix}$$

$$\approx e^{-q\tau_\lambda s}\, Q(e^s)\, \mathbf{V} \begin{pmatrix} F_0^q(s) \\ F_0^{q-1}(s) \\ \cdot \\ \cdot \\ \cdot \\ F_0^2(s) \\ F_0(s) \\ 1 \end{pmatrix} + e^{-q\tau_\lambda s}\, Q(e^s)\, \mathbf{V} \begin{pmatrix} q\,F_0^q(s) \\ (q-1)\,F_0^{q-1}(s) \\ \cdot \\ \cdot \\ \cdot \\ 2\,F_0^2(s) \\ F_0(s) \\ 0 \end{pmatrix} \delta(s)$$

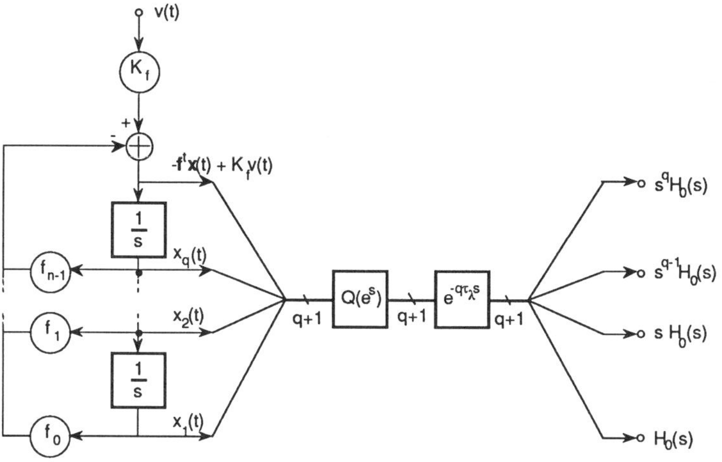

Figure 3.2a : observation filter in controller state-space form with initial condition removing filter and delay.

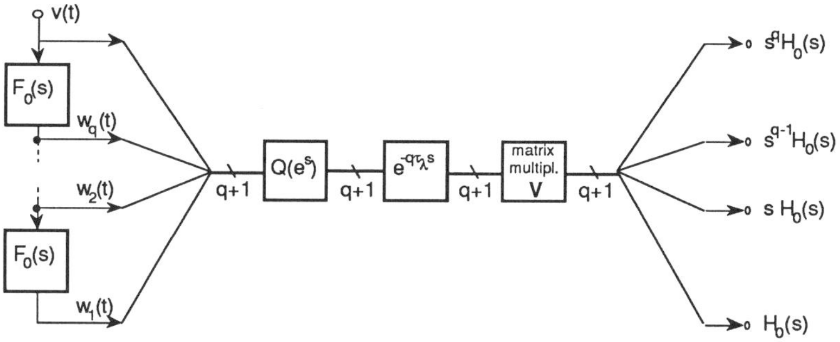

Figure 3.2b : observation filter realized as a cascade of identical first order sections with initial condition removing filter, delay and linear transformation.

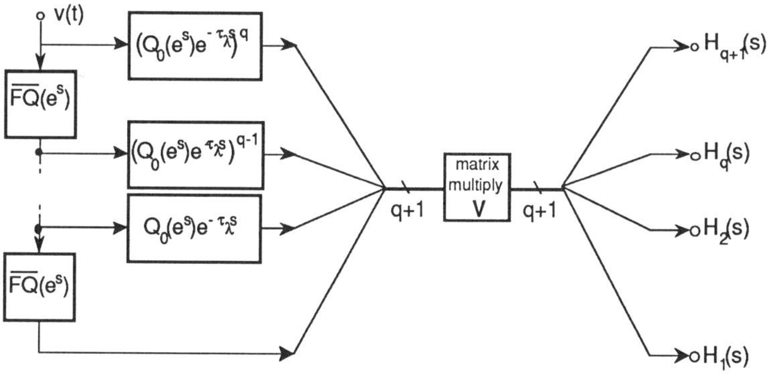

Figure 3.2c : approximation to Fig. 3.2b using digital filtering.

where only the second term involves an approximation. Recalling that $F(s) = K_f/P(s)$ and with (3.2.8) and (3.2.9) we find:

$$H_{i+1}(e^{s}) \approx e^{-q\tau\lambda s} F(s)\, Q(e^{s})\, s^{i} \left(1 + (s+\lambda)\left(\frac{P'(s)}{P(s)} - i\, s^{-1}\right)\delta(s)\right) \qquad i = 0 \ldots q$$

Hence $H_0(e^{s}) = e^{-q\tau\lambda s} F(s)\, Q(e^{s})$ and $\Delta_{i+1}(s) \approx (s+\lambda)\left(\frac{P'(s)}{P(s)} - i\, s^{-1}\right)\delta(s)$.

Finally, with $P(s) = (s+\lambda)^q$, we have

$$\frac{B_d(s)}{B(s)} \approx \left(q - (s+\lambda)\frac{B'(s)}{B(s)}\right)\delta(s)$$

If s_i ($i = 1 \ldots m$) are the roots of $B(s)$, we can write:

$$\frac{B_d(j\omega)}{B(j\omega)} \approx \left((q-m) - \sum_{i=1}^{m} \frac{s_i+\lambda}{j\omega - s_i}\right)\delta(j\omega) \qquad (3.2.11)$$

The relative error has a contribution due to the difference in order of the observation filter and the degree of the polynomial to be identified. Concerning the second term, it is advantageous to choose $\lambda = 0$ if this doesn't imply a loss in $|\delta(j\omega)|$. For $\lambda = 0$, the gain of $\delta(j\omega)$ is reduced to same order of magnitude as for the chain of differentiators.

We will now investigate the influence of λ on the quality of the approximation in (3.2.10). The same testing conditions and filter design techniques as in § 3.2.1 are assumed. The measure of goodness for a first order differenced filter with pole at $-\lambda$ is defined as:

$$\varepsilon_\lambda = \max_{|\omega| < 2\pi f_m} |\delta(j\omega)|$$

Table 3.2 shows the influence of λ on this measure for an 8-th order FIR approximation with $l = 1, 2, 3$ or 4. Higher values of l should not be considered if $f_m = 0.25$ Hz, since Q_0 produces a q-fold zero in the neighborhood of $\omega = \frac{2\pi}{l}$. Particularly for small λ, this degenerates the ability of the setup to measure at these frequencies. Also, correcting the measurement channel to achieve a flat magnitude response and a linear phase response for frequencies between 0 and f_b becomes hard. From Table 3.2 and the considerations regarding (3.2.11), we conclude that the two-sample integrator has optimal properties:

$$\boxed{\begin{array}{c}\lambda = 0 \\ l = 2\end{array}} \qquad (3.2.12)$$

An aspect not apparent from Table 3.2 is that increasing λ or decreasing l has the disadvantage of increasing τ_λ. The use of an IIR filter with a 6-th order numerator and 2-nd order denominator can improve the filter designs for high values of λ and $l \neq 2$, but they remain inferior to the choice (3.2.12). An extra advantage of the integrated sampling approach is that the transformation **V** becomes trivial.

Table 3.2 : worst approximation error of 8-th order FIR halfband first order differenced filters.

λ	l = 1 (dB)	l = 2 (dB)	l = 3 (dB)	l = 4 (dB)
0.0	-70.9	-100.1	-78.4	-96.2
0.1	-70.9	-95.5	-78.6	-91.6
0.2	-70.9	-92.3	-78.7	-87.2
0.5	-71.0	-90.9	-78.4	-88.6
1.0	-71.1	-80.3	-76.8	-83.3
1.5	-71.3	-76.5	-75.4	-76.6

Using similar notations as in (3.2.5), we find with $\lambda = 0$ and $l \leq n_{\varepsilon,0}$

$$\boxed{\begin{aligned} M_{init,0} &= q\, n_{\varepsilon,0} \\ c_0 &= 2q\,(n_{\varepsilon,0}+q+1) \\ \tau_{t,0} &= q\, \max(\tau_0, l) \end{aligned}} \qquad (3.2.13)$$

where $n_{\varepsilon,0}$ and τ_0 are read from Table 3.3 for any given value of ε_0 . Notice that the sample integrator can meet the requirements on $|\delta(j\omega)|$ with lower orders than for the differentiators. Typically, the order of each approximating filter can be reduced by 6. On the other hand, the complexity of the total observation filters is increased due to the extra weighted differences Q_0 that have to be implemented and the need for two observation filters consisting of q $\overline{FQ}$ -filters (compared to m+n units for the chain of differentiators). If m = n < 6, the sample integrator is more efficient to implement than the differentiator. For m < n (or n < m), the chain of differentiators becomes more efficient as m (n) decreases.

Another disadvantage of integrated sampling is that the equalization of the measurement channel (assuring that $G(j\omega)$ has a flat magnitude response and a linear phase response) depends on q via $H_0(e^s)$, whereas $H_0(e^s)$ is already a pure delay for the chain of differentiators. Hence, for time-varying systems, this procedure must be re-executed if the model order is increased. On the other hand, the total estimation delay τ_t is smaller for the sample integrator because of the reduced order of each digital filter and because $\tau_d > \tau_0$ for $n_{\varepsilon,d} = n_{\varepsilon,0}$. This can be an important consideration for on-line application of the method.

Table 3.3 : performance of FIR halfband two-sample integrators.

filter order $n_{\varepsilon,0}$	approx. error ε_0 in dB	optimal delay τ_0
6	-77	2
7	-77	2
8	-95	3
9	-96	3
10	-113	4
11	-113	4
12	-130	5
13	-131	5
14	-148	6
15	-149	6

3.2.3 Chain of second order differenced filters.

In § 3.2.2 we already elaborated on the comparison of the chain of differentiators versus the chain of differenced first order filters. These observation filter structures have in common that they are realized as a cascade of identical discrete-time systems, each simulating a continuous-time system with one pole or one zero, with identical approximation errors. Of course, other structures can be devised. For a cascade of second order differenced filters that introduce two approximation errors, the control of the transfer function estimation error becomes harder. Similarly to § 3.2.2, we choose

$$P(s) = (s^2 + 2\alpha s + \alpha^2 + \beta^2)^{q/2}$$

$$\frac{K_f}{P(s)} = F_0^{q/2}(s)$$

$$F_0(s) = \frac{l^2(\alpha^2 + \beta^2)}{(1-2e^{-\alpha l}\cos(\beta l) + e^{-2\alpha l})(s^2 + 2\alpha s + \alpha^2 + \beta^2)}$$

$$Q(e^s) = Q_0^{q/2}(e^s)$$

$$Q_0(e^s) = 1 - 2e^{-\alpha l}\cos(\beta l)\,e^{-s} + e^{-2\alpha l}\,e^{-2s}$$

where q is the even integer number greater than or equal to the maximum of m and n, α is real and positive and β is real (complex conjugate poles) or imaginary (real poles). Suppose we design a digital system with one input and two outputs that approximates the continuous-time system with transfer functions $F_0(s)\,Q_0(e^s)$ and $(s+\lambda)\,F_0(s)\,Q_0(e^s)$ with λ any real number:

$$\overline{FQ_1}(e^s) = e^{-\tau s}F_0(s)\,Q_0(e^s)\,(1+\delta_1(s)) \qquad \text{(first output)}$$

$$\overline{FQ_2}(e^s) = (s+\lambda)\,e^{-\tau s}F_0(s)\,Q_0(e^s)\,(1+\delta_2(s)) \qquad \text{(second output)}$$

These systems are cascaded by connecting the first output of one system to the input of the next system. The observation filter is formed by cascading of q/2 of these digital systems with addition of appropriate equalizing delays and Q_0-filters in the output lines. The analysis of the measurement channel is similar to the one carried out in § 3.2.2 but is more involved and will therefore be omitted. The result is:

$$\frac{B_d(s)}{B(s)} \approx \frac{s^2 + 2\alpha s + \alpha^2 + \beta^2}{2(s+\alpha)}\left(\frac{P'(s)}{P(s)} - \frac{B'(s)}{B(s)}\right)\delta_1(s)$$
$$+ \frac{\gamma(s)}{B(s)}\left(\frac{s^2 + 2\alpha s + \alpha^2 + \beta^2}{2(s+\alpha)}\,\delta_1(s) + (s+\lambda)\,(\delta_2(s)-\delta_1(s))\right)$$

where $\gamma(s)$ is some polynomial of the same degree as B(s). The ratio $|\gamma(j\omega)/B(j\omega)|$ attains values that are inversely proportional to the product of all roots of B(s), measured in α-units. To avoid problems with DUTs that have poles or zeros close to the origin, the second term must be minimized in the design of the second order differenced filters, which is a complex problem.

Considering the observation filter approximation as the design of a SIMO digital system with a q+1 outputs is a generalization of the above second order formulation. A good observation filter design does not simply minimize the magnitude of the approximation errors of all outputs, but minimizes their joint worst-case effect on the transfer function estimate. This is even more difficult than in the second order case.

3.2.4 Discussion and conclusions.

With the present digital filter design techniques, the previous section shows that it is safest to either use the chain of differentiators or the chain of 2-sample integrators, depending on the comparison of (3.2.6) versus (3.2.13).

For the chain of first order differenced filters we mentioned before that for $\lambda = 0$, the integrated sampling approach is found. It is customary to compute the integrals over multiple samples using classical numerical integration methods. The resulting method is called the numerical integration approach (Sagara and Zhao 1990). The method we suggest here differs from this approach due to the introduction of τ_λ and the fact that $n_{\varepsilon,0} \geq l$. Using the band-limited properties of the input signals, their integral over l samples is computed using samples outside the integration interval aswell. This method will allow a slower sampling rate than the numerical integration approach and the Nyquist frequency can be chosen closer to the bandwidth of the DUT. For a given sampling rate, smaller errors result, but the approach requires more samples and has the disadvantage of the delay inherent to using samples outside the integration interval. In other words, we consider the integration as a filter design problem in the frequency domain instead of resorting to the classical numerical integration techniques that are based on time-domain criteria. The design of the differentiators is treated in the same spirit: as a filter design problem in the frequency domain and not as a numerical problem that is solved using time-domain considerations. An important aspect of the filter design approach is that the observation filter approximation error can be reduced to any level by increasing the digital filter order.

4. Maximum likelihood off-line estimation for time-invariant systems.

Consider the model equation (2.5.1) rewritten as:

$$\mathbf{he}_k^t \, \mathbf{pe} = 0 \qquad \text{for } k = 1 \ldots K \tag{4.0.1}$$

$$\mathbf{he}_k, \mathbf{pe} \in \mathbb{R}^{L\text{x}1}$$

where $\mathbf{he}_k$ are the noise-free (exact) observations and $\mathbf{pe}$ are the exact system parameters of a time-invariant system. We wish to obtain the maximum likelihood estimate (MLE) of $\mathbf{pe}$ from noisy observations

$$\mathbf{g}_k = \mathbf{he}_k + \mathbf{n}_k$$

where $\mathbf{n}_k$ is stationary zero-mean Gaussian observation noise with known autocovariance up to a scalar σ^2. Assume that among the L parameter vector components, L-Q (Q ≤ L and Q ≤ K) are known (suffix k) and Q are unknown (suffix u). Introduce the following notations:

$$\mathbf{p}^t = (\mathbf{pk}^t, \mathbf{pu}^t)$$

$\mathbf{pk} = (\mathbf{pe}_1, \ldots, \mathbf{pe}_{L\text{-}Q})^t \in \mathbb{R}^{(L\text{-}Q)\text{x}1}$ are the known parameters

$\mathbf{pu} \in \mathbb{R}^{Q\text{x}1}$ are the unknown parameters

$$\mathbf{B_p} = \begin{pmatrix} \mathbf{p}^t & \mathbf{0} & \ldots & \mathbf{0} \\ \mathbf{0} & \mathbf{p}^t & \ldots & \mathbf{0} \\ & & \ddots & \\ \mathbf{0} & \mathbf{0} & \ldots & \mathbf{p}^t \end{pmatrix} \in \mathbb{R}^{K\text{x}KL} \tag{4.0.2a}$$

$$\mathbf{g} = (\mathbf{g}_1^t, \ldots, \mathbf{g}_K^t)^t, \ \mathbf{h} = (\mathbf{h}_1^t, \ldots, \mathbf{h}_K^t)^t \text{ and } \mathbf{n} = (\mathbf{n}_1^t, \ldots, \mathbf{n}_K^t)^t \in \mathbb{R}^{KL\text{x}1}$$

$$\mathbf{G} = (\mathbf{g}_1, \ldots, \mathbf{g}_K)^t, \ \mathbf{H} = (\mathbf{h}_1, \ldots, \mathbf{h}_K)^t \in \mathbb{R}^{K\text{x}L}$$

$$\mathbf{G} = (\mathbf{Gk}, \mathbf{Gu}), \quad \mathbf{H} = (\mathbf{Hk}, \mathbf{Hu})$$

with $\mathbf{Gk}, \mathbf{Hk} \in \mathbb{R}^{K\text{x}(L\text{-}Q)}$ and $\mathbf{Gu}, \mathbf{Hu} \in \mathbb{R}^{K\text{x}Q}$

$\mathbf{r_p} = \mathbf{G}\,\mathbf{p} = \mathbf{B_p}\,\mathbf{g} \in \mathbb{R}^{K\text{x}1}$ are the residuals for parameter choice $\mathbf{p}$ (4.0.2b)

$\mathbf{C} = \sigma^{-2}\, E\{\mathbf{n}\,\mathbf{n}^t\}$ nonsingular where σ^2 is an unknown real number (4.0.2c)

$$\mathbf{C}^{-1} = \mathbf{F}^t\mathbf{F} \text{ with } \mathbf{F} \in \mathbb{R}^{KL\text{x}KL} \tag{4.0.2d}$$

and $\mathbf{C_p} = \mathbf{B_p}\mathbf{C}\mathbf{B}_\mathbf{p}^t = \sigma^{-2}\, E\{(\mathbf{r_p} - E\{\mathbf{r_p}\})\,(\mathbf{r_p} - E\{\mathbf{r_p}\})^t\} \in \mathbb{R}^{K\text{x}K}$ (4.0.2e)

4.1 The negative loglikelihood function.

The probability density function (p.d.f.) of the observation $\mathbf{g}$, given $\mathbf{h}$ is:

$$P[\mathbf{g}\,|\,\mathbf{h}] = \frac{1}{\sigma^K \sqrt{2\pi}^K \sqrt{\det(\mathbf{C})}} \exp\left(-\frac{1}{2\sigma^2}(\mathbf{g}-\mathbf{h})^t\,\mathbf{C}^{-1}\,(\mathbf{g}-\mathbf{h})\right) \tag{4.1.1}$$

The maximum likelihood estimates $\mathbf{h}_{\mathrm{ML}}$ and $\mathbf{p}_{\mathrm{ML}}$ are obtained by maximizing $P[\mathbf{g}\,|\,\mathbf{h}]$ over $\mathbf{h}$ and the unknown components of $\mathbf{p}$ with the constraint

$$\mathbf{B}_\mathbf{p}\,\mathbf{h} = 0 \tag{4.1.2a}$$

Equation (4.1.2a) implies that $\mathbf{h}$ must be in the null space of $\mathbf{B}_\mathbf{p}$, or equivalently:

$$\mathbf{h} = \mathbf{D}_\mathbf{p}\,\mathbf{x} \text{ for some } \mathbf{x} \in \mathbb{R}^{K(L-1)x1} \tag{4.1.2b}$$

where $\mathbf{D}_\mathbf{p}$ is a KLxK(L-1) matrix of full rank whose columns span the null space of $\mathbf{B}_\mathbf{p}$. The problem can be reformulated as minimizing the negative loglikelihood function which, within an additive constant, is given by:

$$L_g(\mathbf{x},\,\mathbf{p}) = \frac{1}{2}(\mathbf{g}-\mathbf{D}_\mathbf{p}\,\mathbf{x})^t\,\mathbf{C}^{-1}\,(\mathbf{g}-\mathbf{D}_\mathbf{p}\,\mathbf{x}). \tag{4.1.3}$$

The minimization over $\mathbf{x}$ is a weighted linear least squares problem with solution

$$\mathbf{x}_\mathbf{p} = (\mathbf{D}_\mathbf{p}^t\,\mathbf{C}^{-1}\,\mathbf{D}_\mathbf{p})^{-1}\,\mathbf{D}_\mathbf{p}^t\,\mathbf{C}^{-1}\,\mathbf{g}, \tag{4.1.4}$$

With this expression, the weighted residuals can be written as

$$\mathbf{F}\,(\mathbf{g}-\mathbf{D}_\mathbf{p}\,\mathbf{x}_\mathbf{p}) = (\mathbf{I}-(\mathbf{F}\mathbf{D}_\mathbf{p})\,(\,(\mathbf{F}\mathbf{D}_\mathbf{p})^t\,(\mathbf{F}\mathbf{D}_\mathbf{p}))^{-1}\,(\mathbf{F}\mathbf{D}_\mathbf{p})^t)\,\mathbf{F}\,\mathbf{g}$$

Since the idempotent matrix $\mathbf{P}^\perp_{\mathbf{FD}} = \mathbf{I} - (\mathbf{F}\,\mathbf{D}_\mathbf{p})\,((\mathbf{F}\mathbf{D}_\mathbf{p})^t\,(\mathbf{F}\mathbf{D}_\mathbf{p}))^{-1}\,(\mathbf{F}\mathbf{D}_\mathbf{p})^t$ is the projection onto the orthogonal complement of the column space of $\mathbf{F}\mathbf{D}_\mathbf{p}$ and since $\mathbf{B}_\mathbf{p}\mathbf{D}_\mathbf{p} = \mathbf{0}$, we have

$$\begin{aligned}
\mathbf{P}^\perp_{\mathbf{FD}} &= \mathbf{I} - (\mathbf{F}\mathbf{D}_\mathbf{p})\,((\mathbf{F}\mathbf{D}_\mathbf{p})^t\,(\mathbf{F}\mathbf{D}_\mathbf{p}))^{-1}\,(\mathbf{F}\mathbf{D}_\mathbf{p})^t \\
&= (\mathbf{B}_\mathbf{p}\mathbf{F}^{-1})^t\,((\mathbf{B}_\mathbf{p}\mathbf{F}^{-1})(\mathbf{B}_\mathbf{p}\mathbf{F}^{-1})^t)^{-1}\,(\mathbf{B}_\mathbf{p}\mathbf{F}^{-1})
\end{aligned} \tag{4.1.5}$$

$$\begin{aligned}
\text{so } L_g(\mathbf{x}_\mathbf{p},\,\mathbf{p}) &= \frac{1}{2}\mathbf{g}^t\,\mathbf{F}^t\,(\mathbf{B}_\mathbf{p}\mathbf{F}^{-1})^t\,((\mathbf{B}_\mathbf{p}\mathbf{F}^{-1})(\mathbf{B}_\mathbf{p}\mathbf{F}^{-1})^t)^{-1}\,(\mathbf{B}_\mathbf{p}\mathbf{F}^{-1})\,\mathbf{F}\,\mathbf{g} \\
&= \frac{1}{2}\mathbf{g}^t\,\mathbf{B}_\mathbf{p}^t\,(\mathbf{B}_\mathbf{p}\mathbf{C}\mathbf{B}_\mathbf{p}^t)^{-1}\,\mathbf{B}_\mathbf{p}\,\mathbf{g}
\end{aligned}$$

Finally, given the observations $\mathbf{g}$, we have to minimize

$$K(\mathbf{p},\,\mathbf{g}) = \frac{1}{2}\mathbf{p}^t\,\mathbf{G}^t\,(\mathbf{B}_\mathbf{p}\mathbf{C}\mathbf{B}_\mathbf{p}^t)^{-1}\,\mathbf{G}\mathbf{p} = \frac{1}{2}\mathbf{r}_\mathbf{p}^t\,\mathbf{C}_\mathbf{p}^{-1}\,\mathbf{r}_\mathbf{p} \tag{4.1.6}$$

over the unknown components of $\mathbf{p}$ to find $\mathbf{p}_{\mathbf{ML}}$. This value is then substituted in (4.1.4) from which it is possible to recover $\mathbf{h}_{\mathbf{ML}}$ using (4.1.2b).

Notice that $\mathbf{G}\,\mathbf{p}$ is the equation error for the chosen parameter value $\mathbf{p}$ and that if $\mathbf{n}_k$ is stationary, $\mathbf{C}_\mathbf{p}$ is the Toeplitz, symmetric autocorrelation matrix of this equation error.

4.2 Minimizing the cost function.

The minimization of K($\mathbf{p}$) requires nonlinear programming techniques. The i-th component of the gradient of K is given by:

$$\frac{\partial K(\mathbf{p}, \mathbf{g})}{\partial \mathbf{p}_i} = \mathbf{p}^t \mathbf{G}^t \mathbf{C}_\mathbf{p}^{-1} \mathbf{G}\,\mathbf{e}_i - \mathbf{p}^t \mathbf{G}^t \mathbf{C}_\mathbf{p} (\mathbf{B}_\mathbf{p} \mathbf{C} \mathbf{B}^t_{\mathbf{e}_i})^{-1} \mathbf{C}_\mathbf{p} \mathbf{G}\mathbf{p} \qquad (4.2.1)$$

where $\mathbf{B}_{\mathbf{e}_i}$ is given by (4.2a) with $\mathbf{p}$ replaced by $\mathbf{e}_i$. A positive semi definite approximation of the i,j-th entry of the Hessian matrix of K is:

$$\frac{\partial^2 K(\mathbf{p}, \mathbf{g})}{\partial \mathbf{p}_i \partial \mathbf{p}_j} = \mathbf{e}_j^t \mathbf{G}^t (\mathbf{B}_\mathbf{p} \mathbf{C} \mathbf{B}^t_\mathbf{p})^{-1} \mathbf{G}\,\mathbf{e}_i + O(\mathbf{r}_\mathbf{p}) \qquad (4.2.2)$$

where $O(\mathbf{r}_\mathbf{p})$ denotes a matrix with entries of the order of $\| \mathbf{r}_\mathbf{p} \|$ or smaller. Let $\mathbf{gr}(\mathbf{p}) \in \mathbb{R}^{Qx1}$ and $\mathbf{Hs}(\mathbf{p}) \in \mathbb{R}^{QxQ}$ be the gradient (column) vector and Hessian matrix corresponding to the unknown components of $\mathbf{p}$. The computation of (4.1.6), (4.2.1) and (4.2.2) requires the simultaneous solution of Q+1 symmetrical Toeplitz linear KxK sets of equations, which can be accomplished with a number of floating point operations that is proportional to K^2 using a Levinson-type algorithm (Golub and Van Loan, 1985). Then, provided that the initial guess $\mathbf{p}^{(0)}$ is sufficiently close to $\mathbf{p}_{ML}$, the following iterative scheme will yield the maximum likelihood estimate of the system parameters:

$$\mathbf{p}^{(i+1)^t} = \left(\mathbf{pk}^t, (\mathbf{pu}^{(i)} - \mathbf{Hs}^{-1}(\mathbf{p}^{(i)})\, \mathbf{gr}(\mathbf{p}^{(i)})\,)^t\,\right) \qquad (4.2.3)$$

The initial guess $\mathbf{p}^{(0)}$ is obtained from the total least squares (TLS) solution of the K equations

$$\mathbf{Gu}\ \mathbf{pu} = -\mathbf{Gk}\ \mathbf{pk}, \qquad (4.2.4)$$

i.e. $\mathbf{pu}$ is the solution of the set of linear equations $\mathbf{A}\ \mathbf{pu} = \mathbf{b}$ with $\mathbf{A}$ and $\mathbf{b}$ chosen such that $\| (-\mathbf{Gk\,pk}, \mathbf{Gu}) - (\mathbf{b}, \mathbf{A}) \|_F$ is minimal (Van Huffel, 1987). Taking the covariance of ($-\mathbf{Gk\,pk}$, $\mathbf{Gu}$) into account, TLS generates consistent estimates of $\mathbf{pu}_e$. The maximum likelihood estimator (4.2.3) is used to obtain asymptotically efficient estimates. The details of the TLS estimator are given in lemma 4.1.

4.3 The Cramér-Rao lower bounds.

Consider the problem of estimating $\mathbf{h}_k$ and the unknown components of $\mathbf{p}e$ in (4.0.1). Thanks to (4.1.2b), this is tantamount to the estimation of $\mathbf{x}$ and $\mathbf{pu}e$. The Fisher information matrix for this system is:

$$\mathbf{Fi} = E\left\{\left(\begin{matrix} \dfrac{\partial^2}{\partial \mathbf{x}^2} & \dfrac{\partial^2}{\partial \mathbf{x}\,\partial \mathbf{pu}} \\ \dfrac{\partial^2}{\partial \mathbf{pu}\,\partial \mathbf{x}} & \dfrac{\partial^2}{\partial \mathbf{pu}^2} \end{matrix}\right) L_g(\mathbf{x}, \mathbf{p})\right\}\Bigg|_{\substack{\mathbf{p} = \mathbf{p}e \\ \mathbf{D_p x} = \mathbf{h}e}}$$

where $L_g(\mathbf{x}, \mathbf{p})$ is given by (4.1.3). Partition $\mathbf{Fi}$ as:

$$\mathbf{Fi} = \begin{pmatrix} \mathbf{Fi}_{11} & \mathbf{Fi}_{12} \\ \mathbf{Fi}_{12}^t & \mathbf{Fi}_{22} \end{pmatrix}$$

then

$$\mathbf{Fi}_{11} = \mathbf{D_p^t}\, \mathbf{C}^{-1}\, \mathbf{D_p}\Big|_{\substack{\mathbf{p} = \mathbf{p}e \\ \mathbf{D_p x} = \mathbf{h}e}}$$

$$\mathbf{Fi}_{12} = \mathbf{D_p^t}\, \mathbf{C}^{-1}\, \frac{\partial(\mathbf{D_p x})}{\partial \mathbf{pu}}\Bigg|_{\substack{\mathbf{p} = \mathbf{p}e \\ \mathbf{D_p x} = \mathbf{h}e}}$$

$$\mathbf{Fi}_{22} = \left(\frac{\partial(\mathbf{D_p x})}{\partial \mathbf{pu}}\right)^t \mathbf{C}^{-1}\, \frac{\partial(\mathbf{D_p x})}{\partial \mathbf{pu}}\Bigg|_{\substack{\mathbf{p} = \mathbf{p}e \\ \mathbf{D_p x} = \mathbf{h}e}}$$

The Cramér-Rao inequality states that for any unbiased estimator $\hat{\mathbf{pu}}$:

$$Var\{\hat{\mathbf{pu}}\} = E\left\{(\hat{\mathbf{pu}}-E\{\hat{\mathbf{pu}}\})\,(\hat{\mathbf{pu}}-E\{\hat{\mathbf{pu}}\})^t\right\} \geq \mathbf{Cr_{pu}}$$

where the matrix inequality must be interpreted in a positive definite context and

$$\begin{aligned} \mathbf{Cr_{pu}^{-1}} &= \mathbf{Fi}_{22} - \mathbf{Fi}_{12}^t\, \mathbf{Fi}_{11}^{-1}\, \mathbf{Fi}_{12} \\ &= \left(\frac{\partial(\mathbf{D_p x})}{\partial \mathbf{pu}}\right)^t \mathbf{F}^t\, \mathbf{P}_{\mathbf{FD}}^{\perp}\, \mathbf{F}\left(\frac{\partial(\mathbf{D_p x})}{\partial \mathbf{pu}}\right)\Bigg|_{\substack{\mathbf{p} = \mathbf{p}e \\ \mathbf{D_p x} = \mathbf{h}e}} \end{aligned}$$

With (4.1.5) we find

$$\mathbf{Cr}^{-1}_{\mathbf{pu}} = \left(\frac{\partial(\mathbf{D_p x})}{\partial \mathbf{pu}}\right)^t \mathbf{B}^t_{\mathbf{p}} (\mathbf{B_p C B^t_p})^{-1} \mathbf{B_p} \left(\frac{\partial(\mathbf{D_p x})}{\partial \mathbf{pu}}\right) \Bigg|_{\substack{\mathbf{p} = \mathbf{pe} \\ \mathbf{x} = \mathbf{D}^+_{\mathbf{pe}} \mathbf{he}}}$$

Using $\frac{\partial \mathbf{B_p}}{\partial \mathbf{pu}_i}\mathbf{h} = -\frac{\partial \mathbf{h}}{\partial \mathbf{pu}_i}\mathbf{B_p}$ and (4.1.2b), the i,j-th component of $\mathbf{Cr}^{-1}_{\mathbf{pu}}$ becomes

$$\left(\mathbf{Cr}^{-1}_{\mathbf{pu}}\right)_{ij} = \left(\frac{\partial \mathbf{B_p}}{\partial \mathbf{pu}_i}\mathbf{he}\right)^t (\mathbf{B_p C B^t_p})^{-1} \left(\frac{\partial \mathbf{B_p}}{\partial \mathbf{pu}_j}\mathbf{he}\right) \Bigg|_{\mathbf{p} = \mathbf{pe}}$$

Finally,

$$\begin{aligned} \mathbf{Cr}^{-1}_{\mathbf{pu}} &= \mathbf{Hue}^t (\mathbf{B_{pe} C B^t_{pe}})^{-1} \mathbf{Hue} \\ &= \mathbf{Hue}^t \mathbf{C}^{-1}_{\mathbf{pe}} \mathbf{Hue} \end{aligned} \tag{4.3.1}$$

The Cramér-Rao lower bounds can be calculated as the inverse of the (approximated or true) Hessian matrix of the negative loglikelihood function as given by (4.2.2), evaluated at **pe** and **he**. For a real-life experiment, where **pe** and **he** are not known, an approximation of the Cramér-Rao lower bound can be generated by substituting the estimates in (4.3.1). Assuming that the MLE has reached its asymptotic properties, this result can be used as an approximation of the covariance of the estimates obtained.

Often, the covariance matrix of the parameters is not easily interpreted and it is better to examine the variance of the transfer function estimate. Model (4.0.1) is used to identify the linear system with transfer function:

$$\mathrm{DUTe}(s) = \frac{\sum_{i=0}^{m} be_i\, s^i}{\sum_{i=0}^{n} ae_i\, s^i} = \frac{\mathrm{Be}(s)}{\mathrm{Ae}(s)}$$

The parameters a_i (= α_i) and b_i (= β_i) are arranged in $\mathbf{p}$ with the known coefficients placed ahead of the unknown quantities: $\mathbf{p}^t = (\mathbf{ak}^t, \mathbf{bk}^t, \mathbf{au}^t, \mathbf{bu}^t)$. The variance of the estimated frequency response $\hat{DUT}(j\omega)$ is lower-bounded by

$$Var\{\hat{DUT}(j\omega)\} \geq \frac{\partial DUT(j\omega)}{\partial \mathbf{pu}}^{H} \mathbf{Cr_{pu}} \frac{\partial DUT(j\omega)}{\partial \mathbf{pu}} = \frac{\mathbf{w}^{H}(j\omega)\,\mathbf{Cr_{pu}}\,\mathbf{w}(j\omega)}{Ae(j\omega)} \qquad (4.3.2)$$

where superscript H denotes conjugate transpose and where $\mathbf{w}^t(s) = (-DUTe(s)\,\mathbf{wa}^t(s), \mathbf{wb}^t(s))$. The i-th component of $\mathbf{wa}(s) \in \mathbb{C}^{Qx1}$ is equal to s raised to the power k if the i-th component of **au** equals a_k and **wb** is defined similarly. The proof of (4.3.2) is a generalization to nonscalar **pu** of the proof given in §17.14 of Stuart and Ord (1943).

4.4 Relation to frequency-domain estimation.

Assume that the excitation signal u(t) is periodic with a period of R (integer) samples and that K (the number of samples) is an integer multiple of R: K = S R. The excitation signal is allowed to violate the Nyquist sampling condition. Hence, we write:

$$u(t) = \sum_{k=-\infty}^{\infty} \hat{\mathbf{u}}_k \, e^{j\frac{2\pi}{R}k\,t}$$

where $\hat{\mathbf{u}}_k$ is k-th Fourier coefficient of the periodic signal u(t). The plant and, as usual, all measurement filters have reached their steady state. The observation noise on the input and output are Gaussian and uncorrelated with each other. Both have a power spectral density (p.s.d.) such that the Fourier coefficients of the noise filtered by G[.] are uncorrelated.

Let $\mathbf{E} \in \mathbb{C}^{KxK}$ be the matrix with k,l-th entry

$$\mathbf{E}_{kl} = e^{j\frac{2\pi}{K}(k-1)(l-1)} \qquad k, l = 1 \ldots K$$

which has the property:

$$\mathbf{E}^H \mathbf{E} = \mathbf{E}\,\mathbf{E}^H = K\,\mathbf{I}_{KxK} \qquad (4.4.1)$$

The discrete Fourier transform (DFT) of a record of K samples of the signal x(t) is defined as:

$$\overline{\mathbf{x}} = \mathbf{E}^H \mathbf{x},\ \overline{\mathbf{x}} \in \mathbb{C}^{Kx1} \text{ where } \mathbf{x} = (x(0), x(1), \ldots, x(K-1))^t.$$

Since u(t) is periodic with period R = K/S, all except components 1, S+1, 2S+1, ... , K-S+1 of $\overline{\mathbf{u}}$ will be zero. Furthermore, since u(t) is real, $\overline{\mathbf{u}}_{iS+1} = \overline{\mathbf{u}}^*_{K-iS+1}$ for i = 1 ... R - 1, i.e. $\overline{\mathbf{u}}$ is completely specified by its components 1, S+1,

2S+1, ... , [R/2]S+1 (the symbol [R/2] equals R/2 if R is even and equals (R-1)/2 if R is odd). Furthermore, due to alias we have:

$$\overline{\mathbf{u}}_{iS+1} = K \sum_{k=-\infty}^{\infty} \hat{\mathbf{u}}_{i+kR}$$

The residuals $\mathbf{r_p}$ are defined in (4.0.2.b) as a linear combination of the output of the observation filters. Hence, its DFT shows the same pattern of zeros and symmetry as the entries of $\overline{\mathbf{u}}$, and

$$\mathbf{r_p} = \frac{1}{K} \mathbf{E}\, \overline{\mathbf{r}}_{\mathbf{p}}$$

With (3.0.1), (3.0.2) and (3.0.5), the DFT of $\mathbf{r_p}$ is expressed as:

$$\begin{aligned} \frac{1}{K} \overline{\mathbf{r}}_{\mathbf{p},iS+1} &= (A(j\tfrac{2\pi}{R}i) + A_d(j\tfrac{2\pi}{R}i)) \sum_{k=-\infty}^{\infty} G(j\tfrac{2\pi}{R}(i+kR))\, \hat{\mathbf{y}}_{i+Rk} \\ &- (B(j\tfrac{2\pi}{R}i) + B_d(j\tfrac{2\pi}{R}i)) \sum_{k=-\infty}^{\infty} G(j\tfrac{2\pi}{R}(i+kR))\, \hat{\mathbf{u}}_{i+Rk} \end{aligned}$$

for i = 0 ... [R/2]

This expression and property (4.4.1) allow to rewrite the cost function (4.1.6) as:

$$\begin{aligned} K(\mathbf{p}, \mathbf{g}) &= \frac{1}{2K^2} \overline{\mathbf{r}}_{\mathbf{p}}^{\mathbf{H}} \mathbf{E}^H \mathbf{C}_{\mathbf{p}}^{-1} \mathbf{E}\, \overline{\mathbf{r}}_{\mathbf{p}} \\ &= \frac{1}{2} \overline{\mathbf{r}}_{\mathbf{p}}^{\mathbf{H}} (\mathbf{E}^H \mathbf{C}_{\mathbf{p}} \mathbf{E})^{-1}\, \overline{\mathbf{r}}_{\mathbf{p}} \end{aligned}$$

The matrix $\mathbf{E}^H \mathbf{C_p} \mathbf{E}$ equals, within a multiplicative constant, the covariance matrix of the DFT coefficients of the assumed noise model of the residuals as depicted in Fig. 4.1. The noise source labeled "n_u" is such that its R-point DFT yields uncorrelated values with variance of the i-th component equal to $R\,\mathbf{\Phi}_{u,i}$ (n_y similarly). With lemma 4.2 and taking into account that the noise processes $\{n_u\}$ and $\{n_y\}$ are independent, we find that $\mathbf{E}^H \mathbf{C_p} \mathbf{E}$ is diagonal with iS+1,iS+1-th entry:

$$\mathbf{\Phi}_{p,iS+1} = K\, | B(j\tfrac{2\pi}{R}i)+B_d(j\tfrac{2\pi}{R}i) |^2\, \mathbf{\Phi}_{u,i} + K\, | A(j\tfrac{2\pi}{R}i)+A_d(j\tfrac{2\pi}{R}i) |^2\, \mathbf{\Phi}_{y,i}$$

for i = 0 ... [R/2]

and $\mathbf{\Phi}_{p,K-iS+1} = \mathbf{\Phi}_{p,iS+1}$ for i = 1 ... [R/2]

Finally, removing the terms that are zero, we can write the cost function as:

$$K(\mathbf{p}, \mathbf{g}) = \frac{K}{2}\sum_{i=0}^{[R/2]} \frac{\gamma_i \,|\rho_i|^2\, |C(e^{j\frac{2\pi}{R}i})\, D(e^{j\frac{2\pi}{R}i})\, H_0(e^{j\frac{2\pi}{R}i})|^2}{|B(j\frac{2\pi}{R}i)+B_d(j\frac{2\pi}{R}i)|^2\, \Phi_{u,i} + |A(j\frac{2\pi}{R}i)+A_d(j\frac{2\pi}{R}i)|^2\, \Phi_{y,i}} \tag{4.4.2a}$$

with ρ_i

$$= (A(j\tfrac{2\pi}{R}i) + A_d(j\tfrac{2\pi}{R}i))\sum_{k=-\infty}^{\infty} AA(j\tfrac{2\pi}{R}(i+kR))\,\hat{\mathbf{y}}_{i+Rk}$$

$$- (B(j\tfrac{2\pi}{R}i) + B_d(j\tfrac{2\pi}{R}i))\sum_{k=-\infty}^{\infty} AA(j\tfrac{2\pi}{R}(i+kR))\,\hat{\mathbf{u}}_{i+Rk}$$

and $\gamma_i = 1$ for $i = 0$ and for possible contribution at the Nyquist frequency

$\gamma_i = 2$ else.

If the excitation is spectrally confined to $\omega_m = \omega_b$, the cost function reduces to:

$$K(\mathbf{p}, \mathbf{g}) = G_0^2\frac{K}{2}\sum_{i=0}^{[R/2]} \frac{\gamma_i\, |A(j\frac{2\pi}{R}i)\,\hat{\mathbf{y}}_{i+Rk} - B(j\frac{2\pi}{R}i)\,\hat{\mathbf{u}}_{i+Rk}|^2}{|B(j\frac{2\pi}{R}i)|^2\, \Phi_{u,i} + |A(j\frac{2\pi}{R}i)|^2\, \Phi_{y,i}} \tag{4.4.2b}$$

This cost function equals the one presented in Pintelon and Schoukens (1990a). Hence, (4.4.2a) extends the MLE based on a frequency-domain analysis to its time-domain counterpart where alias is present and non-ideal differentiation is used.

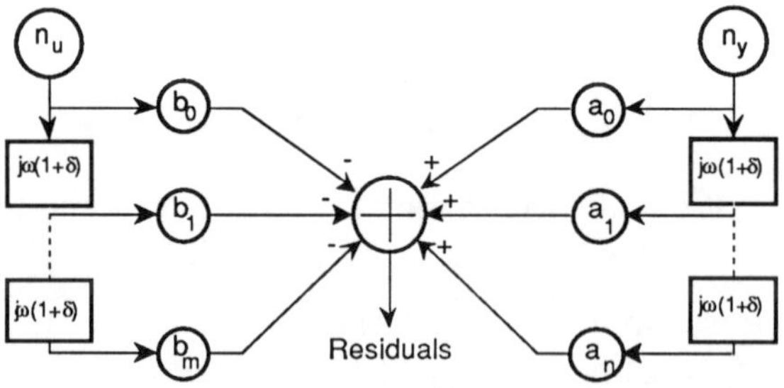

Figure 4.1 : path from noise sources to residuals

We observe that if a spectral component is prominently present in u(t) and y(t), this frequency is weighted strongly in the cost function. Also, the contribution to the total cost of a spectral component at frequency ω is weighted by the variance of the assumed model and by $G(j\omega)$ of which $D(e^{j\omega})$ can be chosen freely for $\omega > \omega_m$. If the D-filter is not included in the noise model, it can be used to deliberately decrease the weight at some frequencies where we expect important errors. This statement will now be illustrated with the aid of Fig. 4.2, where only one channel of the measurement setup is depicted for simplicity.

First assume that the de-emphasis filter is not present as in Fig. 4.2a. At some frequencies, there may be systematic errors due to modeling inaccuracies, time-varying behavior as in § 2.1, the alias phenomenon or the implementation of the continuous-time observation filters by discrete-time approximations. Hence, we want to multiply the variances by a function $|D(e^{j\omega})|^{-2}$ that is large at those frequencies where the systematic errors are important. This is equivalent to reshaping the assumed noise spectrum, as depicted in Fig. 4.2b. From (4.4.2), this is equivalent to the situation shown in Fig. 4.1c. The same goes for $C(e^{j\omega})$ and for $AA(j\omega)$, so in order to maintain the deliberate frequency weighting, the noise properties $\Phi_{u,i}$ and $\Phi_{y,i}$ (or **C** for arbitrary excitations) are those of the noise as measured at the input and output. The advantage of the approach of Fig. 4.2c is its computational simplicity. Its disadvantage is that if a limited number of samples of the signals $u_{aa}(t)$ and $y_{aa}(t)$ are available, more information is lost to initiate the de-emphasis filter.

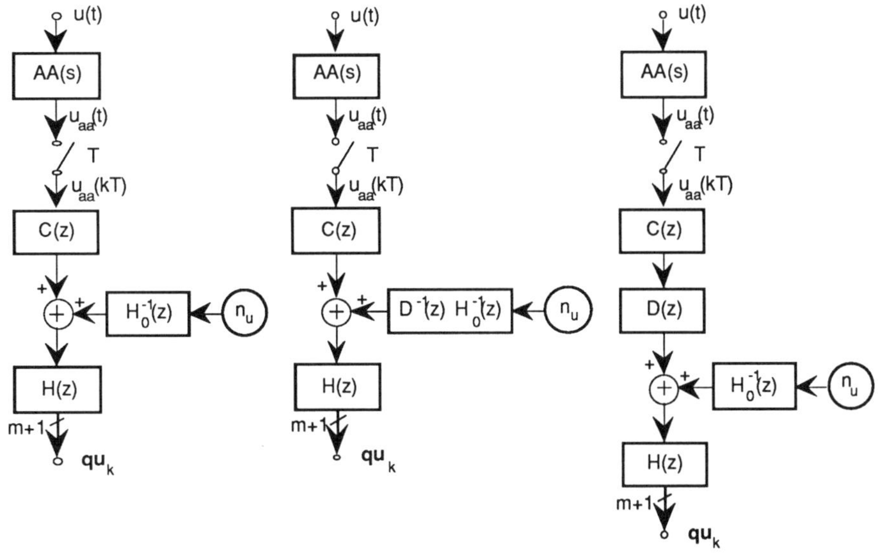

Figure 4.2 : measurement channel with, from left to right :
(a) no de-emphasis
(b) D-filter incorporated in noise model
(c) D-filter placed in signal path.

Appendix 4.A

Lemma 4.1

Assume that $\mathbf{n}_k$ are uncorrelated and stationary over k, i.e.

$$\mathbf{C} = \begin{pmatrix} \mathbf{D} & & \mathbf{0} \\ & \ddots & \\ \mathbf{0} & & \mathbf{D} \end{pmatrix} \text{ where } \mathbf{D} = E\{\mathbf{n}_k\, \mathbf{n}_k^t\} \in \mathbb{R}^{LxL} \tag{4.A.1}$$

and $\mathbf{C_p} = (\mathbf{p}^t\, \mathbf{D}\, \mathbf{p})\, \mathbf{I}_{KxK}$.

The cost function (4.1.6) to be minimized over all unknown components of $\mathbf{p}$ becomes:

$$K(\mathbf{p}, \mathbf{g}) = \frac{\|\mathbf{G}\, \mathbf{p}\|^2}{2\, \mathbf{p}^t\, \mathbf{D}\, \mathbf{p}}.$$

Introduce the matrices

$$\mathbf{E} = \begin{pmatrix} \mathbf{pk} & \mathbf{0}_{(L-Q)xQ} \\ \mathbf{0}_{Qx1} & \mathbf{I}_{QxQ} \end{pmatrix}$$

and $\mathbf{T} = \begin{pmatrix} t_{11} & \mathbf{0}_{1xQ} \\ \mathbf{t}_{21} & \mathbf{T}_{22} \end{pmatrix}$ a Choleski factor of $(\mathbf{E}^t\, \mathbf{D}\, \mathbf{E})^{-1} = \mathbf{T}\, \mathbf{T}^t$

and write $\mathbf{p} = \mathbf{E}\, \mathbf{T}\, \mathbf{y}$ for some $\mathbf{y} \in \mathbb{R}^{(Q+1)x1}$ with the constraint $\mathbf{e}_1^t\, \mathbf{y} = \frac{1}{t_{11}}$, then

$$K(\mathbf{E\,T\,y}, \mathbf{g}) = \|\mathbf{G\,E\,T} \frac{\mathbf{y}}{\|\mathbf{y}\|}\|^2$$

Now find the singular value decomposition of the Kx(Q+1) matrix $(\mathbf{G\,E\,T})$:

$$\mathbf{G\,E\,T} = \mathbf{U} \begin{pmatrix} s_1 & & \mathbf{0} \\ & \ddots & \\ \mathbf{0} & & s_{Q+1} \\ \mathbf{0} & \mathbf{0} & \mathbf{0} \end{pmatrix} \mathbf{V}^t \text{ where } s_1 \geq s_2 \geq \ldots \geq s_Q > s_{Q+1}$$

with $\mathbf{U} \in \mathbb{R}^{KxK}$ and $\mathbf{V} = (\mathbf{v}_1\, , \ldots , \mathbf{v}_{Q+1}) \in \mathbb{R}^{(Q+1)x(Q+1)}$ orthogonal matrices and rewrite $\mathbf{v}_{Q+1}^t = (w_1, \mathbf{w}_2^t)$ with $w_1 \in \mathbb{R}$ and $\mathbf{w}_2 \in \mathbb{R}^{Qx1}$.

The minimizer of $K(\mathbf{E\,T\,y}, \mathbf{g}) = \sum_{i=1}^{Q+1} (s_i\, \mathbf{v}_i^t \frac{\mathbf{y}}{\|\mathbf{y}\|})^2$ over $\mathbf{y}$ is $\mathbf{y_{TLS}} = \frac{\mathbf{v}_{Q+1}}{t_{11}\, w_1}$.

The corresponding estimate for **pu** is recovered as:

$$\mathbf{pu}_{\mathbf{TLS}} = \frac{1}{t_{11}}\left(\mathbf{t}_{12} + \mathbf{T}_{22}\frac{\mathbf{w}_2}{w_1}\right). \tag{4.A.2}$$

which is recognized as the total least squares solution of (4.2.4) (Van Huffel, 1987). The use of the TLS estimator is justified by regarding (4.A.1) as an approximation of the true **C**.

Lemma 4.2

Let $\{w_k\}$ be stationary zero-mean noise that is filtered through a linear discrete-time system to yield the stationary noise $\{v_k\}$. The linear system has impulse response g_k and frequency response $G(e^{j\omega})$. The DFT of a record of K samples of $\{v_k\}$ is:

$$V_K(e^{j\omega}) = \sum_{t=0}^{K-1} v_t\, e^{-j\omega t}$$

$$= \sum_{k=-\infty}^{\infty} g_k \sum_{t=0}^{K-1} w_{t-k}\, e^{-j\omega t}$$

$$= \sum_{k=-\infty}^{\infty} g_k\, e^{-j\omega k} \sum_{\tau=-k}^{K-k-1} w_\tau\, e^{-j\omega \tau}$$

Due to the stationarity of $\{w_k\}$, the statistical moments of the last factor are independent of k. Now compute:

$$E\{V^*_K(e^{j\omega_1})\, V_K(e^{j\omega_2})\} = G^*(e^{j\omega_1})\, G(e^{j\omega_2})\, E\{W^*_K(e^{j\omega_1})\, W_K(e^{j\omega_2})\} \tag{4.A.3}$$

where $W_K(\omega)$ is the DFT of any segment of K points of $\{w_k\}$.

5. Simulation Examples.

In this section we will give a design example of a measurement setup and subsequently use it to simulate the estimation of some plants. It is our intention to model the transfer function with relative systematic errors of the order of magnitude of 10^{-4}. The actual error on transfer function estimate (TFE) will depend on the excitation used and the plant to be estimated.

The design of the measurement channel is a compromise between performance (control of the approximation errors), complexity (of the analog and digital filters), length of impulse response (amount of information that is lost after starting the measurements or a sudden change in the DUT) and realization effort (in particular the analog filters). To achieve a design of reasonable complexity, we set $f_m = f_b = 0.2$ Hz. Above this frequency, the contributions to the estimates will be gradually out-weighted. If we were able to make perfect anti-alias filters and observation filters, we could set $f_m = f_b = 0.5$ Hz (the Nyquist frequency). Hence, with $f_m = f_b = 0.2$ Hz, we require an oversampling rate of 2.5.

Because of its flexibility, we will implement the observation filters as a chain of differentiators. Hence, if the measurement setup is to be applied to TV DUTs like the TI test systems of this section, we will try to realize a flat amplitude response and a linear phase response for $G(j\omega) = AA(j\omega)\,C(e^{j\omega})\,D(e^{j\omega})$.

For the anti-alias filter, we choose a 7-th order inverse Chebyshev with 80 dB attenuation in its stopband. Its frequency response is shown in Fig. 5.1. The choice of this filter is based on its smooth magnitude and phase response in the passband, which makes it easy to compensate. Since the zeros of the filter are located on the imaginary axis, the filter can in practice be realized with the so-called FDNR technique with its known advantages. See Ramakrishna *et al.* (1979) and its references for a treatment of the imperfections of these filter structures. The compensation filter and the de-emphasis filter are compressed into a single 37-th order FIR filter. This filter has sufficient degrees of freedom to accommodate production tolerances in the AA-filter and is designed with the same technique as the one used in Pintelon and Schoukens (1990b). Slightly better results are expected if minimax techniques tailored to the FIR design problem are used. The magnitude response of the D-filter is chosen such that its stopband attenuation is sufficiently large to allow a low-order design of the digital differentiators, without having a prohibitively long impulse response. Fig. 5.2 shows $G(j\omega)$ up to the Nyquist frequency after removing a delay of $\tau = 23.327$ s.

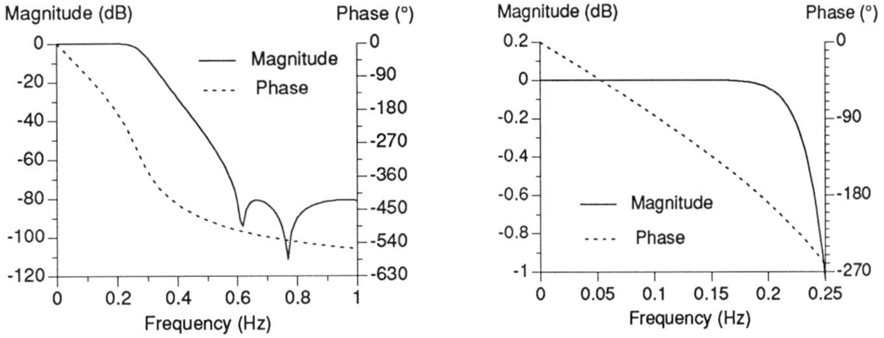

Figure 5.1 : frequency response AA(j2πf) of the anti-alias filter

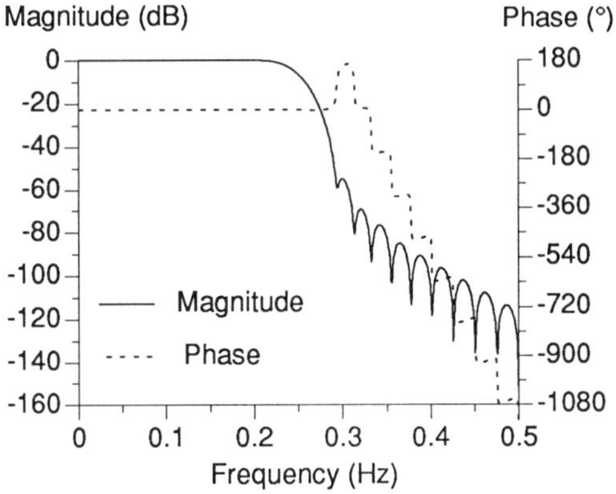

Figure 5.2 : frequency response G(j2πf) of the complete measurement channel

Fig. 5.3 shows how G(jω) controls the model equation approximation errors. The attenuation of components in the input or the output signal with frequency in the range 0.5 Hz to 1.0 Hz can be derived from Fig. 5.3a. The magnitude of Δ(jω) as defined in (2.1.1) is depicted in Fig. 5.3b. Finally, Fig. 5.3c shows the rapidly decaying impulse response of the hybrid linear system G[.] (continuous-time in, discrete-time out). It is possible to reduce the delay of the measurement channel at the expense of larger Δ(jω) (for the same filter orders). If desired, D-filter designs that create an integer global delay τ can be obtained (e.g. for τ = 21 s, similar overall design results are obtained without increasing the filter order).

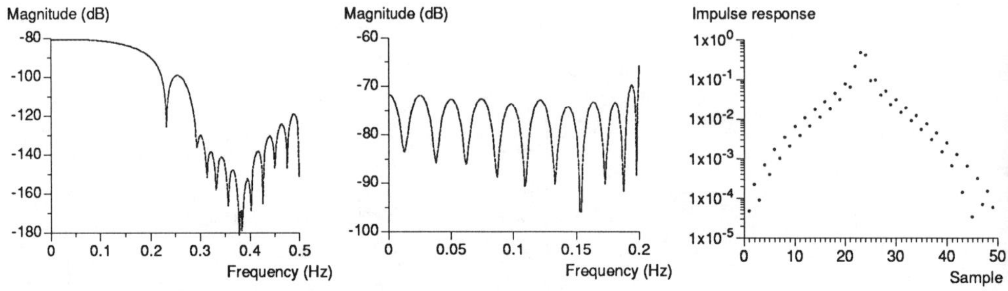

Figure 5.3 : control of various error by design of G[.] : (from left to right)
(a) the alias error due to the first fold, quanified by $AA(j2\pi(1-f))\ D(e^{j2\pi f})\ C(e^{j2\pi f})$
(b) the error due to bringing the band-limited time-varying system parameters outside the G[.]-operator, quantified by $\Delta(j2\pi f)$ in $G(s) = G_0\, e^{-\tau s}\,(1 + \Delta(s))$
(c) the error after a sudden change in the system parameters or the measurement setup, quantified by the impulse response of G[.].

If we expect DUTs with quality factors of the poles and zeros not greater than about 5, an acceptable choice of the order of the digital differentiators is 13. The corresponding optimal delay τ_d equals 6 s. This choice of order is based on (3.2.4) and on the desire to achieve a relative approximation error of -100 dB, i.e. $\varepsilon_d = 10^{-5}$. The difficulty of this design depends on the choice of $G(j\omega)$, in particular the attenuation of anti-alias and de-emphasis filters above f_m. In the design of the differentiators, deviations from the ideal $j\omega$ frequency response are penalized proportional to

$$\omega \sqrt{|G(j\omega)|^2 + |G(j2\pi - j\omega)|^2}.$$

The factor ω is introduced because we desire *relative* approximation errors that are proportional to the second factor, which is derived from consideration of ρ_i and its weight in (4.4.2). For frequencies below f_m, the second factor is approximately constant and above f_m it is proportional to the worst-case weight of the differentiator approximation error $\delta(j\omega)$. In this argument, it is assumed that the useful spectral components in the signals and the folded components are of the same order of magnitude and that only a single fold must be considered.

The relative approximation error

$$|\delta(j\omega)| = \left| \frac{e^{j\tau_d \omega} \, Di(e^{j\omega}) - j\omega}{j\omega} \right|$$

of the resulting design is depicted in Fig. 5.4a. For checking the design of the differentiator, a plot of $G(j\omega)\,\delta(j\omega)$ is also provided, which quantifies the model equation error for a spectral component at ω ($|\omega| < \pi$). The price we pay for not using equiripple design methods becomes apparent in the interval 0.2 Hz to 0.3 Hz, where the design limit of -100 dB is exceeded. The fitted relative differentiator magnitude response is shown in Fig. 5.4b. The differentiator has the desirable property to attenuate above f_m, which reduces the dynamic range of the signals in the observation chain. In § 5.1 we will make an analysis of the systematic errors involved in the method and it will be assumed that all observations are free of noise. The impact of observation noise will be simulated in § 5.2.

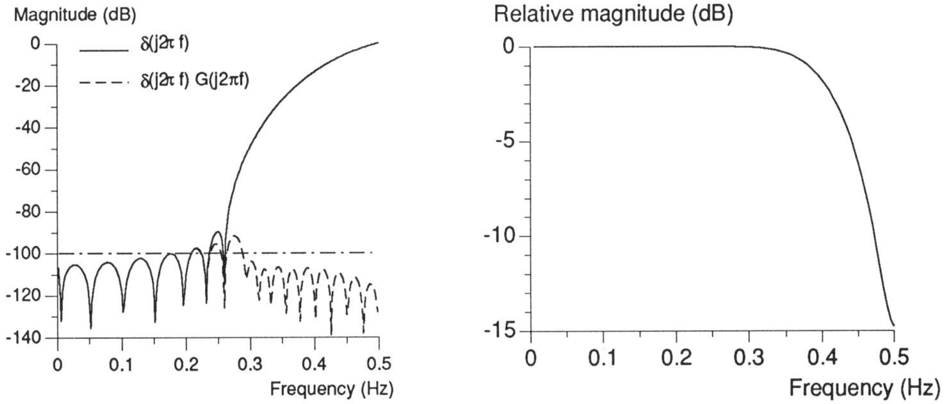

Figure 5.4 : Design of discrete-time approximations to a differentiator. From left to right:
(a) the relative error $\delta(j2\pi f)$ and the weighted relative approximation error $G(j2\pi f)\ \delta(j2\pi f)$
(b) the ratio of the approximated to the true frequency response.

5.1 Systematic estimation errors.

First, we study the influence of the excitation signal on the estimates. The plant is a 4-th order inverse Chebyshev filter with a stopband attenuation of 60 dB for frequencies above 0.4 Hz (Fig 5.5), i.e. it has a transfer function:

$$\frac{Be(s)}{Ae(s)} = \frac{1\,10^{-3}\,s^4 + 5.0532\,10^{-2}\,s^2 + 0.31919}{s^4 + 1.9463\,s^3 + 1.8942\,s^2 + 1.0849\,s + 0.31919} \tag{5.1.1}$$

The highest Q-factor for the poles is 1.4. The excitation signal is a multisine with random phases uniformly distributed between 0 and 2π:

$$u(t) = \sum_{i=0}^{P} \cos\left(\frac{2\pi}{P_m} i\,t + \phi_i\right) \qquad \text{with } P_m = 256. \tag{5.1.2}$$

Also, the number of observation points is chosen such that a total number of 256 observations (2.5.1) are available. All filters and the plant have reached their steady state. We consider three particular signals:

case 1: P = 51, i.e. the signal is spectrally confined to frequencies in the interval 0 Hz to 0.2 Hz. The only approximation in the model equation is due to the observation filters below f_m. The differentiation error was designed to be less than -100 dB (Fig. 5.4a).

case 2: P = 128, i.e. the signal is spectrally confined to frequencies in the interval 0 Hz to 0.5 Hz. The contribution to (5.1.2) for $i \leq 52$ is equal to the signal of case 1. In this case, the approximation due to the observation filters in the whole frequency band is present.

case 3: P = 256, i.e. the signal is spectrally confined to frequencies in the interval 0 Hz to 1.0 Hz. The contribution to (5.1.2) for $i \leq 128$ is equal to the signal of case 2. Here, the approximation involves the fullband differentiation error and the alias error.

Let $A_{ML}(s)$ and $B_{ML}(s)$ denote the maximum likelihood estimates of Ae(s) and Be(s), derived with the prior knowledge that the input and the output observation noise are white with equal variance and independent of each other, that the odd powers of the numerator polynomial have a zero coefficient and that the denominator polynomial is monic. The quantity that we will examine to verify the goodness of a fit is the relative TFE error

$$\eta_{ML}(j\omega) = \frac{\left|\frac{B_{ML}(j\omega)}{A_{ML}(j\omega)} - \frac{Be(j\omega)}{Ae(j\omega)}\right|}{\left|\frac{Be(j\omega)}{Ae(j\omega)}\right|} \tag{5.1.3}$$

Fig. 5.6 shows the magnitude of this quantity for all three cases. Obviously, $G(j\omega)$ was properly chosen to control the fullband differentiation error (case 1 vs. case 2) and the alias error (case 1 vs. case 3).

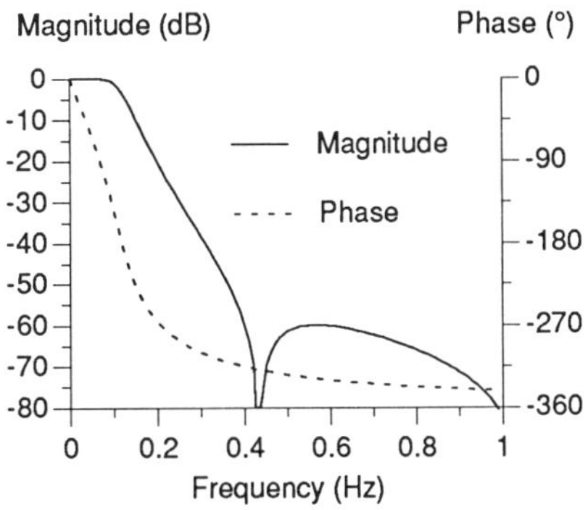

Figure 5.5 : 4-th order inverse Chebyshev filter used as a DUT.

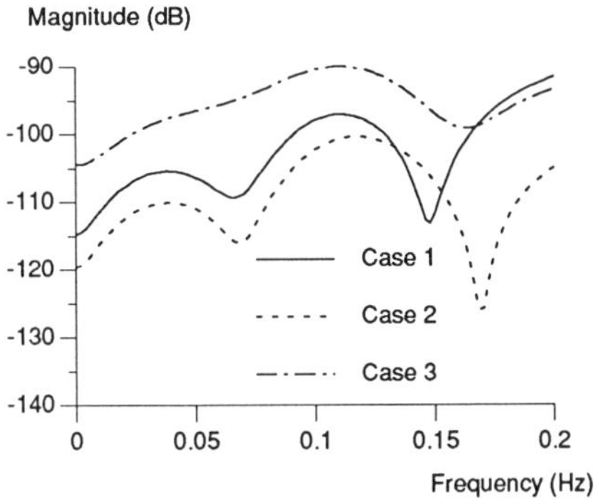

Figure 5.6 : relative TFE error η_{ML} for different excitation signals
case 1 : 0 Hz to 0.2 Hz
case 2 : 0 Hz to 0.5 Hz
case 3 : 0 Hz to 1.0 Hz.

In a second set of simulations, we compare the systematic TFE error in the integrated sampling approach when using numerical integration and when using digital filtering. In particular, we are interested in the sampling rate required in all methods to achieve the same maximal TFE error in the passband of the DUT. The excitation signal is a multisine composed of 50 components each with unit amplitude, with random frequency uniformly distributed in the interval 0 Hz to 0.25 Hz and with random phase uniformly distributed in the interval 0 to 2π. This signal is chosen such that it doesn't require processing by the AA, D or C-filters. The plant is a polynomial Chebyshev filter of the fifth order ($n = 5$, $m = 0$) with a 1 dB passband ripple and a cutoff frequency of 0.18 Hz. The numerical integration approach is found from § 3.2.2 by using the appropriate coefficients for $\overline{FQ}$ and by setting $\lambda = 0$ and $n_{\varepsilon,0} = 1$. For a given integration rule (we will examine the trapezoidal rule and Simpson's rule), it is only possible to reduce the integration error to any desired level by decreasing the sampling period T. Increasing $n_{\varepsilon,0}$ hardly affects the integration error at low frequencies. Moreover, since the frequency response of an integrator over $n_{\varepsilon,0}$ samples exhibits a zero at $f = (Tn_{\varepsilon,0})^{-1}$, it is required in this case that $Tn_{\varepsilon,0} < 4$. Contrarily, in the digital filtering approach, it is possible to increase the accuracy of the integration by taking more samples outside the integration interval into account. The 8-th order FIR filter used in the subsequent simulation is the one given in Schoukens (1990). It approximates the 2-sample integrator for frequencies below 0.25 Hz. In all simulations, the equal input/output variance total least squares estimates as described in lemma 4.1 are used. Enough samples are acquired to generate 256 observations of (2.5.1), regardless of the sampling period. The relative TFE error $\eta_{TLS}(j\omega)$, which is defined as in (5.1.3), is shown in Fig 5.7 for the digital filtering approach (T = 1.0 s) and for the numerical integration approach using Simpson's rule (T = 1.0 s and T = 0.14 s) or the trapezoidal rule (T = 1 s and T = 0.01 s). As suggested in Sagara and Zhao (1990), l is matched to T to have $1 \leq lT \leq 2$ when applying numerical integration. We conclude that digital filtering allows a sampling rate that is 7 times lower than for Simpson integration and about 100 times lower than for the trapezoidal rule. This factor depends on the desired TFE accuracy, on the plant's order and on its highest Q-factor. If the gain of $\delta(s)$ in equation (3.2.11) is increased by a factor of ξ and it is our goal not to increase the maximal TFE error, Table 3.3 reveals that we have to increase the digital filter order by about $2.2\log_{10}\xi$ units, while the sampling period for Simpson's rule must be decreased by a factor of $\sqrt[4]{\xi}$.

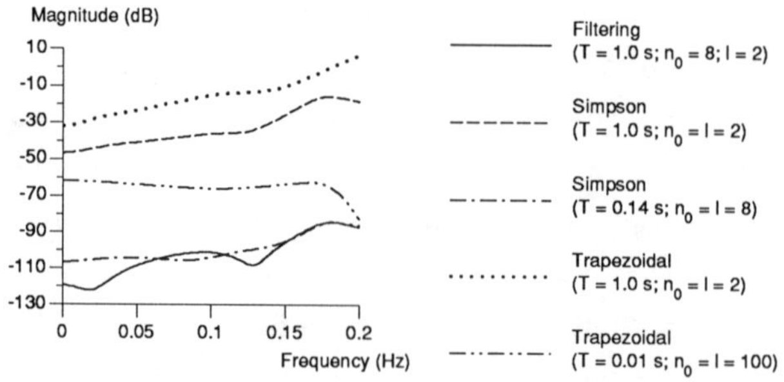

Figure 5.7 : performance of several integrated sampling approaches.

A third issue that will be discussed now is the influence of the plant on the TFE. In all subsequent simulations, the same excitation signal is used: a multisine composed of 50 components with unit amplitude, random frequency uniformly distributed in the interval 0 Hz to 0.35 Hz and with random phase uniformly distributed in thc interval 0 to 2π. The plants are polynomial Chebyshev filters of second up to tenth order ($n = 2 \ldots 10$, $m = 0$). All plants exhibit a 1 dB passband ripple and a cutoff frequency of 0.18 Hz. Again, the plant and the measurement setup have reached the steady state condition before sufficient data is collected to generate 256 observations (2.5.1). The MLE is calculated assuming white, independent, equal-level input/output noise and monicity of the denominator. It is found that the TFE error increases with the order of the DUT, but all $\eta_{ML}(j\omega)$ curves have the same general shape as those in Fig. 5.7: the relative TFE error is maximal in the neighborhood of the cutoff frequency and the slope below this frequency is about 200 dB/Hz. The dependence on the plant order is related to the property of Chebyshev filters of increasing order to have poles with increasing quality factor close to the cut-off frequency. Table 5.1 lists the order of the Chebyshev filter (n), the highest quality factor (Q_{max}) of a pole with imaginary part $2\pi f_{max}$, the expected relative TFE error in dB as predicted from (3.2.4) taking only the pole with the highest Q into account, i.e. $20\log_{10}(2Q_{max}) - 100$ dB, and the observed worst relative TFE error $\eta_{ML}(j2\pi f)$ for $|f| < 0.2$ Hz. It shows an excellent agreement between the simulations and (3.2.4).

Table 5.1: worst TFE error for polynomial Chebyshev filters

n	Q_{max}	f_{max} (Hz)	$20\log_{10}2Q_{max} - 100$ dB (dB)	$\max_{\lvert f\rvert<0.2} \lvert\eta_{ML}(j2\pi f)\rvert$ (dB)
2	0.96	0.1611	-94	-103
3	2.02	0.1739	-88	-88
4	3.55	0.1770	-83	-82
5	5.56	0.1782	-79	-79
6	8.00	0.1788	-76	-76
7	10.9	0.1791	-73	-73
8	14.2	0.1794	-71	-71
9	18.7	0.1795	-68	-68
10	22.3	0.1796	-67	-67

To compare the applicability of this time-domain method with the frequency-domain method discussed in Pintelon and Schoukens (1990a), we use an arbitrary excitation signal. The frequency-domain estimator is based on the assumption that the excitation is periodic and that the measurement time is an exact integer multiple of this period. Hence, for signals like the one used in the first simulation of this section, it would not produce estimation errors for noise-free observations. The excitation signal is a discrete white unit-variance Gaussian noise source that produces a sample every 0.2 s, to which we add a square wave of height 1, a duty cycle 50 % and a period of 100 s. The resulting sequence is fed to a zero-order hold followed by a coloring filter with transfer function 1/(0.8 s + 1). This signal is applied to an 8-th order polynomial Chebyshev filter with 1 dB passband ripple and a cut-off frequency of 0.1 Hz. The magnitude of the TFE error for the MLE (assumptions as in second simulation) based on 256 observations of (2.5.1) is shown in Fig. 5.8. It reveals that the time-domain estimator is still usable in cases where its the frequency-domain counterpart fails to produce a sensible estimate.

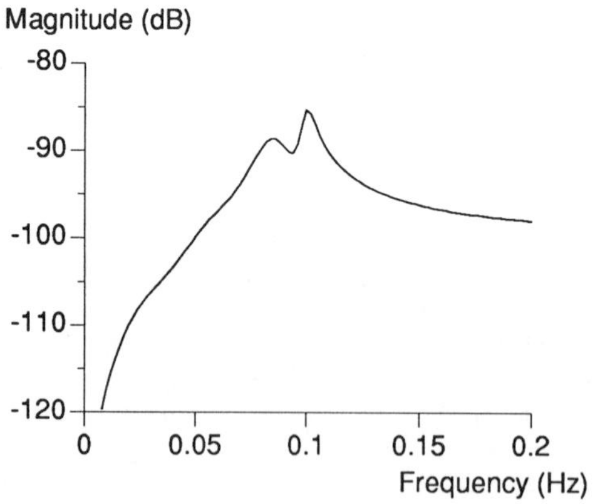

Figure 5.8 : TFE error for an 8-th order polynomial Chebyshev filter with an arbitrary input signal.

The above simulations show that it is possible to reduce the systematic errors to an acceptable level while retaining a slow sampling rate. Alternative design specifications can be met by changing the order of the digital differentiators and, if required, restarting the design of G[.].

5.2 Stochastic estimation errors.

The other aspect of the setup and the associated estimator is its performance in a stochastic framework. In particular, we will disclaim the often mentioned drawback of differentiators that they are noise-accentuating in the continuous-time system identification context. We compare the estimates produced by the total least squares estimator (TLS_t) as explained in lemma 4.1 and those produced by the maximum likelihood estimator (MLE_t) as defined in (4.2.3). The plant is the 4-th order inverse Chebyshev filter defined in (5.1.1). The noise is injected after the anti-alias filters and passes the digital section of the measurement channel. Two data records are considered:

a. 128 samples of $u_d(t)$ and $y_d(t)$ are available, so it is possible to produce 76 (= 128 - $M_{init,d}$) observations of (2.5.1),

b. enough samples of $u_d(t)$ and $y_d(t)$ are available to produce 128 observations of (2.5.1).

The resulting estimators are labelled $TLS_{t,76}$, $TLS_{t,128}$, $MLE_{t,76}$ and $MLE_{t,128}$.

We compare our method to the frequency-domain estimator presented in Pintelon and Schoukens (1990a), which takes the fast Fourier transform (FFT) of the both the input and the output data ($u_d(t)$ and $y_d(t)$ in this case) and then estimates the system parameters that satisfy the spectral input/output relation. Due to noise, this relation will never be satisfied identically. The problem is solved using a total least squares approach (TLS_f) or using a maximum likelihood approach (MLE_f). In order for the FFT to be free of leakage, we are restricted to signals of the form (5.1.2), where we choose P_m = 128 and P = 25 (no sinusoid has a frequency greater than 0.2 Hz). When taking the FFT of the noise-free signals, only the first 26 components will be nonzero. Hence, it is wise to write the frequency-domain relations only at these frequencies. Alternatively, if one is not aware of the spectral confinement of the excitation signal, one might include 64 lines of the FFT. The resulting estimators are labelled $TLS_{f,26}$, $TLS_{f,64}$, $MLE_{f,26}$ and $MLE_{f,64}$.

The mean square errors of all TFEs have been computed based on 100 independent experiments. In each experiment, we add white, zero mean, Gaussian noise with a standard deviation of 0.1 to the noise-free samples of $u_{aa}(t)$ and $y_{aa}(t)$ and filter it by C[.] and D[.]. Fig. 5.9 depicts the mean square error (MSE) of the TFE in dB, together with the Cramér-Rao lower bound as obtained from (4.3.2) when 76 or 128 realizations of (2.5.1) are available (CRB_{76} and CRB_{128} respectively). These bounds are calculated assuming that the differentiators are perfect up to 0.5 Hz and taking the noise coloring by D[.] into account. Under these conditions, CRB_{128} coincides with the Cramér-Rao bound derived for the frequency-domain estimator. The loss of accuracy due to the initialization of the observation filters is found by comparing $TLS_{t,76}$ and $TLS_{t,128}$, or $MLE_{t,76}$ and $MLE_{t,128}$ or CRB_{76} and CRB_{128}.

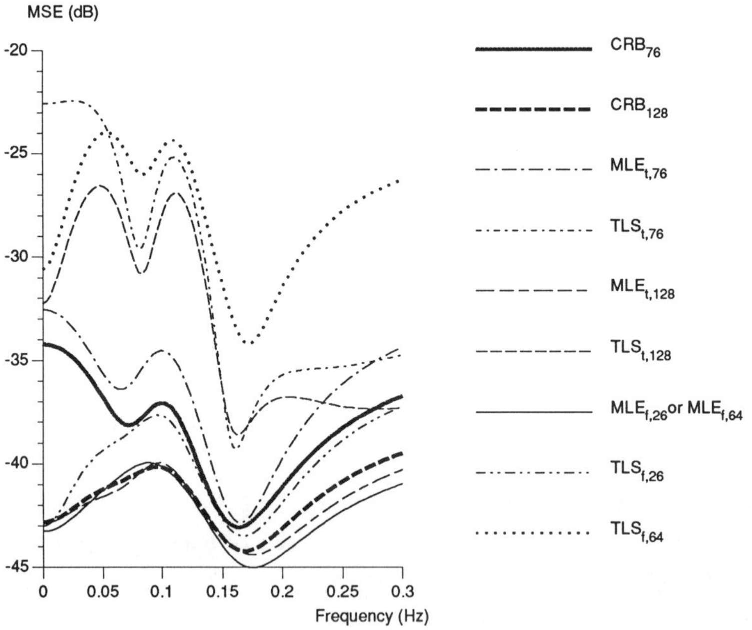

Figure 5.9 : TFE mean square error for the maximum likelihood (MLE) and total least squares (TLS) estimators for a 4-th order inverse Chebyshev filter obtained from noisy input/output measurements.

We observe that $MLE_{t,128}$, $MLE_{f,26}$ and $MLE_{f,64}$ perform equally and attain the CRB within the accuracy of this simulation. $MLE_{t,76}$ is only a couple of dB's above its CRB. The implication of the approximation inherent to the TLS approach is obvious. Replacing **C** (4.0.2c) by (4.A.1) implies that the model equation residuals are white, while they ought to be colored by A(jω) and B(jω) which boost up the high-frequency components of the noise. Hence, TLS weighs the high-frequency components too strongly, which is reflected in an increased MSE of the TFE at low frequencies for $TLS_{f,64}$, $TLS_{t,76}$ and $TLS_{t,128}$. The faulty weighing is not present in $TLS_{f,26}$ because it doesn't use the high-frequency data. Finally, $TLS_{f,26}$ and $TLS_{t,128}$ on the one hand and $MLE_{f,26}$ and $MLE_{t,128}$ on the other hand do not coincide because the time-domain estimator uses the autocorrelation of approximate derivatives.

This simulation shows that MLE outperforms TLS. For some selected plants, differences of up to 40 dB in MSE (not due to thresholding) have been recorded with the same measurement setup and noise properties. On the other hand, it is possible to find examples where the slight improvement in MSE doesn't justify the computational effort involved in the MLE.

6. Conclusions.

We have presented a general framework for converting a linear differential equation with time-varying coefficients to an algebraic equation. The representation retains the original coefficients or system parameters and their physical interpretation. An observation filter is used to reconstruct signals and their derivatives that satisfy this algebraic equation. Classical approaches like Poisson moment functionals, integrated sampling and instantaneous sampling result as special cases and initial conditions can be eliminated by a discrete-time operator. We concentrated on linear SISO systems for simplicity, but the method can be extended to MIMO and nonlinear systems that are linear-in-the-parameters.

When modeling time-varying systems, the measurement setup should exhibit a flat magnitude response and a linear phase response at low frequencies. At higher frequencies, its magnitude response can be used to control systematic modeling errors due to the alias phenomenon and the observation filter approximations. The analysis of the modeling error results in design constraints for all building blocks of the measurement chain. Theoretically, design specifications of any level can be met as long as the DUT remains identifiable from the filtered input/output signals. A chain of identical observation filters, each approximating a first order continuous-time system exhibits a controlled behavior of the systematic transfer function estimation error. The use of digital filters designed in the frequency domain as an alternative to numerical integration allows slower sampling thanks to the use of samples outside the integration interval.

A maximum likelihood errors-in-variables estimator was derived for time-invariant systems. The algorithm is iterative and is used to improve the statistical efficiency of the consistent total least squares estimates. The MLE uses a frequency-domain weighting that is based on statistical considerations at low frequencies, while the weighting function is chosen to conquer the systematic modeling errors at high frequencies.

A design example and simulations show that it is possible to reduce the systematic errors to an acceptable level without increasing the sampling rate. The robustness with respect to the choice of the plant and the excitation signal was verified. The method was found to have a more universal area of application than frequency-domain techniques.

References

Barnett S. and R. G. Cameron (1985). *Introduction to Mathematical Control Theory (second edition)*, Clarendon Press, Oxford.

Golub G. H. and L. F. Van Loan (1985). *Matrix Computations*, The John Hopkins University Press, Baltimore.

Henderson T. L. (1981). Geometric Models for Determining System Poles from Transient Response, *IEEE Transactions on Acoustics, Speech, and Signal Processing*, Vol. ASSP-29, No. 5, pp. 982-988.

Jain V. K., T. K. Sarkar, and D. D. Weiner (1983). Rational Modeling by Pencil-of-Functions Method, *IEEE Transactions on Acoustics, Speech, and Signal Processing*, Vol. ASSP-31, No. 3, pp. 564-573.

Kollár I., R. Pintelon, Y. Rolain, and J. Schoukens (1991). Another Step Towards an Ideal Data Acquisition Channel, *IEEE Transactions on Instrumentation and Measurements*, Vol. IM-40, No. 3, accepted for publication.

Kumaresan R. and D. W. Tufts (1982). Estimating the Parameters of Exponentially Damped Sinusoids and Pole-Zero Modeling in Noise, *IEEE Transactions on Acoustics, Speech, and Signal Processing*, Vol. ASSP-30, No. 6, pp. 833-840.

Pintelon R. and I. Kollár (1991). Exact Discrete Time Representation of Continuous Time Systems, *9th IFAC/IFORS Symposium on Identification and System Parameter Estimation,* Budapest (Hungary), July 8-12, 1991.

Pintelon R. and J. Schoukens (1990a). Robust Identification of Transfer Functions in the s- and z-Domains, *IEEE Transactions on Instrumentation and Measurements,* Vol. IM-39, No. 4, pp. 565-573.

Pintelon R. and J. Schoukens (1990b). Real-Time Integration and Differentiation of Analog Signals by Means of Digital Filtering, *IEEE Transactions on Instrumentation and Measurements,* Vol. IM-39, No. 6, pp. 923-927.

Pintelon R. and L. Van Biesen (1990). Identification of Transfer Functions with Time Delay and Its Application to Cable Fault Location, *IEEE Transactions on Instrumentation and Measurements,* Vol. IM-39, No. 3, pp. 479-484.

Ramakrishna K., K. Soundararajan, and V. K. Aatre (1979). Effect of Amplifier Imperfections on Active Network, *IEEE Transactions on Circuits an Systems,* Vol. CAS-26, No. 11, pp. 922-931.

Schoukens J. (1990). Modeling of Continuous Time Systems Using a Discrete Time Representation, *Automatica,* Vol. 26, No. 3, pp. 579-583.

Stuart A. and J. K. Ord (1986). *Kendall's Advanced Theory of Statistics - Vol. 1,* Charles Griffin & Co, London.

Unbehauen H. and G.P. Rao (1987). *Identification of Continuous Systems,* North-Holland, Amsterdam.

Unbehauen H. and G.P. Rao (1990). Continuous-Time Approaches to System Identification - A Survey, *Automatica,* Vol. 26, No. 1, pp. 23-35.

Van Huffel S. (1987). *Analysis of the Total Least Squares Problem and its Use in Parameter Estimation,* PhD thesis, Katholieke Universiteit Leuven, Belgium.

Wahlberg B. (1987). *On the Identification and Approximation of Linear Systems,* PhD thesis, Linköping University, Sweden.

Whitfield A. H. and N. Messali (1987). Integral-Equation Approach to System Identification, *Int. J. Control,* Vol. 45, No. 4, pp. 1431-1445.

Young P. C. (1979). Parameter Estimation for Continuous-Time Models - A Survey, *Proceedings 5th IFAC/IFORS Symposium on Identification and System Parameter Estimation,* Darmstadt (Germany), September 24-28, pp. 17-41.

The Relationship Between Discrete Time and Continuous Time Linear Estimation

Brett M. Ninness and Graham C. Goodwin
Department of Electrical Engineering & Computer Science
University of Newcastle
Shortland, 2308 AUSTRALIA

Abstract

We examine the problem of discrete time system estimation while not ignoring the underlying continuous time system. This leads to the use of a new discrete time operator, the δ operator, which approximates the continuous time derivative operator $\frac{d}{dt}$. We use this to formulate system estimation algorithms, and discuss their significantly superior numerical properties when compared to the equivalent shift operator formulated algorithms. We provide an overview of this new δ operator and also discuss some practical considerations in recursive least squares parameter estimation.

1 Introduction

Discrete time system analysis is usually done using the q forward shift operator and associated discrete frequency variable z. Unfortunately, the discrete domains resulting are unconnected with the continuous domains which spawn them. This is because the underlying continuous domain descriptions cannot be obtained by setting the sampling period to zero in the discrete domain approximations to them. Furthermore, it is widely known that serious numerical problems arise using shift operator formulations of algorithms at high sampling rates relative to the natural frequencies of the system being estimated.

Here we address this problem by formulating our discrete time description for a process while not ignoring the underlying continuous time process generating it. This leads naturally to the specification of a new discrete time operator, the δ operator, which is a difference operator and thus is the equivalent of the continuous time $\frac{d}{dt}$ operator. The use of this operator will be shown to entail many advantages. Namely:

N. K. Sinha and G. P. Rao (eds.), Identification of Continuous-Time Systems, 79–122.

1. There is a close connection between continuous time results and those formulated using the δ operator in discrete time since setting $\Delta = 0$ in a δ operator result gives the corresponding continuous time result.

2. The δ operator provides more insight into discrete time system analysis because of the similarity between the continuous time description for the process and the δ operator discrete time description.

3. Improved numerical properties are achieved. There are numerous areas relevant to system estimation where δ operator parameterisations of discrete time algorithms are numerically superior to their equivalent conventional q operator implementations. Specifically, the improved numerical properties we consider are:

   - Digital Filtering
     - Finite word length effects
     - Frequency response sensitivity
     - Round-off noise
   - Optimal state estimation: Conditioning of Riccati equations
   - Least squares parameter estimation: Conditioning of covariance matrix

Note that there is a simple linear transformation between δ operator parameterisations and shift operator parameterisations and so no sacrifice in modelling flexibility or statistical efficiency is made in using the δ operator.

We begin our discussion with a presentation of the δ operator, an associated discrete frequency domain transform (the Γ transform), and a discussion of the advantages of the δ operator over the conventional q operator. We also present a new generalised notation that will allow continuous and discrete results to be derived simultaneously. A new derivation and analysis of recursive least squares (RLS) parameter estimation and its variants is later done to illustrate the use of this notation.

Using the δ operator, we also address the problem of state estimation and show how the Kalman filter can be formulated using the δ operator as well as highlighting the advantages of doing this. We conclude with a discussion of some practical aspects associated with RLS parameter estimation.

1.1 A New Discrete Time Operator

When describing a model for a dynamic system it is common to use operator notation. That is, if it is a continuous time system the model for its performance is usually a differential equation, can be more compactly expressed using the operator $\rho \equiv \frac{d}{dt}$. For example:

$$2\frac{d^2x}{dt^2} + \frac{dx}{dt} + 7x = \frac{dy}{dt} + 7y$$

$$\Rightarrow (2\rho^2 + \rho + 7)x = (\rho + 7)y$$

If it is a discrete time system, the model for its performance is usually a difference equation. This can be more compactly expressed using the forward shift operator q which is defined by:

$$qx_k \equiv x_{k+1}$$

For example:

$$\begin{aligned} 2x_{k+2} + x_{k+1} + 7x_k &= y_{k+1} + 7y_k \\ \Rightarrow (2q^2 + q + 7)x_k &= (q + 7)y_k \end{aligned}$$

Unfortunately, the shift operator has no continuous time counterpart. Consequently, discrete time representations using the q operator do not converge smoothly to the underlying continuous time system as the sampling interval goes to zero. This motivates us to present a new discrete time operator called the δ operator that does have a continuous time counterpart. Using this operator continuous time systems are seen as the limiting case of a corresponding discrete time system as the sampling interval tends to zero.

2 The Delta Operator

The δ operator is defined as follows:

$$\begin{aligned} \delta &\triangleq \frac{q-1}{\Delta} \qquad (1) \\ \Rightarrow \delta x_k &= \frac{(q-1)x_k}{\Delta} \\ &= \frac{x_{k+1} - x_k}{\Delta} \end{aligned}$$

This operator has been known in the numerical analysis field as the first divided difference operator (Hilderbrand 1956). From the above we see that the δ operator approximates the derivative:

$$\delta x_k \approx \left. \frac{dx}{dt} \right|_{x=x(k\Delta)}$$

with the approximation becoming better as the sampling interval tends to zero. Therefore, because the δ operator has the continuous time counterpart ρ, models for systems expressed in terms of the δ operator are very similar to models expressed with the differentiation operator ρ, or the Laplace transform variable s. Because of this the use of δ operators permits continuous time intuition and insights to be used in discrete time systems. Furthermore, it provides equivalent flexibility in the context of discrete time modelling as does the shift operator q.

2.1 Obtaining δ Operator Discrete time Models for Systems

Since (1) illustrates that there is a linear transformation between the shift operator and δ operator models for systems, the derivation of the δ operator discrete time models for continuous time systems is as straightforward as for the q operator case.

Specifically, suppose we have a continuous time model for a SISO system expressed in rational proper transfer function form using the differentiation operator ρ:

$$\frac{y(t)}{u(t)} = \frac{B(\rho)}{A(\rho)} = G(\rho)$$

$$\begin{aligned} A(\rho) &= \rho^n + a_{n-1}\rho^{n-1} + \ldots + a_1\rho + a_0 \\ B(\rho) &= b_m\rho^m + b_{m-1}\rho^{m-1} + \ldots + b_1\rho + b_0 \end{aligned}$$

This can be converted to an equivalent state space form using a canonical representation (Goodwin and Sin 1984):

$$\rho\vec{x}(t) = A\vec{x}(t) + Bu(t) \tag{2}$$

$$y(t) = C\vec{x}(t) + b_n u(t) \tag{3}$$

Note that if $G(\rho)$ is strictly proper then $b_n = 0$. Assuming zero order hold sampling, the equivalent shift operator state space description is:

$$q\vec{x}_k = M\vec{x}_k + Nu_k \tag{4}$$

$$y_k = S\vec{x}_k + Tu_k \tag{5}$$

Where M,N,S and T are given by Middleton and Goodwin (1990):

$$M = e^{A\Delta} \quad N = A^{-1}(e^{A\Delta} - I)B \quad S = C \quad T = b_n$$

Here $e^{A\Delta}$ is the matrix exponential (Middleton and Goodwin 1990) and Δ is the sampling period.

Considering the definition in (1) then gives the δ operator state space representation from (4) and (5) by subtracting $\vec{x}_k$,and dividing by Δ on both sides of (4):

$$\delta\vec{x}_k = F\vec{x}_k + Gu_k \tag{6}$$

$$y_k = H\vec{x}_k + Ku_k \tag{7}$$

where

$$F = \Omega A; \quad G = \Omega B; \quad H = C; \quad K = b_n$$

and where we define Ω by:

$$\Omega \triangleq (e^{A\Delta} - I)\frac{A^{-1}}{\Delta} \qquad \text{If A non-singular} \tag{8}$$

$$= I + \frac{A\Delta}{2!} + \frac{A^2\Delta^2}{3!} + \frac{A^3\Delta^3}{4!} + \ldots \tag{9}$$

Therefore, it can be seen that at high sampling rates relative to the system bandwidth, $A\Delta \to 0$ to give $\Omega \approx I$ and so the state space discrete time representation for a system using the δ operator will be very similar to the continuous time representation. Hence, it will be possible to assume $G(\delta) \approx G(s)$ if the sampling rate is sufficiently high. Empirically, sufficiently high has been found to mean sampling rates greater than ten times the plant bandwidth. Of course, it is always possible to work with the exact $G(\delta)$.

Also note that if the calculation of Ω is to be done via a computer, which is almost always the case, then the power series expression up to an appropriate number of terms should be used. This is because finite word length limits on floating point accuracy will more strongly affect the evaluation of the expression (8) then they will the expression (9). Finally, if desired, this δ operator discrete time state space description can be converted into a rational transfer function form by:

$$G(\delta) = \frac{y_k}{u_k} = \frac{det[\delta I - (F - GH)] + (K-1)det(\delta I - F)}{det(\delta I - F)} \tag{10}$$

2.2 Implementation of Filters

There are many ways in which to implement digital filters (Dougherty etc. 1984).

Here we present one way in order to illustrate the implementation of DSP algorithms formulated using the δ operator without converting back to shift operator form. This is essential to realising the improved numerical properties that are possible through δ operator parameterisation.

We assume without loss of generality that the filter is specified in δ operator state space form (6),(7). If it represented in δ operator transfer function form then a canonical representation may be used to obtain a representation as per (6), (7). Multiplying both sides of (6) by the inverse operation δ^{-1} then gives:

$$\vec{x}_k = \delta^{-1}[F\vec{x}_k + Gu_k] \tag{11}$$

$$y_k = H\vec{x}_k + Ku_k \tag{12}$$

This suggests that the basic building block in the implementation of digital filters using δ operators is the function δ^{-1}. This compares with the usual building block, the backward shift function q^{-1}. The function δ^{-1} is defined to operate in the following way:

$$\alpha_k = \delta^{-1}\beta_k \qquad \Rightarrow \qquad \alpha_{k+1} = \alpha_k + \Delta\beta_k \tag{13}$$

The function δ^{-1} is thus seen to act as a discrete time integrator. This δ^{-1} building block can be manipulated in much the same way as integrators are manipulated in the implementation of continuous time models. Additionally, (13) shows that the hardware or software needed to implement this building block is only marginally more complex than that required for the more usual delay building block q^{-1} which it replaces. Considering (11) and (12),then gives that the required recursive updating of the state space equations to implement the filter are:

$$\begin{aligned} \vec{x}_{k+1} &= \vec{x}_k + \Delta F\vec{x}_k + \Delta G u_k \\ y_{k+1} &= H\vec{x}_{k+1} + b_n u_{k+1} \end{aligned} \tag{14}$$

Note that (14) should be implemented exactly. The following difference equation:

$$\vec{x}_{k+1} = (I + \Delta F)\vec{x}_k + \Delta G u_k \tag{15}$$

which is mathematically equivalent to (14) should not be implemented since in practical terms it is not equivalent to the implementation of (14). This is due to finite word length effects in computer implementation which become apparent when sampling rates become high. In this case, Δ becomes small and so:

$$I + \Delta F \approx I \tag{16}$$

If only a finite word length is available, as is always the case in the computer implementation of the recursive update equation (15), then the approximation in (16) may in fact become equality. The required dynamics of the filter will thus not be realised at fast sampling rates. A major exacerbating factor of this problem is that much of the word length in the multiplication involved in (15) is taken up by registering the presence of the identity matrix I.

However, note that in the formulation of the update equation given in (14), the full word length of the computer is available to calculate the possibly small quantity ΔF. Eventually of course, as Δ becomes smaller, finite word length effects will make ΔF appear to be zero and once again the required filter dynamics will not be realised. However, it can be shown (Middleton and Goodwin 1990) that this will occur at much higher sampling rates than occurs for the formulation (14).

Therefore, implementation of the recursive difference equations to realise a digital filter exactly via the formulation (14) rather than the formulation (15) will result in higher sampling rates being possible before numerical difficulties degrade the performance of the filter.

2.3 Stability Region for δ Operator Models

It follows that since the poles of ρ and q domain input - output models for systems must lie in well known areas of the complex plane in order for the response of the system to be asymptotically stable, then a similar result should hold for the poles of δ domain models for systems.

Such a result is seen straightforwardly by noting that the Hurwitz stability region for the poles of q domain models of systems is the open unit disc centred on the origin (Dougherty etc. 1984). Furthermore, the δ operator is defined in terms of the shift operator q via (1). This is a simple linear transformation, under which it is obvious

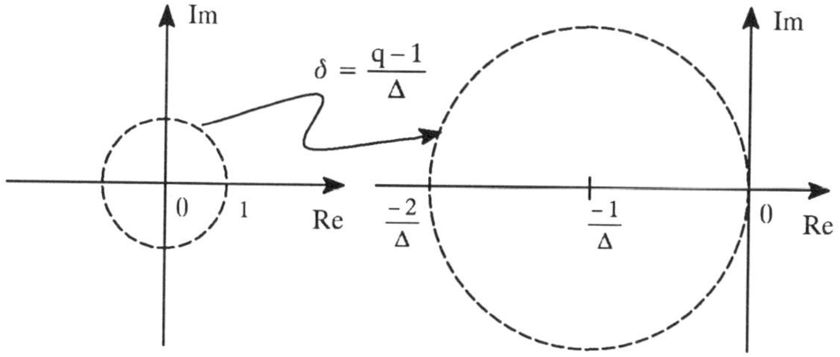

Figure 1: *Hurwitz stability regions for the poles of q and δ operator models*

that the stability region in q maps to the disc radius $\frac{1}{\Delta}$ and centre $\frac{-1}{\Delta}$. This then is the region in the complex plane that the poles of a δ operator model for a system must lie if the response is to be asymptotically stable. This region, and the mapping involved, is shown in figure 1. Note that as $\Delta \to 0$, the δ stability region expands to fill the whole open left half of the complex plane. This, of course, is the continuous time Hurwitz stability region and so once again it is seen that δ discrete time models converge towards continuous time models as the sampling interval tends to zero.

2.4 A Generalised Notation

It has already been mentioned that the δ operator approximates a continuous time derivative operator as the sampling interval tends to zero. This suggests that the formulation of algorithms in terms of delta operators will not only be valid for discrete time implementation, but will tend to the equivalent continuous time formulations as the sampling interval tends to zero. This, in fact, is the case, but to make it clear some specialised notation needs to be introduced. The notation used has been drawn from (Middleton and Goodwin 1990). We present it here for later use in the derivation of the discrete time least squares parameter estimation algorithm. We also present some elementary results formulated using this notation that we will require later.

These results give a flavour for how this new notation is used.

2.4.1 The Generalised Derivative

Definition 1. Up to this point, the symbol ρ has been used to represent continuous time differentiation. This will remain the case, but will also be used to represent the δ operator in discrete time. Therefore, any formulations obtained in terms of ρ can be interpreted as discrete time expressions by replacing ρ with δ. More succinctly, if Δ represents the sampling interval:

$$\rho = \begin{cases} \frac{d}{dt} & \Delta = 0 \\ \delta & \Delta \neq 0 \end{cases}$$

2.4.2 Generalised Integration/Summation

Definition 2. The same symbol $\mathbb{S}$ will be used to represent Riemann integration in continuous time, and summation in discrete time. Specifically the definition is:

$$\mathbb{S}_{t_1}^{t_2} f(\tau)d\tau = \begin{cases} \int_{t_1}^{t_2} f(\tau)d\tau & \Delta = 0 \\ \Delta \sum_{k=\frac{t_1}{\Delta}}^{k=\frac{t_2}{\Delta}-1} f(k\Delta) & \Delta \neq 0 \end{cases}$$

2.4.3 Reciprocity of Integration/Differentiation

Lemma 1. *The results for generalised integration of a generalised derivative and generalised differentiation of a generalised integral follow as per the continuous time cases. Specifically:*

$$\begin{aligned} \rho\left[\mathbb{S}_a^t f(\tau)d\tau\right] &= f(t) \\ \rho\left[\mathbb{S}_t^b f(\tau)d\tau\right] &= -f(t) \\ \mathbb{S}_a^b [\rho f(\tau)]d\tau &= f(b) - f(a) \end{aligned}$$

Proof. These results may be trivially obtained by considering the continuous and discrete time cases separately. ▽▽▽

2.4.4 Differentiation of a Product

Lemma 2. *Consider two real scalar functions of a real scalar variable t: $f(t)$ and $g(t)$. For these functions, the product rule for generalised differentiation is:*

$$\rho(fg) = (\rho f)g + f(\rho g) + \Delta(\rho f)(\rho g)$$

Proof.

$$\begin{aligned} \rho(fg) &= \frac{f[(k+1)\Delta]g[(k+1)\Delta] - f(k\Delta)g(k\Delta)}{\Delta} \\ &= \frac{[f[(k+1)\Delta] - f(k\Delta)]\, g(k\Delta) + f(k\Delta)\,[g[(k+1)\Delta] - g(k\Delta)]]}{\Delta} + \\ &\quad \frac{\Delta\,[f[(k+1)\Delta] - f(k\Delta)]\,[g[(k+1)\Delta] - g(k\Delta)]]}{\Delta^2} \\ &= (\rho f)g + f(\rho g) + \Delta(\rho f)(\rho g) \end{aligned}$$

▽▽▽

Notice that as the sampling interval tends to zero, the result for generalised differentiation of a product of two functions tends to the well known product rule result for continuous time derivatives as expected.

2.4.5 Matrix Differencing Lemma

Lemma 3. *Consider a square invertible matrix A whose generalised derivative can be written via the vector B as:*

$$\rho A = BB^T$$

Then the generalised derivative of A^{-1} is given by:

$$\rho A^{-1} = \frac{-A^{-1}BB^TA^{-1}}{1 + \Delta B^TA^{-1}B}$$

Proof. The generalised derivative of A^{-1} is given by:

$$\begin{aligned}
\rho(A^{-1}) &= \frac{A^{-1}[(k+1)\Delta] - A^{-1}(k\Delta)}{\Delta} \\
&= \frac{A^{-1}[(k+1)\Delta]A(k\Delta)A^{-1}(k\Delta) - A^{-1}[(k+1)\Delta]A[(k+1)\Delta]A^{-1}(k\Delta)}{\Delta} \\
&= \frac{A^{-1}[(k+1)\Delta][A(k\Delta) - A[(k+1)\Delta]]A^{-1}(k\Delta)}{\Delta} \\
&= -A^{-1}[(k+1)\Delta](\rho A)A^{-1}(k\Delta) \\
&= -A^{-1}[(k+1)\Delta]BB^TA^{-1}(k\Delta) \qquad (17)
\end{aligned}$$

However, the generalised derivative of A is also given through definition by:

$$\begin{aligned}
\rho A &= \frac{A[(k+1)\Delta] - A(k\Delta)}{\Delta} \\
\Rightarrow A[(k+1)\Delta] &= A(k\Delta) + \Delta BB^T
\end{aligned}$$

Using the matrix inversion lemma (Goodwin and Payne 1977) and substituting into (17) then gives the result. ∇∇∇

2.5 The Gamma Transform and Frequency Response of δ Models

We now present a new transform, the Γ transform, that can be applied to δ operator models that will give the frequency response of both the continuous time model and the discrete time model for a plant simultaneously; the continuous frequency response being the limit of the discrete frequency response as the sampling interval tends to zero.

A heuristic justification of the Γ transform may be obtained by noting that for zero initial conditions, the z domain transfer function can be obtained by substituting $q = z$. Furthermore, the δ and q operators are related by definition in (1) which may be arranged to:

$$q = \Delta\delta + 1$$

This suggests a definition for the Γ transform, in terms of a new transform variable γ as:

$$\begin{aligned} F_\Delta(\gamma) &= F(z)|_{z=\Delta\gamma+1} \\ &= \sum_{k=0}^{k=\infty} f_k(1+\Delta\gamma)^{-k} \end{aligned}$$

Where $\Gamma[f(\delta)] = F_\Delta(\gamma)$. Middleton and Goodwin (1990) modify this definition in order to ensure that $F_\Delta(\gamma) \to F(s)$ as $\Delta \to 0$ by scaling the above definition by Δ to arrive at the final definition for the Γ transform:

$$F_\Delta(\gamma) = \Delta F(z)|_{z=\Delta\gamma+1}$$

Furthermore, the following shift property holds for the Z transform:

$$Z(qf_k) = \mathrm{Z}\left[zf_k - f_0\right]$$

Therefore, the definition of the Γ transform provides the equivalent differentiation property:

$$\Gamma(\delta f_k) = \gamma\Gamma(f_k) - f_0(1+\Delta\gamma)$$

Applying this result to a δ domain model of a system expressed in state space form as:

$$\begin{aligned} \delta\vec{x}_k &= F\vec{x}_k + Gu_k \\ y_k &= H\vec{x}_k \end{aligned}$$

gives:

$$\begin{aligned} \gamma X(\gamma) - x_0(1+\Delta\gamma) &= FX(\gamma) + GU(\gamma) \\ Y(\gamma) &= CX(\gamma) \end{aligned}$$

Consequently, if we are only interested in the forced response of a system after transients due to initial conditions have died out, then the delta transform model of a plant modelled using the delta operator may be found by replacing δ by γ analagously to the previous cases of replacing q with z and ρ with s. That is:

$$G(\gamma) = \frac{Y(\gamma)}{U(\gamma)} = \left.\frac{B(\delta)}{A(\delta)}\right|_{\delta=\gamma}$$

Finally then, the frequency response of the delta operator model may be found by noting that:

$$\gamma = \frac{z-1}{\Delta} \tag{18}$$

and since the frequency response of z transform models is found by substituting $z = e^{j\omega\Delta}$, so the frequency response of $G(\gamma)$ at the angular frequency ω is found by:

$$\text{Frequency Response} = G(\gamma)\big|_{\gamma=\frac{e^{j\omega\Delta}-1}{\Delta}}$$

Morover, a Taylor series expansion for $\frac{e^{j\omega\Delta}-1}{\Delta}$ about $\omega = 0$ gives:

$$\frac{e^{j\omega\Delta}-1}{\Delta} = j\omega\left(1 + \frac{j\omega\Delta}{2!} - \frac{\omega^2\Delta^2}{3!} + \ldots\right)$$

Therefore, for sampling rates greater than 20 times any frequency of interest (ie. 20 times the system bandwidth) we have:

$$\frac{\omega\Delta}{2} < 0.15$$

and so the frequency response of a δ operator transfer function can be roughly found by substituting $\delta = j\omega$. This is obviously intuitively appealing.

2.6 Relationship between γ Domain Poles and System Response

The derivation of the Γ transform allows the mapping of poles from the s domain to the γ domain. Since the relationship between the location of s domain poles and plant response is well known, this mapping will provide insight into how the location of the γ domain poles of a system affect the response of a discrete time system. Note that (Dougherty etc. 1984) gives the mapping between the s and z domains as $z = e^{s\Delta}$. Considering (18) we find the mapping between the s and γ domains as:

$$\gamma = \frac{e^{s\Delta}-1}{\Delta} \tag{19}$$

Note that this mapping is irrespective of the type of holding circuit used for the discrete time system. This is so because the poles of a system describe the natural response of a system when the input forcing signal is zero. Obviously, then the input holding circuit cannot affect the natural response of the system and therefore has no effect on the poles of the system. Such a simple situation does not hold for the zeroes of a system which describe non zero input forcing signals which cause zero output from the system. Obviously these zeroes will be intimately related to the type of input holding circuit used for the discrete time system, and there will be no simple mapping from continuous time zeroes to discrete time zeroes as there is for continuous time poles to discrete time poles (Middleton and Goodwin 1990). The simple mapping proposed for the poles leads to three important conclusions:

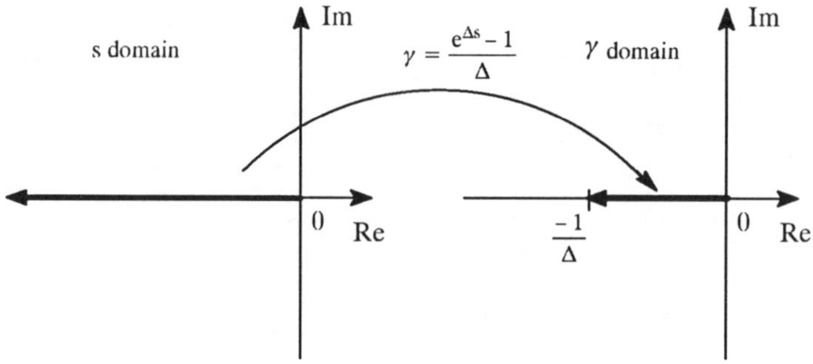

Figure 2: *Mapping of the negative real axis in the s domain to the γ domain*

1. $s = 0 \Rightarrow \gamma = 0$ and as $s \to -\infty$ along the real axis, $\gamma \to \frac{-1}{\Delta}$ along the real axis. This mapping is shown in figure 2. Therefore, poles in the γ domain near the real axis between the origin and the point $\gamma = \frac{-1}{\Delta}$ coincide with a well damped system response, with the response becoming quicker as the poles move to the left, analogous to the continuous time case. Furthermore, this mapping highlights the fact that there is a finite limit to how fast a sampled data system can respond. That is, it is obvious that it can respond no quicker than the sampling interval Δ.

2. Assume a continuous time pole at $s = -\alpha + j\beta$. Substituting this into (19) gives:
$$1 + \Delta\gamma = e^{(-\alpha+j\beta)\Delta} = e^{-\alpha\Delta}(\cos\Delta\beta + j\sin\Delta\beta)$$
Now suppose that γ is a complex number given by $\gamma = x + jy$. In this case:
$$\begin{aligned}\cos\Delta\beta &= e^{\alpha\Delta}(1 + \Delta x)\\ \sin\Delta\beta &= e^{\alpha\Delta}(\Delta y)\end{aligned}$$
$$\Rightarrow \left(x + \frac{1}{\Delta}\right)^2 + y^2 = \frac{1}{(\Delta e^{\alpha\Delta})^2} \qquad (20)$$
Therefore, the straight line locii $z = -\alpha + j\beta$ with α constant in the s domain maps to a circle centre $\frac{-1}{\Delta}$ and radius $\frac{1}{\Delta e^{\alpha\Delta}}$ in the γ domain. This is shown in figure 3. This highlights the interesting result that poles near the real axis in the γ domain can represent a very poorly damped system response if the pole is to the left of $\frac{-1}{\Delta}$. Furthermore, it is interesting to note how the locii of poles with a fixed damping ratio in the s domain map to the γ domain. The loci of poles in the s plane with constant damping ratio ζ is defined by the equation:
$$s = -\omega\cot\phi + j\omega \qquad \zeta = \cos\phi$$
Poles defined by this equation map to poles in the γ domain defined by:
$$\gamma = \frac{e^{-\Delta\omega\cot\phi}e^{j\Delta\omega} - 1}{\Delta}$$

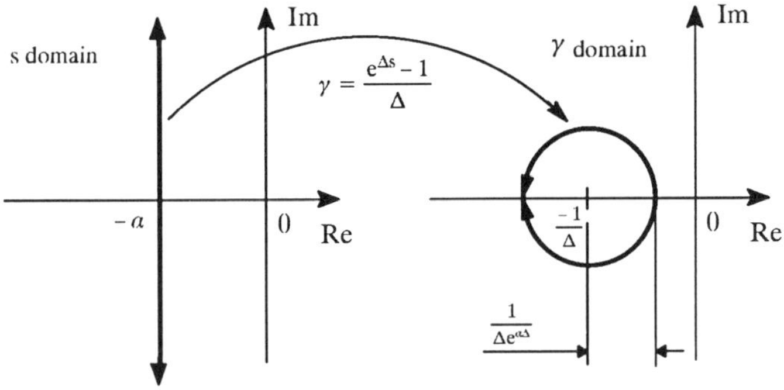

Figure 3: *Mapping the loci of poles with constant real part in the s domain to the γ domain*

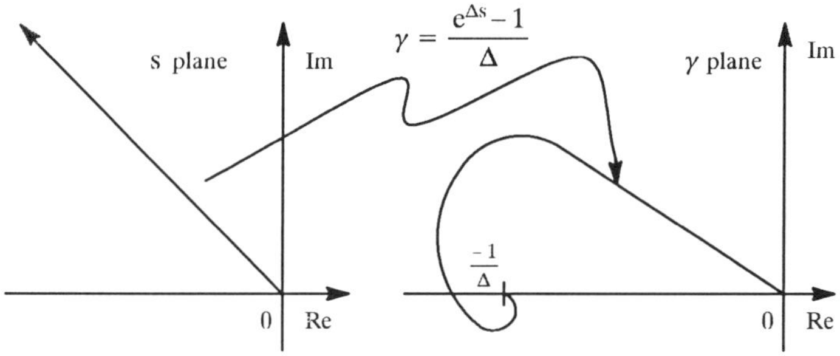

Figure 4: *Loci of poles with constant damping ratio in the s domain, and the loci they map to in the γ domain for a particular case.*

This is an exponentially decaying spiral centred on $\gamma = -\frac{1}{\Delta}$ as shown in figure 4.

3. Finally, by substituting $\alpha = 0$ into (20) the s domain stability boundary is seen to map to the circle shown in figure 1. This provides agreement with the same result found earlier.

2.7 Advantages of δ Operators

By now, the intuitive advantages of using the δ operator to parameterise discrete time approximations to continuous time systems should be apparent. We now proceed to emphasise these advantages, as well as to highlight significant numerical benefits stemming from their use.

2.7.1 Insight Advantages

As has already been shown, the δ operator is a difference operator that approximates the derivative operator ρ. As a result of this, the discrete time model for a sampled continuous time system is very similar to the continuous time system model expressed in terms of the ρ operator. Thus continuous time insights can be used in discrete design. This is not true if we use discrete time models expressed in terms of the shift operator q.

For example, consider the continuous time system modelled using the ρ operator:

$$G(\rho) = \frac{-\rho + 5}{\rho^4 + 23\rho^3 + 185\rho^2 + 800\rho + 2500} \tag{21}$$

The zero order hold discrete time approximation to this system parameterised by the q operator and assuming a 100Hz sampling rate (rounded to 5 significant digits) is:

$$G(q) = \frac{-10^{-6}(0.15537q^3 + 0.41605q^2 - 0.47402q - 0.14200)}{q^4 - 3.7777q^3 + 5.3506q^2 - 3.3674q + 0.79453} \tag{22}$$

This is not at all similar to the continuous time model and the poles and zeroes of the above transfer function are not obviously related to those of the continuous time system. That is, a 'quick pole' in continuous time, i.e. one in the far left hand of the complex ρ plane, does not map to one in the far left hand of the complex q plane. However, the δ operator form of the transfer function describing the discrete time domain response (rounded to 2 significant digits) is:

$$G(\delta) = \frac{-0.80\delta + 4.5}{\delta^4 + 22\delta^3 + 180\delta^2 + 760\delta + 2200} \tag{23}$$

This is quite similar to the continuous time model, with the poles and the zero being quite close to the continuous time poles and zero. This similarity between $G(s)$ and the δ operator discrete time approximation $G(\delta)$ is particularly important in the context of discrete time parameter estimation since it will be particularly easy to relate the estimated transfer function $\hat{G}(\delta)$ to the underlying continuous time process. Such is not the case for a q operator parameterised estimate $\hat{G}(q)$.

2.7.2 Finite Word Length Considerations

Many algorithms are better conditioned numerically using δ operator implementation than shift operator implementation. Most fundamentally, digital filtering operations are less prone to finite word length problems at higher sampling rates. This is due to the fact that as sampling rates increase, the poles and zeroes of models represented using shift operator notation tend to cluster about the point $q = 1$. Thus, the shift operator discrete time state transition matrix M tends to the identity matrix. This may be seen by noting that if A is the continuous time state transition matrix then M is given by:

$$M = e^{A\Delta} = 1 + \frac{A\Delta}{1!} + \frac{(A\Delta)^2}{2!} + \frac{(A\Delta)^3}{3!} + \dots$$

The dynamics of the system will be captured by the fractional part of the entries in M, but in floating point computer implementation much of the available word length will be used recognising the presence of the non-fractional part (i.e. 1) of M. Consequently, at a sufficiently high sampling rate the dynamics of the filter will be lost since the fractional part will be too small to be represented with the word length available.

The δ operator, being defined as $\delta = \frac{q-1}{\Delta}$, avoids this problem by shifting the point $q = 1$ to the point $\delta = 0$ and then scaling by the factor $\frac{1}{\Delta}$. Thus, the poles of δ operator models will tend to their continuous time values as the sampling rate increases. Again, this may be seen by considering the discrete time state transition matrix F. This time for a delta operator model:

$$F = A\left(I + \frac{A\Delta}{2!} + \frac{(A\Delta)^2}{3!} + \ldots\right)$$

Therefore, the discrete time transition matrix will tend towards the continuous time transition matrix. Additionally, using δ operators the states of a filter are updated according to (14). Evidently the dynamics of the filter are contained in the matrix ΔF. This will become very small as the sampling rate increases, but with computer implementation, the full word length of the computer may be used to store the important fractional entries in ΔF. Eventually, as the sampling rate increases, finite word length effects will cause ΔF to appear as the zero matrix, just as F will eventually appear like the identity matrix using shift operators, and the dynamics of the filter will no longer be achieved. However, this will occur at a much higher sampling rate using δ operators (Middleton and Goodwin 1986). In (Middleton and Goodwin 1986) it is shown that for a given word length, samplings rate five to ten times faster than achievable with shift operators can be sustained with δ operator implementation. The motivation for using fast sampling will be detailed presently.

To illustrate this numerical advantage, the step response of the continuous time system given by (21) was simulated over 10 seconds as is shown in figure 5. Also shown superimposed on this is the step response for the shift and δ operator discrete time approximations (22) and (23). The shift approximation is the upper trace in figure 5, and the continuous and δ approximation are the lower traces. As can be seen, the q operator implementation is not a very good approximation to the continuous time system, while the δ operator implementation is so close to the true continuous time response that it is difficult to tell the two apart. Notice too that the shift co-efficient representation in (22) involves 3 more significant digits than for the δ operator representation in (23) !

2.7.3 Frequency Response Sensitivity

There are other numerical advantages. Suppose we have a fixed precision in the representation of the coefficients of the denominator of a discrete time transfer function because of finite word length effects; be it in δ or q form. This means that there is an error ϵ in the representation of the coefficients of a filter. Given this error in the denominator for example, there must also be an error η in the pole positions of the

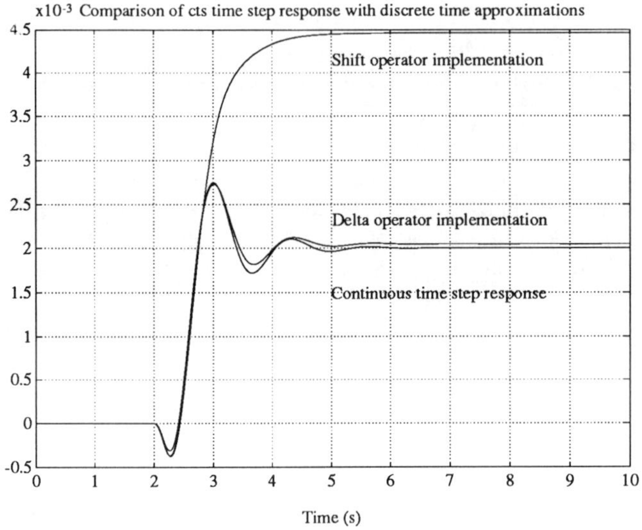

Figure 5: *Comparison of step response of continuous time system to the step responses of discrete time approximating systems using shift and δ operator implementations*

discrete time transfer function. It can be shown (Middleton and Goodwin 1986) that for a given ϵ, the upper bound for η is always smaller using a δ operator implementation than with a shift operator implementation. This is regardless of whether a fixed or a floating point implementation is used. Stated another way, to get within x % of required pole positions requires less bits of accuracy when using δ operator implementations rather than shift operator implementation. In general, to achieve an error of less then x % in pole positions, requires 5 to 7 bits less word length using δ operators (Middleton and Goodwin 1990).

2.7.4 Round Off Noise

In most finite word length implementations of DSP algorithms there will be errors introduced into the system due to the finite word length available for the calculation and storage of intermediate quantities. Under suitable conditions on the input to a discrete time system (namely that there be sufficient noise and/or input variation), the errors introduced may be considered as an almost stochastic process (Wilkinson 1963; Rabiner and Gold 1975). Thus the term 'round off noise' has arisen. It is shown in (Middleton and Goodwin 1990) that at sampling rates high with respect to the system bandwidth, δ operator implementation of filters results in the introduction of less round off noise on the output than for shift operator implementation.

2.7.5 Fast Sampling Rates

Note that the δ operator will always outperform the shift operator in terms of intuitive insights and numerical performance. However, the difference in performance is most appreciable at sampling rates greater than twenty times the system bandwidth where, unless great care is taken, shift operator implementation of filters will simply fail (see figure 5). Furthermore, we have shown how at these frequencies $G(\delta) \approx G(s)|_{s=\delta}$ and the discrete frequency response is $\approx G(\delta)|_{\delta=j\omega}$.

We emphasise these advantages by enumerating other motivations for fast sampling in the context of adaptive control (a large application area of system estimation). Specifically, these advantages are;

1. Aliasing effects due to frequency folding are significantly reduced or eliminated, and consequently the specification of the front end anti-aliasing filters can be relaxed. Since these filters have to be taken into account in system estimation this is a major advantage.

2. There is a smoother progression in control input to the plant. If slow sampling is used then the control input can be a sequence of large step changes (Goodwin and Sin 1984). This can feed significant energy into high frequency mechanical resonances. Rapid sampling ensures a smooth sequence of smaller changes to achieve the same bandwidth.

3. The discrete time response is a better approximation to the desired continuous time response.

4. Higher closed loop bandwidths can be achieved.

Having introduced the δ operator and enumerated its virtues, we now go on to examine system estimation algorithms parameterised with the δ operator.

3 System Estimation

The vast majority of discrete time system estimation literature uses formulations involving the shift operator q. Consequently, the resultant estimates pertain to a discrete time system only, seemingly ignoring the underlying continuous time system. As discussed in (Salgado 1989; Goodwin 1988; Middleton and Goodwin 1990; and Middleton etc. 1988) this causes state estimates to degenerate at high sampling rates to a constant irregardless of the underlying continuous time system, namely the trivial model $y_{k+1} = y_k$.

Consequently, we begin with a consideration of this problem. We will use the δ operator to establish a direct connection between the continuous and discrete formulations of the state estimation problem and show that this eliminates the Kalman filter degeneracy problem at high sampling rates. We will also compare the numerical robustness of the Kalman filter using the q and δ operators.

All the results here will, for the sake of brevity, be presented without proof. For readers interested in these proofs, and a more detailed discussion of the results, then (Salgado 1989; Goodwin 1988; Middleton and Goodwin 1990; and Middleton etc. 1988) are the appropriate references.

3.1 State Estimation

Consider the continuous time SISO linear time invariant stochastic system given by:

$$dx(t) = Ax(t)dt + d\nu(t) \tag{24}$$
$$dz(t) = Cx(t)dt + d\omega(t) \tag{25}$$

where $\nu(t)$ and $\omega(t)$ are Wiener processes with incremental covariances:

$$\mathcal{E}\left\{\begin{bmatrix} d\nu(t) \\ d\omega(t) \end{bmatrix}\begin{bmatrix} d\nu^T(t) & d\omega^T(t) \end{bmatrix}\right\} = \begin{bmatrix} Q & 0 \\ 0 & R \end{bmatrix} dt \tag{26}$$

and $\mathcal{E}$ denotes expectation over the underlying probability space. If we wish to find a state estimate $\hat{x}(t)$ minimising

$$J(t) = \mathcal{E}\{[x(t) - \hat{x}(t)][x(t) - \hat{x}(t)]^T\} \tag{27}$$

then the solution is well known (Astrom 1970) to be the continuous time Kalman filter

$$d\hat{x}(t) = A\hat{x}(t)dt + H(t)[dz(t) - C\hat{x}(t)] \tag{28}$$
$$H(t) = P(t)C^T R^{-1} \tag{29}$$

Where $P(t)$ satisfies the continuous time Riccati differential equation (CRDE):

$$\dot{P}(t) = P(t)A^T + AP(t) - P(t)C^T R^{-1} CP(t) + Q \tag{30}$$

Formulating the discrete time equivalent to the system in (24) and (25) presents the problem that sampling the output $\frac{dz}{dt}$ generates a system having output noise of infinite variance. This is overcome (Middleton etc. 1988) by accounting for low pass filtering prior to sampling to arrive at a discrete time approximation to (24),(25):

$$x(k+1) = A_q x(k) + \nu_q(k) \tag{31}$$
$$y(k) = C_q x(k) + \omega_q(k) \tag{32}$$

with

$$\mathcal{E}\left\{\begin{bmatrix} \nu_q(k) \\ \omega_q(k) \end{bmatrix}\begin{bmatrix} \nu_q^T(k) & \omega_q^T(k) \end{bmatrix}\right\} = \begin{bmatrix} Q_q & S_q \\ S_q^T & R_q \end{bmatrix}$$

The optimal state estimate for a system described via this joint Markov model is well known (Goodwin and Sin 1984) as:

$$\hat{x}(k+1) = A_q\hat{x}(k) + H_q(k)[y(k) - C_q\hat{x}(k)] \quad (33)$$

$$H_q(k) = [A_qP_q(k)C_q^T + S_q][C_qP_q(k)C_q^T + R_q]^{-1} \quad (34)$$

Where $P_q(k)$ satisfies the following discrete Riccati difference equation (DRDE):

$$P_q(k+1) = Q_q + A_qP_q(k)A_q^T - H_q(k)[C_qP_q(k)C_q^T + R_q]H_q^T(k) \quad (35)$$

However, it is noted (Middleton etc. 1988) that the Kalman gain H_q in this solution has the property

$$\lim_{\Delta\to 0} H_q = 0$$

where Δ is the sampling interval. This is seen to stem from the fact that:

$$\lim_{\Delta\to 0} A_q = I \qquad \lim_{\Delta\to 0} Q_q = 0 \qquad \lim_{\Delta\to 0} R_q = \infty \quad (36)$$

These results apply for a large class of pre-sampling filters and irregardless of the dynamics of underlying continuous time process in (24),(25). This is the problem of Kalman filter degeneracy mentioned in the introduction and will obviously lead to numerical problems at high sampling rates. These problems can be overcome by formulating the discrete time state space approximation to (24) and (25) using the δ operator:

$$\delta x(k) = A_\delta x(k) + \nu_\delta(k) \quad (37)$$

$$y(k) = C_\delta x(k) + \omega_\delta(k) \quad (38)$$

where

$$A_\delta = \frac{1 - A_q}{\Delta} \qquad \nu_\delta(k) = \frac{\nu_\delta(k)}{\Delta}$$

$$C_\delta = C_q \qquad \omega_\delta(k) = \omega_q(k)$$

Furthermore, if we define the δ operator covariances as:

$$Q_\delta = \frac{Q_q}{\Delta} \qquad R_\delta = \Delta R_q \qquad S_\delta = S_q \quad (39)$$

then the δ formulation for the optimal filter may be obtained from (33),(34) and (35) as:

$$\delta\hat{x}(k) = A_\delta\hat{x}(k) + H_\delta(k)[y(k) - C_\delta\hat{x}(k)] \quad (40)$$

where

$$H_\delta(k) = \frac{H_q(k)}{\Delta} = [P_\delta(k)C_\delta^T + \Delta A_\delta P_\delta(k)C_\delta^T + S_\delta][R_\delta + \Delta C_\delta P_\delta(k)C_\delta^T]^{-1} \quad (41)$$

and $P_\delta(k)$ satisfies:

$$\delta P_\delta(k) = Q_\delta + A_\delta P_\delta(k) + P_\delta A_\delta^T - H_\delta(k)[R_\delta + \Delta C_\delta P_\delta(k)C_\delta^T]H_\delta^T(k) + \Delta A_\delta P A_\delta^T \quad (42)$$

There are several points to note about this formulation.

1. This δ formulation for the optimal filter can be derived directly without going through the shift operator formulation first (Middleton etc. 1988).

2. The formulation of the δ covariances in (39) implies that they converge as $\Delta \to 0$ to the spectral densities of the continuous time processes. That is:

$$\lim_{\Delta \to 0} Q_\delta = Q \quad \lim_{\Delta \to 0} R_\delta = R \quad \lim_{\Delta \to 0} S_\delta = 0$$

3. The specification of the discrete time approximation to the continuous time process converges to the continuous time process as $\Delta \to 0$. That is

$$\lim_{\Delta \to 0} A_\delta = A \quad \lim_{\Delta = 0} C_\delta = C$$

4. Because of the above, the discrete Riccati difference equation (DRDE) in (42) converges to the CRDE in (30) and hence the discrete gain vector $H_\delta(k)$ tends to the continuous gain vector $H(t)$.

5. Because the δ operator formulation converges to the continuous time expressions as $\Delta \to 0$ we can use our generalised notation to express the Kalman filter for both continuous and discrete systems in a unified manner as:

$$\rho\hat{x} = A\hat{x} + H(y - C\hat{x}) \tag{43}$$

where

$$H = [(\Delta A + I)PC^T + S][\Delta CPC^T + R]^{-1}$$

By substituting (41) into (42) we get that P satisfies

$$\rho P = Q + PA^T + AP + PC^T(\Delta CPC^T + R)^{-1}CP + \mathcal{O}(\Delta)$$

Here, A,P,C,S, and R are the δ subscripted versions defined earlier and we note that setting $\Delta = 0$ gives the continuous time solution.

Therefore, the δ operator formulation achieves the aim of preserving the innate link between continuous and discrete time in the context of state estimation. Furthermore, because of this smooth transition from discrete time results to continuous time results, at high sampling rates the solution of the DRDE and DRAE are better numerically conditioned when formulated using the δ operator rather than the shift operator q (Middleton etc. 1988).

For example, consider a system having transfer function:

$$G(s) = \frac{-s + 2}{(s + 2)(s + 10)}$$

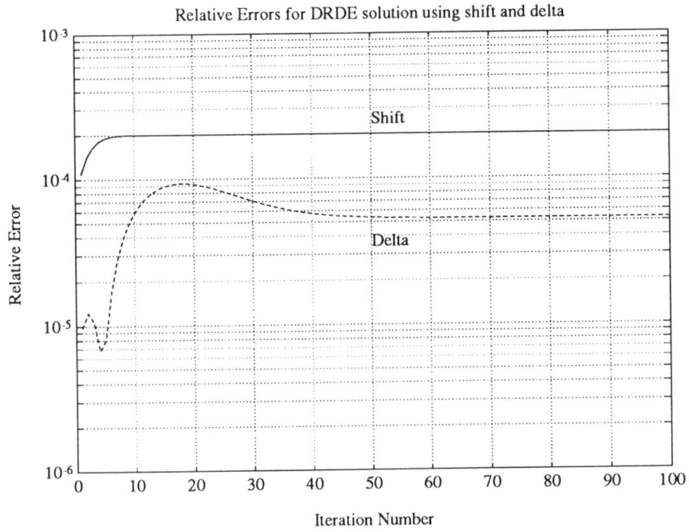

Figure 6: *Comparison of errors for solution of DRDE using δ and q operators*

with continuous state space model in observer form. The DRDE for this system was solved with A_δ and A_q rounded to 4 decimal places and $R = 1$, $Q = I$, $\Delta = 0.02$ and $P_0 = I$. Figure (6) shows the propagation of the relative error defined as:

$$\zeta_k = \frac{\|P_{FP}(k) - P(k)\|_F}{\|P(k)\|_F}$$

where $P_{FP}(k)$ and $P(k)$ denote the floating point and 'infinite' precision solution to the Riccati equation and $\|.\|_F$ denotes the Fröbenius norm. Obviously the δ formulation leads to a significant improvement in the relative error in the computation of the DRDE. Furthermore, the levelling off of the error suggests that the solution of the DARE for this system will also be more precise using δ operator implementation. A different 2nd order example is considered in (Middleton etc. 1988) where a more thorough discussion of these numerical considerations is given.

3.2 ARMAX Modelling

We now go on to motivate ARMAX modelling by showing it to be a convenient way of expressing the innovations form of the Kalman Filter given in (43). Defining the innovations process $\{\xi_t\}$ by:

$$\xi_t = y_t - C\hat{x}_t$$

then the steady state Kalman Filter of (43) can be written as:

$$\rho\hat{x}_t = A\hat{x}_t + Bu_t + K\xi_t \quad (44)$$

$$y_t = C\hat{x}_t + \xi_t \tag{45}$$

Where K is the steady state value of H_t. Note that the innovations process $\{\xi_t\}$ is a white 'noise' process with spectral density D given by:

$$D = \Delta CPC^T + R$$

To convert (44) and (45) to a more convenient form for black box modelling we may assume, without loss of generality, that the model is in observer form. That is A,B and K are of the form:

$$A = \begin{bmatrix} -a_{n-1} & 1 & 0 & \cdots & 0 \\ -a_{n-2} & 0 & 1 & & \vdots \\ \vdots & \vdots & & \ddots & 0 \\ \vdots & \vdots & & & 1 \\ -a_0 & 0 & \cdots & \cdots & 0 \end{bmatrix} \tag{46}$$

$$B^T = [b_{n-1}, b_{n-2}, \cdots, b_0] \tag{47}$$

$$K^T = [k_{n-1}, k_{n-2}, \cdots, k_0] \tag{48}$$

With this model structure, we can successively eliminate the state vector $\hat{x}_t$ to yield a model expressed purely in terms of input and output quantities. This yields the following left matrix fraction representation (Goodwin and Sin 1984; Kailath 1980):

$$A(\rho)y_t = B(\rho)u_t + C(\rho)\xi_t \tag{49}$$

where

$$A(\rho) = a_n\rho^n + a_{n-1}\rho^{n-1} + \cdots + 1 \tag{50}$$

$$B(\rho) = b_{n-1}\rho^{n-1} + \cdots + b_0 \tag{51}$$

$$C(\rho) = c_n\rho^n + c_{n-1}\rho^{n-1} + \cdots + c_0 \tag{52}$$

with

$$c_i = k_i + a_i \qquad i \in [0, n-1]$$

This is the familiar ARMAX model for a plant (Ljung 1987) except for the oddity that we have chosen to normalise $A(\rho)$ from the right so that it is not monic as is usually the case. The reasons for this are discussed in the last section of this chapter.

3.3 Fractional Representations

It is common to write models for systems using fractional representations (Francis 1987)

$$M(\rho)y_t = N(\rho)u_t + \nu_t \tag{53}$$

Here $M(\rho)$ and $N(\rho)$ belong to some desired class of transfer functions. A commonly desired class are those analytic and bounded in the right half plane so that $M(s)$ and $N(s)$ are in H_∞. We note that the ARMAX representation (49) representation just derived will not have $A(\rho)$ and $B(\rho)$ in this class. However, if we choose a Hurwitz polynomial $E(\rho)$ of the same order as $A(\rho)$ and also normalised from the right:

$$E(\rho) = e_n\rho^n + e_{n-1}\rho^{n-1} + \cdots + e_1\rho + 1 \tag{54}$$

then the simple choice

$$M(\rho) = \frac{A(\rho)}{E(\rho)} \qquad N(\rho) = \frac{B(\rho)}{E(\rho)}$$

will give a fractional representation for the ARMAX model (49) of the form (53) with $M(\rho)$ and $N(\rho)$ in H_∞. Other restrictions on $E(\rho)$ allow fractional representations for other classes of transfer functions to be found.

The advantage of this fractional representation is that it allows the innovations process $\{\xi_t\}$ to be expressed as a function of the plant input and output sequences $\{y_t\}$ and $\{u_t\}$ (Francis 1987). This allows the characterisation of all linear, unbiased estimates of the state of a plant to be expressed as an affine function of a free design variable (Francis 1987; Middleton and Goodwin 1990). Consequently, the design of observers subject to min-max or H_∞ type constraints (as opposed to the quadratic constraint just used) is possible (Middleton and Goodwin 1990). Because the topic of the existence of many linear unbiased estimates for the state has been brought up, we should note that the solution given in (43) is the best (in the quadratic sense of (27)) linear estimate of the state, and if the noise distributions are Gaussian is well known to be the best estimate of the state (linear or non-linear).

3.4 Parameter Estimation

Now that we have considered the problem of state estimation we will go on to consider parameter estimation. We will begin by deriving a parameter estimator from the Kalman filter state estimate just presented and then go on to consider least squares estimation.

3.4.1 Derivation from Kalman Filtering

For the sake of generality, we will include $E(\rho)$ in our ARMAX description of the Kalman filter (53) to give:

$$\frac{A(\rho)}{E(\rho)}y_t = \frac{B(\rho)}{E(\rho)}u_t + \frac{C(\rho)}{E(\rho)}\xi_t \tag{55}$$

Because of its connection with the Kalman filter $E(\rho)$ is commonly referred to as the observer polynomial. Notice that in discrete time theory, it is common to use $E = q^n$

to give $M(q)$ and $N(q)$ only involving backward time shifts. However, we will argue below that the choice $E = q^n$ is, in general, a poor one. It is easy to rearrange (55) into linear regression form by:

$$\begin{aligned} y_t &= \left(\frac{E(\rho) - A(\rho)}{E(\rho)}\right) y_t + \frac{B(\rho)}{E(\rho)} u_t + \frac{C(\rho)}{E(\rho)} \xi_t & (56) \\ \Rightarrow \quad y_t &= \phi_t^T \theta_0 + \eta_t & (57) \end{aligned}$$

where

$$\begin{aligned} \phi_t^T &= \left[\frac{\rho^n y_t}{E(\rho)}, \cdots, \frac{\rho y_t}{E(\rho)}, \frac{\rho^m u_t}{E(\rho)}, \cdots, \frac{u_t}{E(\rho)}\right] & (58) \\ \theta_0^T &= [e_n - a_n, \cdots, e_1 - a_1, b_m, \cdots, b_0] & (59) \end{aligned}$$

Note due to the introduction of the observer polynomial $E(\rho)$ the elements of the regression vector ϕ_t are filtered derivatives. The condition that the parameter vector is time invariant may be expressed as;

$$\delta\theta_0 = 0 \tag{60}$$

Equations (60) and (57) are then precisely in the joint Markov model form of (37) and (38) with:

$$A_\delta = 0 \quad C_\delta = \phi_t^T \quad \omega_\delta(t) = \eta_t \quad \nu_\delta(t) = 0 \quad x(t) = \theta_0$$

For the moment, we assume that $C(\rho) = E(\rho)$ so that η_t represents a white noise sequence. The more general case of coloured noise will be taken up in section 3.7.3. The definitions on the noise imply:

$$Q_\delta = S_\delta = 0$$

and

$$R_\delta = \Delta\mathcal{E}\{\nu_t^2\} \triangleq \sigma_t^2 \tag{61}$$

Therefore, we can use the δ operator formulation of the optimal filter in (40), (41) and (42) to recursively calculate the minimum variance linear unbiased estimate of θ by;

$$\begin{aligned} \delta\hat{\theta}_k &= \frac{P_k \phi_k (y_k - \phi_k^T \hat{\theta}_k)}{\sigma_k^2 + \Delta\phi_k^T P_k \phi_k} & (62) \\ \delta P_k &= \frac{-P_k \phi_k \phi_k^T P_k}{\sigma_k^2 + \Delta\phi_k^T P_k \phi_k} & (63) \end{aligned}$$

We note that since $\hat{\theta}$ is an unbiased estimate of θ_0, then the variance of the prediction error

$$\varepsilon_t = y_t - \hat{y}_t = y_t - \phi_t^T \hat{\theta}$$

will be given by:

$$\mathcal{E}\{\varepsilon_t^2\} = \phi_t^T P_t \phi_t - y_t^2$$

Therefore, since the Kalman filter gives the minimum variance estimate of $\hat{\theta}$, then for $\hat{\theta}'$ any other linear unbiased estimate with covariance P_t' we have

$$P_t \leq P_t'$$

in a matrix sense and hence the Kalman filter estimate $\hat{\theta}$ gives the predictor of minimum error variance.

3.4.2 Recursive Least Squares

We now approach the problem of parameter estimation from a different viewpoint. We formulate this by using our generalised notation and by defining a new cost function to be minimised:

$$J(\hat{\theta}) = \frac{1}{2}\left\{\mathop{\mathrm{S}}_0^t \frac{1}{\sigma_\tau^2}(y_\tau - \phi_\tau^T\hat{\theta})^2 d\tau + (\hat{\theta} - \hat{\theta}_0)^T P_0^{-1}(\hat{\theta} - \hat{\theta}_0)\right\} \tag{64}$$

where

$$\begin{aligned} \hat{\theta}_0 &= \text{Some } \textit{a-priori} \text{ estimate for } \theta_0 \\ \{\sigma_t^2\} &= \text{Sequence of positive scalars} \\ P_0^{-1} &= \text{Positive definite symmetric matrix} \end{aligned}$$

The last term in (64) is a term reflecting *a-priori* information about what θ_0 might be and σ_t^2 in the first term in (64) weights the importance of the data in the cost function. If we wish to find the estimate $\hat{\theta}$ which minimises this cost function then it is appropriate to find the partial derivative of $J(\hat{\theta})$ with respect to $\hat{\theta}$;

$$\frac{\partial J(\hat{\theta})}{\partial \hat{\theta}} = \mathop{\mathrm{S}}_0^t \frac{\phi_\tau y_\tau}{\sigma_\tau^2} d\tau + P_0^{-1}\hat{\theta}_0 - \left(P_0^{-1} + \mathop{\mathrm{S}}_0^t \frac{\phi_\tau \phi_\tau^T}{\sigma_\tau^2} d\tau\right)\hat{\theta}$$

Setting this to zero then gives the modified least squares estimate in generalised notation as:

$$\hat{\theta}_t = P_t\left(P_0^{-1}\hat{\theta}_0 + \mathop{\mathrm{S}}_0^t \frac{\phi_\tau y_\tau}{\sigma_\tau^2} d\tau\right) \tag{65}$$

$$P_t^{-1} = P_0^{-1} + \mathop{\mathrm{S}}_0^t \frac{\phi_\tau \phi_\tau^T}{\sigma_\tau^2} d\tau \tag{66}$$

A recursive formulation may be found by applying the generalised derivative to both sides of (65):

$$\rho\hat{\theta}_t = (\rho P_t)\left(P_0^{-1}\hat{\theta}_0 + \mathop{\mathrm{S}}_0^t \frac{\phi_\tau y_\tau}{\sigma_\tau^2} d\tau\right) + P_t\frac{\phi_t y_t}{\sigma_t^2} + \frac{\Delta}{\sigma_t^2}(\rho P_t)\phi_t y_t \tag{67}$$

However, noting that

$$\rho P_t^{-1} = \frac{\phi_t \phi_t^T}{\sigma_t^2}$$

and applying Lemma 3 gives:

$$\rho P_t = \frac{-P_t \phi_t \phi_t^T P_t}{\sigma_t^2 + \Delta \phi_t^T P_t \phi_t} \tag{68}$$

Substituting this into (67) and rearranging then gives;

$$\rho \hat{\theta}_t = \frac{P_t \phi_t (y_t - \phi_t^T \hat{\theta}_t)}{\sigma_t^2 + \Delta \phi_t^T P_t \phi_t} \tag{69}$$

Some points to note about this solution are:

1. (68) and (69) are exactly the same as (62) and (63) that were derived from the Kalman filter if the sequence $\{\sigma_t^2\}$ is chosen as per (61).

2. If we have no *a-priori* information about θ_0, then we should set $\hat{\theta}_0 = 0$ and $P_0^{-1} = 0$. If we have no *a-priori* information about the noise process $\{\nu_t\}$ then we can arbitrarily set $\sigma_t^2 = 1 \quad \forall t \in R^+$. With these choices (65) and (66) become:

$$\hat{\theta}_t = P_t \mathrm{S}_0^t \phi_\tau y_\tau d\tau \tag{70}$$

$$P_t = \mathrm{S}_0^t \phi_\tau \phi_\tau^T d\tau \tag{71}$$

which in discrete time notation becomes:

$$\hat{\theta}_n = \left(\sum_{k=0}^{n-1} \phi_k \phi_k^T \right)^{-1} \sum_{k=0}^{n-1} \phi_k y_k \tag{72}$$

This is easily recognised as the standard least squares estimate of $\hat{\theta}$.

This completes our formulation of parameter estimation using our generalised notation. we now go on to consider the numerical properties of discrete time least squares estimation using δ operator formulation.

3.5 Conditioning of Least Squares Estimation

The expression (72) shows that least squares estimation implicitly involves calculating the solution to a linear system of equations which will be ill conditioned if the condition number of the covariance matrix $P_n = \sum_{k=0}^{n-1} \phi_k \phi_k^T$ is large. It is shown in (Middleton and Goodwin 1990) that in many cases, if the regression vector is derived from a δ operator ARMAX model rather than a shift operator one, then the condition number of the associated covariance matrix is lower. This effect is exacerbated for high model orders and for high sampling rates relative to the bandwidths of the plant input and output signals y_t and u_t.

As an example, consider the continuous time system used in the discussion on Kalman Filtering:

$$G(s) = \frac{-s+2}{(s+2)(s+10)}$$

Least squares as per (72) was then used to fit a fixed denominator model to this system using both q and δ parameterisation. In this case the regression vector was:

$$\phi_t^T = \left[\frac{\hat{b}_0 u_t}{\hat{A}(\xi)}, \frac{\hat{b}_1 \xi u_t}{\hat{A}(\xi)}, \ldots, \frac{\hat{b}_m \xi^m u_t}{\hat{A}(\xi)}\right]$$

Where ξ is either the δ of q operator. The sampling rate was ranged from 0.5 Hz to 50 Hz, the numerator model order m was chosen as 3, the input signal u_t was a 0.1 Hz fundamental square wave, and the observation record was 20 seconds long. Finally the fixed denominator for the shift and δ operator forms was chosen to be:

$$\hat{A}(\delta) = (\delta+4)^2(\delta+8)^2 \qquad \hat{A}(q) = (q-(1-4\Delta))^2(q-(1-8\Delta))^2$$

For each sampling frequency the condition numbers of the q and δ operator covariance matrices were calculated and are shown vs. sampling frequency in figure 7. Note that the P matrices were scaled to give 1's along the diagonal so that the condition numbers where more meaningful comparisons of the fixed point difficulty of inversion. As can be seen from figure 7, the normal equations are much better conditioned using δ operator implementation, especially at higher sampling rates.

3.6 Calculation of Regressors

In order to implement the recursive solution to parameter estimation it is necessary to have a method for calculating the elements of the regression vector ϕ, which was defined in (58). This is easily achieved by noting that the definition of $E(\rho)$ given in (54) implies:

$$E(\rho)y \quad = \quad e_n\rho^n y + e_{n-1}\rho^{n-1}y + \ldots + e_1\rho y + y$$

to give (dropping the explicit dependence on ρ):

$$\frac{\rho^n y}{E} = \left[-\left(\frac{e_{n-1}}{e_n}\right)\left(\frac{\rho^{n-1}y}{E}\right) - \ldots - \left(\frac{e_1}{e_n}\right)\left(\frac{\rho y}{E}\right) - \left(\frac{1}{e_n}\right)\left(\frac{y}{E}\right)\right] + \frac{y}{e_n}$$

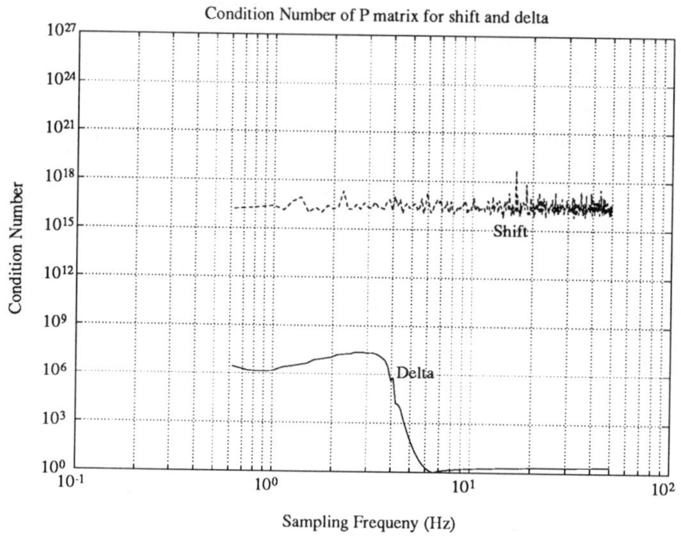

Figure 7: *Comparison of condition numbers for shift operator and δ operator covariance matrices.*

Defining ψ and ϕ_y by:

$$\begin{aligned}\psi^T &\triangleq \left[\frac{-e_{n-1}}{e_n}, \ldots, \frac{-e_1}{e_n}, \frac{-1}{e_n}\right]\\ \phi_y^T &\triangleq \left[\frac{\rho^{n-1}y}{E}, \ldots, \frac{\rho y}{E}, \frac{y}{E}\right]\end{aligned}$$

then gives:

$$\frac{\rho^n y}{E} = \psi^T \phi_y + \frac{y}{e_n}$$

Performing this similarly for the input $u(t)$ gives:

$$\frac{\rho^n u}{E} = \psi^T \phi_u + \frac{u}{e_n}$$

Where ϕ_u is defined similarly to ϕ_y. These equations can be written in state space form as:

$$\begin{aligned}\rho\phi_y &= \Upsilon\phi_y + \Sigma y\\ \rho\phi_u &= \Upsilon\phi_u + \Sigma u\end{aligned}$$

With the following definitions for Υ and Σ:

$$\Upsilon = \begin{bmatrix} 0 & 1 & 0 & \cdots & & \cdots & 0 \\ 0 & 0 & 1 & 0 & & \cdots & 0 \\ \vdots & & \ddots & \ddots & & & \vdots \\ \vdots & & & \ddots & & \ddots & 0 \\ 0 & \cdots & \cdots & \cdots & & 0 & 1 \\ -\frac{1}{e_n} & -\frac{e_1}{e_n} & \cdots & \cdots & & -\frac{e_{n-2}}{e_n} & -\frac{e_{n-1}}{e_n} \end{bmatrix}$$

$$\Sigma^T = \left[0, \ldots, 0, \frac{1}{e_n}\right]$$

In discrete time, we interpret ρ as representing the discrete time operator δ to get the discrete time recursive update equations that are necessary to calculate the filtered derivatives of $y(t)$ and $u(t)$ that are required in the regression vector ϕ:

$$\begin{aligned} \delta\phi_y &= \Upsilon\phi_y + \Sigma y \\ \delta\phi_u &= \Upsilon\phi_u + \Sigma u \\ \frac{\delta^n y}{E} &= \psi^T\phi_y + \frac{y}{e_n} \end{aligned}$$

Which gives the difference equations:

$$\phi_y^{k+1} = \phi_y^k + \Delta\Upsilon\phi_y^k + \Delta\Sigma y_k \tag{73}$$

$$\phi_u^{k+1} = \phi_u^k + \Delta\Upsilon\phi_u^k + \Delta\Sigma u_k \tag{74}$$

$$\left(\frac{\delta^n y}{E}\right)_{k+1} = \psi^T\phi_y^{k+1} + \frac{y_{k+1}}{e_n}$$

All the filtered derivatives of the regression vector ϕ are thus known from these equations and so ϕ may be formed from the calculated elements by:

$$\phi_k^T = \left[\left(\frac{\delta^n y}{E}\right)_k, \phi_y^k(1), \ldots, \phi_y^k(n-1), \phi_u^k(n-m-1), \ldots, \phi_u^k(n)\right]$$

Note that not all the terms in ϕ_y and ϕ_u are used in forming the regression vector ϕ. Specifically, the term $\frac{y}{E}$ in ϕ_y is not used in ϕ and the terms of higher order than $\frac{\rho^m u}{E}$ in ϕ_u are not used in ϕ.

3.7 Altering the Dynamic Behaviour of RLS

Consideration of (68) shows that $P_k = 0$ is a solution of the difference equation for the update of P_k. Consequently, for RLS of P_k given by (68) we get $P_k \to 0$. Consideration of (69) shows that this means that the algorithms will not be able to track time varying parameter changes. In order to overcome this, we need to modify the dynamic behaviour of RLS. Here we discuss two algorithms for doing this namely the gradient and constant trace schemes.

3.7.1 Gradient Algorithm

The most obvious solution to the problem is to fix the covariance matrix so that it cannot tend to $\vec{0}$. The simplest constant to use for the covariance matrix is some multiple of the identity matrix. This then leads to the following recursive update scheme (we have dropped the explicit dependence on time):

$$\begin{aligned} \rho\hat{\theta} &= \frac{\alpha\phi e}{1+\Delta\alpha\phi^T\phi} \\ P &= \alpha I \end{aligned}$$

This solution is only a crude one to the problem of tracking time varying plants. Specifically, it slows down the rate of parameter convergence. To see this consider first the RLS update scheme (ie. not the gradient scheme). In this case the covariance matrix P is updated. Considering the inverse of this matrix and the parameter estimation error $\tilde{\theta}$:

$$\tilde{\theta} \triangleq \hat{\theta} - \theta_0$$

Lemma 2 can be used to find the generalised derivative of their product:

$$\rho(P^{-1}\tilde{\theta}) = (\rho P^{-1})\tilde{\theta} + P^{-1}\rho\tilde{\theta} + \Delta(\rho P^{-1})(\rho\tilde{\theta}) \tag{75}$$

Noting that

$$\begin{aligned} \rho\tilde{\theta} &= \frac{-P\phi\phi^T P}{1+\Delta\phi^T P\phi} && (76) \\ \rho P^{-1} &= \phi\phi^T && (77) \end{aligned}$$

and substituting into (75) gives:

$$\rho(P^{-1}\tilde{\theta}) = \phi\phi^T\tilde{\theta} + P^{-1}\left(\frac{-P\phi\phi^T P}{1+\Delta\phi^T P\phi}\right) + \Delta\phi\phi^T\left(\frac{-P\phi\phi^T P}{1+\Delta\phi^T P\phi}\right) \tag{78}$$

consider also the quadratic form:

$$V(t) = \tilde{\theta}^T(t)P^{-1}\tilde{\theta}(t)$$

Using Lemma 2 gives the generalised derivative of $V(t)$ as:

$$\rho V = (\rho\tilde{\theta}^T)P^{-1}\tilde{\theta} + \tilde{\theta}^T\rho(P^{-1}\tilde{\theta}) + \Delta(\rho\tilde{\theta}^T)\rho(P^{-1}\tilde{\theta})$$

Using (77),(78) then results in:

$$\rho V = -\tilde{e}_t^2 \tag{79}$$

Where $\tilde{e}_t$ is the normalised prediction error defined as:

$$\tilde{e}_t \triangleq \frac{y-\phi^T\hat{\theta}}{\sqrt{1+\Delta\phi^T P\phi}}$$

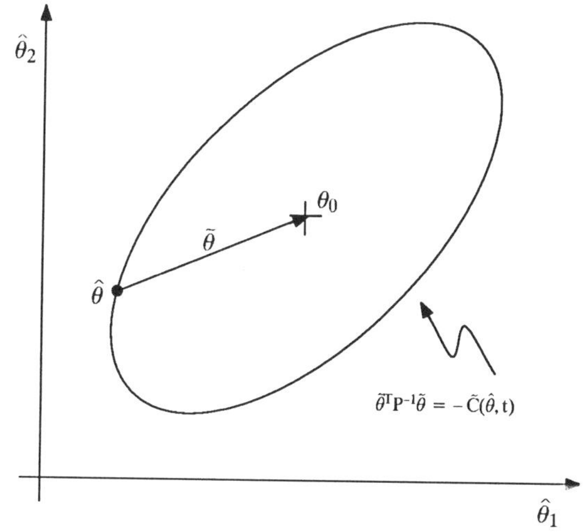

Figure 8: *Ellipse of possible estimates giving the same cost function value*

Defining an associated normalised cost function:

$$\tilde{C}(\hat{\theta}, t) \triangleq \mathcal{S}_0^t \tilde{e}_t^2$$

and applying the generalised integral to both sides of (79) gives:

$$V(t) = -\mathcal{S}_0^t \tilde{e}_t^2 = -\tilde{C}(\hat{\theta}, t)$$

Note that as $\Delta \to 0$ the normalised cost function tends towards the least squares cost function $C(\hat{\theta}, t)$. Therefore, the equation:

$$V(t) = (\hat{\theta} - \theta_0)^T P^{-1} (\hat{\theta} - \theta_0) = -\tilde{C}(\hat{\theta}, t) \tag{80}$$

describes a hyper-ellipsoid centred around the true parameter values. Furthermore, the 'size' of the hyper-ellipsoid is proportional to the value of the

normalised cost function. That is, for a certain normalised least squares cost function value, all the possible estimates of the plant that could achieve that cost function value must lie on the surface of the hyper-ellipsoid. The two dimensional case is shown diagrammatically in figure 8. Note that the vector:

$$\tilde{\theta} = \left(\sqrt{\frac{\tilde{C}(\hat{\theta}, t)}{\phi^T P \phi}} \right) P\phi$$

is a solution of the ellipsoid equation (80). That is, given a particular parameter estimate $\hat{\theta}$, projection should be made in the direction $P\phi$ in order to arrive at the true parameter value θ_0. Projection such as this would require a parameter update of the form:

$$\rho\hat{\theta} = \alpha P \phi \qquad \alpha \text{ a scalar}$$

Note that this is precisely what the unmodified least squares algorithm does. Because P is not updated with the gradient scheme the projection of $\hat{\theta}$ is not in the correct straight line direction towards θ_0 and hence convergence is slower than for RLS.

3.7.2 Constant Trace Algorithm

A solution to the problem of the covariance matrix going to zero, whilst not fixing it and thus losing valuable search direction information, is to simply multiply it by an appropriate constant. Specifically, a possible solution is to update the covariance matrix using the basic RLS, and then multiply the matrix by an appropriate scalar determined so as to fix the trace of the matrix at a specified value. This scheme will give the constant trace algorithm which may be written

$$\begin{aligned}
\hat{\theta}_{k+1} &= \hat{\theta}_k + \frac{\Delta P_k \phi_k e_k}{1 + \Delta \phi_k^T P_k \phi_k} \\
P_{k+1} &= \xi \left(P_k - \frac{\Delta P_k \phi_k \phi_k^T P_k}{1 + \Delta \phi_k^T P_k \phi_k} \right) \\
\xi &= \frac{\text{Trace}(P_0)}{\text{Trace}(P_{k+1})} \\
e_k &= y_k - \phi_k^T \hat{\theta}_k
\end{aligned}$$

Note that scaling P by a scalar does not distort the search direction information P contains, but it does prevent P tending to zero. Other related algorithms are described in (Salgado etc. 1988).

3.7.3 Dealing with Coloured Noise

The linear regression model (57) that was derived from the Kalman filter motivated ARMAX model (49) involves a measurement noise process $\{\eta_t\}$. This is derived from the white 'noise' innovations process $\{\xi_t\}$ via:

$$\eta_t = \frac{C(\rho)}{E(\rho)} \xi_t$$

Therefore, $\{\eta_t\}$ will in general not be a white noise process. Consequently, since both ϕ_t and ξ_t will then depend on past data, they will be correlated, Therefore, the normal least squares estimate $\hat{\theta}_t$ will not be consistent (Goodwin and Payne 1977). There are a number of ways of overcoming this bias in $\hat{\theta}$.

1. If $C(\rho)$ is known then putting $E(\rho) = C(\rho)$ gives the linear regression form (57) as:
$$y_t = \left(\frac{C-A}{C}\right) y_t + \frac{B}{C} u_t + \xi_t \tag{81}$$
This gives the white measurement noise process we require.

2. If $C(\rho)$ is not known, then $C(\rho)$ can be estimated from the regression model (81). However, this will be a non-linear optimisation problem. This implies non-unique minima to the least squares cost function (64). Therefore $\hat{\theta}$ may converge to a local minima and give worse bias than if no steps were taken. Furthermore, it is usually necessary to project $\hat{C}(\rho)$ to ensure that it is stable (Goodwin and Sin 1984).

3. If appropriate instruments are known, then the instrumental variable method may be used to obtain a consistent estimate (Soderstrom and Stoica 1989).

4. A Pseudo-Linear regression may be used. That is, the ARMAX model (49) may be written in the regression form:
$$y_t = \left(\frac{E-A}{E}\right) y_t + \frac{B}{E} u_t + \left(\frac{C-E}{E}\right) \xi_t + \xi_t$$
This is a linear regression model of the form (57) with white noise where
$$\begin{aligned} \phi^T &= \left[\frac{\rho^n y_t}{E}, \cdots, \frac{\rho y_t}{E}, \frac{\rho^m u_t}{E}, \cdots, \frac{u_t}{E}, \frac{\rho^n \xi_t}{E}, \cdots, \frac{\xi_t}{E}\right] \\ \theta^T &= [e_n - a_n, \cdots, e_1 - a_1, b_m, \cdots, b_0, c_n - e_n, \cdots, c_0 - e_0] \end{aligned}$$
This is not in a form suitable for parameter estimation since ϕ depends on the unmeasured innovations process $\{\xi_t\}$. The pseudo-linear regression method circumvents this by replacing ξ_t by an on-line estimate $\hat{\xi}_t$:
$$\hat{\xi}_t = y_t - \phi_t^T \hat{\theta}_t$$
This method provides a consistent estimate for θ so long as $E(\rho)$ and $C(\rho)$ satisfy a positive real condition (Goodwin and Sin 1984; Soderstrom and Stoica 1989; Goodwin 1988).

5. We can try to linearize the estimation of $C(\rho)$ proposed in 2. This can be done by writing $C^{-1}(\rho)$ as a power series and then approximating $C^{-1}(\rho)$ by truncation of the series at r terms:
$$C^{-1}(\rho) \approx \sum_{k=1}^{r} c_k \rho^{-k} \tag{82}$$

Operating on both sides of (49) when $\rho = q$ then shows this to be the well known idea of modelling MA processes with AR approximations:

$$A'(q^{-1})y_k \approx B'(q^{-1})u_k + \xi_k$$

Where

$$A'(q^{-1}) = A(q^{-1})q^n \sum_{k=1}^{r} c_k q^{-k} \qquad B'(q^{-1}) = B(q^{-1})q^n \sum_{k=1}^{r} c_k q^{-k}$$

We can now perform estimation with approximately white measurement errors. Note that the resultant model estimate is non-minimal.

We intend to pursue this last method here, but suggest the use of *a priori* knowledge of $C(\rho)$ in order to improve the performance of the method. Specifically, we note that the convergence of the power series (82) depends on the location of the zeroes of $C(\rho)$. If these are close to the stability boundary, then the convergence will be slow and large orders of $A'(\rho)$ and $B'(\rho)$ will be required to provide approximately white errors.

Therefore, we propose that $C(\rho)$ be expanded not in terms of ρ^{-1}, but in terms of $(\rho + e)^{-1}$, where $-e$ is chosen by prior knowledge to be close to the zeroes of $C(\rho)$ so that the expansion will converge quickly. This idea is summarised in the following Lemma.

Lemma 4. *Consider the stochastic operator model (55). Provided we know an* $e \in (0, \frac{1}{\Delta}]$ *such that the zeroes* γ_i *of* $C(\rho)$ *satisfy:*

$$|\gamma_i + e| < e \qquad \forall i \in [1, n] \tag{83}$$

then $\forall \varepsilon > 0$ *there exists a stable operator* $E'(\rho)$ *such that (55) can be expressed as*

$$\frac{A'(\rho)}{E'(\rho)} y_t = \frac{B'(\rho)}{E'(\rho)} u_t + \xi_t + \xi'_t$$

where $\{\xi_t\}$ *is a white noise process and* ξ'_t *has variance less than* ε.

Proof. For the sake of simplicity, assume that $C(\rho)$ has no repeated zeroes so that by partial fraction expansion we may write:

$$\begin{aligned} \frac{E(\rho)}{C(\rho)} &= \frac{E(\rho)}{\prod_{i=1}^{n}(\rho + \gamma_i)} \\ &= 1 + \sum_{i=1}^{n} \left(\frac{\alpha_i}{\rho + \gamma_i} \right) \\ &= 1 + \sum_{i=1}^{n} \left(\frac{\alpha_i}{\rho + e} \right) \left[\frac{1}{1 + \left(\frac{\gamma_i - e}{\rho + e} \right)} \right] \end{aligned} \tag{84}$$

The term in square brackets may be expanded via a power series as (Kreyszig 1983):

$$\frac{1}{1+\left(\frac{\gamma_i - e}{\rho+e}\right)} = \sum_{j=0}^{\infty}\left(\frac{e-\gamma_i}{\rho+e}\right)^j$$

By the Ratio Test (Kreyszig 1983), this power series is convergent for:

$$\left|\frac{e-\gamma_i}{\rho+e}\right| < 1 \Rightarrow |\gamma_i - e| < e \quad \forall i \in [1,n] \quad \forall \rho \in [0, \frac{1}{\Delta}]$$

That is, e must be 'closer to' the γ_i's than to the origin. This may be satisfied if e is bigger than all the γ_i's are. The closer e is to the γ_i's, the faster the power series will converge. If we truncate this power series at m terms, then we may write:

$$\begin{aligned}\frac{1}{1+\left(\frac{\gamma_i - e}{\rho+e}\right)} &= \sum_{j=0}^{m}\left(\frac{e-\gamma_i}{\rho+e}\right)^j + R(\rho) \\ &\triangleq \frac{F(\rho)}{(\rho+e)^m} + R(\rho)\end{aligned}$$

Where $R(\rho)$ is a remainder term. Substituting this into (84) gives:

$$\frac{1}{C(\rho)} = \frac{1}{E(\rho)}\left(1 + \frac{F(\rho)}{(\rho+e)^m} + R(\rho)\right)$$

Motivated by this, define:

$$\frac{F'(\rho)}{E'(\rho)} \triangleq \frac{1}{E(\rho)}\left(1 + \frac{F(\rho)}{(\rho+e)^m}\right) \approx \frac{1}{C(\rho)}$$

In this case, $E'(\rho) \triangleq E(\rho)(\rho+e)^m$ and $F'(\rho)$ is of order m. Operating on both sides of (57) by this stable operator gives:

$$A'\left(\frac{y}{E'}\right) = B'\left(\frac{u}{E'}\right) + \nu - \left(\frac{HR}{E}\right)\nu$$

Where $A'(\rho) \triangleq A(\rho)F'(\rho)$ and $B'(\rho) \triangleq B(\rho)F'(\rho)$. $\nabla\nabla\nabla$

This Lemma suggests a paradigm of adding extra zeroes to the observer polynomial $E(\rho)$, appropriately chosen by *a-priori* knowledge and (83), and then by fitting an appropriately higher order model to the process. Some points to note about this method are;

1. The model used is non-minimal. However, because of the normalisation of $A(\rho)$ and $E(\rho)$ from the right, the extraction of an appropriate minimal order estimate is easy since the high order terms go to zero as the power series converges. Such is not the case if normalisation from the left is used to force the high order term to be 1.

2. The convergence condition in (83) is precisely the positive real condition necessary for the pseudo-linear regression method 4 to converge (Goodwin and Sin 1984; Soderstrom and Stoica 1989). This highlights the fact that this expansion method essentially involves the estimation of $C(\rho)$.

To illustrate this discussion of noise an example is now presented. The following discrete time system was simulated:

$$\begin{aligned}(\delta+2)y_k &= 2u_k + (\delta+2)\nu_k \\ \Rightarrow y_k &= \phi^T\theta + \left(\frac{\delta+2}{E(\delta)}\right)\nu_k\end{aligned}$$

The sampling rate used was 30 Hz and ν_k was a white Gaussian distributed process with variance $\sigma^2 = 0.01$. A Constant Trace estimator was run for 5 seconds with the trace set to 100. Initially the observer polynomial was set to $E(\delta) = (\delta+5)$ so that the noise error was coloured. As expected, this produced a bias in the estimation of the parameters with the identified model being:

$$\hat{G}(\delta) = \frac{0.55}{0.14\delta+1}$$

This is not very close to the true model of $\frac{1}{0.5\delta+1}$. Figure 9 shows the evolution of the estimates together with a comparison between the true and estimated frequency responses. As can be seen, there is a large bias error. Note also that the parameter estimates vary with input changes. This is typical of undermodelled behaviour.

The same estimator was then run with $E(\delta) = (\delta+2)$ so that the discrete time simulation became:

$$\begin{aligned}y_k &= \phi^T\theta + \left(\frac{\delta+2}{\delta+2}\right)\nu_k \\ &= \phi^T\theta + \nu_k\end{aligned}$$

In this case the noise error is white, and so by the Kalman optimal filter properties, the estimated plant should be unbiased. This was found to be the case with the estimates being:

$$\hat{G}(\delta) = \frac{0.83}{0.41\delta+1}$$

This is quite close to the true plant as the frequency response curves in figure 10 show. This illustrates the use of method 2. If the spectral properties are not known, then the expansion method of Lemma 4 may be if the order of $E(\delta)$ and consequently $\hat{G}(\delta)$ are extended by m. The case $m = 1$ was simulated with $E(\delta) = (\delta+5)^2$ and the estimated plant arrived at was:

$$\hat{G}(\delta) = \frac{-0.012\delta+0.87}{0.009\delta^2+0.43\delta+1} \approx \frac{0.87}{0.43\delta+1}$$

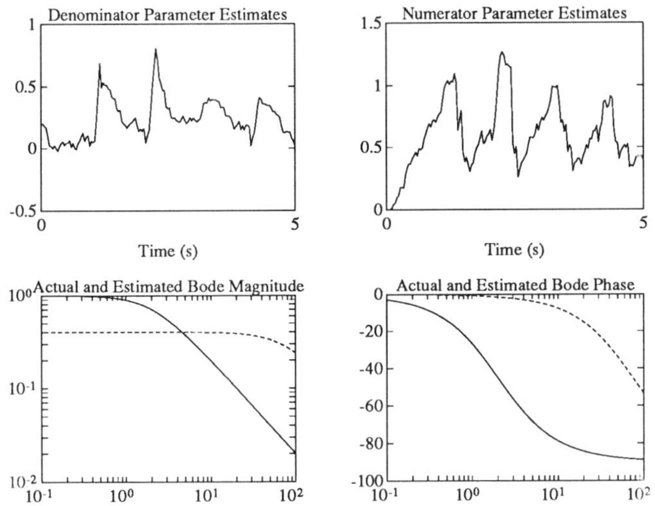

Figure 9: *Evolution of Estimates, and comparison of frequency responses of true and estimated plants for* $E(\delta) = (\delta + 5)$

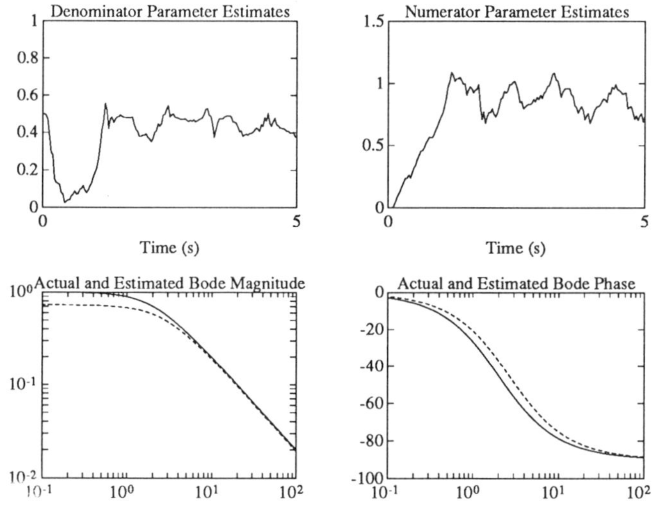

Figure 10: *Evolution of Estimates, and comparison of frequency responses of true and estimated plants for* $E(\delta) = (\delta + 2)$

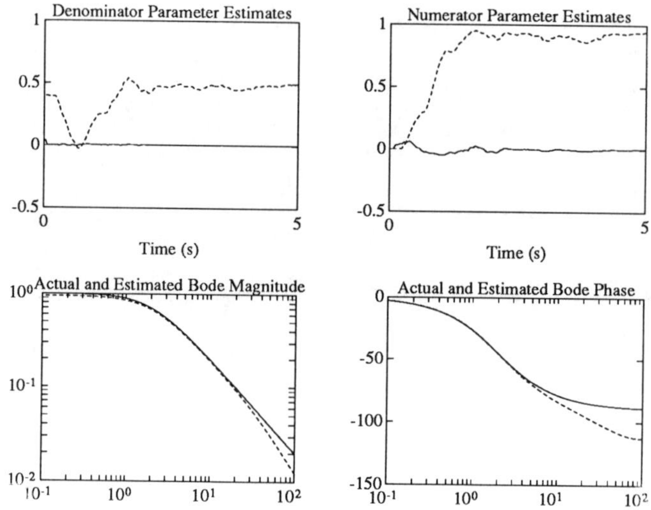

Figure 11: *Evolution of Estimates, and comparison of frequency responses of true and estimated plants for* $E(\delta) = (\delta + 5)^2$, $m = 1$

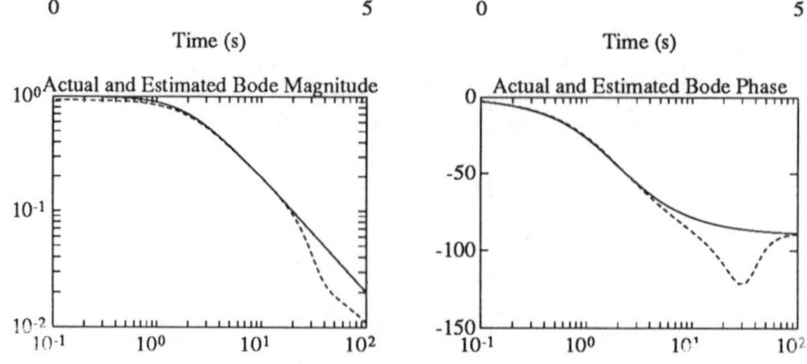

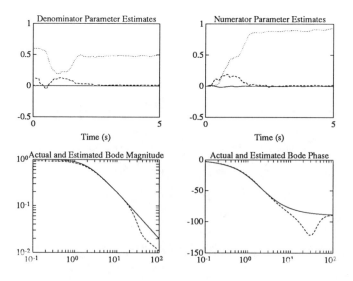

Figure 12: *Evolution of Estimates, and comparison of frequency responses of true and estimated plants for* $E(\delta) = (\delta + 5)^3$, $m = 2$

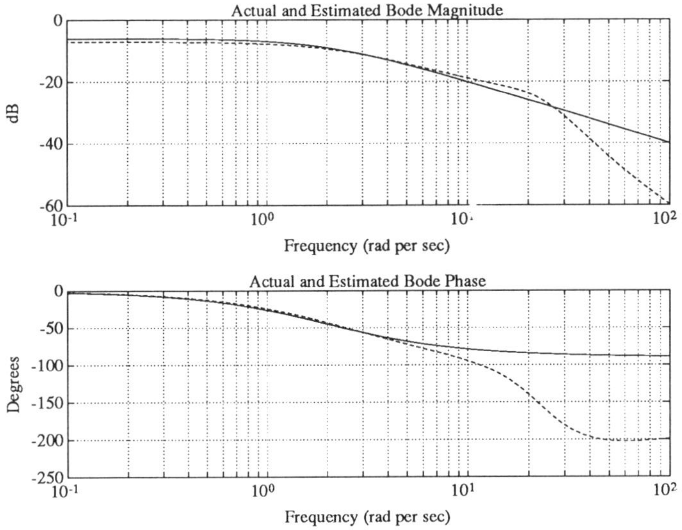

Figure 13: *Results of estimation for* $G_1(s)$.

This estimate is quite close to the true plant as the frequency response comparison in figure 11 shows. The case $m = 2$ was also simulated with $E(\delta) = (\delta + 5)^3$ to give an estimate of:

$$\hat{G}(\delta) = \frac{0.0007\delta^2 + 0.0086\delta + 0.87}{0.0007\delta^3 + 0.019\delta^2 + 0.46\delta + 1} \approx \frac{0.87}{0.46\delta + 1}$$

The frequency response comparison in figure 12 shows a negligible improvement over the case $m = 1$ for this example. Note how, with normalisation from the right, it is easy to extract the minimal order subsystem from the non-minimal estimate.

3.7.4 Effect of Normalising the Plant Model from the Right

In the analysis presented so far, it has been assumed that, apart form measurement noise, the system response can be exactly described by an nth order system parameterised by θ_0. In practice, however, all systems are infinite dimensional, and all we can hope to do is find and order n which allows approximate modelling of the system. The practice of normalising the plant model from the right is particularly amenable to this problem since it allows a paradigm of fitting a high order model to the system. If this order turns out to be too high then the high order parameters are estimated as zero. Such is not the case if the nominal model is normalised from the left and the highest order parameter is fixed at 1.

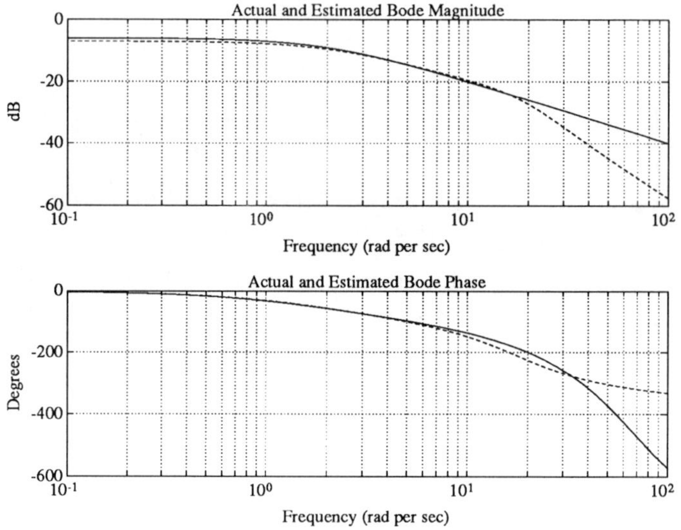

Figure 14: *Results of estimation for* $G_2(s)$.

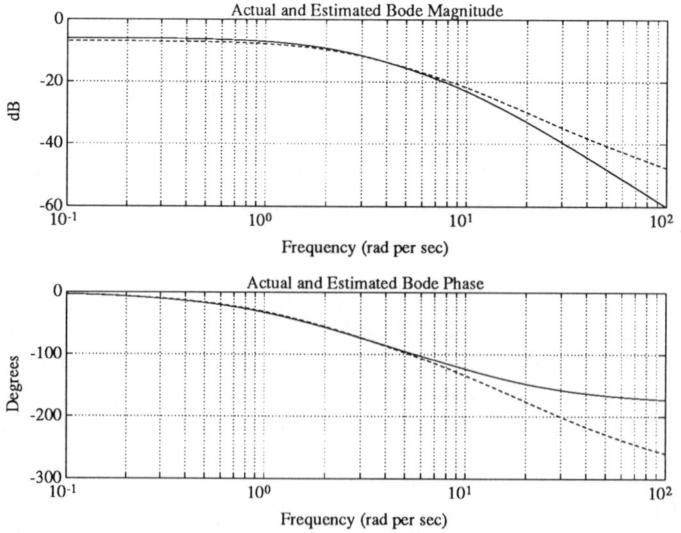

Figure 15: Results of estimation for $G_3(s)$.

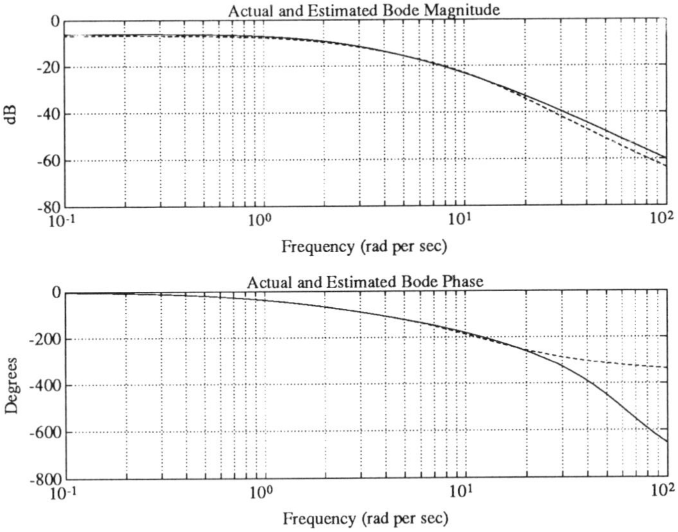

Figure 16: *Results of estimation for* $G_4(s)$.

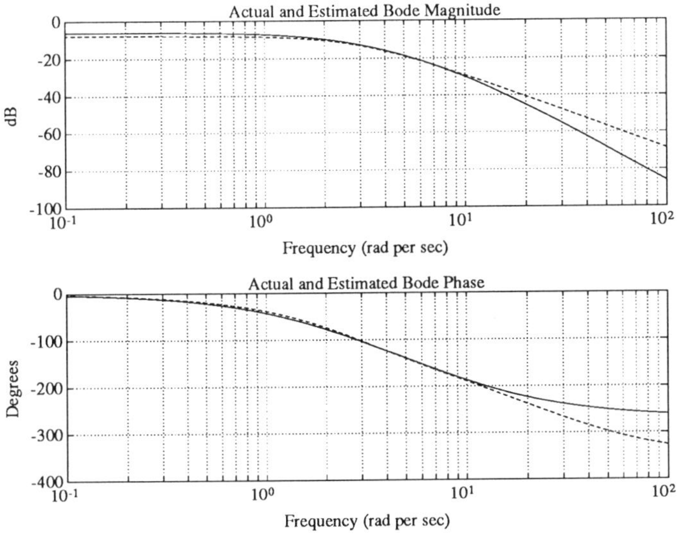

Figure 17: *Results of estimation for* $G_5(s)$.

To illustrate this paradigm the following model:

$$G(\delta) = \frac{\hat{b}_1\delta + \hat{b}_0}{\hat{a}_3\delta^3 + \hat{a}_2\delta^2 + \hat{a}_1\delta + 1}$$

was fitted to the following five plants:

$$G_1(s) = \frac{2}{s+2} \quad G_2(s) = \frac{2e^{0.1s}}{s+2} \quad G_3(s) = \frac{2}{(s+2)(0.1s+1)}$$

$$G_4(s) = \frac{2e^{0.1s}}{(s+2)(0.1s+1)} \quad G_5(s) = \frac{2}{(s+2)(0.1s+1)(0.2s+1)}$$

A sampling frequency of 30 Hz was used in the simulations. The plant excitation signal used was a 1.5 radian per second square wave. A Constant Trace Least Squares identification algorithm was used with the trace set to 100 and an observer polynomial of $E(\delta) = (\delta + 10)^3$ was used. The results are shown in figures 13 to 17. In each of the figures the true and estimated Bode magnitude and phase plots are shown. As can be seen the fixed third order model is very successfully fitted to all five different order plants. The comparison of frequency responses show that in all cases a good fit is found over a 10 radian per second range. This suggests that a method of fitting a high order model to a plant response would be useful in finding a smooth estimate of the frequency response of an unknown order plant. This could replace the common method of dividing the sample cross correlation between the plant output signal and the plant input signal by the sample autocorrelation function of the plant input signal since this method suffers from providing non-smooth estimates of frequency response and also of being sensitive to noise (Ljung 1987).

4 Conclusion

We have provided a new perspective on linear estimation by formulating our algorithms in a new generalised notation that provides both the discrete time, and the appropriate continuous time result simultaneously. Central to this has been the introduction of a new discrete time operator, the δ operator that is derived from a linear transformation of the familiar shift operator q. The major conceptual advantage of this new operator is that, unlike the q operator case, discrete time formulations do not ignore the fact that they stem from an underlying continuous time process. Consequently, the problem of having to juggle two different domains, the continuous and discrete, disappears.

We have shown that as well as the conceptual benefit of linking the continuous and discrete domains more closely, the δ operator allows much more numerically robust formulations of some key digital signal processing algorithms. In particular, we highlighted the improved numerical conditioning of Kalman filter state estimation calculations, and the improved conditioning of the normal equations involved in least squares parameter estimation.

We concluded with a discussion of some practical aspects of parameter estimation, not limited to δ operator formulations. We showed how the dynamic properties of *RLS* could be improved to track time varying plants. We showed how the problem of measurement noise correlated with the regression vector could be overcome to avoid biased estimates, and we showed how the problem of parameter estimation on systems of unknown order could be approached with a simple paradigm.

References

Astrom, K. J. (1970), "*Introduction to Stochastic Control Theory*", Academic Press, New York.

Dougherty, G.R., Stanley, W.D., and Dougherty, R. (1984), "*Digital Signal Processing*", Reston Publishing Company, Inc., Reston, Virginia, second edition.

Francis, B.A. (1987), "*A Course in H_∞ Control Theory*", Springer-Verlag, Berlin.

Goodwin, G.C. (1988), "Some observations on robust stochastic estimations", *8th IFAC Symposium on System Identification and System Parameter Estimation*, Beijing, vol.1, pp.22-32.

Goodwin, G.C. and Payne, R.L. (1977), "*Dynamic System Identification*", Academic Press, New York.

Goodwin, G.C. and Sin, K.W. (1984), "*Adaptive Filtering Prediction and Control*, Prentice-Hall, Inc., New York.

Hilderbrand, F.B. (1956), "*Introduction to Numerical Analysis*", McGraw Hill, New York.

Kailath, T. (1980), "*Linear Systems*", Prentice-Hall, Inc., New York.

Kreyszig, E. (1983), "*Advanced Engineering Mathematics*", John Wiley and Sons, London.

Ljung, L. (1987), "*System Identification: Theory for the User*", Prentice-Hall, Inc., New Jersey.

Middleton, R.H. and Goodwin, G.C. (1986), "Improved finite word length characteristics in digital control using delta operators", *IEEE Transaction on Automatic Control*, AC-31, pp.1015-1021.

Middleton, R.H. and Goodwin, G.C. (1990), "*Digital Estimation and Control: A Unified Approach*", Prentice-Hall, Inc., New Jersey.

Middleton, R.H., Salgado, M.E., and Goodwin, G.C. (1988), "Connection between continuous and discrete riccati equations with applications to Kalman filtering", *IEE Proceedings, Part D*, vol.135.

Rabiner, L.R. and Gold, B. (1975), "*Theory and Application of Digital Signal Processing*", Prentice-Hall, Inc., New Jersey.

Salgado, M.E. (1989), "*Issues in Robust Identification*", Ph.D. thesis, University of Newcastle.

Salgado, M.E., Goodwin, G.C., and Middleton, R.H. (1988), "A modified least squares algorithm incorporating exponential setting and forgetting", *International Journal of Control*, vol.47, pp.477-491.

Soderstrom, T. and Stoica, P. (1989), "*System Identification*", Prentice-Hall, Inc., New Jersey.

Wilkinson, J.H. (1963), "*Rounding Errors in Algebraic Processes*", Prentice-Hall, Inc., New Jersey.

Transformation of discrete-time models

N.K. Sinha
Department of Electrical & Computer Engineering
McMaster University
Hamilton, Canada L8S 4L7

and

G.J. Lastman
Department of Applied Mathematics
University of Waterloo
Waterloo, Canada N2L 3G1

Abstract

Several approaches to estimating the parameters of a continuous-time model of a linear multivariable system from an equivalent discrete-time model are presented. It is assumed that a suitable discrete-time model has been obtained from the samples of input and output observations using techniques that are already well established. Here, our emphasis is on transformations which will lead to a suitable continuous-time model from the identified discrete-time model. Several algorithms for such transformation are critically compared. Finally, a straightforward procedure for determining a continuous-time model from the δ-model is described.

1. Introduction

The identification of the process parameters for control purposes must often be done, using a digital computer, from samples of input-output observations. The dynamical model of the process is, on the other hand, usually described in terms of state equations in the continuous-time domain. Thus, the problem may be stated as estimation of the parameters of a continuous-time model from the samples of the input-output observations for a multivariable system.

One way to solve the problem is to divide it into two subproblems, (i) the estimation of the parameters of a discrete-time model from the samples of the observations, and (ii) the determination of a continuous-time model corresponding to the discrete-time model thus obtained. The main advantage of this approach is that considerable literature is already available on the first part of the problem (Tse and Wienert 1975, Sinha and Kwong 1979, El-Sherief and Sinha 1979, Sinha and Kuszta 1983). On the other hand, relatively less has been published about the second subproblem (Sinha 1972, Hsia 1972, Strmčnik and Bremšak 1979, Sinha and Lastman

N. K. Sinha and G. P. Rao (eds.), Identification of Continuous-Time Systems, 123–137.

1982). Another advantage is that only the first subproblem requires considerations of the stochastic nature of measurement noise, while the second subproblem is essentially deterministic.

It will be assumed that the sampling interval has been selected carefully and the discrete-time model obtained from the data has been duly validated. Furthermore, the choice of the sampling interval is often very critical, especially for input-output data contaminated with noise (Sinha and Puthenpura 1985). A procedure for selecting the optimum sampling interval has been described by Puthenpura and Sinha (1985). This topic will be discussed in more detail in Section 4.

2. Background

Consider a continuous-time linear multivariable system described by the state equations

$$\dot{\boldsymbol{x}}(t) = \boldsymbol{A}\boldsymbol{x}(t) + \boldsymbol{B}\boldsymbol{u}(t) \tag{1}$$

$$\boldsymbol{y}(t) = \boldsymbol{C}\boldsymbol{x}(t) + \boldsymbol{D}\boldsymbol{u}(t) \tag{2}$$

where $\boldsymbol{x}(t) \in \Re^n$, $\boldsymbol{u}(t) \in \Re^m$, and $\mathbf{y}(t) \in \Re^p$ are the state, input and output vectors, respectively, while $\boldsymbol{A}$, $\boldsymbol{B}$, $\boldsymbol{C}$ and $\boldsymbol{D}$ are matrices of appropriate dimensions.

Assuming that the state of the system is known at $t = t_1$, its value at $t_2 > t_1$ is readily obtained as

$$\boldsymbol{x}(t_2) = e^{A(t_2 - t_1)}\boldsymbol{x}(t_1) + \int_{t_1}^{t_2} e^{A(t_2 - \tau)}\boldsymbol{B}\boldsymbol{u}(\tau)\,d\tau \tag{3}$$

If the sampling interval is T, and the values of t_1 and t_2 are selected such that

$$t_1 = kT \tag{4}$$

$$t_2 = (k + 1)T \tag{5}$$

where k is any non-negative integer, equation (3) may be written as

$$\boldsymbol{x}_{k+1} = e^{AT}\boldsymbol{x}_k + \int_{kT}^{(k+1)T} e^{A(kT+T-\tau)}\boldsymbol{B}\,\boldsymbol{u}(\tau)\,d\tau \tag{6}$$

where

$$\boldsymbol{x}_i \triangleq \boldsymbol{x}(iT), \quad \boldsymbol{u}_i \triangleq \boldsymbol{u}(iT) \tag{7}$$

and $$y_i \triangleq y(iT) = Cx_i + Du_i \tag{8}$$

In order to evaluate the integral in equation (6), we must have the precise mathematical expression for **u**(t). If the only available information is the value of **u**(t) at the sampling instants *kT*, we can obtain an approximation based on some simplifying assumption. The problem is considerably simplified if it is assumed that the input to the system is held constant between the sampling instants. In many practical situations, this is true. For example, the block diagram of a control system using a digital computer as part of the control loop is shown below, where *ZOH* represents a zero-order hold.

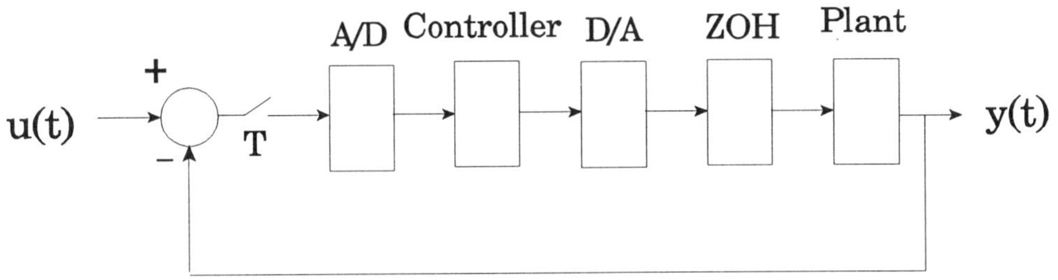

Fig. 1 A typical computer-controlled system

In this case, we obtain the *step-invariant* transformation between the continuous-time system and its equivalent discrete-time model, represented by the equations

$$x_{k+1} = Fx_k + Gu_k \tag{9}$$

$$y_k = Cx_k + Du_k \tag{10}$$

where $$F = e^{AT} \tag{11}$$

$$G = \int_0^T e^{A\tau}\, d\tau\, B \tag{12}$$

It is well known that F and G are easily calculated from the following power series expansions

$$F = \sum_{i=0}^{\infty} \frac{(AT)^i}{i!} \tag{13}$$

and

$$G = \left[\sum_{i=0}^{\infty} \frac{(AT)^i}{(i+1)!} \right] BT \tag{14}$$

where the series can be truncated after a certain number of terms to obtain a desired accuracy. It can be shown that if the sampling interval is selected so that $\rho(\boldsymbol{A})T < 0.5$, where $\rho(\boldsymbol{A})$ is the spectral radius of $\boldsymbol{A}$, one can truncate the series after 12 terms to with an error of less than 10^{-12} in the elements of $\boldsymbol{F}$ and $\boldsymbol{G}$.

In many other cases, a zero-order hold is not used, but we are still required to obtain the system model from the samples of the input-output data. It is evident that although the assumption that the input is held constant between the sampling instants is no longer valid, the error will be small if the sampling interval is very small. Further improvement is possible if it is assumed that the input varies linearly between the sampling instants. This leads to the *ramp-invariant* transformation between the continuous-time system and its discrete-time transformation. As shown in a recent paper (Bingulac and Sinha 1989), the equivalent discrete-time model is similar to that for the step-invariant case, but equation (9) is modified to take the form

$$\boldsymbol{x}_{k+1} = \boldsymbol{F}\boldsymbol{x}_k + \boldsymbol{G}_0\boldsymbol{u}_k + \boldsymbol{G}_1\boldsymbol{u}_{k+1} \tag{15}$$

where $\boldsymbol{F}$ is as given in equation (11) while $\boldsymbol{G}_0$ and $\boldsymbol{G}_1$ can be calculated from the following equations

$$\boldsymbol{G}_0 = \boldsymbol{M}_0\, \boldsymbol{B}T \tag{16}$$

and

$$\boldsymbol{G}_1 = \boldsymbol{M}_1\, \boldsymbol{B}T \tag{17}$$

where

$$\boldsymbol{M}_0 \triangleq \sum_{i=0}^{\infty} (i+1)\frac{(AT)^i}{(i+2)!} \tag{18}$$

and

$$\boldsymbol{M}_1 \triangleq \sum_{i=0}^{\infty} \frac{(AT)^i}{(i+2)!} \tag{19}$$

Thus, our problem may be stated as the evaluation of the matrices $\boldsymbol{A}$ and $\boldsymbol{B}$, from the estimates of the matrices $\boldsymbol{F}$ and $\boldsymbol{G}$ (for the step-invariant transformation), or from the estimates of $\boldsymbol{F}$, $\boldsymbol{G}_0$ and $\boldsymbol{G}_1$ (for the ramp-invariant case). It follows from equation (11) that

$$AT = \ell n\, \boldsymbol{F} \tag{20}$$

As the sampling interval is known, this problem is essentially that of finding the natural logarithm of $\boldsymbol{F}$. After $\boldsymbol{A}$ has been obtained, it is fairly straightforward to evaluate $\boldsymbol{G}$, $\boldsymbol{G}_0$ or $\boldsymbol{G}_1$. For example, we have

$$\boldsymbol{B} = \boldsymbol{R}^{-1}\boldsymbol{G} \tag{21}$$

where

$$\boldsymbol{R} = T\left[\sum_{i=0}^{\infty} \frac{(AT)^i}{(i+1)!}\right] \tag{22}$$

and can be easily calculated. It may be noted that nonsingularity of $\boldsymbol{R}$ is guaranteed if the sampling interval is selected so that the spectral radius of AT is less than 0.5. Similarly, we can calculate $\boldsymbol{G}_0$ and $\boldsymbol{G}_1$ from equations (16) and (17) since nonsingularity of the matrices $\boldsymbol{M}_0$ and $\boldsymbol{M}_1$ are guaranteed.

In the following section, we shall study various methods for evaluating the natural logarithm of a square matrix.

3. Evaluation of the natural logarithm of a square matrix

During the last 13 years, several methods have been proposed by different authors for evaluating the natural logarithm of a square matrix. These are based on the following approaches:

Transformation to diagonal or Jordan canonical forms:

The main idea here is that any square matrix $\boldsymbol{F}$ can be transformed to the diagonal or Jordan canonical form through the relationship

$$\boldsymbol{J} = \boldsymbol{P}^{-1}\boldsymbol{F}\boldsymbol{P} \tag{23}$$

where the columns of $\boldsymbol{P}$ are the eigenvectors of $\boldsymbol{F}$ (generalized eigenvectors if the matrix $\boldsymbol{F}$ cannot be diagonalized). The elements on the main diagonal of $\boldsymbol{J}$ are the eigenvalues of $\boldsymbol{F}$. Furthermore, the eigenvalues of AT are the natural logarithms of the eigenvalues of $\boldsymbol{F}$, while the (generalized) eigenvectors of the AT and $\boldsymbol{F}$ are identical. Consequently, we have

$$AT = \boldsymbol{P}\,\boldsymbol{Q}\,\boldsymbol{P}^{-1} \tag{24}$$

where $\boldsymbol{Q}$ is a matrix with diagonal elements

$$Q_{\text{ii}} = \ell n\, J_{\text{ii}} \tag{25}$$

While this approach is direct, it requires computation of the eigenvalues and eigenvectors of $\boldsymbol{F}$. Further difficulties arise when $\boldsymbol{F}$ has complex eigenvalues. Since the elements of $\boldsymbol{F}$ are real, such eigenvalues must occur in conjugate pairs. Let these be denoted by $\alpha\, e^{\pm j\beta}$. The corresponding eigenvalues of $\boldsymbol{A}$ are, then, given by

$$\lambda = \frac{1}{T}\,\ell n\,\alpha \pm j\frac{1}{T}(\beta + 2\pi k) \tag{26}$$

where k is an integer. The nonuniqueness of the logarithm of a complex number is the cause of our difficulty, and is related to the spectral density of the sampled signals. In view of the low-pass nature of the system, we are justified in making $k = 0$, especially if the sampling interval has been selected to make the spectral radius of $\boldsymbol{A}$ less than 0.5.

A related problem arises when $\boldsymbol{F}$ has a real negative eigenvalue. This may happen due to a slight error in the estimates of the elements of $\boldsymbol{F}$. Such an eigenvalue of $\boldsymbol{F}$ is possible only if $\boldsymbol{A}T$ has complex eigenvalues of the form $c \pm j(2k+1)\pi$. For this complex pair, $\boldsymbol{F}$ must have two identical negative real roots. Consequently, whenever the characteristic polynomial for $\boldsymbol{F}$ has negative real roots, these must occur in complex conjugate pairs. If $\boldsymbol{F}$ is diagonalizable, the corresponding eigenspace of $\boldsymbol{F}$ is spanned by two real and linearly independent eigenvectors $\boldsymbol{u}$ and $\boldsymbol{v}$. But $\boldsymbol{u} \pm j\boldsymbol{v}$ are also eigenvectors of $\boldsymbol{F}$. Hence, the eigenvectors of $\boldsymbol{A}T$ corresponding to the eigenvalues $c \pm j(2k+1)\pi$, are $\boldsymbol{u} \pm j\boldsymbol{v}$ and these are the vectors that must be used in $\boldsymbol{P}$ for determining $\boldsymbol{A}$ in equation (24). It should be noted that this problem will not arise if T has been selected properly so that the spectral radius of $\boldsymbol{A}T$ does not exceed 0.5. It can be easily shown that if the eigenvalues of the matrix $\boldsymbol{A}T$ are $\alpha \pm j\beta$, then the corresponding eigenvalues of $\boldsymbol{F}$ are given by $e^{\alpha}(\cos\beta \pm j\sin\beta)$ and since $|\beta| \leq 0.5$, the real part of the eigenvalue must be a positive number between $e^{-0.5}$ and 1. If we set $a = e^{\alpha}\cos\beta$, $b = e^{\alpha}\sin\beta$, then with $\alpha = \mu\cos\Theta$, $\beta = \mu\sin\Theta$, where $0 \leq \mu \leq 0.5$, and $\pi/2 \leq \Theta \leq 3\pi/2$,

$$a^2 + b^2 = e^{2\mu\cos\Theta} \tag{27}$$

Using the bounds for μ and Θ, we can easily determine the region in the z-plane in which the eigenvalues of $\boldsymbol{F}$ must lie (where $z = a + jb$). This is shown in Figure 2. Conversely, if the eigenvalues of $\boldsymbol{F}$ lie inside the lens-shaped region in Figure 2, then each eigenvalue of $\boldsymbol{A}T$ has a negative real part and magnitude less than 0.5. On the other hand, if the sampling frequency is made five times faster, all poles must lie in a much smaller lens-shaped region in the z-plane, bounded by the unit circle and the curve for $\rho(A_m)T = 0.1$, as indicated in the figure.

In addition to being computationally tedious, this approach suffers from the drawback that it depends upon accurate evaluation of the eigenvalues and (generalized) eigenvectors of $\boldsymbol{F}$. In some cases, even a slight error in the estimation of the parameters of the discrete-time model, as will often be the case with noise contaminated input-output data, we may get complex or negative eigenvalues although they should actually be real and positive. In such cases some of the difficulties mentioned above will arise. We shall now consider some other methods which do not require evaluation of the eigenvalues of the matrix $\boldsymbol{F}$.

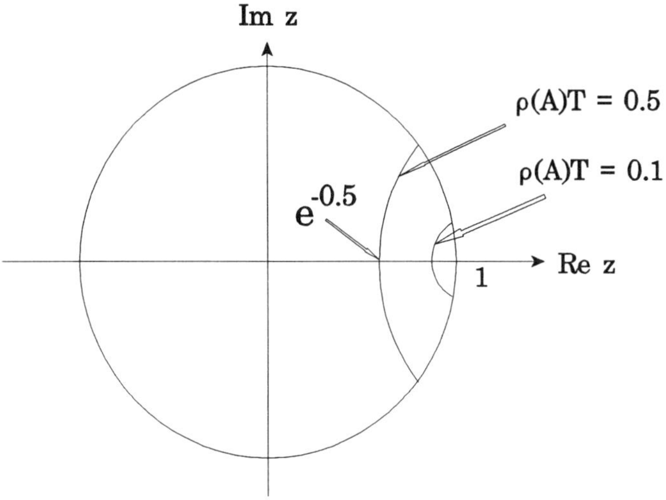

Use of Sylvester's Interpolation Formula

If we can write $\ell n\ \boldsymbol{F}$ as a power series in terms of a matrix $\boldsymbol{L}$, such as in equation (33), then we can use the following expansion

$$\ell n\ \boldsymbol{F} = \alpha_0\boldsymbol{I} + \alpha_1\boldsymbol{L} + \alpha_2\boldsymbol{L}^2 + \cdots + \alpha_{m-1}\boldsymbol{L}^{m-1} \tag{28}$$

where m is the degree of the minimal polynomial of $\boldsymbol{L}$. If the eigenvalues of $\boldsymbol{L}$ are distinct, the degree of the minimal polynomial is equal to the order of $\boldsymbol{L}$, and the constants α_i, $i = 0, 1, \cdots m-1$, are obtained by solving the following set of linear equations:

$$\begin{bmatrix} 1 & \lambda_1 & \lambda_1^2 & & \lambda_1^{m-1} \\ 1 & \lambda_2 & \lambda_2^2 & \cdots & \lambda_2^{m-1} \\ \vdots & \vdots & \vdots & \cdots & \vdots \\ 1 & \lambda_m & \lambda_m^2 & \cdots & \lambda_m^{m-1} \end{bmatrix} \begin{bmatrix} \alpha_0 \\ \alpha_1 \\ \alpha_2 \\ \vdots \\ \alpha_{m-1} \end{bmatrix} = \begin{bmatrix} \ell n\ \mu_1 \\ \ell n\ \mu_2 \\ \ell n\ \mu_3 \\ \vdots \\ \ell n\ \mu_m \end{bmatrix} \tag{29}$$

where λ_i is an eigenvalue of $\boldsymbol{L}$ and μ_i is an eigenvalue of $\boldsymbol{F}$. Equation (29) must be modified if the minimal polynomial of $\boldsymbol{L}$ involves multiple roots. If the multiplicity of a certain eigenvalue, say λ_p is r, then the corresponding r rows of the matrix on the left-hand side of equation (29) will need to change.

It will be seen that this method is computationally laborious, since one must first determine the eigenvalues of $\boldsymbol{L}$ and then determine the constants α_i by solving a set of linear simultaneous equations. Furthermore, as in the previous method, there will be a problem if some of the eigenvalues are found to be negative, as may be the case when the parameters are estimated from measurements contaminated with noise.

We shall now discuss some methods which avoid the computation of eigenvalues and consequently, are less affected by small errors in estimation caused by the presence of noise in the data.

An iterative algorithm for computing the logarithm of a matrix

One approach to evaluating $\boldsymbol{AT}$ from $\boldsymbol{F}$ without calculating the eigenvalues is to use a fixed-point iteration algorithm (Sinha and Lastman 1981). The main idea behind the method is based on the fact that if the spectral radius of a matrix is less than 0.5, it is possible to calculate its exponential efficiently by summing the power series for the exponential which may be truncated after 12 terms with an error of less than 10^{-12}. Thus, if we start with a suitable guess for $\boldsymbol{AT}$, we may use the following iterative algorithm to successfully improve it.

$$(\boldsymbol{AT})^{(k+1)} = (\boldsymbol{AT})^{(k)} + \boldsymbol{F}^{-1}(\boldsymbol{F} - \boldsymbol{F}^{(k)}) \tag{30}$$

where $(\boldsymbol{AT})^{(k)}$ is the guess for $\boldsymbol{AT}$ at the kth iteration and $\boldsymbol{F}^{(k)} = e^{(\boldsymbol{AT})^{(k)}}$, calculated by using the power series. A good initial guess for $\boldsymbol{AT}$ is

$$(\boldsymbol{AT})^{(0)} = \frac{\boldsymbol{F} - \boldsymbol{F}^{-1}}{2} \tag{31}$$

Note that the algorithm requires only one matrix inversion for the given matrix $\boldsymbol{F}$. Finally, the matrix $\boldsymbol{B}$ can be calculated using the following relationship obtained from (12) by assuming that $\boldsymbol{A}$ is nonsingular

$$\boldsymbol{B} = \boldsymbol{A}(\boldsymbol{F} - \boldsymbol{I})^{-1}\boldsymbol{G} \tag{32}$$

Results of simulation (Sinha and Lastman 1982) indicate that with the initial guess given by equation (32), the algorithm of equation (31) converges very fast if the spectral radius of $\boldsymbol{AT}$ is less than or equal to 0.5.

Algorithm using a series expansion of the logarithm of a matrix

Although the iterative algorithm described above works quite well, there may be some difficulty in inverting the matrix $\boldsymbol{F}$, if it is ill-conditioned. An algorithm (Lastman, Puthenpura and Sinha 1984) based on the following power series expansion eliminates this problem and is computationally more efficient,

$$AT = \ell n(\boldsymbol{I}+\boldsymbol{L}) = \sum_{i=1}^{\infty} \frac{(-1)^{i+1}\boldsymbol{L}^i}{i} = \boldsymbol{P}_m(\boldsymbol{L}) + \boldsymbol{E}_m(\boldsymbol{L}) \tag{33}$$

where

$$\boldsymbol{L} = \boldsymbol{F} - \boldsymbol{I} \tag{34}$$

$$\boldsymbol{P}_m(\boldsymbol{L}) = \sum_{i=1}^{m} \frac{(-1)^{i-1}\boldsymbol{L}^i}{i} \tag{35}$$

and $\boldsymbol{E}_m(\boldsymbol{L})$ is the error introduced by truncating the series after m terms.

It is easily shown that the spectral radius of $\boldsymbol{L}$ is less than one if $\rho(\boldsymbol{A})T \leq 0.5$. Furthermore, the error due to truncation (Lastman, Puthenpura and Sinha 1984) is given by

$$\|\boldsymbol{E}_m\| \leq \frac{|\rho(\boldsymbol{L})|^{m+1}}{(m+1)(1-\rho(\boldsymbol{L}))} \tag{36}$$

where $\rho(\boldsymbol{L})$ denotes the spectral radius of $\boldsymbol{L}$.

Since $\boldsymbol{L} = \boldsymbol{F} - \boldsymbol{I}$, is known, the above equation can be utilized to determine the value of m required for a given accuracy. Thus, the following procedure can be utilized for calculating $\boldsymbol{A}T$ and $\boldsymbol{B}$ from $\boldsymbol{F}$ and $\boldsymbol{G}$ (Raol, Puthenpura and Sinha 1987):

(i) Determine $\boldsymbol{L} = \boldsymbol{F} - \boldsymbol{I}$.
(ii) Set $\boldsymbol{D}^{(1)} = \boldsymbol{I}$, $\boldsymbol{M}^{(1)} = \boldsymbol{I}$ and $i = 1$.
(iii) Perform the following computations recursively,

$$\boldsymbol{M}^{(i+1)} = \frac{-i\boldsymbol{L}\boldsymbol{M}^{(i)}}{i+1}$$

$$\boldsymbol{D}^{(i+1)} = \boldsymbol{D}^{(i)} + \boldsymbol{M}^{(i+1)}$$

Replace i with $i+1$

until $i = m$ (some predetermined value), or, the change in $\boldsymbol{D}^{(i)}$ is negligible.
(iv) Finally, $\boldsymbol{A}T = \boldsymbol{D}\boldsymbol{L}$ and $\boldsymbol{B}T = \boldsymbol{D}\boldsymbol{G}$

Note that this procedure does not require any matrix inversion. If desired, the value of m can be determined from an estimate of the spectral radius of $\boldsymbol{L}$, which does not require evaluation of the eigenvalues of $\boldsymbol{L}$, (Puthenpura and Sinha 1987) although this is not necessary if one is monitoring the change in $\boldsymbol{D}^{(i)}$. As pointed out in an earlier paper (Raol, Puthenpura and Sinha 1987), this is an efficient algorithm for determining $\boldsymbol{A}$ and $\boldsymbol{B}$ for given $\boldsymbol{F}$ and $\boldsymbol{G}$.

4. Choice of the sampling interval

In our discussions so far, we have assumed that the sampling interval has been selected carefully so that the spectral radius of $\boldsymbol{A}T$ is less than 0.5. Although this choice is important for the computation of the exponential or the logarithm of a matrix, the main reason is the need for the input to varying sufficiently fast to persistently excite the system to be identified. Furthermore, computational experiments (Sinha and Puthenpura 1985) indicate that even with a small amount of noise contaminating the data, there is a considerable error in the estimates of the parameters of the discrete-time model unless the sampling interval is selected so that the spectral radius of $\boldsymbol{A}T$ is between 0.1 and 0.5. In practice, the problem is complicated because one does not know *á priori* the location of the eigenvalues of the system model to be identified. If the selected value of the sampling interval is low, there is the danger of missing the faster natural frequencies of the system model. On the other hand, making the sampling interval very small causes numerical problems due to the fact that the poles of the discrete-time model are forced in a very small region of the z-plane, as was shown in Figure 2. Puthenpura and Sinha (1985) have described a method for determining the optimum sampling interval for estimating the discrete-time model of a given continuous-time system.

There may be situations where the $\boldsymbol{F}$ and $\boldsymbol{G}$ may be known for a system where the sampling interval is larger than required by the condition stated above. In such cases, the iterative algorithm is unsuitable and the series for calculating the logarithm of $\boldsymbol{F}$ may not converge. Cooper and Bingulac (1990) have suggested the successive use of square roots of $\boldsymbol{F}$ to reduce the spectral radius of $\boldsymbol{L}$. Otherwise, the only alternatives are the methods based on calculating the eigenvalues of $\boldsymbol{F}$. It has been claimed (Feliu 1986, Feliu, Cerrada and Cerrada 1988) that the method based on using the Jordan form gives satisfactory results as long as the spectral radius of $\boldsymbol{A}T$ is less than π.

On the other hand, if may often be necessary to sample the input-output data at a much faster rate, especially if the data contain a significant amount of noise with the result that the spectral radius of $\boldsymbol{A}T$ is less than 0.1. In such cases, it is more desirable to estimate the discrete-time model based on the δ-operator (Middleton and Goodwin 1986, 1990) and then obtain the corresponding continuous-time model by transformation, This will be described in the next section.

5. Transformation of models obtained using the δ-operator

The δ operator is defined as

$$\delta \triangleq \frac{q-1}{T} \tag{37}$$

where q is the unit advance operator and T is the sampling interval, as before. It will be seen to correspond to the first divided difference operator in the field of numerical analysis (Hildebrand 1956).

From its definition, it is evident that this operator can be used as an approximation to the derivative operator p, so that one may write

$$\delta \boldsymbol{x}_k \approx p\boldsymbol{x}\big|_{x=x(kT)} \tag{38}$$

and the approximation improves as $T \rightarrow 0$.

Consequently, the state equations of a linear multivariable system, described by equations (1) and (2) can be discretized as (Ninness and Goodwin 1991)

$$\delta \boldsymbol{x}_k = \boldsymbol{F}_\delta \boldsymbol{x}_k + \boldsymbol{G}_\delta \boldsymbol{u}_k \tag{39}$$

$$y_k = \boldsymbol{C}\boldsymbol{x}_k + \boldsymbol{D}\boldsymbol{u}_k \tag{40}$$

Substituting for $\boldsymbol{x}_{k+1}$ from equation (9) and utilizing equations (11) and (12), we can derive the following expressions for $\boldsymbol{F}_\delta$ and $\boldsymbol{G}_\delta$

$$\boldsymbol{F}_\delta = \frac{\boldsymbol{F}-\boldsymbol{I}}{T} = \boldsymbol{\Omega}\boldsymbol{A} \tag{41}$$

$$\boldsymbol{G}_\delta = \boldsymbol{\Omega}\boldsymbol{B} \tag{42}$$

where

$$\boldsymbol{\Omega} = \sum_{i=0}^{\infty} \frac{(AT)^i}{(i+1)!} \tag{43}$$

It may be noted that equations (41) and (42) are essentially the same as equations (13) and (14).

It is understood that the infinite series for $\boldsymbol{\Omega}$ can be truncated after 12 terms with fairly high accuracy since the spectral radius of AT is less than 0.5.

As pointed out by Ninness and Goodwin (1991) estimation of the matrices $\boldsymbol{F}_\delta$ and $\boldsymbol{G}_\delta$ does not suffer from the problem of ill-conditioning for smaller sampling intervals that is encountered the estimation of the discrete-time model described by equation (9). In particular, the δ-model approaches the continuous-time model in the limit as $T \rightarrow 0$, and the eigenvalues are not crowded into a small region as in the model using the shift operator. They have also described algorithms for estimating these matrices from samples of the input-output data. Consequently, we shall only address the problem of determining the matrices $\boldsymbol{A}$ and $\boldsymbol{B}$ from the estimates of $\boldsymbol{F}_\delta$ and $\boldsymbol{G}_\delta$.

It follows from equations (11) and (41) that

$$\boldsymbol{A}T = \ell n\ (\boldsymbol{I} + \boldsymbol{F}_\delta T) \tag{44}$$

Hence, to determine $\boldsymbol{A}$, we must calculate the natural logarithm of $(\boldsymbol{I} + \boldsymbol{F}_\delta T)$. Although all the algorithms described in section can be used for this purpose, we recommend using the most efficient algorithm, described by equations (33) and (34). The only difference is that here we make

$$\boldsymbol{L} = \boldsymbol{F}_\delta T \tag{45}$$

After $\boldsymbol{A}$ has been determined. it is now quite straightforward to calculate $\boldsymbol{\Omega}$ and then $\boldsymbol{B}$ using equation (42), since $\boldsymbol{\Omega}$ will be nonsingular.

It is possible to develop an efficient computer program which would calculate $\boldsymbol{A}$ and $\boldsymbol{B}$ for given values of $\boldsymbol{F}_\delta$, $\boldsymbol{G}_\delta$ and T, on the lines suggested by Raol, Puthenpura and Sinha (1987) for discrete-time models based on the shift operator.

6. Conclusions

In this chapter, we have described several methods for obtaining an equivalent continuous-time model from the discrete-time model of a multivariable linear system. Two of the methods require evaluation of the eigenvalues of the system model. Since the usual discrete-time model is highly dependent on the sampling interval, it is very important that this be selected very carefully. Moreover, the input-output data used for estimating the discrete-time model usually contains some noise, which affects the accuracy of the estimates. Consequently, the methods based on the evaluation of eigenvalues may lead to incorrect results, especially if the sampling interval is not selected properly. This difficulty is overcome to a large extent if the samples of the input and output are used for estimating a model based on the δ-operator.

It has been shown that in all cases, the problem of determination of the equivalent continuous-time model boils down to calculating the natural logarithm of a square matrix. In our opinion, this can be done most efficiently by using the series expansion of the natural logarithm, which can be truncated after a suitable number of terms for a desired accuracy. An advantage with this algorithm is that one can use the spectral radius of a known matrix to decide how many terms in the series are required for a specified error bound.

References

Bingulac, S., and Sinha, N.K. (1989), "On the identification of continuous time multivariable systems from samples of input-output data", Proc. Seventh Int.Conf. on Mathematical and Computer Modelling, (Chicago, Ill, August 1989), pp. 231-239.

Bingulac, S. and Cooper, D.L. (1990), "Derivation of discrete and continuous time ramp invariant representations", *Electronics Letters*, vol. 26, pp. 664-666.

Cooper, D. and Bingulac, S. (1990), "Computational improvement in the calculation of the natural log of a square matrix", *Electronics Letters*, vol. 26, pp. 861-862.

El-Sherief, H., and Sinha, N.K. (1979), "Choice of models for the identification of linear multivariable discrete-time systems", *Proc. I.E.E.*, vol. 126, pp. 1326-1330.

Feliu, V. (1986), "A transformation algorithm for estimating system Laplace transform from sampled-data", *IEEE Transactions on Systems, Man and Cybernetics*, vol. SMC-16, pp. 168-173.

Feliu, V., Cerrada, J.A., and Cerrada, C. (1988), "An algorithm to compute the continuous state model from its equivalent discrete model", *Control -Theory and Advanced Technology*, vol. 4, pp. 231-241.

Guidorzi, R. (1975), "Canonical structures in the identification of multivariable systems", *Automatica*, vol. 11, pp. 113-116.

Haykin, S.S. (1972), "A unified treatment of recursive digital filtering", *IEEE Transactions on Automatic Control*, vol. AC-17, pp. 113-116.

Hildebrand, F.B. (1956), "*Introduction to Numerical Analysis*", McGraw Hill, New York, 1983.

Hsia, T.C. (1972), "On sampled-data approach to parameter identification of continuous linear systems", *IEEE Transactions on Automatic Control*, vol. AC-17. pp. 247-249.

Lastman, G.J., Puthenpura, S, and Sinha, N.K. (1984), "Algorithm for the identification of continuous-time multivariable systems from their discrete-time models", *Electronics Letters*, vol. 20, pp. 918-919.

Middleton, G.H. and Goodwin, G.C. (1986), "Improved finite word length characteristics digital control using delta operators", *IEEE Transactions on Automatic Control*, vol. AC-31, pp. 1015-1021.

Middleton, G.H. and Goodwin, G.C. (1990), "*Digital Estimation and Control : A Unified Approach*", Prentice-Hall, New Jersey.

Niness, B.M. and Goodwin, G.C. (1991), "The relationship between discrete time and continuous time linear estimation", Chapter 3 of this book.

Puthenpura, S., and Sinha, N.K. (1984), "Transformation of continuous-time model of a linear multivariable system from its discrete-time model", *Electronics Letters*, vol. 20, pp. 737-738.

Puthenpura, S., and Sinha, N.K. (1985), "A procedure for determining the optimum sampling interval for system identification using a digital computer", *Can. Elec. Eng. J.*, vol. 10, pp. 152-157.

Puthenpura, S., and Sinha, N.K. (1987), "Extended power method and stability of linear discrete time systems", *Electronics Letters*, vol.23, pp. 4-5.

Raol, J.R., Puthenpura, S.C., and Sinha, N.K. (1987), "Algorithms for transformation of multivariable discrete-time models to continuous-time models", *Advances in Modelling and Simulation*, (AMSE Press), Vol. 6, pp. 52-62.

Sinha, N.K. (1972), "Estimation of transfer function of continuous-time systems from sampled data", *Proceedings IEE*, vol. 126, pp. 612-614.

Sinha, N.K., and Kuszta, B. (1983), "*Modeling and Identification of Dynamic Systems*", Von-Nostrand Reinhold, New York.

Sinha, N.K., Kwong, Y.H., (1979), "Recursive identification of the parameters of linear multivariable systems", *Automatica*, vol. 15, pp. 471-475.

Sinha, N.K., and Lastman, G.J. (1981), "Transformation algorithm for identification of continuous-time multivariable systems from discrete data", *Electronics Letters*, vol. 21, pp. 779-780.

Sinha, N.K., and Lastman, G.J. (1982), "Identification of continuous-time multivariable systems from sampled data", *International Journal of Control*, vol. 35, pp. 117-126.

Sinha, N.K., and Puthenpura, S. (1985), "Choice of the optimum sampling interval for the identification of continuous-time systems from samples of input/output data". *IEE Proceedings*, vol. 132, Pt. D., pp. 263-267.

Strmčnik, S., and Bremšak, F, (1979), "Some new transformation algorithms in the identification of continuous-time multivariable systems using discrete identification methods", Preprints 5th IFAC Symposium on Identification and System Parameter Estimation (Darmstadt, West Germany), pp. 397-405.

Tse, E., and Weinert, H. (1975), "Structure determination and parameter identification for multivariable stochastic linear systems", *IEEE Transactions on Automatic Control*, vol. AC-20, pp. 603-613.

Methods using Walsh functions

E.V. Bohn
Department of Electrical Engineering
University of British Columbia
Vancouver, Canada V6T 1W5

Abstract

Analytical expressions for integral functions of linear system output signals and their Walsh function coefficients are derived. Personal computer signal processing algorithms are then developed. Methods for Walsh coefficient evaluation of integral functions are discussed. Walsh function representation of zero-order hold digital controller outputs and application to system parameter identification are reviewed.

1. Introduction

The properties of piecewise constant orthogonal functions and their application to signal processing of continuous-time systems have been extensively investigated (Harmuth 1971, Beauchamp 1984, Rao 1983). Since physical processes are continuous in nature it is often important to identify the parameters of continuous-time models. The underlying basic models for continuous-time systems are differential equations. However, derivatives cannot be accurately estimated from sampled input-output data. Consequently, parameter identification requires that differential equation models be transformed into some form of integral equation models. Many forms of parametric and non-parametric models have been extensively investigated.

Analytical expressions for integrals of Walsh functions that have found application in continuous-time parameter identification were originally derived by Fine (1949). These integrals allow repeated integrations of continuous-time signals to be performed by multiplication of operational matrices (Corrington 1973, Rao and Palanisamy 1983). Bohn (1982) has shown that a *sal-cal* ordering of Walsh functions results in a particularly simple form of operational matrix suitable for sampled-data signal processing operations.

This chapter places emphasis on the signal processing properties of Walsh functions that are particularly useful in the case of time-limited rapidly sampled continuous-time signals. Walsh transforms in this regard are complimentary to Fourier transforms which are more useful in the case of frequency-limited signals. Walsh functions appear to have an advantage in the continuous-time case when fast sampling of time-limited signals is used. Efficient use of fast sampling of continuous-time signals requires the use of some form of data compression. Such data processing operations are easily realized when Walsh functions are used. The standard additive-type noise models used in the discrete-time case are not appropriate when fast sampling is used in conjunction with data compression. It would appear that a differential equation is an appropriate noise model. System order and noise model parameter identification are then similar problems. Because of space limitations these problems are not discussed. Emphasis is placed in

N. K. Sinha and G. P. Rao (eds.), *Identification of Continuous-Time Systems*, 139–157.

this chapter on Walsh function analysis useful in the development of a personal computer (P
Walsh function signal processing toolbox. Since no single signal processing or identificati
method is "best", the availability of PC toolboxes to perform comparative studies is
considerable importance. The PC software MATLAB and its associated toolboxes are chos
as a standard. All algorithms in this chapter are realized by use of standard MATLA
statements.

2. Integral equation models

An integral equation model is obtained from a differential equation model by repeat
integrations. Bohn (1982) defines the integral function of $y(x)$ by

$$y^1(x) = \int_0^x y(t)dt - a\int_0^a y(t)dt, \quad (2.$$

where the binary factor $a=1/2$ is used for its notational convenience. Definition (2.1) is bas
on the periodicity condition

$$y(x) = -y(x+a) \quad (2.$$

for periodic functions of period 1 and odd symmetry. Definition (2.1) is, however, al
applicable to non-periodic functions. For example, the integral function for a non-periodic fir
derivative is

$$\frac{dy^1}{dx} = y(x) - f_0\, sal(1,x) \quad (2.$$

where

$$f_0 = a(y(0) + y(a)) \quad (2.$$

An integral equation model for a second order system based on (2.1) has the form

$$y(x) + a_2 y^1(x) + a_2 y^2(x) = b_2 u^1(x) + b_1 u^2(x) + r(x) \quad (2.$$

where $y(x)$ and $u(x)$ are the input and output respectively, and where

$$r(x) = f_0\, sal(1,x) + (a_2 f_0 + f_1)\, sal^1(1,x)$$

contains initial condition terms. By use of (2.1) integral functions can be approximate
numerically evaluated from fast sampled data. System parameters and initial conditions can the
be identified from data equation (2.5).

Numerous numerical methods are available for evaluating repeated integrations. Th
choice of a Walsh function approach is natural when input $u(x)$ is the output of a zero-order hol
digital controller. Orthogonality of Walsh functions can be used to reduce the number of term
in (2.5). Furthermore, different sets of equations for the system parameters are easily ' obtaine
and data compression is easily realized.

It is desirable in computer simulation studies of approximate numerical methods to hav
an analytically exact method for evaluating integral functions. This is possible in the case c
linear systems where the output is a superposition of complex exponentials. Consider

$$y(x) = \exp(sx) \tag{2.6}$$

The integral functions of (2.6) have the analytic forms

$$y^1(x) = \frac{\exp(sx)}{s} - a(1+\exp(sa))\frac{sal(1,x)}{s},$$

$$y^2(x) = \frac{\exp(sx)}{s^2} - a(1+\exp(sa))\left[\frac{sal(1,x)}{s^2} + \frac{sal^1(1,x)}{s}\right] \tag{2.7}$$

$$\cdots$$

It should be noted that if input is *sal(1,x)* then the integral functions are polynomials in x:

$$p_1(x) = sal^1(x) = x \quad a^2$$

$$p_2(x) = sal^2(x) = ax(x-a) \tag{2.8}$$

3. Walsh coefficients of *exp(sx)*

It is seen from (2.7) that analytical evaluation of Walsh coefficients of integral functions requires the analytical evaluation of Walsh coefficient of (2.6). The desired expressions can be derived in a recursive form by use of the defining properties of *sal-cal* functions.

Let $N=2^{n-1}$, $M=2^n$. The *sal-cal* functions of sequency k, $N \leq k < M$, are defined by

$$\begin{aligned} sal(k,x) &= sal(M,x)cal(M\text{-}k,x), \\ cal(k,x) &= sal(M,x)sal(M\text{-}k,x), \\ sal(M,x) &= r_n(x). \end{aligned} \tag{3.1}$$

In (3.1), $r_n(x)$ are the Rademacher functions defined by

$$r_n(x) = r_{n-1}(2x). \tag{3.2}$$

For notational convenience in discussing sequency relations induced by (3.1) a simplified notation is used:

$$s(k,x) = sal(k,x),$$

$$c(k,x) = s(M,x)s(M-k,x) = s(M,M-k,x) \tag{3.3}$$

Let $M_i=2^{n_i}$, $M>M_{i+1}>M_i$, be chosen such that k has the unique representation

$$k = (M-M_1)+\cdots+(M_{J-1}-M_J). \tag{3.4}$$

By recursive use of (3.1) and (3.4) it is seen that

$$c(k,x) = s(M,M1,\cdots,MJ,x). \tag{3.5}$$

Consider

$$k' = M-k-1 = (M_1-M_2)+\cdots+(M_J-1). \tag{3.6}$$

It is seen from (3.4), (3.5), and (3.6) that

$$c(k',x) \;=\; s(M_1,M_2,\cdots,M_J,1,x). \tag{3.7}$$

Equations (3.1), (3.4), and (3.7) yield

$$\begin{aligned} c(k,x)\,s(1,x) \;&=\; [s\,(M,M_1,\cdots,M_J,x)]\,s(1,x) \\ &=\; s\,(M,x)\,c(k',x) \;=\; s(k+1,x). \end{aligned} \tag{3.8}$$

Since $s(1,x)=1$ when $(0<x<a)$, it follows from (3.8) that

$$s\,(k+1,x) \;=\; c\,(k,x). \tag{3.9}$$

Equation (3.9) allows the discussion to be restricted to a set of *cal* functions and occurs, fc example, when (2.2) is satisfied. The Walsh coefficients of an odd function are given by

$$F_w(k) = \;=\; \int_0^1 cal(k,x)\,f(x)\,dx \;=\; 2\int_0^a cal(k,x)\,f(x)\,dx, \tag{3.10}$$

where $w=c$ and $w=s$ for odd/even values of k, respectively (see (3.9)). (It is easily seen th the restriction to the interval $(0<x<a)$ and the use of (2.2) does not result in loss of generalit The use of (3.9) does, however, simplify the discussion).

The points of discontinuity where $cal(k,x)$ changes sign are referred to as zeros. Th transformation $v=x/x_1$, where $x_1=a^{n+1}$ is the first zero, maps the open interval $(0<x<a)$ ont the open interval $I[n]=(0<v<M)$. (Note that the midpoint of $I[n]$ is $v=N$). Define a left/rigl partition of $I[n]$ by the open sets

$$L[n] \;=\; (0<v<N) \tag{3.11}$$

$$R[n] \;=\; I[n]-L[n] \;=\; (N<v<M) \tag{3.12}$$

Let $v \in I[n]$ and define

$$Z_K \;=\; [v_1,\cdots,v_K] \tag{3.13}$$

as the ordered set of zeros of $c(K,v)$. Note that sequency k gives the number of zeros in set $I[n$

Consider (3.10), where $f(x) \;=\; exp(sx)$. It is easily seen that

$$Fc(k) \;=\; -\frac{2c\,(k)}{s} \tag{3.14}$$

where $c(k)$ has the polynomial form

$$c(k) \;=\; 1-2\alpha^{v_1}+2\alpha^{v_2}\mp\ldots+b_k\alpha^M \tag{3.15}$$

where

$$\alpha \;=\; exp(sx_1) \tag{3.16}$$

(Note that the -2 and +2 coeficients are the jump in function value at the zeros).

Bohn (1983) has derived a recursive algorithm for evaluating (3.15) based on th periodicity of zero patterns and their association with a factorization of polynomial $c(k)$. simpler and more rigorous algebraic proof based on (3.1) and (3.2) is given below.

To understand the algebraic basis for factorization of (3.15) consider the case c Rademacher functions, $n=1,2$. It is easily seen that

$$c(1) = 1 - 2\alpha + \alpha^2, \qquad c(3) = 1 - 2\alpha + 2\alpha^2 - 2\alpha^3 + \alpha^4 = c(1)(1 + \alpha^2) \tag{3.17}$$

The generalization of (3.17) is given by the following theorem:

Theorem 1. Let $M = 2^n$, $N = 2^{n-1}$. Then

$$c(K) = c(k)(1 + \alpha^N) \tag{3.18}$$

where $K = 2k+1 = M\text{-}1$.

Consider a fixed n and let $v \in I[n]$. It is easily seen by geometric inspection of Rademacher function zeros that

$$Z_K = [Z_K(L), N, Z_K(R)] \tag{3.19}$$

where $Z_K(L) = Z_k = [v_1, \ldots, v_k]$, are the zeros in $L[n]$, $Z_K(R) = N + Z_k$, are the zeros in $R[n]$, and where $v_{k+1} = N$ is the midpoint zero, respectively. The distribution of zeros on the v-axis given by (3.19) in conjunction with definition (3.15) shows that

$$c(K) = c(k) + \alpha^N c(k)$$

providing a geometric-type proof of (3.18).

In the general case an algebraic proof using (3.1) and (3.2) is simpler. Let $v = Mx = aMx'$. It is seen that

$$r_n(v) = r_{n-1}(v) = r_{n-1}(v + N) \tag{3.20}$$

It follows from (3.1) and (3.9) that

$$r_n(v) = s(M,v) = c(K,v) = r_{n-1}(v) = s(aM,v) = c(k,v) \tag{3.21}$$

where $K = 2k+1 = M\text{-}1$. Equation (3.21) shows that $Z_K[L] = Z_k$. Let $v' = v + N$. It is seen from (3.20) and (3.21) that

$$c(K,v') = c(k,v) \tag{3.22}$$

Equation (3.22) shows that $Z_K[R] = N + Z_k$. Noting that the midpoint zero is $v_{k+1} = N$ completes the proof of (3.19).

The generalization of (3.18) is given by the following theorem:

Theorem 2. Let $aN \le k < N$ and consider the two cases: (a) $K = 2k$, (b) $K = 2k+1$. Then

$$Z_k = \begin{cases} [Z_k,\ N, N + Z_k] & \text{for odd } K \\ [Z_k,\ N + Z_k] & \text{for even } K \end{cases} \tag{3.23}$$

$$c(K) = c(k)(1 + P_k \alpha^N) \tag{3.24}$$

where

$$P_k = \begin{cases} 1 & \text{for even } K-k \\ -1 & \text{for odd } K-k \end{cases} \tag{3.25}$$

Proof of (3.23). Case (a).

It follows from (3.1) and (3.5) that

$$c(K,v) = s(M, M_1, \cdots, M_J, v)$$

is a product of Rademacher functions. Use of (3.20) and $aM=2^{n-1}$ yields

$$c(K,v) = s(aM, aM_1, \cdots, aM_J, v) = c(k,v)$$

The proof is then completed by noting that there is no midpoint zero when K is even.

Proof of (3.23). Case (b).

It follows from (3.1) and (3.9) that

$$c(K,v) = s(M,v)s(M-K,v) = s(M,v)c(M-2k-2,v)$$

Use of (3.20) yields

$$c(K,v) = s(aM,v)c(aM-k-1,v) = s(k+1,v) = c(k,v)$$

The proof is then completed by noting that N is a midpoint zero when K is odd. (See (3.19)).

Proof of (3.24) and (3.25).

Let B_K be the coefficient of α^N associated with midpoint zero $v_{k+1}=N$. There is n
midpoint zero when K is even. Hence $B_K=0$, K even.

Consider K odd. It follows by expanding (3.24) that $B_K=b_k+P_K=2q_k$, where $q_k=\pm1$
Since B_K gives the function value jump at midpoint and since k is the number of zeros in $L[n$
it is seen that

$$q_k=1,\ k \text{ odd and } q_k=-1,\ k \text{ even.}$$

Table 1, where L is an integer, summarizes the above relations and in conjunction with (3.23
completes the proof of (3.24) and (3.25).

Table 1

k	K	b_k	B_k	P_k
$2L$	$4L$	-1	0	1
$2L$	$4L+1$	-1	-2	-1
$2L+1$	$4L+2$	1	0	-1
$2L+1$	$4L+3$	1	2	1

A MATLAB program for evaluating (3.14) using (3.24) recursively is given in Appendix *A*. In the numerical evaluation of $c(k)$ it should be noted that (see (3.16)

$$\alpha_n = \exp(sa^{n+1}), \quad (\alpha_{n+1})^2 = \alpha_n.$$

4. Sampled data Walsh coefficients

There are numerous ways for representing sampled Walsh functions by binary bits. The method used by Bohn (1982) is easily implemented on MATLAB. The sampling instants are taken at the zeros of $cal(N_s,x)$, where N_s is the maximum sequency. The mapping $(1,-1) \leftrightarrow (0,1)$ is used to convert function values to binary bits. With this choice the multiplications in (3.1) are replaced by exclusive or binary operations and can be realized in MATLAB by a not equal ($\sim =$) relational operator.

An algorithmic approach for obtaining a Walsh binary matrix Wb is essential in order to reduce the number of input operations. The bit doubling method (Bohn 1982) can be initialized by choosing N_s=16 and using a hexadecimal representation for the binary values of $cal(k,x)$. The hexadecimal input matrix is

$$Cal = [C1;C2;C3;C4],$$

where the k-th row of *Cal* is associated with *cal(k-1,x)* and where

$$C1 = [0\ 0\ 0\ 0;0\ 0\ 15\ 15;0\ 15\ 15\ 0;0\ 15\ 0\ 15],$$

$$C2 = [3\ 12\ 3\ 12;3\ 12\ 12\ 3;3\ 3\ 12\ 12;3\ 3\ 3\ 3],$$

$$C3 = [6\ 6\ 6\ 6;6\ 6\ 9\ 9;6\ 9\ 9\ 6;6\ 9\ 6\ 9],$$

$$C4 = [5\ 10\ 5\ 10;5\ 10\ 10\ 5;5\ 5\ 10\ 10;5\ 5\ 5\ 5].$$

The following MATLAB program converts Cal to a temporary Walsh binary matrix TWb:

```
    b8=[8 4 2 1];v4=[1 2 3 4];v=v4;w=[];
    for k=1:16
    hn=cal(k,:);
            for l=1:4
            hc=hn(l);
                    for i=1:4
                    w(i)=(hc>=b8(i));
                    if w(i)==1 hc=hc-b8(i);end;
                    end
            TWb(k,v)=w;v=v+4;
            end
v=v4;
end
```

Consider updating TWb(N_s=16) to Wb(N_s=32). Binary bit doubling, based on (3.1), i used to obtain one half of the Wb entries:

```
Wb(:,1:2:31)=TWb,
Wb(:,2:2:32)=TWb.
```

The remaining entries are obtained by the following binary realizations of (3.1):

```
Wb(17,:)=[TWb(9,:) TWb(9,:)],
Wb(25,:)=[TWb(13,:) TWb(13,:)],
Wb(29,:)=[TWb(15,:) TWb(15,:)],
Wb(32,:)=[TWb(16,:) TWb(16,:)],
```

```
for i=2:8
Wb(16+i,:)=(Wb(17,:)~=Wb(i,:));
end;

for i=2:4
Wb(24+i,:)=(Wb(25,:)~=Wb(i,:));
end;
```

```
Wb(30,:)=(Wb(29,:)~=Wb(2,:));

v10=[1 0 1 0];v10=[v10 v10];v10=[v10 v10];
Wb(31,:)=[TWb(16,:) v10].
```

Updating Wb for N_s=64 is achieved by repeating the above procedure.

It is convenient to decompose Wb into *sal-cal* binary bit matrices:

```
Sb=Wb(1:2:31,:) ; Cb=Wb(2:2:32,:) .
```

The numerical evaluation of Walsh coeficients from sampled values of a function $y(x)$ is straightforward in MATLAB:

```
for k=1:16
LN=Sb(k,:);LP=~LN;
ys(k)=sum(y(LP))-sum(y(LN));
end;ys=(1/32)*ys;
```

Note that in MATLAB logic row vector *LN* consists of zeros and ones and that *LP* is the logic complement of *LN*. The operation *sum*(*y*(*L*)) sums those columns of *y* where corresponding elements of *L* are non-zero. A similar operation generates $yc(k)$.

The inverse Walsh transform

$$y(x) = S(x)ys' + C(x)yc', \tag{4.1}$$

where
$$S(x)=[s(1,x),\ s(3,x),\ \ldots.,\ s(N_s\text{-}1,\ x)],$$
$$C(x)=[c(1,x),\ c(3,x),\ldots.,\ c(N_s\text{-}1,\ x)],$$

is evaluated by the following algorithm:

```
for i=1:32
Zs=Sb(:,i);Zc=Cb(:,i);
LNs=Zs';LPs=~LNs;
LNc=Zc';LPc=~LNc;
y(i)=sum(ys(LPs))-sum(ys(LNs))+
    sum(yc(LPc))-sum(yc(LNc));
end;
```

Note that the resulting $y(x)$ can be considered as the output of a zero-order hold digital controller with the original sampled values as input.

The computational simplicity of the Walsh function approach is evident from these algorithms. The fact that no multiplications are required is of importance when fast sampling and data compression of continuous-time signals is required. Blachman (1974) has compared the computational requirements of Walsh and Fourier coefficients. Fourier transform methods have an advantage when signals are frequency-limited and when a large number of Fourier coefficients are required. This is the case, for example, when a power spectrum has to be identified. Walsh transform methods have an advantage when signals are time-limited and when a small number of Walsh (or Fourier) coefficients are required (Tadokoro and Higuchi 1978). This is the case, for example, when parameters of an integral equation system model are to be identified.

5. Evaluation of integral functions

There are numerous numerical methods available for evaluating repeated integrations of $y(t)$. For example, a standard fourth-order Runge-Kutta method could be used. Methods more specific to the required integrations, however, seem more appropriate. Consider the two differential equations

$$f_1'(x) = y(x),\ f_1(0) = 0, \tag{5.1}$$

and

$$f_2''(x) = y(x),\ f_2(0) = 0, f_2'(0) = 0, \tag{5.2}$$

Let $ym(i)$ be the sampled value of $y(x)$ at sampling point $x=(i\text{-}1)h/2$, $i=1{:}1{:}2N_s+1$, and let $y(k)$ be the sampled value of $y(x)$ at sampling point $x=(k\text{-}1)h$, $k=1{:}1{:}N_s+1$, where h is the stepsize. Simpson's rule

$$f_1(k) = f_1(k-1) + \frac{h}{6}(y(k) + 4ym(i) + y(k-1)) \tag{5.3}$$

where $i=2(k\text{-}1)$, can be used to integrate (5.1). It then follows from (2.1) that

$$y^1(k)=f_1(k)\text{-}af_1(N_s+1). \tag{5.4}$$

The method of Numerov, based on matching the coefficients of a Taylor expansion to fourth order in h of the expression

$$f_2(k+1) = 2f_2(k) - f_2(k-1) + \frac{h^2}{12}(y(k+1) + 10y(k) + y(k-1)) \tag{5.5}$$

can be used to integrate (5.2). A Taylor expansion of $y(x)$ to second order in h is used to obtain initial values for (5.5):

$$f_2(1) = 0; \quad f_2(2) = \frac{13y(1)+14y(2)-3y(3)}{48}. \tag{5.6}$$

It then follows from (2.1) and (2.8) that

$$y^2(k) = f_2(k) - a(f_2(N_s+1) + f_1(N_s+1)p_1(k)) \tag{5.7}$$

Walsh coefficients are then determined from the above data samples by use of the Wb matrix (see section 4).

A different approach is based on the use of operational matrices defined by (see (4.1))

$$y_s^{n+1} = -A_c y_c^n, \quad y_c^{n+1} = -A_s y_s^n \tag{5.8}$$

(See Appendix B for a discussion of the algebraic and numerical properties of operational matrices). The use of operational matrices avoids the necessity of first evaluating integral functions numerically before finding Walsh coefficients. The use of (4.1) in conjunction with (5.8) results in a direct data compression of sampled data.

Example. Consider (2.5) and the case of a periodic input $u(x)=sal(1,x)$. Let $s=[\text{-}1\ \text{-}3];b_1=2$, $b_2=1$. Table E shows numerical results for Walsh coefficients based on analytic formulas (2.7) and the algorithm given in Appendix A.

Table E

Ysal	s(1,x)	s(3,x)	s(5,x)	s(7,x)
10Ysal(1,:)	0.3569	-0.1318	-0.0340	-0.0663
100Ysal(2,:)	-1.8670	0.6927	0.1770	0.3479
1000Ysal(3,:)	-0.8904	0.3647	0.0752	0.1759

Ycal	c(1,x)	c(3,x)	c(5,x)	c(7,x)
10Ycal(1,:)	-1.1149	-0.5673	-0.0100	-0.2850
100Ycal(2,:)	-0.5561	-0.2372	0.0411	-0.1133
1000Ycal(3,:)	2.9117	1.2407	-0.2161	0.5927

In Table E the j-th row of matrix Ysal/Ycal has the sal/cal coefficients of $y^{j-1}(x)$ as elements.

Table Wb shows numerical results based on integration formulas (5.4), (5.7) and the use of Walsh binary matrix Wb (see section 4). The step-size is $h=1/64$ and maximum sequency is $N_s=32$.

Table Wb

Ysal	s(1,x)	s(3,x)	s(5,x)	s(7,x)
10Ysal(1,:)	0.3570	-0.1318	-0.0340	-0.0663
100Ysal(2,:)	-1.8679	0.6926	0.1770	0.3479
1000Ysal(3,:)	-0.8907	0.3648	0.0753	0.1766

Ycal	c(1,x)	c(3,x)	c(5,x)	c(7,x)
10Ycal(1,:)	-1.1149	-0.5673	-0.0099	-0.2849
100Ycal(2,:)	-0.5563	-0.2373	0.0411	-0.1134
1000Ycal(3,:)	2.9128	1.2412	-0.2161	0.5930

Table A_sA_c gives numerical results based on operational matrices (5.8)., where $dim(A_s)=dim(A_c)=16\times16$.

Table A_sA_c

Ysal	s(1,x)	s(3,x)	s(5,x)	s(7,x)
10Ysal(1,:)	0.3570	-0.1318	-0.0340	-0.0663
100Ysal(2,:)	-1.8651	0.6927	0.1770	0.3479
1000Ysal(3,:)	-0.8993	0.3643	0.0751	0.1757

Ycal	c(1,x)	c(3,x)	c(5,x)	c(7,x)
10Ycal(1,:)	-1.1149	-0.5673	-0.0099	-0.2849
100Ycal(2,:)	-0.5558	-0.2371	0.0411	-0.1132
1000Ycal(3,:)	2.9070	1.2383	-0.2162	0.5915

It is seen from these tables that for the values of step-size and maximum sequency chosen both forms of discrete approximations result in comparable numerical accuracy.

6. Parameter identification

Definition (2.1) is based on a normalized time $x=t/T$, where t is a real-time variable and T is a fixed data interval. (In the periodic case T is the period). To illustrate parameter identification of an integral equation model consider (2.5) subject to periodicity condition (2.2). Let y, $y1$, $y2$, $u1$, $u2$ be column vectors of dimension R having sampled values of $y(x)$, $y^1(x)$, $y^2(x)$, $u^1(x)$, $u^2(x)$ as elements, respectively. The parameter identification equation has the form

$$M\theta = -y \tag{6.1}$$

where
$$\theta = [a_2 \;\; a_1 \;\; -b_2 \;\; -b_1]', \tag{6.2}$$

is the normalized parameter vector and

$$\text{M} = [y1 \;\; y2 \;\; u1 \;\; u2], \tag{6.3}$$

is the data matrix. Note that the un-normalized parameter vector is $[Ta_2 \;\; T^2a_1 \;\; -Tb_2 \;\; -T^2b_1]'$.

MATLAB solves (6.1) by Householder orthogonalization with column pivoting. This gives the least-squares solution when the set of parameter equations is overdetermined. This occurs when $R>4$. The solution is given by the MATLAB statement

$$\theta = \text{M}\backslash\text{-y} \,. \tag{6.4}$$

Experiment design for parameter identification deals with choice of a suitable data interval T and input $u(x)$. Consider the case of the **Example** of section 5. All methods to be described have a MATLAB output

$$\theta = [4 \;\; 3 \;\; -1 \;\; -2]'. \tag{6.5}$$

Method E. Method E is based on analytical formulas (2.7),(2.8), for the integral functions. There are 32 sampling points. Hence $R=32$. Equation (6.4) yields (6.5).

Method I. Method I is based on discrete integration formulas (5.4) and (5.7). Thirty two approximate sampled values of integral functions are used. Equation (6.4) yields (6.5).

Method Wb. Method Wb uses the 32 Walsh coefficients computed from the sampled data of Method 1 and the Walsh binary matrix. It is seen that Wb is a non-singular transformation of the data matrix M and output y used in Method I. Consequently, the least-squares solution in unchanged.

Method A,A$_c$. In this method sets of Walsh coefficients of (2.5) are used. The parameter equation has the form

$$M\theta = -y_w, \qquad (6.6)$$

where

$$M = [y^1_w \quad y^2_w \quad u^1_w \quad u^2_w\ ],$$

and where w represents a *sal* or *cal* Walsh function. A full set of Walsh coefficients (see (4.1)) results in R=32 and yields a non-singular matrix transformation of (6.1). A choice $4<R<32$ results in a weighted-least-squares solution. This approach has been used by Bohn (1982) for parameter identification of a servomotor position control system.

Orthogonality of Walsh functions can be used to reduce the number of parameters in a data equation. Walsh methods 1 and 2 illustrate this approach.

Walsh method 1. Making use of the fact that $u^2(x)$ is orthogonal to $cal(k,x)$, k odd, yields a parameter vector equation of dimension 3 (see (6.6)):

$$\mathrm{M}\theta =- y_c,$$

where

$$\mathrm{M} = [y^1_c \quad y^2_c \quad u^1_c],$$

$$\theta = [a_2 \quad a_1 \quad -b_1]' = M\backslash -y_c = [4 \quad 3 \quad -1]'.$$

The parameter b_2 is the found by evaluating the *sal*(1,*x*) coefficient of (2.5):

$$b_1 = (y_{s1} + 4\mathrm{y}^1{}_{s1} + 3y^2{}_{s1})/u^2{}_{s1} = 2.$$

Walsh method 2. Making use of the fact that the Walsh functions *sal*(1,*x*)+*csal*(3,*x*) and *cal*(1,*x*)-2*cal*(3,*x*), where c=$-u^1{}_{s1}/u^2{}_{s3}$, are orthogonal to $u^1(x)$ and $u^2(x)$ results in a parameter vector equation of dimension 2:

$$M = [y^1_{s1} + cy^1_{s3} \quad y^2_{s1} + cy^2_{s3}\ ;\ y^1_{c1} - 2y^1_{c3} \quad y^2_{c1} - 2y^1_{c3}];$$

$$z = -[y_{s1} + cy_{s3};\ y_{c1} - 2y_{c3}];$$

$$\theta = [a_2 \ a_1]' = M\backslash z = [4 \ 3]'.$$

The parameters b_1 and b_2 can then be determined by choosing the *sal* and *cal* coefficients of (2.5).

An extensive literature survey of related block pulse methods and computer simulation studies is given by Rao (1983).

7. Walsh function inputs

Persistently exciting inputs are generally required to obtain well-conditioned data matrices M. Pseudo-random binary inputs are often chosen for this purpose. It is clear that over a limited time interval such inputs can be represented by Walsh functions. This requires that the previous restriction to odd functions satisfying periodicity condition (2.2) and the choice of $sal(1,x)$ as input be removed. Let $y(x)$ and $y(x,s1)$ be the system responses to a general Walsh function input $u(x)$ and an input $sal(1,x)$, respectively. It is shown in this section that $y(x,s1)$ can be expressed as a linear combination of time-shifted $y(x)$. Consequently, Walsh function data processing methods based on $y(x,s1)$ are applicable to the case of general Walsh function inputs.

To illustrate the Walsh function properties required in dealing with the general case let $v=a^3$ and consider a persistently exciting pulse input

$$u(x) = u_o(x)+u_e(x) = \begin{cases} 7, & 0 < x < v \\ -1, & v < x < a \end{cases} \tag{7.1}$$

where

$$u_o(x) = sal(1,x) + sal(3,x) + cal(1,x) + cal(3,x)$$

$$= \begin{cases} 4, & 0 < x < v \\ 0, & v < x < a \end{cases} \tag{7.2}$$

$$u_e(x) = sal(2,x) + cal(2,x) + sal(4,x)$$

$$= \begin{cases} 3, & 0 < x < v \\ -1, & v < x < a \end{cases} \tag{7.3}$$

The functions (7.2) and (7.3) are odd and even, respectively:

$$u_o(x) = -u_o(x+a), \quad u_e(x) = u_e(x+a). \tag{7.4}$$

Ohta (1976) has shown that Walsh functions can be expressed in terms of time-shifted $sal(1,x)$:

$$\begin{aligned} cal(1,x) &= sal(1,x+2v) \\ sal(3,x) &= sal(1,x)+sal(1,x+3v)+sal(1,x-3v) \\ cal(3,x) &= sal(3,x-2v) \end{aligned} \tag{7.5}$$

The output response to input (7.1) can be resolved into odd-even components:

$$y(x) = y_o(x) + y_e(x), \tag{7.6}$$

where

$$y_o(x) = a(y(x) - y(x+a)),$$
$$y_e(x) = a(y(x) + y(x+a)). \tag{7.7}$$

Making use of the fact that $4v=a$ it is seen from (7.2),(7.3),(7.4) and (7.5) that

$$\begin{aligned} y_o(x) &= 2y(x,s1)+y(x+3v,s1)+y(x-3v,s1)+y(x+2v,s1) \\ &\quad +y(x-2v,s1)+y(x+v,s1)+y(x-5v,s1) \qquad (7.8) \\ &= 2(y(x,s1)-y(x-v,s1)) \end{aligned}$$

Equation (7.8) shows that an odd system output can be expressed in terms of time-shifted $y(x,s1)$. For identification purposes, however, the inverse relation is required. Replacing x with x-v in (7.8) and adding the two equations yields

$$y(x,s1) - y(x\text{-}2v,s1) = a(y_o(x) + y_o(x\text{-}v). \tag{7.9}$$

Replacing x with x-$2v$ in (7.9), adding the two equations, and using (7.4) yields the inverse of (7.8):

$$y(x,s1) = a^2(y_o(x) + y_o(x\text{-}v) + y_o(x\text{-}2v) + y_o(x\text{-}3v)). \tag{7.10}$$

For the even input (7.3) a scale change $x'=2x$ associated with a period decrease $T'=2T$ gives

$$u_e(x') = sal(1,x') + cal(1,x') + sal(2,x'). \tag{7.11}$$

It is seen that a scale change resolves $u_e(x)$ into odd-even components with respect to T'. Repeating the procedure used to obtain (7.10) it is seen that a sequence of scale changes can be used to express the sequency response components of $y(x)$ in terms of time-shifted $y(x)$. Consequently, the parameter identification methods discussed in section 6 are applicable to the case of general Walsh function inputs.

8. Summary

Input-output data processing of continous-time signals for identification of continous-time models requires the use of fast sampling and some form of data compression. Such data processing requirements are easily realized when Walsh functions are used. Analytic formulas for Walsh coefficients of linear system output signals are derived. A MATLAB implementation of a recursive algorithm for computing the Walsh coefficients of *exp(sx)* is given. The analytic formulas allow a comparison with various discrete approximation methods for numerical evaluation of repeated integrations. MATLAB algorithms are given for the computation of Walsh coefficients using a Walsh binary matrix and operational matrices. Applications to the identification of continous-time integral equation models are discussed. It is shown that Walsh function properties allow sequency components to be determined by additive time-shift operations on output *y(x)*.

References

Beauchamp, K.G. (1984), "*Applications of Walsh and Related Functions*", Academic Press, New York.

Blachman, N.M. (1974), "Sinusoids versus Walsh functions", *Proc. IEEE*, Vol .62, No. 3, pp. 346-354.

Bohn, E.V. (1982), "Estimation of continous-time linear system parameters from periodic data", *Automatica*, Vol. 18, No. 2, pp. 27-36.

Bohn, E.V. (1982), "Measurement of continous-time linear system parameters via Walsh functions", *IEEE Trans. Ind. Electron. Control Instrum.*, Vol. 29, No. 1, pp.38-46.

Bohn, E.V. (1983), "Recursive expressions for evaluating Walsh coefficients for linear dynamic systems", *Intl. J. Systems Sci.*, Vol. 14, pp. 673-682.

Bohn, E. V. (1984), "Recursive evaluation of Walsh coefficients for multiple integrals of Walsh series", *Automatica*, Vol. 20, No. 2, pp. 243-246.

Corrington, M.S, (1973), "Solution of differential and integral equations with Walsh functions", *IEEE Trans. Circuit Theory*, Vol. CT-20, No. 5, pp. 470-476.

Fine, N.J. (1949), "On Walsh functions", *Trans. Amer. Math Soc.*, Vol. 65, pp. 373-414.

Harmuth, H.F. (1971), "*Transmission of Information by Orthogonal Functions*", Springer-Verlag, Berlin.

Ohta, T. (1976), "Expansion of Walsh functions in terms of shifted Rademacher functions and its applications to the processing and the radiation of electromagnetic waves", *IEEE Trans. Electromagn. Compat.*, EMC-18.

Rao, G.P. (1983), "*Piecewise Constant Orthogonal Functions and Their Application to Systems and Control*", Springer-Verlag, Berlin.

Rao, G.P., and Palanisamy, K.R. (1983), " Improved algorithms for parameter identification in continous systems via Walsh functions", *Proc. IEE*, 130,9.

Tadokoro, Y. and Higuchi, T. (1978), "Discrete Fourier transform computation via the Walsh transform", *IEEE Trans. Acoustics, Speech, & Signal Proc.*, ASSP-26,295.

Appendix A

Walsh coefficients of $exp(s(j)x)$

Let $s=[s(1)\ s(2)....s(N)]$. Consider (3.10). Define $cal(0,x)=sal(1,x)$ and let $EW(j,k+1)=E_W(k)$ where $f(x)=exp(s(j)x)$.

```
        for i=1:M
        b2(i)=(1/2)^i;
        end
        for j=1:N
        a=exp(s(j)*b2);
        EW1=(-2/s(j))*(1-a(1));
        v=[1];TN=v*(1-a(2));
        c=TN*(1-a(2));
                for n=2:M-1
                TP=v*(1+a(n));
                i=n-1;
                TNR=TN(:,i:-1:1); % reverse order of elements
                v=[TNR TP];
                TN=v*(1-a(n+1));
                c=[c TN*(1-a(n+1))];
                end
EW(j,:)=[EW1 (-2/s(j))*c];
end
```

Example. s = [-1 -3]; M = 4.

Appendix B

Operational Matrices

Operational matrices are defined by representing analytical series for integrals of Walsh functions in a vector-matrix form (see (4.1) and Bohn 1984),

$$S^1(x)=A_cC(x)\ ,\ C^1(x)=A_sS(x) \tag{B.1}$$

where

(B.2)

It is seen that A is an infinite-dimensional matrix that must be truncated for numerical data processing. It was shown by Bohn (1984), however, that when (4.1) is sequency limited residual truncated terms have infinite geometric series forms that are analytically summable. To illustrate this property consider

$$F^{n+1} = AF^n, \tag{B.3}$$

and partition A into submatrices A_{jk} of dimension $n_T \times n_T$, and partition F into corresponding subvectors F^n_K. Define a reversed unit matrix

$$I_R = [e_n\ .\ .\ .\ .\ .\ .\ e_1], \tag{B.3}$$

where $I=[e_1.\ .\ .\ .\ .\ .\ e_n]$ is a unit matrix. The vector $G^n{}_K$ obtained by reversing the order of the elements of $F^n{}_K$ is given by

$$G^n_K = I_R F^n_K\ . \tag{B.4}$$

(Note that $F^n{}_K=I_RG^n{}_K$). Choosing $n_T=2^J$ it is seen from (B.2) that the submatrices satisfy the following conditions:

$$A_{kk} = 0\ ,\quad k > 1$$

$$A_{jk} = -A_{kj}\ ,\quad j \neq k$$

$$A_{(j+1)(k-1)} = A_{jk}\ ,\quad j < k$$

$$A_{1k} = \begin{cases} \alpha a^{L-1} I_R\ , & k = 2^L \\ 0, & k \neq 2^L \end{cases} \tag{B.6}$$

where $\alpha=a^{J+4}$. The case $F_K=0$, $K>1$, occurs when (4.1) has n_T Walsh coefficients. It is seen from (B.5) and (B.6) that the non-zero subvectors of F^1 are

$$F^1{}_1 = A_{11}F_1,$$

$$F^1{}_2 = A_{21}F_1 = -A_{12}F_1 = -\alpha I_R F_1 = -\alpha G_1,$$

$$F^1{}_4 = -\alpha a G_1,$$

$$F^1{}_8=-\alpha a^2 G_1, \tag{B.7}$$

$$\cdots\cdots\cdots$$

Consider the case $n=1$. It is seen from (B.5), (B.6), and (B.7) that

$$F^2{}_1 = A_{11}F^1{}_1 + R^2{}_1, \tag{B.8}$$

where the residual term due to truncation is given by

$$\begin{aligned} R^2{}_1 &= A_{12}F^1{}_2 + A_{14}F^1{}_4 + A_{18}F^1{}_8 + \ldots. \\ &= \alpha(G^1_2 + aG^1_4 + a^2G^1_8 + \ldots.) \\ &= -\alpha^2(1 + a^2 + a^4 + \ldots.)F_1 \\ &= -\alpha^2\gamma F_1, \end{aligned} \tag{B.9}$$

and where $\gamma=1/(1-a^2)$. The subvectors $F^2{}_K$ are given by

$$\begin{aligned} F^2_2 &= A_{21}F^1_1 = -A_{12}F^1_1 = -\alpha G^1_1 \\ F^2_4 &= -\alpha a G^1_1 \\ F^2_8 &= -\alpha a^2 G^1_1 \\ &\cdots\cdots\cdots\cdots \end{aligned} \tag{B.10}$$

Consider the case $n=2$. Repeating the above procedure it is seen that

$$\mathrm{F}^3_1=\mathrm{A}_{11}\mathrm{F}^2_1+\mathrm{R}^3_1, \tag{B.11}$$

where the residual term is given by (see (B.10))

$$\begin{aligned} R^3_1 &= A_{12}F^2_2 + A_{14}F^2_4 + A_{18}F^2_8 + \ldots. \\ &= \alpha(G^2_2 + aG^2_4 + a^2G^2_8 + \ldots.) \\ &= -\alpha^2\gamma F^1_1. \end{aligned} \tag{B.12}$$

General expressions for the residuals are given by Bohn (1984).

Use of the block pulse operator

Shien-Yu Wang
Department of Automatic Control & Electronic Engineering
East China University of Chemical Technology
Shanghai, 200237, China

Abstract

A block pulse operator (BPO) and its applications to continuous model identification are introduced in this chapter. It includes some research works by the authors on BPO in the recent years. The applications of BPO to the identification of nonlinear and distributed parameter systems are given. The BPO method of optimal input design for identifying parameters in continuous dynamic systems are presented. Numerical examples is presented to illustrate the utility of this method.

1. Introduction

In recent years aspects of continuous model identification (CMI) have been discussed in a multitude of papers, in several books (Unbehauen and Rao 1987, Saha and Rao 1983, Sinha and Kuszta 1983, Rao 1983). Several activity techniques for CMI have been suggested in recent years. As stated by Prof. Unbehauen and Prof. Rao in their books, the methods of parameter estimation in the CMI may in general be seen to consist of two stages. These are the primary stage and secondary stage. The primary stage involves converting a dynamic system described by defferential equation (or transfer function) into a system described by algebraic equation which has two important properties: it is realizable in the parameter and contains only realizable functions of the data. The secondary stage involves estimating unknown parameter from the resulted equation by means of obvious technique (linear algebra, least squares method and instrumental variable method etc.).

In fact, the above procedure of the CMI generally has double approximation meanings. Firstly, the original CMI is approximated by a new identification problem. Secondly, an approximation technique (recursive or iterative) is used to solve the latter problem. Therefore, it will be an important factor in evaluating the methods that whether the approximation solution to the approximation problem is equivalent (or converges) to the trued solution for the original problem. In the procedure, it is obvious that the primary stage will be the key to solving the CMI.

The purpose of this chapter is to introduce a method based on the block pulse operator (BPO) which is used to solve the CMI. The BPO proposed by Wang (1983).

N. K. Sinha and G. P. Rao (eds.), Identification of Continuous-Time Systems, 159–203.

It was perfected, developed and applied in references (Wang and Jiang 1984, 1985, 1989, Wang 1984, 1989, 1990, Wu and Wang, 1988, 1989, 1990, Zhu and Lu, 1988 etc.). The study of the BPO originated in an attempt to analyse the convergence of expansion methods based on block pulse functions (BPFs) for use in various fields. This operator made proper mathematics frame for the block pulse function analysis method, so some common law could be drawn and the strict theory basis could be set up.

The basic idea of the method based on the BPO is to convert the problems of system identification in original function space into the equivalent approximation problems in the image space of BPO. In the image space, the algorithm for solving the approximation problems may be considerably simplified. Then, the solutions of original problems will be obtained if the convergence (or equivalent) of the approximating solutions to the trued solution is proved. The process of system identification via BPO is shown in Fig. 1.

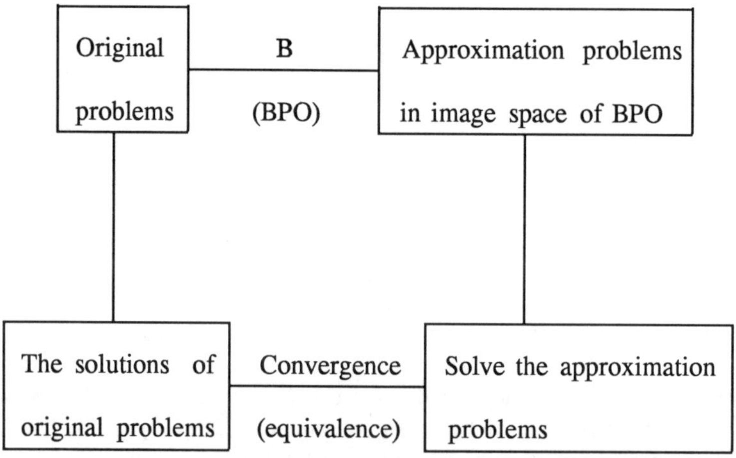

Fig. 1. The system identification process via BPO

The main advantages of the BPO method are that:

(1) This approximate method has been strictly established based on the theory of convergence in the mean square.
(2) It is not only convenient in operation which is similar to Laplace operator, but also simple in numerical computation which is similar to finite difference method.
(3) Using the BPO method, we can find it easy to derive the system of parameter estimation equations from the dynamical model form of the system to be identified.
(4) The convergence (or equivalence) above mentioned can be discussed in the proper mathematics frame.

We now briefly review the contents of this chapter. The definitions and properties of BPO are given in section 2. The operational rules of BPO are introduced in section 3. While section 2 and 3 can be considered as introductory, the main theme is taken up in the following sections. BPO methods of system identification for the estimation of the unknown parameters of a dynamic system described by nonlinear differential equation are discussed and applications are given in section 4. This section presents the key theorems which are required in order to analyze the equivalent relation between original problem and approximation problem. The applications of BPO to identification of distributed parameter systems are developed in section 5. The BPO methods of optimal inputs design for system identification are introduced in section 6.

Throughout this chapter we shall use $L_2^k(a,b)=L_2([a,b], R^k)$ to denote the customary Lebesque space of R^k - valued functions on [a, b] whose components are square integrable. Whenever n=1 we shall write just $L_2[a, b]$. The usual space of continuous functions on [a, b] with value in R^k space will be denoted by $C_k[a, b]$. Let $C_k^r[a, b]$ be the space of rth differentiable functions on [a, b] with value in R^k. The vector space of n×m matrices will be denoted by $R^{n\times m}$. For norms of elements in Banach spaces, we use $|\cdot|$, whereas $\|\cdot\|$ is used for operator norms. Furthermore we shall occasionally use a subscript such as $|\cdot|_B$ to distinguish a norm in the space B from other norms. Finally, $<\cdot, \cdot>_{R^n}$ and $<\cdot, \cdot>_{L_2}$ will denote the standard inner products in R^n and L_2, respectively, and $|\cdot|_B = (<\cdot, \cdot>_B)^{1/2}$. The transpose of matrix A is given by A^T. R^+ is a set of positive real number.

2. Definition of Block-Pulse Operator

Consider a set of orthonormal block pulse functions (BPFs) over [0,T]

$$\varphi_1(t) = \begin{cases} \sqrt{\dfrac{m}{T}} & 0 \le t \le \dfrac{T}{m} \\ 0 & otherwise \end{cases}$$

$$\varphi_i(t) = \begin{cases} \sqrt{\dfrac{m}{T}} & \dfrac{(i-1)T}{m} < t \le \dfrac{iT}{m} \\ 0 & otherwise \end{cases} \quad i = 2,\cdots,m$$

As $m \to \infty$ the BPFs is a set of complete (Kwong and Chen 1981, Rao and Srinirasan 1978). By means of the method presented by Courant and Hilbert (1953), we can construct an n-dimensional BPFs over the set E_n as follows:

$$\varphi_{i_1 i_2 \cdots i_{j-1} 1 i_{j+1} \cdots i_n}(t_1, t_2, \cdots, t_n) = \begin{cases} \sqrt{\dfrac{m}{T}} & t_j \in \left[0, \dfrac{T_j}{m_j}\right];\ t_k \in \begin{cases} \left[0, \dfrac{T_k}{m_k}\right], & i_k = 1 \\ \left(\dfrac{(i_k-1)T_k}{m_k}, \dfrac{i_k T_k}{m_k}\right], & i_k > 1 \end{cases} \\ & i_k = 1,2,\cdots,m_k,\quad k = 1,2,\cdots,j-1,j+1,\cdots n,\ j = 1,\cdots,n \\ 0 & otherwise \end{cases}$$

$$\varphi_{i_1 i_2 \cdots i_n}(t_1, t_2, \cdots, t_n) = \begin{cases} \sqrt{\dfrac{m}{T}} & t_k \in \left(\dfrac{(i_k-1)T_k}{m_k}, \dfrac{i_k T_k}{m_k}\right], \quad \begin{matrix} i_k = 2,\cdots,m_k \\ k = 1,2,\cdots,n \end{matrix} \\ 0 & otherwise \end{cases}$$

where $E_n \mathrel{\hat{=}} \{ t = (t_1, \cdots, t_n)^T : 0 \le t_k \le T_k,\ k = 1, 2, \cdots, n,\ T_k \in R^+ \}$

$$m = \prod_{k=1}^{n} m_k, \qquad T = \prod_{k=1}^{n} T_k$$

as $m_k \to \infty$ $(k=1,\cdots,n)$ the n-dimensional BPFs is complete and orthonormal set. Another proof of the completeness of n-dimensional BPFs is given by Nath and Lee (1983).

Let

$$N_m = span\left\{\varphi_{i_1 \cdots i_n}(t_1, \cdots, t_n),\ i_k = 1,2,\cdots,m_k,\ k = 1,2,\cdots,n\right\}$$

N_m is a subspace of $L_2(E_n)$. If a function $f(t_1,\cdots,t_n) \in L_2(E_n)$, by the projection Theorem (Curtain and Pritchard 1977), there exists a unique $\hat{f}_m(t_1,\cdots,t_n) \in N_m$ given by

$$\hat{f}_m(t_1, \cdots, t_n) = \sum_{i_1=1}^{m_1} \cdots \sum_{i_n=1}^{m_n} f_{i_1 \cdots i_n} \varphi_{i_1 \cdots i_n}(t_1, \cdots, t_n)$$

$$\mathrel{\hat{=}} F^T \Phi_m(t_1, \cdots, t_n)$$

where

$$F^T \hat{=} (f_{1\cdots 1}, \cdots, f_{1\cdots m_n}, \cdots, f_{m_1\cdots m_{n-1}1}, \cdots, f_{m_1\cdots m_n})$$

$$f_{i_1\cdots i_n} = \int_{E_n} f(t)\varphi_{i_1\cdots i_n}(t)\, dt \qquad (1)$$

$$\Phi_m(t_1, \cdots, t_n) = \left(\varphi_{1\cdots 1}(t_1, \cdots, t_n), \cdots, \varphi_{1\cdots m_n}(t_1, \cdots, t_n), \cdots, \varphi_{m_1\cdots m_n}(t_1, \cdots, t_n)\right)^T$$

such that $\hat{f}_m(t_1, \cdots, t_n)$ is the closest approximation to $f(t_1, \cdots, t_n)$ in N_m. Now, we introduce the following definition:

Definition 1: (Wang 1983) An operator $B_n: L_2(E_n) \to R^m$ is said to be a n-dimensional M-BPO if

$$B_n f(t) = F^T \qquad (2)$$

where $f \in L_2(E_n)$. Specifically, if n=1 and $f(t) \in L_2[o,T]$, we have

$$\hat{f}(t) = F^T\Phi(t) \quad , \qquad Bf(t) = F^T$$

where

$$F^T = (f_1, f_2, \cdots, f_m)^T$$

$$f_i = <f(t), \varphi_i(t)>_{L_2} = \int_0^T f(t)\varphi_i(t)dt = \sqrt{\frac{m}{T}}\int_{\Delta_i} f(t)dt$$

$$\Phi(t) = \left(\varphi_1(t), \varphi_2(t), \cdots, \varphi_m(t)\right)^T$$

$$\Delta_1 = \left[0, \frac{T}{m}\right]; \quad \Delta_i = \left(\frac{(i-1)T}{m}, \frac{iT}{m}\right], \quad i=2,\cdots n$$

such that $\hat{f}(t)$ is the closest approximation to $f(t)$ in N_m. we define

$$\overline{B}_n f(t) = \sqrt{m/T} B_n f(t)$$

The operator $\overline{B}_n$ is said to be a non-normal BPO.

Definition 2: For a matrix-valued function $A(t) \hat{=} (a_{ij}(t))_{p\times q}$, $a_{ij}(t) \in L_2(E_n)$. We define

$$B_n A(t) = (B_n a_{ij}(t))_{p\times qm}$$

Definition 3: For a vector-value function $f(t) \in R^k$, we define

$$\omega_f = \limsup_{|s-v|\le\delta} |f(s) - f(v)|_{R^k}, \quad \delta = \frac{T}{m}$$

The BPO has the following properties (Wang 1983, Wang and Jiang 1989):

Property 1: The BPO is a linear, continuous and bounded operator.

Property 2: If $f(t) \in L_2^r(E_n)$, then

$$| (B_n f(t))\Phi(t) - f(t) |_{L_2^r} \to 0 \qquad as\ m\to\infty$$

$$\limsup_{t\in E_n} | (B_n f(t))\Phi(t) - f(t) |_{R^r} \le \omega_f$$

Property 3: $\|B\| = 1$

property 4: $|Bf(t)\Phi(t)|_{L_2} = |Bf(t)|_{R^m} \quad \forall f \in L_2[0, T]$

3. Operational Rules of the BPO

First of all, we introduce the following definitions and symbols.

Definition 4: Vector α and vector β are said to be asymptotic approximation if

$$|\alpha - \beta|_{R^m} \to 0 \qquad (m \to \infty) \qquad (3)$$

where $\alpha, \beta \in R^m$. we shall denote this relation (3) by $\alpha \sim \beta$. For example, $Bf(t) \sim Bg(t)$, it means:

$$\lim_{m\to\infty} | Bf(t) - Bg(t) |_{R^m} = 0$$

We now introduce the following operational Rules (Wang 1983, Wang and Jiang 1989):

(A) Operational Rules for one-dimensional BPO

Rule 1: (Product theorem) If f(t), g(t), f(t)g(t) $\in L_2[0, T]$ and f(t) and g(t) are bounded, then

$$B(f(t)g(t)) \sim \sqrt{m/T}(Bf(t))\circ(Bg(t))$$

and

$$| f(t)g(t) - [(\overline{B}f(t))\circ(\overline{B}g(t))]\overline{\Phi}(t) |_{L_2} \to 0 \quad as\ m \to \infty$$

where we define the Hadamard products of matrix $A = (a_{ij})_{n\times m}$ and matrix $B = (b_{ij})_{n\times m}$ as

$$A\circ B = (a_{ij}b_{ij})_{n\times m}$$

Rule 2: (First shift theorem) Let $\lambda = (q+\mu)T/m$, q is integer and $1\leq q\leq m$, $0 \leq\mu< 1$. Then

$$Bf(t-\lambda) \sim (Bf(t))(I_m-\mu Q)W_1^q \quad (t>\lambda)$$

and

$$| f(t-\lambda) - (Bf(t))(I_m-\mu Q)W_1^q\Phi(t) |_{L_2} \to 0 \quad (t>\lambda) \quad as\ m\to\infty$$

where $Q = I_m - W_1$, I_m is m×m identity matrix and

$$W_1 = \begin{bmatrix} 0 & 1 & 0 & 0 & \cdots & 0 \\ 0 & 0 & 1 & 0 & \cdots & 0 \\ \cdot & \cdot & \cdot & \cdot & \cdots & \cdot \\ 0 & 0 & 0 & 0 & \cdots & 1 \\ 0 & 0 & 0 & 0 & \cdots & 0 \end{bmatrix}_{m\times m}$$

Rule 3: (Second shift theorem) Let $\lambda=(q+\mu)T/m$, q is integer and $1\leq q\leq m$, $0\leq\mu<1$. Then

$$Bf(t+\lambda) \sim (Bf(t))(I_m-\mu Q^T)(W_1^q)^T \quad (0\leq t\leq T-\lambda)$$

and

$$| f(t+\lambda) - (Bf(t))(I_m-\mu Q^T)(W_1^q)^T\Phi(t) |_{L_2} \to 0 \quad (0\leq t\leq T-\lambda) \quad as\ m\to\infty$$

Rule 4: (Integral Theorem) If f(t) $\in L_2[0, T]$ and f(t) is bounded, i.e.

$$|f(t)| \leq K \qquad \forall t \in [0, T] \text{ and } K \in R^+$$

then

$$B\int_0^t f(s)ds \sim (Bf)P_m$$

$$B\int_t^T f(s)ds \sim (Bf)P_m^T$$

and

$$\left\| \int_0^t f(\tau)d\tau - (Bf)P_m\Phi(t) \right\|_{L_2} \leq C\frac{1}{m}$$

$$\max_{0\leq t\leq T}\left| \int_0^t f(\tau)d\tau - (Bf)P_m\Phi(t) \right| \leq \frac{2KT}{m}$$

$$where\ C = T\sqrt{T}[K + \frac{1}{2}\omega_f(T/m)]$$

The matrix P_m is called the operational matrix of integration and given by

$$P_m = \frac{T}{m}\begin{bmatrix} ½ & 1 & 1 & \cdots & 1 & 1 \\ 0 & ½ & 1 & \cdots & 1 & 1 \\ \cdot & \cdot & \cdot & \cdots & \cdot & \cdot \\ 0 & 0 & 0 & \cdots & ½ & 1 \\ 0 & 0 & 0 & \cdots & 0 & ½ \end{bmatrix}_{m\times m}$$

For the operational matrix P_m, one obtain the following Lemma:

Lemma 1: (Sannuti 1977) Let A is a n×m matrix, then

$$(AP_m)_1 = \frac{T}{2m}(A)_1$$

$$(AP_m)_i = \frac{T}{m}\sum_{j=1}^{i-1}(A)_j + \frac{T}{2m}(A)_i$$

$$(AP_m)_{i+1} = (AP_m)_i + \frac{T}{2m}\left[(A)_{i+1} + (A)_i\right]$$

where $(A)_i$ and $(AP_m)_i$ are the ith columns of the n×m matrices A and AP_m, respectively.

Definition 5: A vector $\Delta(\alpha)$ is said to be vector difference of a vector $\alpha = (\alpha_1, \alpha_2, \cdots, \alpha_m)$ if α_0 is known real number and

$$\Delta(\alpha) = \left(\alpha_1-\alpha_0,\ \alpha_2-\alpha_1,\ \cdots,\ \alpha_m-\alpha_{m-1}\right)$$

Rule 5: (First differential theorem) If $f(t) \in C^1[0, T]$ and $f_0 = \sqrt{T/m}\ f(0)$, $Bf(t) = (f_1, f_2, \cdots, f_m)$, then

$$B\dot{f}(t) \sim \frac{m}{T}\Delta(Bf(t))$$

and

$$\left|\dot{f}(t) - \frac{m}{T}\Delta(Bf(t))\Phi(t)\right|_{L_2} \to 0 \quad as\ m\to\infty$$

Furthermore, if $|\ddot{f}(t)| \le G_1\ \ \forall\ t \in [0, T],\ G_1 \in R^+$, *then*

$$\left|\dot{f}(t) - \frac{m}{T}\Delta(Bf(t))\Phi(t)\right|_{R^m} = O(m^{-1/2}) \quad as\ m \to \infty$$

Rule 6: (Second differential theorem)

(1) If $f(t) \in C^2[0, T]$, then

$$B\dot{f}(t) \sim d^T$$

and

$$\max_{0 \le t \le T} | \dot{f}(t) - d^T\Phi(t) | \to 0 \quad as\ m \to \infty$$

where $d^T = (d_1, d_2, \ldots, d_m)$ and

$$\left. \begin{array}{ll} d_1 = (2m/T)(Bf)_1 - 2\sqrt{m/T}f(0) & \\ d_{i+1} = -d_i + (2m/T)[(Bf)_{i+1} - (Bf)_i] & i = 2, \cdots, m-1 \end{array} \right\}$$

(2) If the 3rd derivative of f(t) with respect to t is bounded, then

$$\max_{0 \le t \le T} |\dot{f}(t) - d^T\Phi(t)| = O(m^{-1}) \quad as\ m \to \infty$$

Rule 7: (Convolution theorem) If f(t), g(t) and f(t)g(t) $\in L_2[0, T]$ and f(t) and g(t) are bounded, then

(1) $$B\int_0^t f(x)g(t-x)dx \sim \frac{1}{2}\sqrt{T/m}(Bf)J_{B(g)}W_2$$

where

$$J_y \triangleq \begin{bmatrix} y_1 & y_2 & y_3 & \cdots & y_m \\ 0 & y_1 & y_2 & \cdots & y_{m-1} \\ \cdot & \cdot & \cdot & \cdots & \cdot \\ 0 & 0 & 0 & \cdots & y_1 \end{bmatrix}_{m \times m} \qquad y = (y_1, y_2, \cdots, y_m)$$

and $W_2 = I_m + W_1$.

(2) $$\left| \int_0^t f(x)g(t-x)dx - \frac{1}{2}\sqrt{T/m}(Bf)J_{B(g)}W_2\Phi(t) \right|_{L_2} \to 0 \quad as\ m \to \infty$$

(B) Operational rules for m-BPO

For convenience, our disscussions are limited to the n=2 case.

Rule 8: If f(x, t) and g(x, t) are bounded, f, g, fg $\in L_2(E_2)$. Then

$$B_2[f(x, t)g(x, t)] \sim \left[\frac{m_1 m_2}{T_1 T_2}\right]^{1/2} B_2 f(x, t) \circ B_2 g(x, t)$$

Rule 9: If $f(x) \in L_2[0, T_1]$, $g(t) \in L_2[0, T_2]$, $fg \in L_2(E_2)$, then

$$B_2[f(x)I(t)] = [B_1 f(x)] \otimes e_2^T$$

$$B_2[I(x)g(t)] = e_1^T \otimes [B_1 g(t)]$$

$$B_2[f(x)g(t)] = [B_1 f(x)] \otimes [B_1 g(t)]$$

where $\otimes$ indicates the Kronecker product and

$$I(x) = \begin{cases} 1, & x \in [0, T_1] \\ 0, & otherwise \end{cases} \qquad I(t) = \begin{cases} 1, & t \in [0, T_2] \\ 0, & otherwise \end{cases}$$

$$e_i^T \hat{=} \left[\frac{T_i}{m_i}\right]^{1/2} (1, 1, \cdots, 1)_{1 \times m_i}, \quad i = 1, 2.$$

Rule 10: (First shift theorem) Let $\lambda_i = q_i T_i/m_i + \tau_i$, q_i are natural numbers, $1 \leq q_i \leq m_i$, $0 \leq \tau_i < T_i/m_i$, $i = 1,2$, $f(x,t)$, $f(x-\lambda_1,t)$, $f(x,t-\lambda_2)$ and $f(x-\lambda_1,t-\lambda_2) \in L_2(E_2)$. Then

$$B_2 f(x-\lambda_1, t) \sim [B_2 f(x, t)][H_1 \otimes I_{m_2}]$$

$$B_2 f(x, t-\lambda_2) \sim [B_2 f(x, t)][I_{m_1} \otimes H_2]$$

$$B_2 f(x-\lambda_1, t-\lambda_2) \sim [B_2 f(x, t)][H_1 \otimes H_2]$$

where I_{m_1} is an $m_1 \times m_1$ identity matrix,

$$H_i \hat{=} \left[I_{m_i} - \frac{m_i \tau_i}{T_i} Q_i\right] W_i^{q_i}$$

$$Q_i \hat{=} I_{m_i} - W_i$$

$$W_i \hat{=} \begin{bmatrix} 0 & | & I_{m_i-1} \\ -- & + & -- \\ 0 & | & 0 \end{bmatrix}_{m_i \times m_i}, \quad i = 1, 2$$

Rule 11: (Integral theorem) Let f(x, t) ϵ $L_2(E_2)$, Then

$$B_2 \int_0^x \int_0^t f(x', t') dt' dx' \sim [B_2 f(x, t)] \left[P_{m_1} \otimes P_{m_2}\right]$$

$$B_2 \int_0^x f(x', t) dx' \sim [B_2 f(x, t)] \left[P_{m_1} \otimes I_{m_2}\right]$$

$$B_2 \int_0^t f(x, t') dt' \sim [B_2 f(x, t)] \left[I_{m_1} \otimes P_{m_2}\right]$$

where

$$P_{m_i} = \frac{T_i}{m_i} \begin{bmatrix} ½ & 1 & 1 & \cdots & 1 & 1 \\ 0 & ½ & 1 & \cdots & 1 & 1 \\ \cdot & \cdot & \cdot & \cdots & \cdot & \cdot \\ 0 & 0 & 0 & \cdots & ½ & 1 \\ 0 & 0 & 0 & \cdots & 0 & ½ \end{bmatrix}$$

I_{m_i} is an $m_i \times m_i$ identity matrix. i = 1, 2.

The other Rules of BPO and the proofs of the Rlues above mentioned will be omitted. The interested reader is referred to Wang (1983) and Wang and Jiang (1989) for detail discussion of BPO.

4. Parameter identification of continuous nonlinear systems via block pulse operator

Consider the following system

$$\left.\begin{aligned} \dot{x}(t) &= f(x(t), u(t), t, \theta) \\ x(0) &= x_0 \end{aligned}\right\} \qquad (4)$$

$$y(t) = g(x(t), u(t), t) \qquad (5)$$

where $x(t) \epsilon C_n^1[0,T]$, $u(t) \epsilon L_r^2[0,T]$, $y \epsilon R^l$, $\theta \epsilon R^k$. Let f and g be continuous and f satisfy a Lipschitz condition

$$|f(x, u, t, \theta) - f(y, v, t, \theta)|_{R^n} \leq L[|x-y|_{R^n} + |u-v|_{R^r}]$$

where L is a constant.

Suppose now that $z(t_i)$ is the vector of observations at time t_i, $y(t_i)$ is the output response at time t_i, and N is the number of observation times. The performance criterion for least-squares estimation is given by

$$J(\theta) = \sum_{i=1}^{N} |\ g(x(t_i), u(t_i), t_i) - z(t_i)\ |_{R^l}^2$$

The parameter identification problem can be formulated as follows:

Problem $\wp$: Find parameter θ so that minimize the criterion J and satisfy constraint (4).

By means of the BPO, we can solve the problem. First of all, by BPO, the equations (4) - (5) can be approximated by a sequence of matrix algebraic equations (Wang 1989). Then, the approximation problems can be constructed. Three different schemes of the identification algorithms can be presented. Finally, we discuss the convergence of the approximation problems.

Consider the following matrix equation

$$X = [x_0 \mid \cdots \mid x_0] + FP_m \qquad (6)$$

where

$$X = [X_1 \mid \cdots \mid X_m], \qquad X_i \in R^n$$

$$F = [F_1 \mid \cdots \mid F_m]$$

$$F_i = f(X_i, (\bar{B}u)_i, \bar{t}_i, \theta) \tag{7}$$

$$\bar{t}_i = \frac{(2i-1)T}{2m}$$

Eq. (6) may be rewritten recursively as

$$\left.\begin{aligned} X_1 &= \xi_1(X_1) \hat{=} x_0 + \frac{T}{2m}F_1 \\ X_i &= \xi_i(X_i) \hat{=} X_{i-1} + \frac{T}{2m}(F_i + F_{i-1}),\ i=2,\cdots,m \end{aligned}\right\} \tag{8}$$

by means of the Lemma 1.

Solving (8) is equivalent to finding the fixed point X_i of $\xi_i(X_i)$ on R^n if input $U = (U_1, \cdots, U_m)$ is given. Thus, the Contraction Mapping Theorem can be used to generate a method of successive approximations for solving equation (8) numerically (Wang and Jiang 1989).

We will define $\hat{x}(t) = X\bar{\Phi}(t)$ to be a BPO approximation solution of (4), if X satisfies Eq (6) or (8), where $\bar{\Phi}(t) = \sqrt{m/T}\Phi(t)$. The convergence of the BPO solution to the exact solution is proved by Wang (1989a).

We now introduce the convergence theorem of the BPO solution to Eq. (4):

Theorem 1: (Wang 1989a) Suppose that x(t) is a unique solution of (4) and X is a unique solution of (6), then

$$\max_{0 \le t \le T} |x(t) - X\bar{\Phi}(t)|_{R^n} \le O(s_m) + O(m^{-1})$$

where

$$s_m = \omega_f + L\omega_u, \quad \overline{\Phi}(t) = \sqrt{m/T}\Phi(t).$$

Now three kinds of the approximation problems can be formulated as follows:

Problem $\wp_1$: Find parameter θ_m so that minimize the following criterion $J_m(u)$

$$J_m(\theta) = \sum_{j=1}^{N} | g(X_{k_j}, (\overline{B}u)_{k_j}, \overline{t}_{k_j}) - z(t_j) |_{R^l}^2$$

subject to the constraints (6) or (8) and

$$t_j \in \Delta_{k_j} \hat{=} \left(\frac{(k_j-1)T}{m}, \frac{k_jT}{m} \right]$$

Problem $\wp_2$: Using the input-output set $\{u(t_i), z(t_i), i = 1, \cdots, N, N \geq m\}$, we solve the following two sub-problems:

$\wp_{21}$: Find matrix $\hat{X} \in R^{n \times m}$, so that minimize the criterion $S_m(X)$:

$$S_m(X) = \sum_{j=1}^{N} | g(X\overline{\Phi}(t_j), u(t_j), \overline{t}_{k_j}) - z(t_j) |_{R^l}^2$$

$\wp_{22}$: Find θ_{mN} so that minimize the criterion $\overline{J}_m(\theta)$:

$$\overline{J}_m(\theta) = \sum_{j=1}^{m} | g(x_0+(FP_m)_j, (\overline{B}u)_j, t_j) - z(t_j) |_{R^l}^2$$

where

$$F = [F_1 | \cdots | F_m]$$

$$F_i = f(\hat{X}_i, (\overline{B}u)_i, \overline{t}_i, \theta)$$

$$\overline{t}_i = \frac{(2i-1)T}{2m}$$

It is note that criterion $\overline{J}_m(\theta)$ is a direct expression of θ and the sub-problem $\wp_{22}$ is unconstrained optimization problem.

Problem $\wp_3$: Using the input-output set $\{u(t_i), z(t_i), i=1,\cdots,N, N \geq m \}$, we solve the following two sub-problems:

$\wp_{31}$: Same $\wp_{21}$ as.
$\wp_{32}$: Find parameter θ_{mN}, so that minimize the criterion $W_{mN}(\theta)$:

$$W_{mN}(\theta) = \sum_{i=1}^{m} |\varepsilon_i|^2_{R^n}$$

where

$$\varepsilon_1 = \hat{X}_1 - x_0 - \frac{T}{2m} f(\hat{X}_1, (\bar{B}u)_1, \bar{t}_1, \theta)$$

$$\varepsilon_i = \hat{X}_i - \hat{X}_{i-1} - \frac{T}{2m}\left[f(\hat{X}_i, (\bar{B}u)_i, \bar{t}_i, \theta) + f(\hat{X}_{i-1}, (\bar{B}u)_{i-1}, \bar{t}_{i-1}, \theta)\right] \quad i=2,\cdots,m$$

The problem $\wp_{32}$ is also unconstrained optimization problem and the ε_i is equation error of (8).

From above, it is obvious that the procedure to solve problem $\wp_1$, $\wp_2$ and $\wp_3$ is simpler than that to solve problem $\wp$; $\wp_2$ and $\wp_3$ is simpler than $\wp_1$ and problem $\wp_3$ is the simplest. However, a natural question arises in the approximation method: what relations are between problem $\wp$ and $\wp_1$, $\wp_2$ and $\wp_3$? Do the approximation solutions of problem $\wp_1$, $\wp_2$ and $\wp_3$ converge to exact solution of problem $\wp$? In the following, we will answer these questions and several Theorems will be given here.

The theorem of the convergence for Problem $\wp_1$ can be given as follows:

Theorem 2: (Wang and Jiang 1989) Let u(t) is continuous. If
(a) There exists a unique solution $\hat{\theta}$ to the problem $\wp$.
(b) There exists a unique solution $\hat{\theta}_m$ to the problem $\wp_1$ and there exists a unique limit point of sequence $\{\hat{\theta}_m\}$.

then, there exists a subsequence θ_{m_k} such that

$$\hat{\theta}_{m_k} \to \hat{\theta}, \quad as\ m_k \to \infty$$

Now we discuss the relation between Problem $\wp$ and Problem $\wp_2$. To simplify, we consider the following system:

$$\left.\begin{aligned} \dot{x}(t) &= f(x, u, \theta) \\ x(0) &= x_0 \end{aligned}\right\} \qquad (9)$$

$$y(t) = g(x(t)) \qquad (10)$$

Suppose that f satisfy a Lipschitz condition:

$$| f(x, u, \theta) - f(y, v, \theta) |_{R^n} \leq L_1[|x-y|_{R^n} + |u-v|_{R^r}]$$

Function f, g and u are continuous.

The theorem of the convergence for Problem $\wp_2$ can be given as follows:

Theorem 3: (Wang and Jiang 1989) Consider the system (9)-(10). If
(a) $\hat{\theta}_N$ is a unique solution of $\wp$, $\hat{\theta}_{mN}$ is a unique solution of $\wp_{22}$ and θ^* is a unique minimum point of $J_0(\theta)$

$$J_0(\theta) = \lim_{N\to\infty} J_N(\theta)$$

(b) $\hat{X}^{mN}$ is a unique solution of $\wp_{21}$. The sequence $\{\hat{X}^{mN}, N = 1, 2, \cdots\}$ is a consistent sequence of estimators of $\bar{B}x(t)$, i.e.,

$$\hat{X}^{mN} \to \bar{B}x \quad as\ N \to \infty$$

(c) $$\lim_{m\to\infty} \lim_{N\to\infty} \hat{X}^{mN} = \lim_{N\to\infty} \lim_{m\to\infty} \hat{X}^{mN} < \infty$$

$$\lim_{m\to\infty} \lim_{N\to\infty} \bar{J}_m(\theta) = \lim_{N\to\infty} \lim_{m\to\infty} \bar{J}_m(\theta) < \infty$$

then, there exists the subsequence $\hat{\theta}_{m_k N_k}$ and $\hat{\theta}_{N_\lambda}$ such that:

$$\lim_{m_k\to\infty} \lim_{N_k\to\infty} \hat{\theta}_{m_k N_k} = \lim_{N_\lambda\to\infty} \hat{\theta}_{N_\lambda} = \theta^*$$

Theorem 4: (Wang and Jiang 1989) Consider the system (9)-(10). If
(a) $\hat{\theta}_N$ is a unique solution of $\wp$ and θ^0 is a unique minimum point of $J_0(\theta)$. $x^0(t)$ is a solution of (9) corresponding to θ^0 and u(t).
(b) $\hat{X}^{mN}$ is a unique solution of $\wp_{31}$. The sequence $\{\hat{X}^{mN}, N = 1, 2, \cdots\}$ is a consistent sequence of estimators of $\bar{B}x^0(t)$, i.e.,

$$\hat{X}^{mN} \to \bar{B}x^0(t) \qquad as\ N \to \infty$$

(c) $\hat{\theta}_{mN}$ is a unique solution of $\wp_{32}$ and

$$\lim_{m\to\infty}\lim_{N\to\infty} W_{mN}(\theta) = W(\theta)$$

θ^* is the minimum point of $W(\theta)$.

(d) The derivative of u(t) is bounded.

then, there is a subsequence $\hat{\theta}_{m_k N_k}$ such that

$$\lim_{m_k\to\infty}\lim_{N_k\to\infty}\hat{\theta}_{m_k N_k} = \theta^* = \theta^0$$

Example 1: Parameter identification problem of penicillin fermentation process.

The following model represents a batch penicillin fermentation process which has been derived by Constantinides, Spencer and Gaden (1970). The state variables are cell growth and penicillin synthesis, and the control variable is the temperature.

$$\dot{x}_1 = b_1 x_1 - \frac{b_1}{b_2}x_1^2, \qquad x_1(0) = 0.0294$$

$$\dot{x}_2 = b_3 x_1 - b_4 x_2, \qquad x_2(0) = 0$$

$$0 \le t \le T$$

$$x_1 \le b_2$$

$$b_1 = k_1 \frac{1 - 0.005(u - 30)^2}{1 - 0.005(25 - 30)^2}$$

$$b_2 = k_2 \frac{1 - 0.005(u - 30)^2}{1 - 0.005(25 - 30)^2}$$

$$b_3 = k_3 \frac{1 - 0.005(u - 20)^2}{1 - 0.005(25 - 30)^2}$$

$$b_4 = k_4 \exp\left\{-\frac{12210}{1.987}\left(\frac{1}{u + 273.1} - \frac{1}{298.1}\right)\right\}$$

where k_1, k_2, k_3, k_4 are unknown model parameters.
x_1 = cell growth
x_2 = penicillin synthesis
u = temperature in degrees centigrade.

The above models were based on averaged nondimensionalized data, which are taken at 25^0C. For this reason the models are independent of the system of units used for cell and penicillin determinations and of the period of this fermentation. The measurements are generated by

$$h_1(t) = x_1^0(t)[1 + k \cdot GAUSS(0, 0.1)]$$

$$h_2(t) = x_2^0(t)[1 + k \cdot GAUSS(0, 0.1)]$$

where $x_1^0(t)$ and $x_2^0(t)$ are the trajectories of $x_1(t)$ and $x_2(t)$ at 25^0C, GAUSS(0, 0.1) is a normally distributed random variable with mean 0 and standard devistion 0.1 and k = 0, 0.1, 0.2, 0.3. Using the BPO method to solve problem $\wp_2$, we can obtain the estimation value of k_i by means of recursive least-squares algorithm.

The algorithm has a simple form and is very convenient for computation. The comparison of the estimation values and the true values which are obtained in Constantinides et al (1970) is shown in Table 1. It can be seen that the estimation values of the unknown parameters are very close to the true values.

We examine the effect of the level of measurement error and the results for m = 20 are shown in Table 2 for k = 0, 0.1, 0.2 and 0.3. The results are relatively unaffected by the level of measurement error.

The method based on BPO can be also applied to the identification of time varying parameters. Numerical example is presented as follows:

Example 2: Identification of time-varying parameters.

Consider the following scalar system:

Table 1. Comparison of Estimation Values and True Values for k_1, k_2, k_3, k_4.

	m	estimation value	true value
k_1	5	13.39592	13.10
	10	13.16598	
	20	13.11585	
k_2	5	0.9380824	0.9426
	10	0.9420791	
	20	0.9427760	
k_3	5	4.625232	4.66
	10	4.666653	
	20	4.662251	
k_4	5	4.403931	4.4555
	10	4.462688	
	20	4.457976	

Table 2: Effect of Level of Measurement Error on k_i ($i = 1, 2, 3, 4$).

k	k_1	k_2	k_3	k_4
0	13.11585	0.942776	4.662251	4.457976
0.1	13.10194	0.921512	4.662290	4.561039
0.2	13.08739	0.900249	4.662205	4.668977
0.3	13.07211	0.878986	4.662175	4.782143

$$\dot{y}(t) = a(t)y(t) + b(t)y(t)^2 + c(t)u(t)$$

$$0 \le t \le 1$$

with a(t) = t, b(t) = t(t - cost) and c(t) = sint. The function u(t) is input and function y(t) is output.

Using the BPO technique and three different input t·sint (2.6), 2t (.1), t·exp(-t) (4.3) (the numbers in parentheses being the initial conditions), the problem $\wp_3$ can be solved. In such a case, $\theta(t) = (a(t), b(t), c(t))^T$ and the estimation value of $\theta(t)$ is $\theta_i = (a(t_i), b(t_i), c(t_i))^T$ at $t_i = iT/m$. The results are shown in Table 3 and the plots of a(t), b(t) and c(t) for m = 10 are illustrated in Fig. 2.

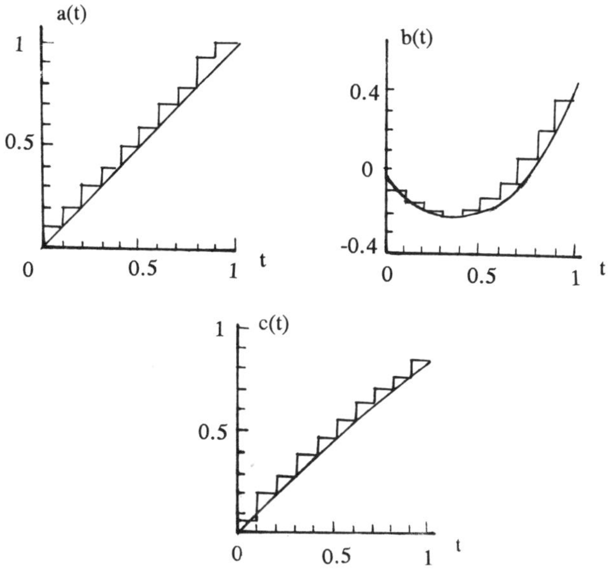

Fig. 2. Plots of a(t), b(t) and c(t) in Example 2

5. Identification of distributed parameter systems via block pulse operator

In this section, we consider the applications of BPO to identification of distributed parameter systems (DPS) (Wang and Jiang 1984, 1985).

Consider a nonlinear distributed parameter system described by the following

Table 3: Comparison of Estimation Values and True Values for Example 2

t_i	estimation values			true values
	m=10	m=50	m=100	
				a(t)
0.1	0.104906	0.100234	0.100258	0.1
0.2	0.200426	0.199985	0.200034	0.2
0.3	0.302079	0.300065	0.299990	0.3
0.4	0.390289	0.399409	0.399842	0.4
0.5	0.499122	0.500231	0.500168	0.5
0.6	0.582153	0.598954	0.600520	0.6
0.7	0.699737	0.700214	0.701050	0.7
0.8	0.786056	0.799576	0.801028	0.8
0.9	0.941241	0.902536	0.901611	0.9
1.0	1.229999	1.017630	1.005460	1.0
				b(t)
0.1	-0.095427	-0.089738	-0.089575	-0.089500
0.2	-0.015697	-0.156048	-0.156031	-0.156013
0.3	-0.201639	-0.196794	-0.196608	-0.196601
0.4	-0.205261	-0.208245	-0.208400	-0.208424
0.5	-0.192972	-0.189040	-0.188851	-0.188791
0.6	-0.130674	-0.134927	-0.135377	-0.135201
0.7	-0.053103	-0.045778	-0.045723	-0.045389
0.8	0.076825	0.082320	0.082254	0.082634
0.9	0.216107	0.248804	0.249912	0.250551
1.0	0.363654	0.453644	0.457991	0.459697
				c(t)
0.1	0.063978	0.098390	0.099680	0.099833
0.2	0.198227	0.198667	0.198623	0.198669
0.3	0.283546	0.295033	0.295482	0.295520
0.4	0.390434	0.389505	0.389403	0.389418
0.5	0.472025	0.479074	0.479348	0.479426
0.6	0.566705	0.564793	0.564493	0.564643
0.7	0.637391	0.643883	0.643906	0.644218
0.8	0.717243	0.717307	0.716996	0.717356
0.9	0.762062	0.782199	0.782716	0.783327
1.0	0.752335	0.834847	0.839346	0.841471

partial differential equation

$$a_1\frac{\partial u^{n_1}(x, t)}{\partial t^2} + a_2\frac{\partial u^{n_2}(x, t)}{\partial x \partial t} + a_3\frac{\partial u^{n_3}(x, t)}{\partial x^2} + a_4\frac{\partial u^{n_4}(x, t)}{\partial t}$$

$$+ a_5\frac{\partial u^{n_5}(x, t)}{\partial x} + a_6 u^{n_6}(x, t) = f^{n_7}(x, t) \tag{11}$$

$$0 \le x \le T_1,\ 0 \le t \le T_2,\ n_i > 0,\ i = 1, 2, \cdots, 7$$

where n_i $(i = 1, 2, \cdots, 7)$ are integers and a_i are unknown parameters. The systems involve the first-order and second-order linear DPS when the n_i $(i = 1, \cdots, 6)$ are chosen appropriate number. Given a record of output u(x, t) and f(x, t), the problems treated here are to solve the following identification problems:

(a) Estimate the unknown parameters a_i $(i = 1, \cdots, 6)$, when the initial and boundary conditions are known.
(b) Estimate a_i $(i = 1, \cdots, 6)$ and the initial conditions, when the boundary conditions are known.
(c) Estimate a_i $(i = 1, \cdots, 6)$ and boundary conditions, when the initial conditions are known.
(d) Estimste a_i $(i = 1, \cdots, 6)$ and the initial and boundary conditions.

In this section, we will first consider the deterministic observations and then noisy observations. The noisy measurements are given by

$$z(x, t) = u(x, t) - e(x, t)$$

where e(x, t) is the measurement error.

Using BPO method, an approximate identification problem described by algebraic equations in the image space of the BPO can be obtained. The unknown parameters characterizing the algebraic equations consist of the unknown model parameters and the unknown images of the initial and boundary conditions under the action of the BPO. This algebraic system is then solved to yield an recursive least square algorithm for identifying the parameters.

Method of Identification

Let

$$U^T \hat{=} B_2 u(x,\ t) = (U_1^T\ U_2^T\ \cdots\ U_{m_1}^T)$$

$$U_i^T = (u_{i1},\ \cdots,\ u_{im_2})$$

$$F^T \hat{=} B_2 f(x,\ t) = (F_1^T\ F_2^T\ \cdots\ F_{m_1}^T)$$

$$F_i^T = (f_{i1},\ \cdots,\ f_{im_2})$$

$$Z^T \hat{=} B_2 z(x,\ t) = (Z_1^T\ Z_2^T\ \cdots\ Z_{m_1}^T)$$

$$Z_i^T = (z_{i1},\ \cdots,\ z_{im_2})$$

Integrating (11) twice with respect to t and twice with respect to x, we have

$$a_1\int_0^x\int_0^x u^{n_1}(x,\ t)dxdx - a_1\int_0^x\int_0^x (u^{n_1}(x,\ 0) + t\frac{\partial u^{n_1}(x,\ t)}{\partial t}\bigg|_{t=0})dxdx$$

$$+ a_2\int_0^x\int_0^t u^{n_2}(x,\ t)dtdx - a_2 x\int_0^t u^{n_2}(0,\ t)dt - a_2\int_0^x\int_0^t\int_0^x \frac{\partial u^{n_2}(x,\ t)}{\partial x}\bigg|_{t=0} dxdtdx$$

$$+ a_3\int_0^t\int_0^t u^{n_3}(x,\ t)dtdt - a_3\int_0^t\int_0^t (u^{n_3}(0,\ t) + x\frac{\partial u^{n_3}(x,\ t)}{\partial x}\bigg|_{x=0})dtdt$$

$$+ a_4\int_0^x\int_0^x\int_0^t u^{n_4}(x,\ t)dtdxdx - a_4 t\int_0^x\int_0^x u^{n_4}(x,\ 0)dxdx$$

$$+ a_5\int_0^x\int_0^t\int_0^t u^{n_5}(x,\ t)dtdtdx - a_5 x\int_0^t\int_0^t u^{n_5}(0,\ t)dtdt$$

$$+ a_6\int_0^x\int_0^x\int_0^t\int_0^t u^{n_6}(x, t)dtdtdxdx = \int_0^x\int_0^x\int_0^t\int_0^t f^{n_7}(x, t)dtdtdxdx \qquad (12)$$

If $\partial u^{n_2}(x, t)/\partial x$ is continuous at the point $t = 0$, then

$$\int_0^x\int_0^t\int_0^x \left.\frac{\partial u^{n_2}(x, t)}{\partial x}\right|_{t=0} dxdtdx = t\int_0^x u^{n_2}(x, 0)dx - xtu^{n_2}(0, 0)$$

Let

$$h^T \hat{=} B_1 u(0, t) = (h_1, h_2, \cdots, h_{m_2})$$

$$g^T \hat{=} B_1 u(x, 0) = (g_1, g_2, \cdots, g_{m_1})$$

$$k^T \hat{=} B_1 \left.\frac{\partial u^{n_3}(x, t)}{\partial x}\right|_{x=0} = (k_1, k_2, \cdots, k_{m_2})$$

$$w^T \hat{=} B_1 \left.\frac{\partial u^{n_1}(x, t)}{\partial t}\right|_{t=0} = (w_1, w_2, \cdots, w_{m_1})$$

$$Y_{ij} \hat{=} (B_2 f^{n_7}(x, t)P_t^2 P_x^2)_{ij}, \quad P_x \hat{=} P_{x_1}, \quad P_t \hat{=} P_{x_2}$$

$$G_{ij}^{(1)} \hat{=} (B_2 u^{n_1}(x, t)P_x^2)_{ij}; \quad G_{ij}^{(2)} \hat{=} (B_2 u^{n_2}(x, t)P_t P_x)_{ij}$$

$$G_{ij}^{(3)} \hat{=} (B_2 u^{n_3}(x, t)P_t^2)_{ij}; \quad G_{ij}^{(4)} \hat{=} (B_2 u^{n_4}(x, t)P_t P_x^2)_{ij}$$

$$G_{ij}^{(5)} \hat{=} (B_2 u^{n_5}(x, t)P_t^2 P_x)_{ij}; \quad G_{ij}^{(6)} \hat{=} (B_2 u^{n_6}(x, t)P_t^2 P_x^2)_{ij}$$

$$H_i^{(1)} \hat{=} \sqrt{T_2/m_2}(B_1 u^{n_1}(x, 0)P_{m_1}^2)_i$$

$$H_{ij}^{(2)} \hat{=} (T_2/m_2)^{3/2}\frac{2j-1}{2}\left(B_1\left.\frac{\partial u^{n_1}(x,\ t)}{\partial t}\right|_{t=0} P_{m_1}^2\right)_i$$

$$H_{ij}^{(3)} \hat{=} (T_2/m_2)^{3/2}\frac{2j-1}{2}\left(B_1 u^{n_2}(x,\ 0)P_{m_1}\right)_i$$

$$H_{ij}^{(4)} \hat{=} (T_2/m_2)^{3/2}\frac{2j-1}{2}\left(B_1 u^{n_4}(x,\ 0)P_{m_1}^2\right)_i$$

$$L_{ij}^{(1)} \hat{=} (T_1/m_1)^{3/2}\frac{2i-1}{2}\left(B_1 u^{n_2}(0,\ t)P_{m_2}\right)_j$$

$$L_{ij}^{(2)} \hat{=} (T_1/m_1)^{1/2}\left(B_1 u^{n_3}(0,\ t)P_{m_2}^2\right)_j$$

$$L_{ij}^{(3)} \hat{=} (T_1/m_1)^{3/2}\frac{2i-1}{2}\left(B_1 u^{n_5}(0,\ t)P_{m_2}^2\right)_j$$

$$L_{ij}^{(4)} \hat{=} (T_1/m_1)^{3/2}\frac{2i-1}{2}\left(B_1\left.\frac{\partial u^{n_3}(x,\ t)}{\partial x}\right|_{x=0} P_{m_2}^2\right)_j$$

$$\alpha_j^T \hat{=} -\sqrt{T_1/m_1}(T_2/m_2)^2 k_0^{n_3-1}(j-1,\ j-2,\ \cdots,\ 1,\ 1/4)$$

$$\beta_j^T \hat{=} -(T_1/m_1)^2\sqrt{T_2/m_2}k_0^{n_1-1}(i-1,\ i-2,\ \cdots,\ 1,\ 1/4)$$

$$\gamma_{ij}^T \hat{=} -(T_1/m_1)^2(T_2/m_2)^{3/2}\frac{2j-1}{2}(i-1,\ i-2,\ \cdots,\ 1,\ 1/4)$$

$$\sigma_{ij}^T \hat{=} -(T_1/m_1)^{3/2}(T_2/m_2)^2\frac{2i-1}{2}(j-1,\ j-2,\ \cdots,\ 1,\ 1/4)$$

$$d_{ij} \hat{=} k_0^3 \frac{(2i-1)(2j-1)}{4}$$

$$\tilde{g}_i \hat{=} a_1 g_i^{n_1}, \quad \tilde{u}_0 \hat{=} a_2 u^{n_2}(0, 0)$$

$$\tilde{w}_i \hat{=} a_1 w_i - a_2(\mathrm{B}_1 u^{n_2}(x, 0) P_{m_1}^{-1})_i - a_4(\mathrm{B}_1 u^{n_4}(x, 0))_i$$

$$\tilde{h}_j \hat{=} a_3 h_j^{n_3}$$

$$\tilde{k}_j^T \hat{=} a_2(\mathrm{B}_1 u^{n_2}(0, t) P_{m_2}^{-1})_j - a_3 k_j - a_5(\mathrm{B}_1 u^{n_5}(0, t))_j$$

$$i = 1, 2, \cdots, m_1, \; j = 1, 2, \cdots, m_2$$

Using the rule 8, we have

$$\mathrm{B}_2 u^{n_i}(x, t) = k_0^{n_i - 1}(u_{11}^{n_i}, \cdots, u_{m_1 m_2}^{n_i}), \; i = 1, \cdots, 7 \qquad (13)$$

By means of Rule 8, Rule 9, Rule 12 and above relations, we may take the block pulse operator B_2 of (12) to yield:

(a) If the initial and boundary conditions are known, then

$$A_{ij}^T \theta^{(1)} \dot{=} Y_{ij}, \quad i=1,\cdots,m_1; \quad j=1,\cdots,m_2 \qquad (14)$$

where

$$\theta^{(1)} \hat{=} (a_1, a_2, \cdots, a_6)^T \qquad A_{ij}^T \hat{=} (A_{1ij}, A_{2ij}, \cdots, A_{6ij})$$

$$A_{1ij} = G_{ij}^{(1)} - H_{ij}^{(1)} - H_{ij}^{(2)} \qquad A_{2ij} = G_{ij}^{(2)} - L_{ij}^{(1)} - H_{ij}^{(3)} - d_{ij} u^{n_2}(0, 0)$$

$$A_{3ij} = G_{ij}^{(3)} - L_{ij}^{(2)} - L_{ij}^{(4)} \qquad A_{4ij} = G_{ij}^{(4)} - H_{ij}^{(4)}$$

$$A_{5ij} = G_{ij}^{(5)} - L_{ij}^{(3)} \qquad\qquad A_{6ij} = G_{ij}^{(6)}$$

Utilizing Lemma 1 and (13), the A_{kij}, Y_{ij} ($k = 1, \cdots, 6, i = 1, \cdots, m_1$, $j=1,\cdots,m_2$) can be calculated recursively.

(b) If the boundary conditions are known, then

$$B_{ij}^T\theta_i^{(2)} \doteq Y_{ij}, \quad i=1,\cdots,m_1; \quad j=1,\cdots,m_2 \qquad (15)$$

where

$$\theta_i^{(2)} \hat{=} (a_1, a_2, \cdots, a_6, \tilde{g}_1, \cdots, \tilde{g}_i, \tilde{w}_1, \cdots, \tilde{w}_i, \tilde{u}_0)^T$$

$$B_{ij}^T \hat{=} (E_{ij}^T, \beta_i^T, \gamma_{ij}^T, d_{ij})$$

$$E_{ij}^T \hat{=} (G_{ij}^{(1)}, G_{ij}^{(2)}-L_{ij}^{(1)}, G_{ij}^{(3)}-L_j^{(2)}-L_{ij}^{(4)}, G_{ij}^{(4)}, G_{ij}^{(5)}-L_{ij}^{(3)}, G_{ij}^{(6)})$$

(c) If the initial conditions are known, then

$$C_{ij}^T\theta_j^{(3)} \doteq Y_{ij}, \quad i=1,\cdots,m_1; \quad j=1,\cdots,m_2 \qquad (16)$$

where

$$\theta_j^{(3)} \hat{=} (a_1, a_2, \cdots, a_6, h_1, \cdots, h_j, k_1, \cdots, k_j, u_0)^T$$

$$C_{ij}^T \hat{=} (Q_{ij}^T, \alpha_j^T, \sigma_{ij}^T, d_{ij})$$

(d) If the initial and boundary are unknown, then

$$D_{ij}^T\theta_{ij}^{(4)} \doteq Y_{ij}, \quad i=1,\cdots,m_1; \quad j=1,\cdots,m_2 \qquad (17)$$

where

$$\theta_{ij}^{(4)} \hat{=} (a_1, a_2, \cdots, a_6, g_1, \cdots, g_i, h_1, \cdots, h_j, w_1, \cdots, w_i, k_1, \cdots, k_j, u_0)^T$$

$$D_{ij}^T \hat{=} (G_{ij}^{(1)}, \cdots, G_{ij}^{(6)}, \beta_i^T, \alpha_j^T, \gamma_{ij}^T, \sigma_{ij}^T, d_{ij})$$

The equations (17) can be written in the form:

$$D \cdot \theta^{(4)}_{m_1 m_2} = Y \qquad (18)$$

where Y is a column vector of dimension $m_1 m_2$ consisting of elements from the right-hand side of equations (17); D is a matrix of dimension $m_1 m_2 \times (7+2m_1+2m_2)$ formed from elements of the left-hand side of equations (17).

The above equations (14), (15), (16) and (18) can be solved for $\theta^{(1)}$, $\theta^{(2)}_{m_1}$, $\theta^{(3)}_{m_2}$ and $\theta^{(4)}_{m_1 m_2}$ using a least-squares technique or recursive least-squares algorithm, respectively. Once these values are determined, then u(0, t), u(x, 0), $\left.\frac{\partial u^{n_3}(x, t)}{\partial x}\right|_{x=0}$, $\left.\frac{\partial u^{n_1}(x, t)}{\partial t}\right|_{t=0}$ *are given by*

$$u(0, t) \doteq h^T \cdot \Phi_1(t) \qquad \left.\frac{\partial u^{n_3}(x, t)}{\partial x}\right|_{x=0} \doteq k^T \cdot \Phi_1(t)$$

$$u(x, 0) \doteq g^T \cdot \Phi_1(x) \qquad \left.\frac{\partial u^{n_1}(x, t)}{\partial t}\right|_{t=0} \doteq w^T \cdot \Phi_1(x)$$

where $\Phi_1(t)$ or $\Phi_1(x)$ is one-dimensional block-pulse vector (Wang and Jiang 1984).

Numerical examples

The computational examples are presented here to illustrate the method proposed in this section.

Example 3: Consider the following DPS:

$$\left.\begin{aligned} & a_1 \frac{\partial u}{\partial x} + a_2 \frac{\partial u}{\partial t} + a_3 u = f(x, t) \\ & u(0, t) = sint \\ & u(x, 0) = 0 \end{aligned}\right\} \qquad (19)$$

$$f(x, t) = t + \sin(t+\pi/6) + (1+\sqrt{3}t)x/2 \qquad (20)$$

$$0 \le x \le 1,\ 0 \le t \le 4$$

We want to estimate the dimensionless unknown parameters a_i $(i = 1, 2, 3)$ from measurements. The measurements are generated by

$$z(x, t) = u^*(x, t)(1 + kGAUSS(0, 0.1)) \qquad (21)$$

where $u^*(x, t)$ is the exact solution of (19) with $a_1=0.5$, $a_2=1$, $a_3=0.866$ and GAUSS(0, 0.1) is a normally distributed random variable with mean 0 and standard deviation 0.1.

By means of the BPO method and recursive least squares method, the $\theta = (a_1, a_2, a_3)^T$ can be estimated recursively. First, we examine the effect of the level of measurement error. The results for $m_1=5$ and $m_2=10$ are shown in Table 4 for $k=0, 0.1, 0.2$ and 0.3. The effect of the number of locations was studied next for $k=0$, $N_2=m_2=10$, and $m_1=5$, the results of which are shown in Table 5. The results are relatively unaffected by the number of locations. The effect of number of measurement times for $N_1=m_1=2$, $k=0$ and $N_2=m_2$, is shown in Table 6. It can be seen that as N_2 decreases the error increases.

Table 4: Effect of Level of Measurement Error on θ

k	$\hat{a}_1$	$\hat{a}_2$	$\hat{a}_3$
0	0.4949407	0.9992491	0.8687934
0.1	0.5192197	1.042198	0.8714733
0.2	0.545137	1.087581	0.8742046
0.3	0.5727175	1.135325	0.8772891

Table 5: Effect of the Number of Locations on θ

N_1	$\hat{a}_1$	$\hat{a}_2$	$\hat{a}_3$
1	0.4941427	1.000572	0.8662676
2	0.4942382	1.000401	0.8670865
3	0.4941898	0.9996961	0.8683941
4	0.4949407	0.9992491	0.8687934
5	0.4949407	0.9992491	0.8687934

Table 6: Effect of the Number of Measurement Times on θ

N_2	$\hat{a}_1$	$\hat{a}_2$	$\hat{a}_3$
5	0.4800166	0.9980812	0.8761826
10	0.4952089	0.9997671	0.8682427
20	0.4988141	0.9999559	0.8665432

Example 4: Consider the following model encountered in heat transfer systems (Yoshimura and Campo, 1982)

$$\left.\begin{aligned} a_1\frac{\partial u}{\partial t} &= a_2\frac{\partial^2 u}{\partial x^2} + a_3 u + f(x, t) \\ u(0, t) &= 1 \\ u(x, 0) &= 1 \\ \left.\frac{\partial u}{\partial x}\right|_{x=0} &= 0 \end{aligned}\right\} \quad (22)$$

$$f(x, t) = (1-3t)\sin^2(\frac{\pi x}{2}) - \pi^2 t\cos(\frac{\pi x}{2}) - 3 \quad (23)$$

$$0 \le x \le 2,\ 0 \le t \le 4$$

Suppose we are going to solve for the unknown parameters a_1, a_2 and a_3, knowing only the input f(x, t) and the measurement z(x, t). As before, the noisy observations are generated by

$$z(x, t) = u^*(x, t)(1 + kGAUSS(0, 0.1)) \quad (24)$$

where u^* is the solution of (22) with $a_1=1$, $a_2=2$, and $a_3=3$.

The effect of the level of measurement error was studied with $N_1=m_1=15$, $N_2=m_2=10$. The results are shown in Table 7.

The effect of the number of locations, for $N_1=m_1$, $N_2=m_2=10$ and $k=0$, is shown in Table 8. It can be seen that as N_1 decreases the estimate error increase. The effect of measurement times, for $N_1=m_1=10$, $k=0$ and $N_2=m_2$, is shown in Table 9. The results are relatively unaffected by the number of measurement times.

Table 7: Effect of Level of Measurement Error on θ

k	$\hat{a}_1$	$\hat{a}_2$	$\hat{a}_3$
0	1.031756	1.955256	3.015132
0.1	1.077308	2.028167	3.09947
0.3	1.248432	2.18627	3.302645

Table 8: Effect of the Number of Locations on θ

N_1	$\hat{a}_1$	$\hat{a}_2$	$\hat{a}_3$
5	1.305004	1.60741	3.13469
10	1.072874	1.89969	3.033085
20	1.017726	1.974799	3.008495

Table 9: Effect of the Number of Measurement Times on θ

N_2	$\hat{a}_1$	$\hat{a}_2$	$\hat{a}_3$
5	1.073479	1.899668	3.034357
20	1.072746	1.899694	3.034148
30	1.072722	1.89969	3.034142

In order to examine the effect of nonlinear, we consider the quasilinear DPS.

Example 5: Consider the DPS

$$a_1\frac{\partial u}{\partial x} + a_2\frac{\partial u}{\partial t} + a_3u^2 = f(x, t) \qquad (25)$$

$$u(0, t) = sint, \; u(x, 0) = 0 \qquad (26)$$

$$f(x, t) = 2t + (1+4t)x + cost + 4sint \qquad (27)$$

with $a_1=1$, $a_2=2$ and $a_3=4$. As before, we examine the effect of the level of measurement error for $N_1=m_1=N_2=m_2=5$. The results are shown in Table 10. ($T_1=2$, $T_2=4$).

Table 10: Effect of Level Measurement Error on θ

k	$\hat{a}_1$	$\hat{a}_2$	$\hat{a}_3$
0	0.9675305	2.003835	4.00395
0.1	1.002559	2.076308	4.184279
0.3	1.081075	2.236225	4.583341

Example 6: Consider the following quasilinear DPS

$$\left.\begin{aligned} & a_1\frac{\partial u(x,t)}{\partial x} + a_2\frac{\partial u(x,t)}{\partial t} + a_3u^2(x,t) = f(x,t) \\ & u(0,t) = 0 \qquad 0\le t\le 4 \\ & u(x,0) = 0 \qquad 0\le x\le 2 \end{aligned}\right\} \qquad (28)$$

$$f(x,t) = 2x + 4x^2t^2 + t \qquad (29)$$

We want to estimate the dimensionless unknown parameters a_i ($i = 1, 2, 3$) from measurements. The noisy measurements are generated by

$$z(x,t) = u^*(x,t)(1 - kGAUSS(0, 0.1)) \qquad (30)$$

where $u^*(x, t)$ is the exact solution of (28)-(29) with $a_1 = 1, a_2 = 2, a_3 = 4$.

By means of the BPO method proposed in this section, the $\theta = (a_1, a_2, a_3)^T$ can be estimated recursively. First, we examine the effect of the level of measurement error. The results for $m_1 = m_2 = 10$ are shown in Table 11. The effect of the number of locations was studied next for $k=0, m_1=m_2=10$, the results of which are shown in Table 12. The effect of number of measurement times, for $N_1 = m_1 = 10, N_2 = m_2$ and $k = 0$, is shown in Table 13.

Table 11: Effect of Level of Measurement Error on θ

k	$\hat{a}_1$	$\hat{a}_2$	$\hat{a}_3$
0	1.021467	2.042939	4.00831
0.1	1.045089	2.090186	4.195851
0.2	1.06983	2.13967	4.396869
0.3	1.095773	2.191552	4.612684

true value: $a_1=1, a_2=2, a_3=4$

Table 12: Effect of the Number of Locations on θ

N_1	$\hat{a}_1$	$\hat{a}_2$	$\hat{a}_3$
2	1.034298	1.882539	4.109793
4	1.032482	1.962098	4.030409
6	1.029828	1.997942	4.015731
8	1.026164	2.022643	4.01065

true value: $a_1=1$, $a_2=2$, $a_3=4$

Table 13: Effect of the Number of Measurement Times on θ

N_2	$\hat{a}_1$	$\hat{a}_2$	$\hat{a}_3$
5	0.9922123	2.219785	4.02183
10	1.021467	2.042939	4.00831
15	1.028673	2.004618	4.006006

true value: $a_1=1$, $a_2=2$, $a_3=4$

Example 7: Consider the hyperbolic nonlinear system (Bellman and Roth, 1979)

$$\left.\begin{array}{l} \dfrac{\partial u(x,\ t)}{\partial t} = au(x,\ t)\dfrac{\partial u(x,\ t)}{\partial t} \\ \\ 0 \le x \le 1,\ 0 \le t \le 3 \end{array}\right\} \qquad (31)$$

$$u(x,\ 0) = g(x) \qquad (32)$$

$$u(0,\ t) = 0 \qquad (33)$$

Given a record of u(x, t), the problem is to estimate unknown initial function g(x) and the parameter a.

Using the procedure outlined above, from (15) the estimates of a and g(x) can be obtained recursively. For simulation purposes, we let initial condition be g(x) = 0.3x and a = 1. In this case the exact solution is

$$u(x,\ t) = \frac{x}{(10/3\ -\ t)} \qquad (34)$$

The comparison of the estimation value and the true value for a is shown in Fig. 3, which shows that the result obtained by the BPO method is quite satisfactory.

The results of the initial condition are shown in Fig. 4 and Fig. 5, respectively. As shown by the curves in Fig. 4 and Fig. 5 the estimate error decrease as m_1 increases.

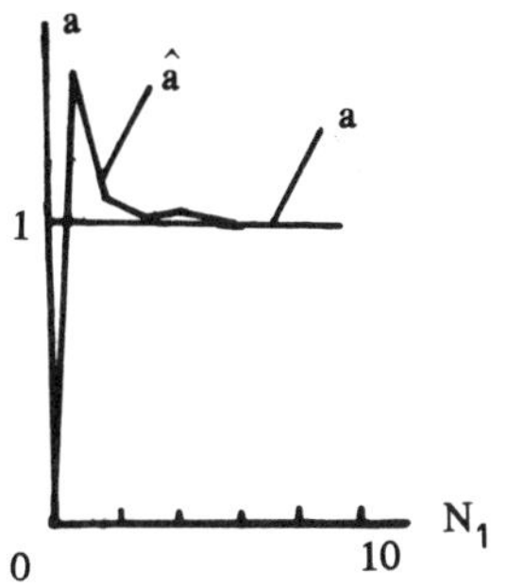

Fig. 3. Comparison of the estimation value and true value for a ($m_1=m_2=10$)

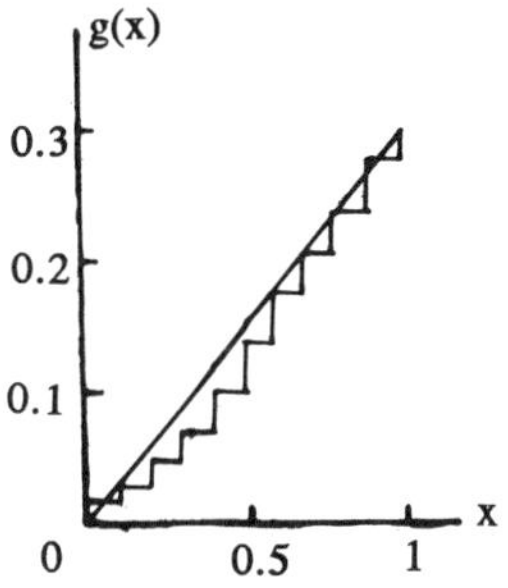

Fig. 4. Comparison of the estimation value and exact value for initial function g(x) ($m_1=m_2=10$)

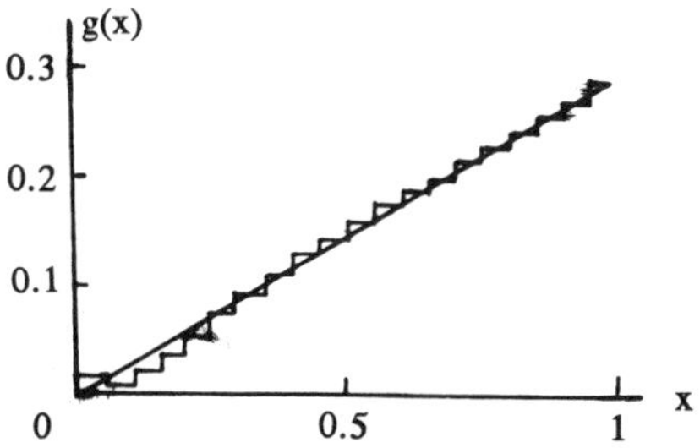

Fig. 5. Comparison of the estimation value and exact value for initial function g(x) ($m_1=20$, $m_2=10$)

6. Optimal inputs design for identifying parameter in dynamic systems via BPO

System parameters are usually identified by applying a known input signal and observing the system response. The estimation accuracy is enhanced by the use of optimal inputs. In recent years, the theory of the optimal inputs design (OID) for identifying parameters in dynamic systems has been successfully developed, both in the frequency domain and the time domain (Goodwin and Payne 1977, Mehra 1979, Kalaba and Spingarn 1982). The frequency domain method is easy to use but gives inputs which are useful only for long experiment durations since the thory is based on the steay-state assumption. In the time domain, the method leads to an iterative algorithm which computation is so complex that the application of the method is limited to low order time invariant systems. And so, the simplifications in the time domain synthesis are necessary. A large body of literature dealing with this problem is summarized in the survey papers by Mehra (1979), Goodwin (1982), and Rafajlowicz (1984).

In this section a new method of OID based on BPO is considered. The basic idea of this method is to convert the problem of OID for identifying parameters in original input space into the approximation problem corresponding with the OID in the image space of BPO. In the image space, the algorithems of solving for approximation problem may be considerably simplified and that it can be used both in long and short experiment durations. The convergence theorem of BPO solution of the approximation problem to the exact solution of original problem is given too.

It is well known, frequency domain method is based on Laplace operator and a set of orthogonal functions--sinusoidal functions while the method developed in this section is based on BPO and a set of orthogonal functions -- block pulse functions. In the frequency domain, optimal inputs should be expressed in terms of sinusoidal functions while the inputs discussed in this section will be expressed by means of block pulse functions. Hence, in a sense, this method is also a "frequency domain" design method and that may be called "generalized frequency domain method".

Consider a nonlinear system

$$\left.\begin{aligned} \dot{x}(t) &= f(x(t), u(t), t, \theta) \qquad t \in [0, T] \\ x(0) &= x_0 \end{aligned}\right\} \qquad (35)$$

$$y(t) = D(\theta)x(t) + v(t) \qquad (36)$$

where $x(t) \in R^n$, $u(t) \in R^r$, $\theta \in R^p$ and $y(t) \in R^N$. $D(\theta)$ is Nxn matrix function of unknown parameter θ and differentiable. v(t) is zero mean Gaussian white noise process with known covariance H. Let f be continous with respect of all variations and differentiable with respect of x and θ and that satisfy a Lipschitz condition

$$| f(x, u, t) - f(y, v, t) |_{R^n} \le L_1(| u-v |_{R^r} + | x-y |_{R^n})$$

The information matrix M for the unknown parameter θ can be shown to be

$$M = \int_0^T \frac{\partial[D(\theta)x(t)]^T}{\partial\theta} H^{-1} \frac{\partial[D(\theta)x(t)]}{\partial\theta} dt \qquad (37)$$

Here we shall employ the standard design criterion

$$J(u) = -\ln(detM) \qquad (38)$$

The input constraint is

$$\int_0^T u^T(t)u(t)dt \le 1 \qquad (39)$$

The input design problem $\wp$ can be formulated as follows:

Problem $\wp$: Find u(t) so that minimize the criterion J and satisfy constraint (35)-(36) and (39).

In order to simplify the optimal input design, we convert the problem $\wp$ in original function space into the equivalent approximating problem $\wp_m$ in the image space of BPO. By means of BPO, the equations (35)-(36) can be approximated by a sequence of matrix algebraic equations and constraint (39) can be approximated by a algebraic relation. Then, the approximation problem $\wp_m$ can be constructed. By means of BPO, it is represented as a sum of block pulse series that

$$u(t) \cong U\overline{\Phi}(t) = \sum_{i=1}^{m} U_i \varphi_i(t)$$

and so the input constraint can be shown to be

$$\Omega \hat{=} \left\{U: \ \frac{T}{m}\sum_{i=1}^{m} U_i^T U_i \le 1\right\} \qquad (40)$$

Now the approximation problem can be formulated as follows:

Problem $\wp_m$: Find $U = (U_1 \mid U_2 \mid \cdots \mid U_m) \in R^{r \times m}$ so that minimize the

$$\min\{J_m(U) = -\ln(\det\overline{M})\}$$

subject to the constraints (8) or (6) and (40). Where $\overline{M}$ is an approximation information matrix corresponding to information matric M.

Representative of Approximation Information Matrix

Let

$$x_{\theta_i} \hat{=} \frac{\partial x}{\partial \theta_i}, \quad f_{\theta_i} = \frac{\partial f}{\partial \theta_i}, \quad f_x = \frac{\partial f}{\partial x}$$

$$z(t) = (x^T(t) \;\; x_{\theta_1}^T \;\; \cdots \;\; x_{\theta_p}^T)^T$$

$$= (z_1(t), \cdots, z_{n(p+1)}(t))^T$$

The derivatives of various matrices are evaluated at a nominal value of θ. From (35) we have that

$$\frac{d}{dt}(x_{\theta_i}) = f_x x_{\theta_i} + f_{\theta_i} \qquad (41)$$

and so

$$\left.\begin{aligned} \dot{z}(t) &= g(z(t), u(t), t, \theta) \\ z(0) &= z_0 = e_1 \otimes x_0 \end{aligned}\right\} \qquad (42)$$

where $e_1^T = (1, 0, \cdots, 0)_{n(p+1)\times 1}$

$$g(z(t), u(t), t, \theta) = \begin{bmatrix} f(x, u, t, \theta) \\ f_x x_{\theta_1} + f_{\theta_1} \\ \cdots \\ f_x x_{\theta_p} + f_{\theta_p} \end{bmatrix}$$

and function g satisfies a Lipschitz condition

$$| g(z, u, t, \theta) - g(w, v, t, \theta) |_{R^{n(p+1)}} \leq L_2(| u-v |_{R^r} + | z-w |_{R^{n(p+1)}})$$

Suppose m_{ij} is the ijth element of the information matrix M. From (37) we have

$$m_{ij} = \int_0^T \frac{\partial[D(\theta)x(t)]^T}{\partial\theta_i} H^{-1} \frac{\partial[D(\theta)x(t)]}{\partial\theta_j} dt$$

$$= \int_0^T z^T(t) E_i H^{-1} E_j^T z(t) dt \qquad (43)$$

where

$$E_i \hat{=} (D_{\theta_i} \mid 0 \mid \cdots 0 \mid \underset{\substack{\uparrow \\ i+1}}{D(\theta)} \mid 0 \mid \cdots \mid 0)^T \in R^{n(p+1)\times r}$$

Suppose **Z** is the solution of the following matrix equation

$$\left.\begin{aligned} Z_1 &= z_0 + \frac{T}{2m} G_1 \\ Z_i &= Z_{i-1} + \frac{T}{2m}(G_i + G_{i-1}) \quad i=2,\cdots,m \end{aligned}\right\} \qquad (44)$$

where

$$Z = [Z_1 \mid Z_2 \mid \cdots \mid Z_m],\ Z_i \in R^{n(p+1)},\ i=1,2,\cdots m$$

$$G = [G_1 \mid G_2 \mid \cdots \mid G_m],\ G_i = g(Z_i, U_i, \bar{t}_i, \theta)$$

Substituting $z(t) = Z\bar{\Phi}(t)$ in (43), we obtain a representative of approximation information matrix M:

$$\bar{m}_{ij} = \int_0^T \bar{\Phi}(t)^T Z^T E_i H^{-1} E_j^T Z \bar{\Phi}(t) dt$$

$$= \frac{T}{m} \sum_{k=1}^{m} Z_k^T Y_{ij} Z_k \qquad (45)$$

where

$$Y_{ij} \hat{=} E_i H^{-1} E_j^T$$

Theorem 5: (Wang 1989b) Consider the nonlinear system given by (42). If input u(t) is given and $U = \bar{B}u(t)$. Let z(t) be a unique solution of (42) and **Z** is a unique solution of (44). Then

$$\lim_{m\to\infty} \bar{m}_{ij} = m_{ij} \qquad \forall\ i, j$$

Corollary: (Wang 1989b) Under assumptions of Theorem 5 and

$$\left|\frac{dg}{dt}\right|_{R^n} < +\infty, \quad \left|\frac{du}{dt}\right|_{R^r} < +\infty$$

then

$$|m_{ij} - \bar{m}_{ij}| = O(m^{-1})$$

Convergence of Approximation solution

We first note that the Problem $\wp_m$ is equivalent to the following roblem:

Problem $\tilde{\wp}_m$: Find $U \hat{=} (U_1 \mid U_2 \mid \cdots \mid U_m) \in R^{r\times m}$ so that minimize the

$$\min\{J_m(U) = -\ln(\det\bar{M})\}$$

where

$$\bar{M} = (\bar{m}_{ij})_{p\times p}, \qquad \bar{m}_{ij} = \frac{T}{m}\sum_{k=1}^{m} Z_k^T Y_{ij} Z_k$$

subject to

$$\frac{T}{m}\sum_{i=1}^{m} U_i^T U_i \le 1$$

$$\left.\begin{aligned} Z_1 &= z_0 + \frac{T}{2m} G_1 \\ Z_i &= Z_{i-1} + \frac{T}{2m}(G_i + G_{i-1}) \quad i=2,\cdots,m \end{aligned}\right\}$$

$$G_i = g(Z_i, U_i, \bar{t}_i, \theta)$$

The problem may be solved by nonlinear programming method.

Theorem 6: (Wang 1989b) Consider the problem $\wp$ and problem $\tilde{\wp}_m$. Suppose that

(a) There exists a unique solution $u^*(t)$ to the problem $\wp$.

(b) There exists a unique solution U^* to the problem $\tilde{\wp}_m$.

Then (1) $J_m(U^*) \rightarrow J(u^*(t))$ as $m \rightarrow \infty$

(2) there exists a subsequence $\hat{u}_{m_k}(t) = (U^*_1 \mid \cdots \mid U^*)\Phi_{m_k m_k}(t)$ such that

$$|u^*(t) - \hat{u}_{m_k}(t)|_{L_2^r} \rightarrow 0 \quad as \ m_k \rightarrow \infty$$

For linear time-varying system (LTVS), we can obtain simple results. The OID can be reduced to find an eigenvector of the "response matrix" corresponding to its maximum eigenvalue. This method leads to a tractable computation algorithm for the LTVS and its convergence has also been proven. Without going into detail, the interested reader is referred to Wang (1984) and Wang and Jiang (1989) for details of the application of BPO method to the input design for identifying parameters in the LTVS.

Examples

The computational examples are presented here to illustrate the BPO method proposed in this section.

Example 8: Consider the following linear system:

$$\dot{x}(t) = -x(t) + \theta u(t) \qquad 0 \leq t \leq 1$$

$$x(0) = 0$$

$$y(t) = x(t) + v(t)$$

where x and u are scalars and θ is the unknown gain. v is a white Gaussian noise with $E\{v\} = 0$ and $E\{v(t)v(s)\} = r\delta(t-s)$.

Let the input be energy constrained as follows:

$$\int_0^T u^2(t)dt \leq 1$$

By means of the BPO method outlined above, the solution of the approximation problem $\wp_m$ can be obtained numerically. The exact solntion of the problem $\wp$ given by Mehra (1979). The comparisons of the BPS solution and the exact solution for the OID are shown in Fig. 6 (a), which show that the BPS solution are quite satisfactory.

Example 9: Consider the following linear time varying system:

$$\dot{x}(t) = -tx(t) + \theta u(t) \qquad 0 \le t \le 1$$

$$x(0) = 0$$

$$y(t) = x(t) + v(t)$$

All assumptions are the same as Example 8. As before, by BPO, the BPO solution can be obtained. The computational results are shown in Fig. 6 (b).

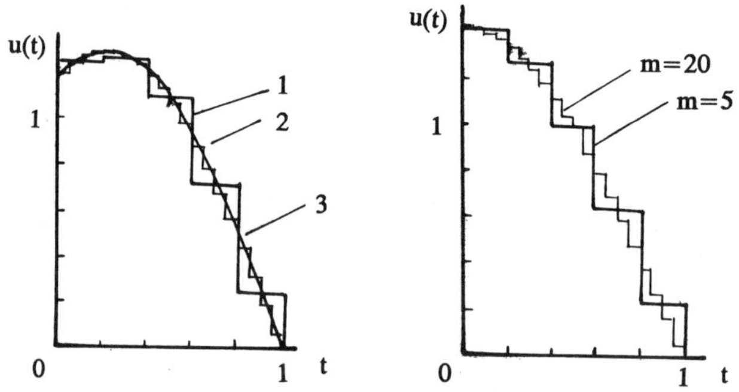

(a) Example 8 (b) Example 9

Fig. 6. The results of Example 8 and Example 9
1 -- BPO solution (m = 5)
2 -- BPO solution (m = 20)
3 -- exact solution

References

Bellman, R. and Roth, R.S. (1979), "Systems identification with partial information", *J. Math. Anal. Appl.*, Vol. 68, pp. 321-333.

Chen, W.L. and Lee, C.L. (1982), "On the convergence of the block-pulse series solution of a linear time-invariant system", *Int. J. Systems Sci.*, Vol. 13, pp. 491-498.

Constantinides, A., Spencer, J.L. and Gaden, E.L. (1970), "Optimization of batch fermentation processes (I), (II)", *J. Biotech. Bioeng.*, Vol. 12, pp. 803-830;

pp. 1081-1098.

Courant, R.F. and Hilbert, D. (1953), "*Methods of Mathematical Physics*",Vol. I, John Wiley & Sons, New York.

Curtain, R.F. and Pritchard, A.J. (1977), "*Functional Analysis in Modern Applied Mathematics*", Academic Press, New York.

Goodson, R.E. and Polis, M.P. (1976), "Parameter identification in distributed systems: a synthetizing overview", *Proc. IEEE*, Vol. 64, pp.45.

Goodwin, G.C. (1982), "An overview of the system identification problem experiment design", Proc. 6th IFAC Symp. Ident. Syst. Parameter Eastimation, (Washington D.C., 1982), pp. 919-924.

Goodwin, G.C. and Payne, R.L. (1977), "*Dynamic System Identification: Experiment Design and Analysis*", Academic Press.

Kalaba, R. and Spingarn, K. (1982). "*Control, Identification and Input Optimization*", Plenum Press. New York and London.

Kwong, C.P. and Chen, C.F. (1981), "The convergence properties of block-pulse series", *Int. J. Systems Sci.*, Vol. 12, pp. 745-751.

Mehra, R.K. (1979), "Choice of input signals", in "*Trends and Progress in System Identification*", Edited by P. Eykhoff.

Nath, A.K. and Lee, T.T. (1983), "On the multidimensional extension of block pulse functions and their completeness", *Int. J. Syst. Sci.*, Vol.14, pp. 201-208.

Rafajlowicz, E. (1984), "Optimum experiment design for parameter identification in distributed systems: brief survey and new results", Proc. 9th World Congress of IFAC, Vol. X, pp. 134-138.

Rao, G.P. (1983), "*Piecewise Constant Orthogonal Functions and Their Application to Systems and Control*", Springer-verlag, New York.

Rao, G.P. and Srinivasan, T. (1978), "Remarks on "author's reply" to "comments on 'Design of piecewise constant gains for optimal control via Walsh functions'", *IEEE Trans. Automat. Contr.*, Vol. 23, pp.762.

Saha, D.C. and Rao, G.P. (1983), "*Identification of Continuous Dynamical Systems*", Springer-Verlag, Berlin, Heidelberg.

Sannuti, P. (1977), "Analysis and synthesis of dynamic systems via BPFs", *Proc. IEE*, Vol. 124, pp.569-571.

Sinha, N.K. and Kuszta, B. (1983), "*Modeling and Identification of Dynamic Systems*", Von-Nostrand Reinhold, New York.

Unbehauen, H. and Rao, G.P. (1987), "*Identification of Continuous Systems*", Ameterdam: North-Holland.

Wang, S.Y. (1983), "Block pulse operator and its application in control theory (I), (II)", *Journal of East China Institute of Chemical Technology*, Vol 9, pp. 1-15; pp. 501-509.

Wang, S.Y. (1984), "The design of suboptimal inputs for identifyiny parameter in linear time-varying system", *IEEE Trans. Automat. Contr.*, Vol 29, pp. 633-636.

Wang, S.Y. (1989a), "The convergence of block pulse series approximation solution for nonlinear systems", Proc. Int. AMSE Conference "Signals & Systems", Vol. 2, pp. 83-94.

Wang, S.Y. (1989b), "A new method of optimal input design for identifing parameter in dynamic systems", Proc. International 89' Dalian Conference "Signal & Systems", (AMSE Press), Vol. 2, pp. 131-142.

Wang, S.Y. (1989c), "The convergence of block pulse operator algorithms for parameter identification problems in nonlinear continuous systems", 13th National Systems Conference, (Kharagpur, Dec. 13-15, Indian, 1989).

Wang, S.Y. (1990a), "Convergence of block pulse series approximation solution for optimal control problem", *Int. J. Systems Sci.*, Vol. 21, pp. 1355-1368.

Wang, S.Y. (1990b), "On multidimesional block pulse operator", Int. 90' Chengdu Conference "Signal & Systems", AMSE, (Oct. 8-10, 1990).

Wang, S.Y. (1990c), "On the block pulse operator method to solving nonlinear control system with time delay", IECON'90, IEEE Industrial Electronics Socicty, 1990.

Wang, S.Y. and Jiang, W.S. (1984). "The application of block pulse operator in identification of distributed parameter systems", A Bridge Between Control Science and Technology, 9th World Congress of IFAC, Vol. 2, pp. 655-660.

Wang, S.Y. and Jiang, W.S. (1985a), "Identification of nonlinear distributed parameter systems using block pulse operator", Proc. 7th IFAC Symposium on

Identification System Parameter Estimation, Vol. 1, pp. 803-808.

Wang, S.Y. and Jiang, W.S. (1985b), "Identification and optimal control of penicillin fermentation process -- an application of block pulse operator", Proc. International Conference on Industry Process Modelling and Control, Vol. II, pp. 163-170.

Wang, S.Y. and Jiang, W.S. (1989), "*Block Pulse Operator and its Applications*", Press of ECUCT, Shanghai, China.

Wu, J.M. and Wang, S.Y. (1988a), "An approach to joint problem of system identification and optimization for linear time delay system with unknown parameters via block pulse operator", Proc. Int. Confer. "Modelling & Simulation", AMSE, (Shenzhen, China, Nov. 7-9, 1988).

Wu, J.M. and Wang, S.Y. (1988b), "Optimal control for nonlinear time delay systems via block pulse operator", Proc. Int. Confer. "Modelling & Simulation", AMSE, (Shenzaen, China, Nov. 7-9, 1988).

Wu, J.M. and Wang, S.Y. (1989), "A new approach to model reference adaptive system using block pulse operator (BPO)", 13th National Systems Conference, (Kharagpur, Dec. 13-15, Indian, 1989).

Wu, J.M. and Wang, S.Y. (1990), "Analysis and optimal control of singular systems via bolck pulse operator (BPO)", Int. 90' Chengdu Conference "Signal & Systems", AMSE, (Oct. 8-10, 1990).

Yoshimura, Y. and Campo, A. (1982), "Identification on a heat transfer system using the Galerkin method", *Int. J. Syst. Sci.*, Vol. 13, pp. 247-255.

Zhu, J.M. and Lu, Y.Z. (1988), "Hierarchical optimal control for distributed parameter systems via block pulse operator", *Int. J. Control*, Vol. 48, pp. 685-703.

Recursive block pulse function method

Z. H. Jiang and W. Schaufelberger
Project Center IDA
Swiss Federal Institute of Technology (ETH)
CH-8092 Zürich, Switzerland

Abstract

A recursive block pulse function method for continuous-time model identification is presented. Based on the block pulse difference equations corresponding to the differential equation models of single-input, single-output linear systems, recursive algorithms developed in discrete-time model identification can be applied directly to estimate the parameters of the original differential equations without much modification. The recursive block pulse function method developed in the single-input, single-output case can also be applied flexibly and conveniently to the identification of certain other continuous-time systems, e.g. multi-input, multi-output linear systems, linear systems with time delays and Hammerstein model nonlinear systems.

1. Introduction

In the continuous-time model identification, methods based on various orthogonal functions and orthogonal polynomials, such as Walsh functions, block pulse functions, Chebyshev polynomials, Legendre polynomials and Laguerre polynomials, have been proposed. A common goal of these methods is to avoid the direct measure of time derivatives of the input and output signals which appear in the differential equation models of continuous-time systems. After expanding the continuous input and output signals by their approximate series and after introducing the integral operational matrices, linear algebraic equation sets can be obtained and the parameters of the differential equations can be estimated. Using these methods, satisfactory results for continuous-time model identification can be obtained.

But as a disadvantage, these methods can only be applied off-line in practice, because data preparation for parameter estimations becomes time-consuming. In establishing the algebraic equation sets, large computations have to be done either in the series expansions of continuous signals or in the multiplications of integral operational matrices. Due to this computational problem, these methods are essentially non-recursive, especially when the period of time for identification is long.

In order to avoid this disadvantage, block pulse difference equations corresponding to the differential equation models of single-input, single-output (SISO)

N. K. Sinha and G. P. Rao (eds.), Identification of Continuous-Time Systems, 205–225.

linear systems are derived in this paper. Using the difference equation forms to represent the systems, recursive algorithms well-developed in the discrete-time model identification can directly be applied to the continuous-time model identification problem without much modification. Following the simple procedures for the block pulse series expansions of continuous signals, the parameters of the original differential equations can easily be estimated from sampled data by means of digital computers. Therefore this recursive block pulse function method is suitable and efficient for continuous-time model identification.

The block pulse difference equations can also be extended to include certain systems which have kernels similar to the SISO linear systems, e.g. multi-input, multi-output linear systems, linear systems with time delays and Hammerstein model nonlinear systems. As in the discussion of the single-input, single-output case, these extended block pulse difference equations can also be applied flexibly and conveniently for the parameter estimations of the corresponding differential equation models.

2. Block pulse function method

Block pulse functions are a set of orthogonal functions with piecewise constant values. Usually, the set of block pulse functions is defined in the interval $[0, T)$ as:

$$\phi_i(t) = \begin{cases} 1 & \text{for } (i-1)T/m \leq t < iT/m \\ 0 & \text{otherwise} \end{cases} \qquad (i = 1, 2, \ldots, m) \tag{1}$$

where m is an arbitrary positive integer. For convenience of expressions, the width of block pulses is denoted by $h = T/m$ in this paper.

Compared with other orthogonal functions and orthogonal polynomials, the block pulse functions have a distinct property, i.e. they are disjoined with each other in the interval $[0, T)$:

$$\phi_i(t)\phi_j(t) = \begin{cases} \phi_i(t) & \text{for } i = j \\ 0 & \text{for } i \neq j \end{cases} \qquad (i, j = 1, 2, \ldots, m) \tag{2}$$

Due to this property, formulas about block pulse functions can be simplified considerably.

Based on the orthogonality of block pulse functions, an arbitrary real function $f(t)$, which is square integrable in the interval $[0, T)$, can be expanded into its block pulse series in the sense of minimizing the mean square error between $f(t)$ and its approximation:

$$f(t) \doteq \sum_{i=1}^{m} f_i \phi_i(t) = F^T \Phi(t) \tag{3}$$

where $\Phi(t)$ is the $m \times 1$ block pulse function vector:

$$\Phi(t) = \begin{pmatrix} \phi_1(t) & \phi_2(t) & \cdots & \phi_m(t) \end{pmatrix}^T \tag{4}$$

and F is the $m \times 1$ block pulse coefficient vector of $f(t)$:

$$F = \begin{pmatrix} f_1 & f_2 & \cdots & f_m \end{pmatrix}^T \tag{5}$$

The ith block pulse coefficient f_i can be determined by:

$$f_i = \frac{1}{h}\int_0^T f(t)\phi_i(t)dt = \frac{1}{h}\int f(t)dt\Big|_{(i-1)h}^{ih} \tag{6}$$

Equations (3) and (6) show that the original function $f(t)$ is approximated by a piecewise constant function, and the constant value in each subinterval is the mean value related to the area bounded by the function $f(t)$ in the same subinterval.

In order to simplify computations, the block pulse coefficients can also be approximated by:

$$f_i \doteq \frac{1}{2}\Big(f((i-1)h) + f(ih)\Big) \tag{7}$$

when the width of block pulses h is small enough. This means that each approximate block pulse coefficient is the mean value of the function at the two end points of the corresponding subinterval.

Operations on original functions can be approximated by operations on their block pulse series. In the problem of parameter estimation of differential equations, the operation rule of integration which transforms approximately the integration of a function into an algebraic operation (Wang 1982) plays an important role:

$$\underbrace{\int_0^t \cdots \int_0^t}_{k \text{ times}} f(t)\,dt \cdots dt \doteq F^T P_k \Phi(t) \tag{8}$$

where P_k is the kth generalized integral operational matrix. It has the form:

$$P_k = \frac{h^k}{(k+1)!}\begin{pmatrix} p_{k,1} & p_{k,2} & p_{k,3} & \cdots & p_{k,m} \\ 0 & p_{k,1} & p_{k,2} & \cdots & p_{k,m-1} \\ 0 & 0 & p_{k,1} & \cdots & p_{k,m-2} \\ \vdots & \vdots & \vdots & \ddots & \vdots \\ 0 & 0 & 0 & \cdots & p_{k,1} \end{pmatrix} \tag{9}$$

with entries:

$$p_{k,j} = \begin{cases} 1 & \text{for } j = 1 \\ j^{k+1} - 2(j-1)^{k+1} + (j-2)^{k+1} & \text{for } j = 2, 3, \ldots, m \end{cases} \tag{10}$$

For a brief description of the block pulse function method (Palanisamy and Bhattacharya 1981, Cheng and Hsu 1982), we use the following general form of a differential equation to characterize the single-input, single-output time invariant linear system:

$$\sum_{i=0}^{n} a_i y^{(i)}(t) = \sum_{i=0}^{n} b_i u^{(i)}(t) \qquad (a_n = 1) \tag{11}$$

where $u(t)$, $y(t)$ are input, output signals of the system, and a_i, b_i $(i = 0, 1, \ldots, n)$ are parameters of the model, respectively. After integrating the above differential equation n times successively from 0 to t on both sides, and after expanding the continuous signals into their block pulse series, the operation rule of integration and the disjoined property provide the relation between the block pulse coefficient vectors of the input and output signals:

$$\begin{aligned} &Y^T \sum_{i=0}^{n} a_{n-i} P_i - E^T \sum_{i=0}^{n-1} \left(y_0^{(i)} \sum_{j=i}^{n-1} a_{n+i-j} P_j \right) \\ &\doteq \; U^T \sum_{i=0}^{n} b_{n-i} P_i - E^T \sum_{i=0}^{n-1} \left(u_0^{(i)} \sum_{j=i}^{n-1} b_{n+i-j} P_j \right) \end{aligned} \tag{12}$$

where $y_0^{(i)}$ $(i = 0, 1, \ldots, n-1)$ are the initial values:

$$y_0^{(i)} = \left. \frac{d^i y(t)}{dt^i} \right|_{t=0} \tag{13}$$

and E is the $m \times 1$ constant vector with all its entries ones. From this linear algebraic equation set, the unknown parameters together with the initial values can be determined if the block pulse coefficients of the input and output signals are known.

The disadvantages of the block pulse function method can be seen clearly from equation (12) . Firstly, in the stage of data preparation, the evaluation of every single entry in the matrix products $Y^T P_k$, $U^T P_k$ and $E^T P_k$ $(k = 1, 2, \ldots, n)$ must always begin from the first entry of the vectors Y, U and E respectively due to the upper triangular forms of the generalized integral operational matrices P_k. The higher the position of an entry of these matrix products, the larger the computations that must be done. This difficulty becomes severe especially when a large amount of data is involved in the estimation problem. Therefore, this block pulse function method is not suitable for practical on-line identification, although recursive algorithms can be applied in the stage of parameter estimation. Secondly, the initial values must either be known *a priori* or be estimated together with the system parameters. In many cases, we do not care about these initial values, but their existence complicates the estimation procedures and increases the size of computations. Besides, both the off-line way of estimation and the initial values hinder the estimations to follow the slow changes of parameters, which happen usually in processes running over long time.

In order to avoid these disadvantages of the block pulse function method, we try to develop a recursive identification method via block pulse functions. The kernel of this recursive method is the block pulse difference equation.

3. Block pulse difference equations

Block pulse difference equations are derived using the properties of the generalized integral operational matrices (Jiang and Schaufelberger 1985a, 1985b). The properties involved in this derivation are:

Property 1. For the integers $k = 0, 1, \ldots, n$ and $l = 2, 3, \ldots$, the following equality holds:

$$\sum_{i=0}^{n}(-1)^i \binom{n}{i} p_{k,l+n-i} = 0 \tag{14}$$

Property 2. For the integers $k = 0, 1, \ldots, n$, the following equality holds:

$$\sum_{i=0}^{n}(-1)^i \binom{n}{i} p_{k,n-i+1} = (-1)^{n+k} \tag{15}$$

Property 3. For the integers $k = 0, 1, \ldots, n-1$ and $l = 1, 2, \ldots$, the following equality holds:

$$\sum_{i=0}^{n}\left[(-1)^i \binom{n}{i} \sum_{j=1}^{l+n-i} p_{k,j}\right] = 0 \tag{16}$$

As a basic and simple case, a SISO time invariant linear system is considered first, which is described by the differential equation (11). We start from the relation (12) which can be separated into m equations. If we denote the lth equation as $E_{(l)}$, it has the form:

$$\begin{aligned}
&\sum_{i=0}^{n}\left(\frac{h^i}{(i+1)!}a_{n-i}\sum_{j=1}^{l} y_j p_{i,l+1-j}\right) \\
&\quad -\sum_{i=0}^{n-1}\left(y_0^{(i)}\sum_{j=i}^{n-1}\left(\frac{h^j}{(j+1)!}a_{n+i-j}\sum_{c=1}^{l} p_{j,l+1-j-c}\right)\right) \\
&= \sum_{i=0}^{n}\left(\frac{h^i}{(i+1)!}b_{n-i}\sum_{j=1}^{l} u_j p_{i,l+1-j}\right) \\
&\quad -\sum_{i=0}^{n-1}\left(u_0^{(i)}\sum_{j=i}^{n-1}\left(\frac{h^j}{(j+1)!}b_{n+i-j}\sum_{c=1}^{l} p_{j,l+1-j-c}\right)\right)
\end{aligned} \tag{17}$$

Applying the operation $\sum_{k=0}^{n}(-1)^k \binom{n}{k} E_{(l+n-k)}$ on the $n+1$ successive equations $E_{(l)}$, $E_{(l+1)}, \ldots, E_{(l+n)}$ $(1 \leq l \leq m-n)$, the properties of the generalized integral operational matrices lead to:

$$\sum_{k=0}^{n} A_k y_{l+k} = \sum_{k=0}^{n} B_k u_{l+k} \tag{18}$$

where A_j, B_j $(j = 0, 1, \ldots, n)$ are linear combinations of the original parameters a_i, b_i $(i = 0, 1, \ldots, n)$, respectively. This equation expresses the relation between the input and output signals of a SISO linear system in an approximate piecewise manner, therefore it can be regarded as the block pulse difference equation corresponding to the differential equation (11). But this difference equation contains no derivatives of signals and no initial values, and the relation between the block pulse coefficients of the input and output signals is much simpler than (12) .

The following relation holds for the parameters a_i and A_j:

$$A_j = \sum_{i=0}^{n} \left(\frac{h^i}{(i+1)!} a_{n-i} \sum_{k=0}^{n-j} (-1)^k \binom{n}{k} p_{i,n-k-j+1} \right) \tag{19}$$

The parameters b_i and B_j also obey the same relation if we substitute B_j for A_j and b_i for a_i in (19). For example, the relation between the parameters a_i and A_j for a second order system is:

$$\begin{pmatrix} A_2 \\ A_1 \\ A_0 \end{pmatrix} = \begin{pmatrix} 1 & 1 & 1 \\ -2 & 0 & 4 \\ 1 & -1 & 1 \end{pmatrix} \begin{pmatrix} a_2 \\ \frac{h}{2} a_1 \\ \frac{h^2}{6} a_0 \end{pmatrix} \tag{20}$$

In fact, the relation between a_i and A_j can be obtained more easily without the complicated computations of (19). In the above example of a second order system, each column of the coefficient matrix in (20) can be determined directly from the second order difference of the first three entries of the matrices P_0, P_1 and P_2 respectively, as depicted in figure 1. As a general rule, the relation between a_i and A_j of an nth order system can be obtained from the nth order difference of the entries of the generalized integral operatoral matrices. The ith column of the coefficient matrix like the one in (20) can be determined from the first $n+1$ entries of the generalized integral operatoral matrix $p_{i-1,1}, p_{i-1,2}, \ldots, p_{i-1,n+1}$ as defined in (10).

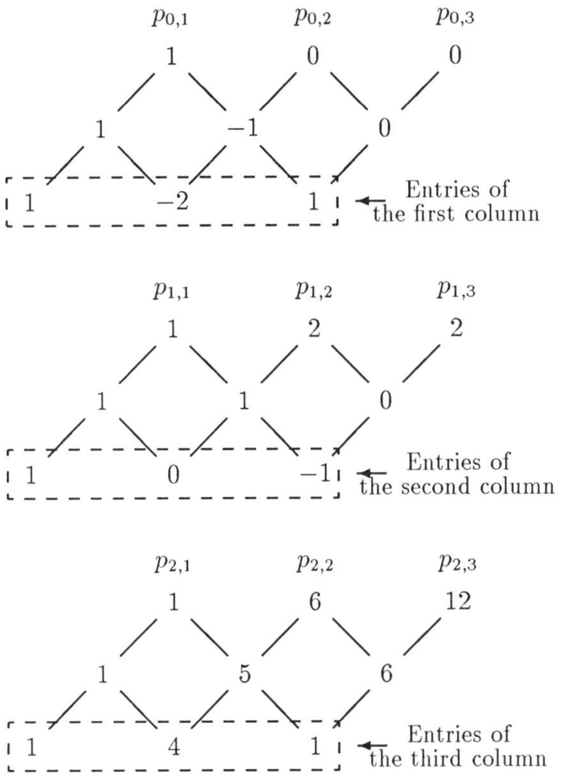

Figure 1: Computation of entries in coefficient matrix of (20).

We can also verify the relation between the parameters A_j $(j = 0, 1, \ldots, n)$ and a_n:

$$(n+1)a_n = \sum_{k=0}^{n} (-1)^{n-k} A_k \Big/ \binom{n}{k} \tag{21}$$

which will be used in the recursive block pulse function method below.

4. Recursive block pulse function method

From the block pulse difference equation derived above, several suitable equation forms can be obtained for solving the problem of identification. A monic difference equation results directly from (18):

$$y_{l+n} = -\sum_{k=0}^{n-1} C_k y_{l+k} + \sum_{k=0}^{n} D_k u_{l+k} \tag{22}$$

where

$$C_k = A_k/A_n, \qquad D_k = B_k/A_n \tag{23}$$

This is the first equation form. With $a_n = 1$, (21) gives:

$$A_n = (n+1) - \sum_{k=0}^{n-1}(-1)^{n-k} A_k \Big/ \binom{n}{k} \tag{24}$$

which leads to a regression equation:

$$x_{n,l} = \sum_{k-0}^{n-1} A_k x_{k,l} + \sum_{k=0}^{n} B_k w_{k,l} \tag{25}$$

In this second equation form, $x_{k,l}$ and $w_{k,l}$ ($k = 0, 1, \ldots, n$) are evaluated from the block pulse coefficients y_{l+j} and u_{l+j} ($j = 0, 1, \ldots, n$):

$$x_{k,l} = \begin{cases} (n+1)y_{l+k} & \text{for } k = n \\ (-1)^{n-k} \Big/ \binom{n}{k} y_{l+n} - y_{l+k} & \text{for } k = 0, 1, \ldots, n-1 \end{cases} \tag{26}$$

and

$$w_{k,l} = u_{l+k} \qquad \text{for } k = 0, 1, \ldots, n \tag{27}$$

Substituting (19) into (18), we obtain another regression equation as the third equation form which contains the parameters of the original differential equation:

$$z_{n,l} = -\sum_{k=0}^{n-1} a_k z_{k,l} + \sum_{k=0}^{n} b_k v_{k,l} \tag{28}$$

In this equation, $z_{k,l}$ and $v_{k,l}$ ($k = 0, 1, \ldots, n$) are linear combinations of the block pulse coefficients y_{l+j} and u_{l+j} ($j = 0, 1, \ldots, n$), respectively. The block pulse coefficients y_{l+j} and $z_{k,l}$ are connected by:

$$z_{k,l} = \frac{h^{n-k}}{(n-k+1)!} \sum_{j=0}^{n} \sum_{i=0}^{n-j} (-1)^i \binom{n}{i} p_{n-k,n-i-j+1} y_{l+j} \tag{29}$$

and the block pulse coefficients u_{l+j} and $v_{k,l}$ also have the same relation if we substitute $v_{k,l}$ for $z_{k,l}$ and u_{l+j} for y_{l+j} in (29). In this equation form, the relation between the block pulse coefficients y_{l+j} and $z_{k,l}$ for an nth order system can also be established using the rule of the nth order difference mentioned above, so that the complicated computations of (29) can be avoided. For example, for a second order system, this relation is:

$$\begin{pmatrix} z_{2,l} \\ \frac{2}{h} z_{1,l} \\ \frac{6}{h^2} z_{0,l} \end{pmatrix} = \begin{pmatrix} 1 & -2 & 1 \\ 1 & 0 & -1 \\ 1 & 4 & 1 \end{pmatrix} \begin{pmatrix} y_{l+2} \\ y_{l+1} \\ y_l \end{pmatrix} \tag{30}$$

Obviously, the coefficient matrix in (30) is the transpose matrix of the one in (20).

Based on these equation forms, the well-developed recursive algorithms for discrete-time model identification can be used without much modification. In the first case of using (22), the block pulse coefficients of the input and output signals u_{l+k}, y_{l+k} can directly be used in the recursive estimation algorithms, but the original parameters a_i and b_i must be computed extra. Once the parameters C_k and D_k are estimated, the original parameters can be determined from the linear algebraic equation set (19). In the second case of using (25), the discrete values $x_{k,l}$ and $w_{k,l}$ must first be computed from the block pulse coefficients of the input and output signals according to (26) and (27), and after estimating the parameters A_k and B_k, they must also be converted to the original parameters a_i and b_i. The third case of using (28) is much simpler. The parameters of the original differential equation can directly be estimated by the recursive estimation algorithms, but the discrete values $v_{k,l}$, $z_{k,l}$ must be computed extra from the block pulse coefficients of the input and output signals according to (29). These three schemes of the recursive block pulse function method are depicted in figure 2, where the computations added to the usual recursive algorithms of discrete-time model identification are marked by their equation numbers. Since the linear algebraic equation sets (19) and (29) are actually as simple as (20) and (30), only little additional computations are involved in these three schemes.

The recursive block pulse function method thus established does avoid the disadvantages of the block pulse function method which were mentioned in Section 2. In each recursion of the estimation, only a small amount of computations must be done, regardless whether the volumn of data involved in the estimation problems is small or large. The elimination of the initial values in the block pulse difference equations reduces the number of the unknowns, so that the computations in the estimation procedures are reduced. Moreover, the forgetting factor can now be introduced in the recursive algorithms. It enables the estimations to follow slow changes of parameters.

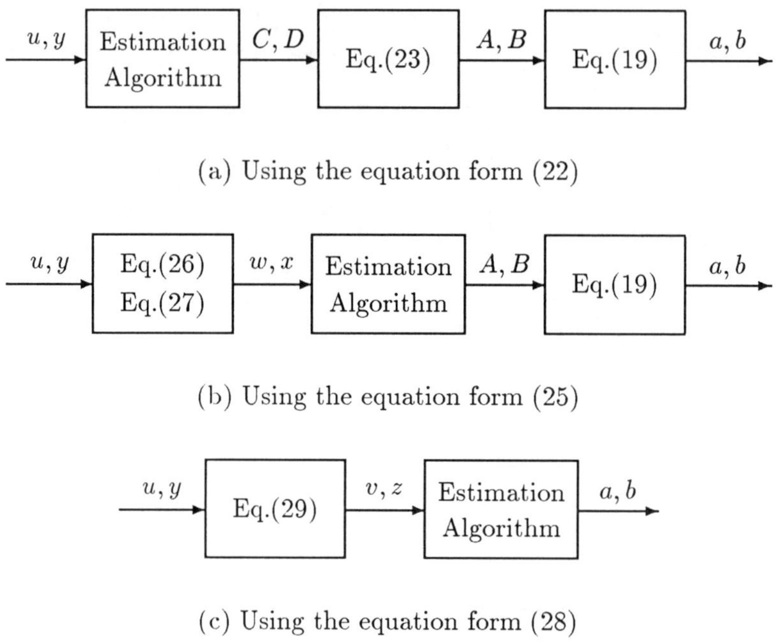

Figure 2: Schemes of recursive block pulse function methods.

In using the third equation form (28), we can directly utilize the *a priori* knowledge about the systems to reduce the number of the unknowns, because this equation form contains the same parameters as the original differential equation. If some terms do not exist in the original differential equation, the corresponding terms in (28) can be eliminated, e.g. the direct coupling between the input and output signals does not exist in most physical systems. If some parameters of the original differential equation are known *a priori* from their physical meaning, the corresponding terms can be included in the left hand side of (28). In these cases, the dimensions of vectors and matrices in the estimation algorithms will be reduced further. But in using the first equation form, we have no such advantage. In the block pulse difference equation (22), all the unknowns C_k $(k = 0, 1, \dots, n-1)$ and D_k $(k = 0, 1, \dots, n)$ always exist, no matter whether we have *a priori* knowledge about the systems or not. The reduction of the unknowns in the case of using the second equation form (25) is possible, but some transformations must be done in this equation. This problem will be discussed in the examples below.

The three equation forms used in the recursive block pulse function method have also different ways of updating data in the algorithms. In the first equation form, the discrete values are fed into the estimation algorithms sequentially. The difference equation form in this case enables us to apply directly ready-made programs which are written for the discrete-time model identification, e.g. the System Identification Toolbox in MATLAB (although only for batch estimations in this software). But such programs must be rewritten before they are used in the second and third equation forms, because the discrete values are fed into the estimation algorithms in parallel in these two cases. As an illustration, figure 3 depicts the different ways of data updating in the algorithms.

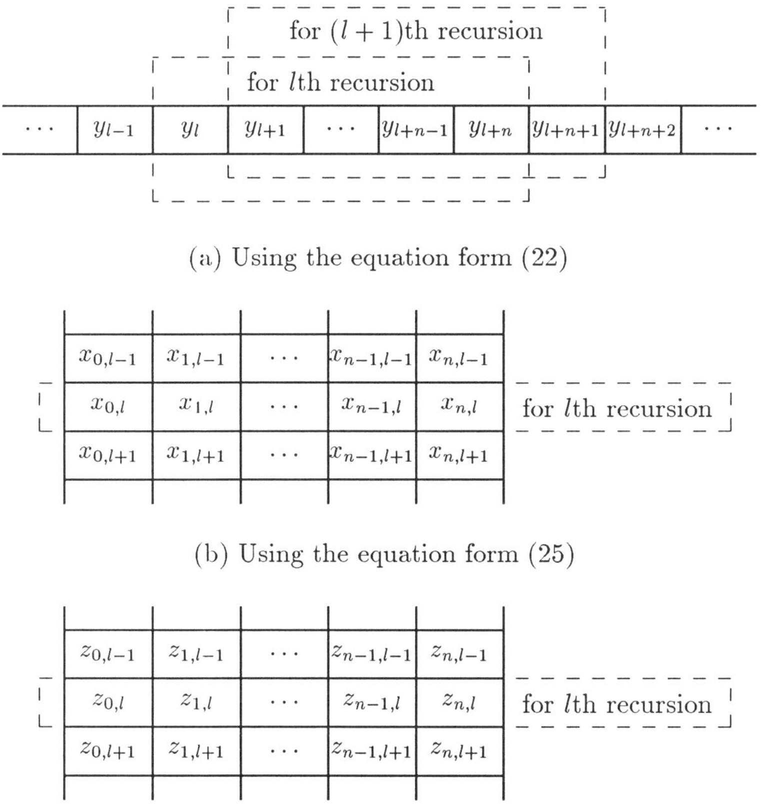

(a) Using the equation form (22)

(b) Using the equation form (25)

(c) Using the equation form (28)

Figure 3: Data updatings in recursive block pulse function methods.

The following three examples show the application of the recursive block pulse function method to continuous-time model identification. As a first example, we consider a second order system with the transfer function:

$$G(s) = \frac{b_0}{s^2 + a_1 s + a_0} \tag{31}$$

where the parameters are:

$$b_0 = 1.0, \qquad a_1 = 3.0, \qquad a_0 = \begin{cases} 2.0 & \text{for } t < 15 \\ 2.5 & \text{for } 15 \leq t < 35 \\ 2.0 & \text{for } t \geq 35 \end{cases} \tag{32}$$

For the identification problem, we assume that the system parameters are unknown, and they should be estimated from the sampled data of the system input and output. In this example, the input signal is $u(t) = \sin(t) + \sin(2t)$, and the sampling period is $h = 0.05$,

We use the second equation form (25) to solve this estimation problem. Since we have some *a priori* knowledge about the system, i.e. $b_2 = 0$ and $b_1 = 0$, we can get the relations $B_1 = 4B_0$ and $B_2 = B_0$ from (20), so that the number of unknowns is reduced to three, which is equal to the number of the original parameters in the transfer function model (31). Based on the regression equation:

$$3y_{l+2} = A_1\left(-\frac{1}{2}y_{l+2} - y_{l+1}\right) + A_0(y_{l+2} - y_l) + B_0(u_{l+2} + 4u_{l+1} + u_l) \tag{33}$$

the recursive least-squares estimation algorithm developed for discrete-time model identification (Ljung 1987) can be applied, in which the block pulse coefficients of the input and output signals are evaluated from the sampled data by the approximation formula (7), and the exponential forgetting factor is set to 0.95. The results of the estimations a_0, a_1 and b_0 are depicted in figure 4. It is clear that these estimations can follow the changes of the actual parameters in the system.

The results above show that the recursive block pulse function method of using the least squares algorithm can provide satisfactory estimations in continuous-time model identification, but in this example, no random noise is involved. The problem of the least squares algorithm in discrete-time model identification, i.e. the estimations become asymptotically biased if the equation residuals are correlated, especially in case of high noise levels, will also occur in the recursive block pulse function method. In order to demonstrate this problem, we give a second example, in which the system model, the input signal and the sampling period are the same as in the first example, but now the parameters are constant:

$$b_0 = 1.0, \qquad a_1 = 3.0, \qquad a_0 = 2.0 \tag{34}$$

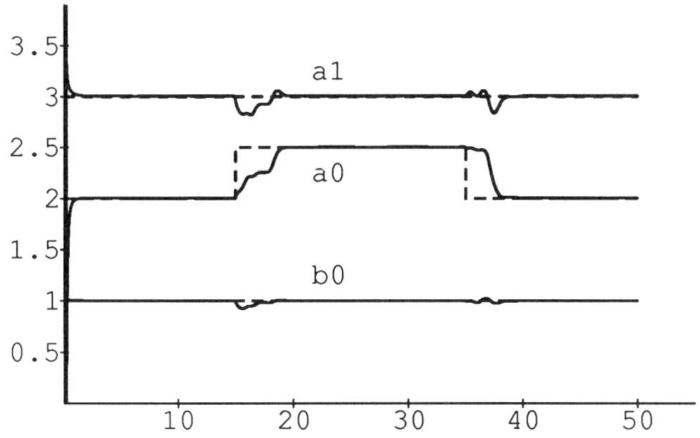

—— : estimated parameters, - - - : true parameters

Figure 4: Estimations tracing the changes of the system parameters.

A random white noise sequence $\{\bar{\epsilon}_k; k = 0, 1, \ldots\}$, which is normally distributed with mean 0.0 and variance 1.0, is added to the sampled output to simulate the noise corrupted output signal $\bar{y}_k$. The level of the noise is characterized by the noise-to-signal ratio:

$$N/S = \left(\frac{\sum \bar{\epsilon}_k^2}{\sum \bar{y}_k^2} \right)^{1/2} = 0.1 \tag{35}$$

We use the third equation form (28) to solve this estimation problem. From the knowledge about b_2 and b_1, it is now reduced to:

$$z_{2,l} = -a_1 z_{1,l} - a_0 z_{0,l} + b_0 v_{0,l} \tag{36}$$

In the lth recursion, the discrete values $(z_{2,l}, z_{1,l}, z_{0,l})$ and $v_{0,l}$ are evaluated respectively from the block pulse coefficients (y_{l+2}, y_{l+1}, y_l) and (u_{l+2}, u_{l+1}, u_l) by relation (30). The recursive least-squares algorithm yields the estimated parameters which are depicted in figure 5. Obviously, the estimations are strongly biased.

In discrete-time model identification, this biased estimation problem can be solved by many techniques, e.g. the different variants of instrumental variable methods (Söderström and Stoica 1983). Such methods can also be used with block pulse difference equations for the continuous-time model identification. Here, we use a simple variant of the instrumental variable method (Wouters 1972), in which the instrumental variables are constructed directly from the data of the input signal. In

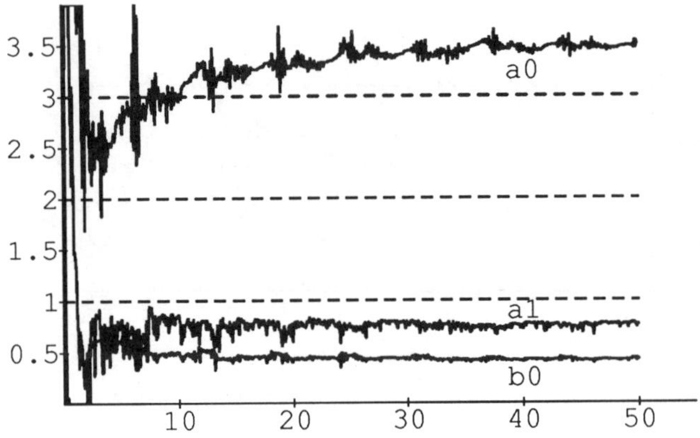

Figure 5: Least-squares algorithm yields the biased estimation results.

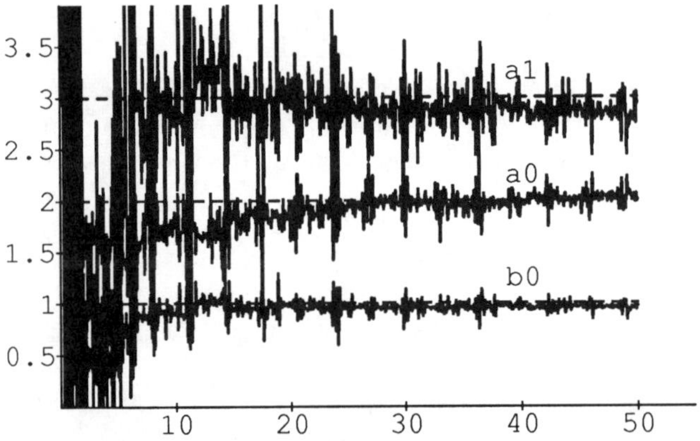

Figure 6: Instrumental variable algorithm improves the estimation results.

the case of using the recursive block pulse function method, the instrumental variables $(\hat{z}_{2,l}, \hat{z}_{1,l}, \hat{z}_{0,l})$ and $\hat{v}_{0,l}$ in the lth recursion are evaluated from the block pulse coefficients (u_{l+2}, u_{l+1}, u_l) and $(u_{l-1}, u_{l-2}, u_{l-3})$ by the relation (30). The parameters thus estimated are depicted in figure 6, in which the asymptotic bias is reduced.

As a prerequisite for applying all the block pulse function methods, the block pulse coefficients of the input and output signals must be known. Strictly speaking, these block pulse coefficients must be evaluated from the formula (6). But such evaluation is possible only when the analytical expressions of the continuous signals are known or when measurements between the sampling instants are available. In the practical identification by means of digital computers, only the data of signals at sampling instants are available. In such cases, the block pulse coefficients of the continuous signals must be approximated. In the above examples, we used the approximation formula (7). In fact, in the absence of any further information about the variations of signals between the sampling instants, this mean value of signals at the two end points of the corresponding subinterval is the simplest approximation.

We can also use other formulas to improve the approximation of block pulse coefficients from the sampled data of continuous signals. As an example, if we interpolate the value of a function $f(t)$ at the middle point of each subinterval via Stirling's interpolation formula with three points, we can improve the approximation of its block pulse coefficient:

$$f_k \doteq \frac{1}{16}\Big(7f(kh) + 10f((k-1)h) - f((k-2)h)\Big) \tag{37}$$

Since approximations like (37) are linear combinations of the samples of continuous signals, they can easily be evaluated in each recursion of the estimation in the recursive block pulse function method.

But weighing the increase of computations and the improvement of estimations resulted by the better approximations of block pulse coefficients of signals, we prefer to use the simplest formula (7) when identification problems are solved by the recursive block pulse function method (Jiang 1990). A more efficient way for improving the identification results via the recursive block pulse function method is to reduce the sampling period. According to the rule proposed by Haykin (1972), for identifying all the modes of a continuous-time system using the bilinear z-transformation method, it is suitable to choose the sampling period h to satisfy $|p_k h| \leq 0.5$, where p_k is the system pole farthest from the origin of the s-plane. Since the estimation results based on the block pulse difference equation method are better than the bilinear z-transformation method (Jiang 1990), this rule is also feasible for choosing the sampling period for the recursive block pulse funtion method.

As a demonstration of the practical application of the recursive block pulse function method, we estimate the parameters of the transfer function of an experimental electrical circuit, which is composed of operational amplifiers, resistances and capacitors, as depicted in figure 7. In the circuit, the resistances are measured as $R_{11} = 997.8\Omega$, $R_{12} = 997.0\Omega$, $R_{21} = 999.2\Omega$, $R_{22} = 1000.0\Omega$ and the capacitances are $C_1 = 35.85\mu F$, $C_2 = 47.95\mu F$, respectively. From the construction of the circuit, a second order system can be obtained which is described by the transfer function (31) with $a_0 = 583.48$, $a_1 = 48.83$ and $b_0 = 583.50$. Now, without measuring the values of the resistances and capacitors, the task of the experiment is to estimate the unknown parameters a_0, a_1 and b_0 from the measured data of the input

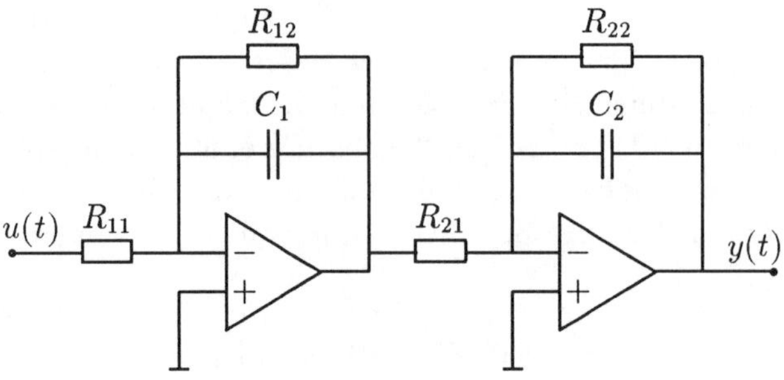

Figure 7: The experimental electrical circuit.

and output signals.

The system input is composed of two sine signals, which are produced by signal generators. They have the frequencies 2Hz, 3Hz and the amplitudes 3.0V, 3.5V, respectively. A SICOMP PC 16-05 personal computer, in which a Burr-Brown PCI-20000 system is installed, is used to fulfill the tasks of data acquisition and parameter estimation. For the A/D convertors, the full range of the analog signal is ±10V. The resolution of these convertors is 12 bits, and it determines the accuracy of the sampled data of the signals. Except for the quantization error in the data acquisition, no extra noise is introduced in this system. The sampling rate is about 75Hz.

Since the noise level is low in this experiment, the recursive least-squares algorithm based on the recursive block pulse function method is used for the parameter estimation. For each step of recursion, the time of about 12ms is needed for data acquisition and computation. The estimated parameters converge quickly as illustrated in figure 8. After about 20 recursions, the estimated parameters converge to the values $\hat{a}_0 = 586$, $\hat{a}_1 = 50$ and $\hat{b}_0 = 587$, with relative errors of about 0.4%, 2.4% and 0.6%, respectively.

5. Extensions of recursive block pulse function method

The recursive block pulse function method discussed in the previous sections can easily be extended to certain systems which have the same kernels as the SISO linear systems, e.g. multi-input, multi-output (MIMO) linear systems, linear systems with time delay and Hammerstein model nonlinear systems. Since these sys-

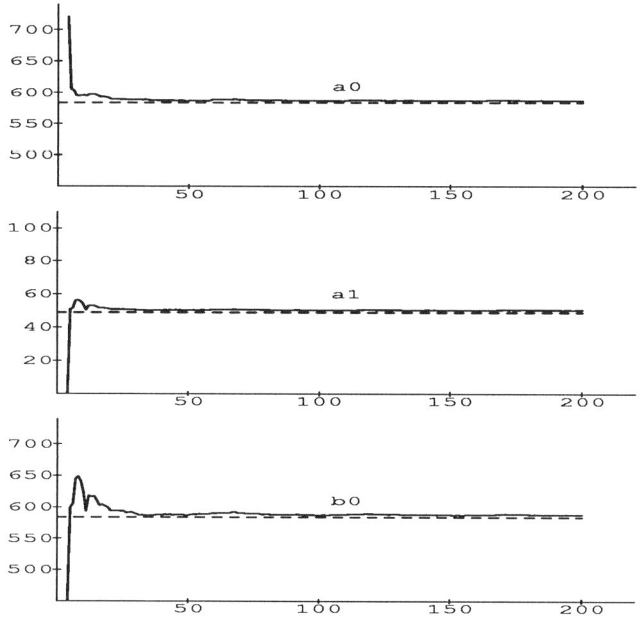

—— : estimated parameters, - - - : true parameters

Figure 8: Estimations of parameters in the experiment.

tems can be discussed in a similar way as the SISO linear systems, we will only mention some particular points about their basic block pulse difference equations. Based on these block pulse difference equations, the three equation forms, as (22), (25) and (28) in the SISO case, can easily be derived and applied to the identification of the corresponding differential equation models.

MIMO linear systems

From the transfer function matrix representation, an M-input, N-output time invariant linear system can be described by N differential equations ($i = 1, 2, \ldots, N$):

$$\sum_{k=0}^{n_i} a_{i,k} y_i^{(k)}(t) = \sum_{j=1}^{M} \sum_{k=0}^{n_i} b_{i,jk} u_j^{(k)}(t) \qquad (a_{i,n_i} = 1) \tag{38}$$

where $u_j(t)$ is the jth input, $y_i(t)$ is the ith output, $a_{i,k}$ and $b_{i,jk}$ ($k = 0, 1, \ldots, n_i; j = 1, 2, \ldots, M$) are system parameters. Noticing that in the derivation of the block pulse difference equation of the SISO case the same operations are carried out separately on the two sides of the original differential equation, and noticing that in each differential equation of (38), the output part and each input part have the same forms, then the block pulse difference equations in the MIMO case can be derived straightforward by the same procedure as discussed in Section 3:

$$\sum_{k=0}^{n_i} A_{i,k} y_{i,l+k} = \sum_{j=1}^{M} \sum_{k=0}^{n_i} B_{i,jk} u_{j,l+k} \tag{39}$$

where $i = 1, 2, \ldots, N$.

Linear systems with time delay

Consider a time invariant linear system with time delay τ in the input, which is described by a differential equation:

$$\sum_{k=0}^{n} a_k y^{(k)}(t) = \sum_{k=0}^{n} b_k u^{(k)}(t - \tau) \tag{40}$$

If we denote the intermediate variable as:

$$g^{(k)}(t) = u^{(k)}(t - \tau) \tag{41}$$

we can immediately obtain the block pulse difference equation from (40) as in the discussion of the SISO case:

$$\sum_{k=0}^{n} A_k y_{l+k} = \sum_{k=0}^{n} B_k g_{l+k} \tag{42}$$

If the width of the block pulses h is chosen small enough, it is reasonable to assume the time delay to be:

$$\tau \doteq dh \tag{43}$$

where d is a positive integer. From the relation between the block pulse coefficients of $u(t - \tau)$ and $g(t)$:

$$g_k = u_{k-d} \tag{44}$$

the block pulse difference equation of the linear system with input time delay can be expressed as:

$$\sum_{k=0}^{n} A_k y_{l+k} = \sum_{k=0}^{n} B_k u_{l+k-d} \tag{45}$$

where $l = d + 1, d + 2, \ldots$.

Hammerstein model nonlinear systems

The structure of a Hammerstein model nonlinear system is a memoryless nonlinear gain followed by a linear subsystem. The linear part of a Hammerstein model nonlinear system is described by the differential equation:

$$\sum_{k=0}^{n} a_k y^{(k)}(t) = \sum_{k=0}^{n} b_k g^{(k)}(t) \tag{46}$$

and its nonlinear part is described by:

$$g(t) = f(u(t)) \tag{47}$$

where $g(t)$ is the intermediate variable. In most cases, the nonlinear gain $f(*)$ can be approximated by a polynomial with an appropriately selected order p:

$$f(u(t)) \doteq \sum_{i=1}^{p} r_i \left(u(t)\right)^i \tag{48}$$

According to the disjoined property of the block pulse functions, we have the relation between the block pulse coefficients of $u(t)$ and $g(t)$:

$$g_k \doteq \sum_{i=1}^{p} r_i u_k^i \tag{49}$$

Since the linear part of a Hammerstein model nonlinear system has the same form as the one of the SISO linear system, we can also obtain its block pulse difference equation directly:

$$\sum_{k=0}^{n} A_k y_{l+k} = \sum_{k=0}^{n} \sum_{i=1}^{p} r_i B_k u_{l+k}^i \tag{50}$$

In all these cases of MIMO linear systems, linear systems with time delays and Hammerstein model nonlinear systems, the relations between the parameters a_i, b_i and A_k, B_k $(i, k = 0, 1, \ldots, n)$ can be determined by (19), only the subscripts in the expressions should be modified to suit these different cases. The rule of the difference as described in figure 1 can also be applied to avoid the complicated computations in these transformations of parameters. After the difference equation or regression equation forms are obtained, the algorithms for the identification of these systems in the continuous-time domain are also similar to those for the discrete-time model identification (Jiang 1987, 1988). Therefore the details will not be discussed further in this paper.

6. Conclusion

Starting from the properties of the block pulse functions and their generalized integral operational matrices, the block pulse difference equations corresponding to the original differential equation models of certain continuous-time systems, such as single-input, single-output linear systems, multi-input, multi-output linear systems, linear systems with time delays and Hammerstein model nonlinear systems, are derived. As the basis of the recursive block pulse function method, these block pulse difference equations can approximately express the relations between input and output without derivatives of signals and without system initial values. In the recursive block pulse function method, the algorithms well-developed for discrete-time model identification can easily be applied to the block pulse difference equations, and continuous signals can easily be approximated from their sampled data; therefore the parameters of the original differential equations can be estimated efficiently by means of digital computers. The satisfactory results in the examples also show that this recursive block pulse function method is a suitable method for the continuous-time model identification.

References

Cheng, B. and Hsu, N.-S. (1982), "Single-input-single-output system identification via block pulse functions", *Int. J. Systems Sci.*, vol.13, pp.697-702.

Haykin, S.S. (1972), "A unified treatment of recursive digital filtering", *IEEE Trans. on Automatic Control*, vol.17, pp.113-116.

Jiang, Z.H. (1987), "Block pulse function approach to the identification of MIMO-systems and time-delay systems", *Int. J. Systems Sci.*, vol.18, pp.1711-1720.

Jiang, Z.H. (1988), "Block pulse function approach for the identification of Hammerstein model non-linear systems", *Int. J. Systems Sci.*, vol.19, pp.2427-2439.

Jiang, Z.H. (1990), "*System identification via block pulse difference equations*", Ph.D. thesis, ETH No.8966, Dept. of Automatic Control, Swiss Federal Institute of Technology, Zürich.

Jiang, Z.H. and Schaufelberger, W. (1985a), "A new algorithm for single-input single-output system identification via block pulse functions", *Int. J. Systems Sci.*, vol.16, pp.1559-1571.

Jiang, Z.H. and Schaufelberger, W. (1985b), "Design of adaptive regulators based on block pulse function identification", Proc. on IEE Int. Conference "Control 85", (Cambridge, July 1985), pp.581-586.

Ljung, L. (1987), "*System identification — Theory for the user*", Prentice-Hall, Englewood Cliffs, New Jersey.

Palanisamy, K.R. and Bhattacharya, D.K. (1981), "System identification via block block functions", *Int. J. Systems Sci.*, vol.12, pp.643-647.

Söderström, T. and Stoica, P.G. (1983), "*Instrumental variable methods for system identification*", Springer-Verlag, Berlin.

Wang, C.-H. (1982), "Generalized block pulse operational matrices and their applications to operational calculus", *Int. J. Control*, vol.36, pp.67-76.

Wouters, W.R. (1972), "On-line identification in an unknown stochastic environment", *IEEE Transactions on Systems, Man and Cybernetics*, vol.SMC-2, pp. 666-668.

Continuous model identification via orthogonal polynomials

K.B.Datta
Department of Electrical Engineering
Indian Institute of Technology
Kharagpur - 721302,India

Abstract

The use of orthogonal polynomials as a class of potential functions for the estimation of parameters of a stipulated model of dynamical systems is outlined in this Chapter. In this process, the system's initial conditions are also determined. Lumped and distributed parameter system models which are linear and time-invariant are considered for our study. The estimates of parameters and initial conditions obtained by orthogonal polynomials are compared with those determined by block-pulse and sine-cosine functions. Results of estimates with noisy output data are also included.

1 Introduction

The history of orthogonal polynomials is very old. The Legendre polynomials originated from determining the force of attraction exerted by solids of revolution (Kline 1972) and their orthogonal properties were established by Adrien Marie Legendre during (1784-90). The problem of sloving ordinary differential equations over infinite or semi-infinite intervals and of obtaining expansion of arbitrary functions over such intervals attracted the attention of famous mathematicians in the nineteenth century and to resolve it, functions known today as Hermite polynomials were introduced (Kline 1972) in 1864 by Charles Hermite (1822-1905). The theory of continued fraction gives rise to all the orthogonal polynomials (Szegö, 1959) and, in fact, Edmond Laguerre (1834-1866) in his effort to convert a divergent power series into a convergent continued fraction in 1879 (Bell 1945) discovered polynomials known today as Laguerre polynomials.

In 1807, Joseph Fourier (1768-1830) while solving the partial differential equa-

N. K. Sinha and G. P. Rao (eds.), Identification of Continuous-Time Systems, 227–258.

tion encountered in connection with conduction of heat in a rod discovered that the solution can be expressed as a seies of exponentially weighted sine functions. Later, he extended this idea to represent any arbitrary function as an infinite sum of sine and cosine functions. Pafnuti L. Tchebycheff (1821-1894) observed that of all polynomials approximation of an arbitrary function in the interval $-1 \leq x \leq 1$, the one that minimizes the maximum error is a linear combination of polynomials known today as the Tchebycheff polynomials (Kline 1972). The genesis of orthogonal polynomials and sine-cosine functions is although different arising as they are in a bid to slove a diversity of problems, they are employed here for the sole purpose of identification of a continuous time dynamical model.These orthogonal functions dated more than one century back but surprisingly enough their potential application to the problems of identification is barely one decade old.

The paper by Corrington (1973) triggered into action a massive effort to identify continuous time models by using piecewise constant basis functions since mid-seventies (Rao 1983). The beginning of eighties witnessed a spurt of activities in continuous model identification via orthogonal polynomials and sine-cosine functions. In the orthogonal function approach, the problem of estimating the parameters and initial conditions is converted into the problem of solving a set of linear simultaneous equations with these parameters and initial conditions as unknowns.Chen and Hsiao (1975) applied Walsh functions to estimate the parameters of linear time-invariant systems with zero initial conditions and simultaneously Rao and Sivakumar (1975) showed that Walsh functions are not only powerful to estimate the parameters but also the initial conditions of the system. These initial conditions in most cases are unknown, in presence of noise they cannot be measured accurately and they are instrumental to give us an estimate of the states of the system. The basic contribution of the last two papers is the discovery of the *integration operational matrix* germinating the seed to develop subsequently a systematic methodology for estimation of system parameters and initial conditions via any class of orthogonal functions such as piecewise constant basis functions, orthogonal polynomials and sine-cosine functions. Later, *one shot operational matrix for repeated integration*, which will be called matrix OSOMRI, was introduced by Rao and Palanisamy (1983) to improve the numerical accuracy for a certain class of orthogonal functions.

A recursive approach for the identification problem was introduced by Palanisamy and Bhattacharyya (1981) using block-pulse functions. Hwang and Shih (1982) initiated lumped parameter system identification via Laguerre polynomials with noisy data. This problem was also studied by Chang and Wang (1982) via Legendre polynomials and Cheng and Hsu (1982) via block pulse functions. Rao and Sivakumar (1982) studied successfully the problem of simultaneously identifying the systems's order as well as parameters. The identification algorithms for parameters of linear time-invariant systems via orthogonal polynomials as mentioned above and published thereafter are all based on the one developed by Hwang and Shih (1982) which is not valid if the system order is more than one and if nonzero initial conditions are to be determined simultaneously.

Paraskevopoulos (1983) considered a model with no derivatives of the forcing function to estimate both parameters and initial conditions via first kind Tchebycheff polynomials, and later using the same model Paraskevopoulos and Kekkeris (1983b) studied parameter estimation via Hermite polynomials and Chung (1987) via sine-cosine functions. A successful solution for estimation of parameters and initial condtions of an n-th order linear time-invariant system containing derivative terms in the input was given in (Mohan and Datta 1988, 1989 and 1991b) via Legendre polynomials, Fourier series and all classes of orthogonal functions employing matrix OSOMRI.

There are many research workers who studied the continuous model identification of distributed parameter systems. Some of the publications in this area are (Paraskevopoulos and Bonus 1978) using Walsh functions, (Paraskevopoulos and Kekkeris 1983a) and (Hrong,Chou and Tsai 1986) using Tchebycheff polynomials, and (Ranganathan, Jha and Rajamani 1984, Jha and Zaman, 1985) using Laguerre polynomials with noisy data. The algorithms given in all these publications are restricted in numerical applications to only first order systems.It may be remarked that the identification of distributed parameter systems can be studied in a general framework using all the classes of orthogonal functions. A numerically viable algorithm to identify parameters and initial conditions in this general framework is developed for a class of second order distributed parameter systems using Legendre polynomials, sine-cosine functions and all classes of orthogonal functions in (Mohan and Datta 1988, 1989 and 1991a) with noisy data.

There are numerous methods to determine the transfer function matrix (TFM) description of a multiinput and multioutput linear time-invariant system from the input-output data (Chen 1987). Among these methods, the one based on the theory of realization which uses the Hankel matrix, and the other based on the generalized Sylvester's resultant matrix are widely used. The advantage of these methods is that the system structure in terms of TFM need not be known a priori but the disadvantage is that the initial conditions can not be determined simultaneously. However, if the structure of the TFM is known, then to estimate the parameters in the transfer function matrix and simultaneously to compute the initial conditons, the orthogonal function approach is obviously the superior method.

Identification of transfer function matrix along with the estimation of initial conditions via Walsh functions is studied in (Paraskevopoulos 1978, Rao and Sivakumar 1981), via block-pulse fucntions in (Jiang 1987) and via Legendre polynomials in (Hwang and Guo 1984). The potential of other orthogonal polynomials and sine-cosine functions is not still explored in the literature for the above problem.

This Chapter provides a brief account of the orthogonal polynomial and sine-cosine function approach for the identification of linear time-invariant continuous-time lumped as well as distributed parameter models with noisy output data. The mathematical methods are developed in Sections 2-4, the lumped parameter model identification is considered in Section 5, the distributed parameter model identification in Section 6, the identification of a transfer function matrix and the estimation of initial conditions of a linear time-invariant multiinput and multioutput system via any kind of orthogonal polynomials such as Legendre, Tchebycheff first and second kind, Laguerre and Hermite, and sine-cosine functions in Section 7.

2 Orthogonal polynomials, integration operational matrix and two dimensional orthogonal polynomials

Orthogonal polynomials

A set of functions $\phi_r(t), r = 0, 1, \ldots, m-1$ is said to be orthogonal with respect to a nonnegative weight function $w(t)$ over the interval $t \in [t_0, t_f]$ if

$$\int_{t_0}^{t_f} w(t)\phi_j(t)\phi_r(t)dt = \begin{cases} 0, & j \neq r, \\ \delta_r, & j = r; \end{cases} \tag{1}$$

where δ_r is a nonzero positive constant.

The set of orthogonal polynomials satisfies a three-term recurrence relation given by [Abramowitz and Stegun, 1964]

$$\phi_{r+1}(t) = (a_r t + b_r)\phi_r(t) + c_r\phi_{r-1}(t) \tag{2}$$

$r = 0, 1, \ldots$ and

$$
\begin{aligned}
\phi_{-1}(t) &= 0 \\
\phi_0(t) &= 1 \\
\phi_1(t) &= a_0 t + b_0
\end{aligned} \tag{3}
$$

where the coefficients a_r, b_r and c_r for some important orthogonal polynomials are given in Table 1.The first eight Legendre, Laguerre, Hermite, Tchebycheff first and second kind polynomials are also included in Tables 2-6.

Table 1: Coefficients of Three-term Recurrence Relation of Orthogonal Polynomials

(a)	Legendre Polynomials:	$P_r(t)$	
	$t_o = -1$	$t_f = 1$	$w(t) = 1$
	$a_r = \frac{(2r+1)}{(r+1)}$	$b_r = 0$	$c_r = -\frac{r}{(r+1)}$
(b)	Laguerre Polynomials:	$L_r(\alpha t)$	
	$t_0 = 0$	$t_f = \infty$	$w(t) = \exp(-\alpha t)$
	$a_r = -\frac{\alpha}{(r+1)}$	$b_r = \frac{(1+2r)}{(r+1)}$	$c_r = -\frac{r}{(r+1)}$
(c)	Hermite Polynomials:	$H_r(\alpha t)$	
	$t_0 = -\infty$	$t_f = \infty$	$w(t) = \exp(-\alpha^2 t^2)$
	$a_r = 2\alpha^2$	$b_r = 0$	$c_r = -2\alpha^2 r$
(d)	Tchebycheff Polynomials	of First Kind,	$T_r(t)$:
	$t_0 = -1$	$t_f = 1$	$w(t) = \frac{1}{\sqrt{(1-t^2)}}$
	$a_0 = 1, a_r = 2(r \geq 1)$	$b_r = 0$	$c_r = -1$
(e)	Tchebycheff Polynomials	of Second Kind,	$U_r(t)$
	$t_0 = -1$	$t_f = 1$	$w(t) = \sqrt{(1-t^2)}$
	$a_r = 2$	$b_r = 0$	$c_r = -1$

Table 2: Legendre Polynomials

$P_0(t)$	1
$P_1(t)$	t
$P_2(t)$	$(3t^2-1)/2$
$P_3(t)$	$(5t^3-3t)/2$
$P_4(t)$	$(35t^4-30t^2+3)/8$
$P_5(t)$	$(63t^5-70^3+15t)/8$
$P_6(t)$	$(231t^6-315t^4+105t^2-5)/16$
$P_7(t)$	$(429t^7-693t^5+315t^3-35t)/16$

Table 3: Laguerre Polynomials

$L_0(t)$	1
$L_1(t)$	$1-t$
$L_2(t)$	$(2-4t+t^2)/2$
$L_3(t)$	$(6-18t+9t^2)-t^3)/6$
$L_4(t)$	$(24-96t+72t^2-16t^3+t^4/24$
$L_5(t)$	$(120-600t+600t^2-200t^3+25t^4-t^5)/120$
$L_6(t)$	$(720-5320t+5400t^2-2200t^3+450t^4-36t^5+t^6)/720$
$L_7(t)$	$(5040-48280t+53920t^2-26800t^3$ $+7150t^4-882t^5+49t^6-t^7)/5040$

Table 4: Hermite Polynomials

$H_0(t)$	1
$H_1(t)$	$2t$
$H_2(t)$	$4t^2-2$
$H_3(t)$	$8t^3-12t$
$H_4(t)$	$16t^4-48t^2+12$
$H_5(t)$	$32t^5-160t^3+120t$
$H_6(t)$	$64t^6-480t^4+240t^2-120$
$H_7(t)$	$128t^7-1344t^5+1968t^3-1680t$

Integration operational matrix

The ordinary differential recurrence relation of the orthogonal polynomials has the form [Abramowitz and Stegun, 1964]

$$\phi_r(t) = A_r\dot{\phi}_{r+1}(t) + B_r\dot{\phi}_r(t) + C_r\dot{\phi}_{r-1}(t) \tag{4}$$

where the coefficients A_r, B_r and C_r are given in Tables 7-8.

Integrating both sides of the differential recurrene relation (4) from t_0 to t we have

$$\int_{t_0}^{t} \phi_r(t)dt = A_r\phi_{r+1}(t) + B_r\phi_r(t) + C_r\phi_{r-1}(t) + D_r\phi_0(t) \tag{5}$$

where $\phi_0(t) = 1$ and

Table 5: Tchebycheff Polynomials of First Kind

$T_0(t)$	1
$T_1(t)$	t
$T_2(t)$	$2t^2 - 1$
$T_3(t)$	$4t^3 - 3t$
$T_4(t)$	$8t^4 - 8t^2 + 1$
$T_5(t)$	$16t^5 - 20t^3 + 5t$
$T_6(t)$	$32t^6 - 48t^4 + 18t^2 - 1$
$T_7(t)$	$64t^7 - 112t^5 + 56t^3 - 7t$

Table 6: Tchebycheff Polynomials of Second Kind

$U_0(t)$	1
$U_1(t)$	$2t$
$U_2(t)$	$4t^2 - 1$
$U_3(t)$	$8t^3 - 4t$
$U_4(t)$	$16t^4 - 12t^2 + 1$
$U_5(t)$	$32t^5 - 32t^3 + 6t$
$U_6(t)$	$64t^6 - 80t^4 + 24t^2 - 1$
$U_7(t)$	$128t^7 - 192t^5 + 80t^3 - 8t$

$$D_r = -[A_r\phi_{r+1}(t_0) + B_r\phi_r(t_0) + C_r\phi_{r-1}(t_0)]$$

for $r = 1, 2, \ldots$ and

$$\int_{t_0}^{t} \phi_0(t)dt = t - t_0 \quad = \quad A_0\phi_1(t) + B_0\phi_0(t) \tag{6}$$

where

$$A_0 = \frac{1}{a_0}, B_0 \quad = \quad -(\frac{b_0}{a_0} + t_0)$$

For Tchebycheff first kind polynomials, if $r = 1$

$$\int_{t_0}^{t} T_1(t)dt \quad = \quad \frac{1}{4}[T_0 + T_2] - \frac{1}{2}t_0^2 T_0$$

Table 7: Coefficients of Differential Recurrence Relation: $r = 0$

Polynomials	A_0	B_0	C_0	D_0
$P_0(t)$	1	1	0	0
$L_0(t)$	-1	1	0	0
$H_0(t)$	$\frac{1}{2}$	0	0	0
$T_0(t)$	1	1	0	0
$U_0(t)$	$\frac{1}{2}$	1	0	0

Table 8: Coefficients of Differential Recurrence Relation:$r \geq 1$

Polynomials	A_r	B_r	C_r	D_r
$P_r(t)$	$\frac{1}{(2r+1)}$	0	$-\frac{1}{(2r+1)}$	0
$L_r(\alpha t)$	-1	1	0	0
$H_r(\alpha t)$	$\frac{1}{2(r+1)}$	0	0	$0, r =$even $-\frac{1}{2}(-\alpha^2)^{\frac{r+1}{2}}\frac{r!}{(\frac{r+1}{2})!}, r =$odd
$T_r(t)$	$\frac{1}{2(r+1)}$ $\frac{1}{4}$	0 0	$-\frac{1}{2(r-1)}$ $\frac{1}{4}$	$\frac{(-1)^{r+1}}{r^2-1}, r > 1$ $-\frac{1}{2}, r = 1$
$U_r(t)$	$\frac{1}{2(r+1)}$	0	$-\frac{1}{2(r+1)}$	$\frac{(-1)^r}{r+1}$

Combining (5) and (6) we can write

$$\int_{t_0}^{t} \phi(t)dt = E\phi(t) \tag{7}$$

where $\phi(t) = [\phi_0(t)\phi_1(t)\ldots\phi_{m-1}(t)]^T$ and E is the integration operational matrix having the form:

$$E = \begin{bmatrix} B_0 & A_0 & 0 & 0 & \cdots & 0 & 0 & 0 \\ D_1 + C_1 & B_1 & A_1 & 0 & \cdots & 0 & 0 & 0 \\ D_2 & C_2 & B_2 & A_2 & \cdots & 0 & 0 & 0 \\ \vdots & \vdots & \vdots & \vdots & \vdots & \vdots & \vdots & \\ D_{m-2} & 0 & 0 & 0 & \cdots & C_{m-2} & B_{m-2} & A_{m-2} \\ D_{m-1} & 0 & 0 & 0 & \cdots & 0 & C_{m-1} & B_{m-1} \end{bmatrix} \tag{8}$$

Any square integrable function $f(t)$ defined over $[t_0, t_f]$ can be approximated as a series of m number of orthogonal functions by

$$f(t) \approx \sum_{r=0}^{m-1} f_r \phi_r(t) \tag{9}$$

where

$$f_r = \frac{a}{\delta_r} \int_{t_0}^{t_f} f(t)\phi_r(t)dt \tag{10}$$

In vector notation this can be written as

$$f(t) = \mathbf{f}^T \phi(t) \tag{11}$$

where $\mathbf{f} = [f_0 f_1 \ldots f_{m-1}]^T$.

The shifted polynomials, orthogonal over any arbitrary interval $\tau \in [\tau_0, \tau_f]$ may be obtained by setting $t = \tau^\star$ such that

$$\tau^\star = \frac{\tau(t_0 - t_f) + (t_f\tau_0 - t_0\tau_f)}{\tau_0 - \tau_f} \quad = \quad a\tau + b \tag{12}$$

where

$$a = \frac{t_f - t_0}{\tau_f - \tau_0}, b \quad = \quad \frac{t_0\tau_f - t_f\tau_0}{\tau_f - \tau_0}$$

if the interval over which the functions are orthogonal is finite.Consequently the orthogonality relation (1) for shifted orthogonal functions becomes

$$\frac{t_f - t_0}{\tau_f - \tau_0} \int_{\tau_0}^{\tau_f} w(\tau^\star)\phi_j(\tau^\star)\phi_r(\tau^\star)d\tau^\star \quad = \quad \begin{cases} 0, & j \neq r, \\ \delta_r, & j = r \end{cases} \tag{13}$$

Substituting from (12) into (2) the three-term recurrence relation of shifted orthogonal polynomials is

$$\psi_{r+1}(\tau) = (\alpha_r \tau + \beta_r)\psi_r(\tau) + \gamma_r \psi_{r-1}(\tau) \tag{14}$$

where

$$\begin{aligned} \phi_r(\tau^*) = \psi_r(\tau), \alpha_r = aa_r, \beta_r &= ba_r + b_r, \gamma_r = c_r \\ \psi_0(\tau) = 1, \psi_1(\tau) = (\alpha_0 \tau + \beta_0) \end{aligned} \tag{15}$$

Again in view of (4) and (12), the differential recurrence relation for shifted orthogonal polynomials becomes,

$$a\psi_r(\tau) = A_r \dot{\psi}_{r+1}(\tau) + B_r \dot{\psi}_r(\tau) + C_r \dot{\psi}_{r-1}(\tau)$$

which on integrating from τ_0 to τ yields

$$\int_{\tau_0}^{\tau} \psi_r(\tau) d\tau = \frac{1}{a}[A_r \psi_{r+1}(\tau) + B_r \psi_r(\tau) + C_r \psi_{r-1}(\tau) + \bar{D}_r \psi_0(\tau)] \tag{16}$$

where

$$\bar{D}_r = -A_r \psi_{r+1}(\tau_0) - B_r \psi_r(\tau_0) - C_r \psi_{r-1}(\tau_0) \tag{17}$$

$$\int_{\tau_0}^{\tau} \psi_0(\tau) d\tau = \tau - \tau_0 = \frac{1}{a}[A_0 \psi_1(\tau) + \bar{B}_0 \psi_0(\tau)] \tag{18}$$

where

$$\bar{B}_0 = -(b + \frac{b_0}{a_0}) - a\tau_0 \tag{19}$$

combining (16) and (17) we can write

$$\int_{\tau_0}^{\tau} \psi(\tau) d\tau = E_s \psi \tag{20}$$

where

$$\psi(\tau) = [\psi_0(\tau)\psi_1(\tau) \ldots \psi_{m-1}(\tau)]^T$$

and E_s is the integration operational matrix for shifted orthogonal polynomials given by

$$E_s = \frac{1}{a}\begin{bmatrix} \bar{B}_0 & A_0 & 0 & 0 & \cdots & 0 & 0 & 0 \\ \bar{D}_1 + C_1 & B_1 & A_1 & 0 & \cdots & 0 & 0 & 0 \\ \bar{D}_2 & C_2 & B_2 & A_2 & \cdots & 0 & 0 & 0 \\ \vdots & \vdots & \vdots & \vdots & \vdots & \vdots & \vdots & \vdots \\ \bar{D}_{m-2} & 0 & 0 & 0 & \cdots & C_{m-2} & B_{m-2} & A_{m-2} \\ \bar{D}_{m-1} & 0 & 0 & 0 & \cdots & 0 & C_{m-1} & B_{m-1} \end{bmatrix} \tag{21}$$

Any square integrable function $g(\tau)$ defined over $[\tau_0, \tau_f]$ can be approximated as a series of m number of shifted orthogonal functions by

$$g(\tau) = \sum_{r=0}^{m-1} g_r \psi_r(\tau) = \mathbf{g}^T \boldsymbol{\psi}(\tau) \tag{22}$$

where

$$g_r = \frac{a}{\delta_r}\int_{\tau_0}^{\tau_f} g(\tau)\psi(\tau)d\tau \tag{23}$$

$$\mathbf{g} = [g_0 g_1 \cdots g_{m-1}]^T$$

and putting (12) into (1) we have

$$\delta_r = a\int_{\tau_0}^{\tau_f} \omega(\tau)\psi_r^2(\tau)d\tau \tag{24}$$

where

$$\omega(\tau) = w(\tau^*)$$

Two dimensional square-integrable polynomials

The construction of a complete system of functions of two or more variables is based on the following result (Sansone, 1959)

Let $\psi_0(x), \psi_1(x), \cdots$, be a complete system of functions in the interval $x_0 \le x \le x_f$, and let $\phi_0(t), \phi_1(t), \cdots$, be a similar system in the interval $t_0 \le t \le t_f$.Then the functions $\psi_i(x)\phi_j(t), i = 0, 1, 2, \cdots$ and $j = 0, 1, 2, \cdots$, form a complete orthogonal system of functions in x and t in the rectangle $x_0 \le x \le x_f, t_0 \le t \le t_f$.

Based on the above theorem any square integrable function $f(x,t)$ in the region $x_0 \leq x \leq x_f, t_0 \leq t \leq t_f$ can be represented as a finite series of orthogonal functions in the form

$$f(x,t) \approx \sum_{i=0}^{m-1} \psi_i(x) \sum_{j=0}^{n-1} f_{ij}\phi_j(t) = \boldsymbol{\psi}^T(x) F \boldsymbol{\phi}(t) \tag{25}$$

where

$$\boldsymbol{\psi}(x) = [\psi_0(x), \psi_1(x), \cdots, \psi_{m-1}(x)]^T \tag{26}$$

an $m-$dimensional basis vector in x,

$$\boldsymbol{\phi}(t) = [\phi_0(t), \phi_1(t), \cdots, \phi_{n-1}(t)]^T \tag{27}$$

an $m-$dimensional basis vector in t, and

$$F = \begin{bmatrix} f_{00} & f_{01} & \cdots & f_{0,n-1} \\ f_{10} & f_{11} & \cdots & f_{1,n-1} \\ \vdots & \vdots & \vdots & \vdots \\ f_{m-1,0} & f_{m-1,1} & \cdots & f_{m-1,n-1} \end{bmatrix} \tag{28}$$

an $m \times n$ coefficient matrix of $f(x,t)$ with respect to the orthogonal system employed.

Let $w_1(x)$ and $w_2(t)$ be the weight functions of orthogonal systems chosen. Then the coefficients f_{ij} in the expansion of any arbitrary function $f(x,t)$ in terms of shifted orthogonal functions can be expressed as outlined below.

For the Legendre polynomials the coefficients f_{ij} can be expressed as

$$f_{ij} = \frac{(2i+1)(2j+1)}{(x_f - x_0)(t_f - t_0)} \int_{t_0}^{t_f} \int_{x_0}^{x_f} f(x,t)\psi_i(x)\phi_j(t)\,dx\,dt \tag{29}$$

The coefficients f_{ij} for Laguerre polynomials are

$$f_{ij} = \int_{t_0}^{t_f} \int_{x_0}^{x_f} w_1(x) w_2(t) f(x,t)\psi_i(x)\phi_j(t)\,dx\,dt \tag{30}$$

with

$$\begin{aligned} w_1(x) &= e^{-(x-x_0)}, \\ w_2(t) &= e^{-(t-t_0)} \end{aligned} \tag{31}$$

The Laguerre polynomials defined with weight functions as above will be called as the shifted Laguerre polynomials. The coefficients f_{ij} for Hermite polynomials are

$$f_{ij} = \frac{1}{\pi 2^{i+j} i! j!} \int_{t_0}^{t_f} \int_{x_0}^{x_f} w_1(x) w_2(t) f(x,t) \psi_i(x) \phi_j(t) dx dt \tag{32}$$

with

$$\begin{aligned} w_1(x) &= e^{-(x-x_0)^2} \\ w_2(t) &= e^{-(t-t_0)^2} \end{aligned} \tag{33}$$

The Hermite polynomials defined with the above weight functions are called shifted Hermite polynomials. The coefficients f_{ij} for Tchebycheff first kind polynomials are

$$f_{ij} = c_1 c_2 \int_{t_0}^{t_f} \int_{x_0}^{x_f} w_1(x) w_2(t) f(x,t) \psi_i(x) \phi_j(t) dx dt \tag{34}$$

with

$$c_1 = \begin{cases} 2/[\pi(x_f - x_0)], & \text{for } i = 0 \\ 4/[\pi(x_f - x_0)], & \text{for } i = 1, 2, \cdots, (m-1) \end{cases}$$

$$c_2 = \begin{cases} 2/[\pi(t_f - t_0)], & \text{for } j = 0 \\ 4/[\pi(t_f - t_0)], & \text{for } j = 1, 2, \cdots, (n-1) \end{cases}$$

$$w_1(x) = \frac{(x_f - x_0)}{2\sqrt{(x - x_0)(x_f - x)}} \tag{35}$$

and

$$w_2(t) = \frac{(t_f - t_0)}{2\sqrt{(t - t_0)(t_f - t)}} \tag{36}$$

The coefficients f_{ij} for Tchebycheff second kind polynomials are

$$f_{ij} = \frac{16}{\pi^2 (x_f - x_0)(t_f - t_0)} \int_{t_0}^{t_f} \int_{x_0}^{x_f} w_1(x) w_2(t) f(x,t) \psi_i(x) \phi_j(t) dx dt \tag{37}$$

where

$$w_1(x) = 2[(x - x_0)(x_f - x)]^{0.5} / (x_f - x_0) \tag{38}$$

and

$$w_2(t) = 2[(t - t_0)(t_f - t)]^{0.5}/(t_f - t_0) \tag{39}$$

3 One shot operational matrix for repeated integration (OSOMRI)

To improve accuracy of the computational procedure of identification, the matrix OSOMRI was introduced by Rao and Palanisamy (1983). It is evident from (7) that if the basis vector ϕ is integrated k times then

$$I_k[\phi(t)] \quad = \quad \int_{t_0}^{t}\int_{t_0}^{t_1}\cdots\int_{t_0}^{t_{k-1}} \phi(t_k)dt_k\cdots dt_2dt_1 \approx E^k\phi(t)$$

by repeated application of (7). On the other hand, if the function, obtained from integrating $\phi(t)$ k-times, is approximated as an m-term series of orthogonal function we have

$$I_k[\phi(t)] \approx E_k\phi(t)$$

For Laguerre and Hermite polynomials, it can be shown that $E_k = E^k$.

The k times repeated integration of ith degree shifted Legendre polynomial from t_0 to t is given by

$$I_k[\phi_i(t)] = \frac{1}{(2c)^k}\sum_{j=0}^{k}\frac{(-1)^{j+k}\ {}^kC_{i+2j-k}(t)}{\prod_{\ell=0,\ell\neq k-j}^{k}[2(i+j-\ell)+1]}$$

and

$$\phi_{-i}(t) = -\phi_{i-1}(t)$$

for all i. The matrix OSOMRI E_k for Legendre polynomials can be derived from the above relation. For Tchebycheff polynomials similar kind of expression can be obtained to compute matrix OSOMRI.

The distinction between the integration operational matrix E^3 and matrix OSOMRI E_3 using the first five Legendre polynomials is shown below.

$$E_3 = \frac{(t_f - t_0)^3}{8}\begin{bmatrix} 1/3 & 3/5 & 1/3 & 1/15 & 0 \\ -1/5 & -1/3 & -1/7 & 0 & 1/105 \\ 1/15 & 3/35 & 0 & -1/45 & 0 \\ -1/105 & 0 & 1/63 & 0 & -2/315 \\ 0 & -1/315 & 0 & 2/405 & 0 \end{bmatrix}$$

$$E^3 = \frac{(t_f - t_0)^3}{8} \begin{bmatrix} 1/3 & 3/5 & 1/3 & 1/15 & 0 \\ -1/5 & -1/3 & -1/7 & 0 & 1/105 \\ 1/15 & 3/35 & 0 & -1/45 & 0 \\ -1/105 & 0 & 1/63 & 0 & -3/385 \\ 0 & -1/315 & 0 & 1/165 & 0 \end{bmatrix}$$

4 Sine-cosine functions(SCF)

Integration operational matrix

The Fourier series approximation of the square integrable function $f(t)$ defined in the interval $t_0 \le t \le t_f$ is given by

$$\begin{aligned} f(t) &\approx \hat{f}_o \hat{F}_o(t) + \sum_{r=1}^{m} [\hat{f}_r \hat{F}_r + \tilde{f}_r \tilde{F}_r(t)] \\ &= \mathbf{f}^T \mathbf{F}(t) \end{aligned}$$

where

$$\begin{aligned} \hat{F}_r(t) &= cos\{r\pi[2t - (t_f + t_0)]c\}, \\ r &= 0, 1, 2, \cdots, m \\ \tilde{F}_r(t) &= sin\{r\pi[2t - (t_f + t_0)]c\}, \\ r &= 1, 2, \cdots, m \\ c &= \frac{1}{t_f - t_0} \\ \hat{f}_o &= \int_{t_0}^{t_f} f(t) dt, \\ \hat{f}_r &= 2c \int_{t_0}^{t_f} f(t) \hat{F}_r(t) dt, \\ \tilde{f}_r &= 2c \int_{t_0}^{t_f} f(t) \tilde{F}_r(t) dt \\ \mathbf{f} &= [\hat{f}_o, \hat{f}_1, \cdots, \hat{f}_m, \tilde{f}_1, \cdots, \tilde{f}_m]^T, \\ \mathbf{F}(t) &= [\hat{F}_o(t), \hat{F}_1(t), \cdots, \hat{F}_m(t), \tilde{F}_1(t), \cdots, \tilde{F}_m(t)]^T \end{aligned}$$

From these equations the integration operational matrix associated with the Fourier sine-cosine functions can be easily derived (Mohan and Datta, 1989) and is represented by

$$E_s = \frac{1}{c} \begin{bmatrix} \frac{1}{2} & \mathbf{0}^T & \frac{\mathbf{e}^T}{\pi} \\ \mathbf{0} & O & \frac{U}{2\pi} \\ -\frac{\mathbf{e}}{2\pi} & -\frac{U}{2\pi} & O \end{bmatrix}$$

where

$$\mathbf{e} = \left[\begin{array}{ccccc} 1 & -\frac{1}{2} & \frac{1}{3} & \cdots & \frac{(-1)^{m+1}}{m} \end{array}\right]^T,$$
$$U = \text{diag}\left[\begin{array}{cccc} 1 & \frac{1}{2} & \cdots & \frac{1}{m} \end{array}\right],$$

0 is an m-vector with zero elements, and O an $m \times m$ null matrix.

OSOMRI of sine-cosine functions

The matrix OSOMRI $E_k = [e_{ijk}]$ for sine-cosine functions can be recursively generated employing the following relations for $i, j = 1, 2, \cdots, (m-1)$ and $k = 2, 3, \cdots,$.

$$\begin{aligned}
e_{1,1,k} &= [1/c(k+1)]e1,1,k-1 \\
e_{1,j+1,k} &= -(1/2j\pi c)e_{1,m+j,k-1} \\
e_{1,m+j,k} &= [(-1)^{j+1}/j\pi c]e_{1,1,k-1} + (T/2j\pi)e_{1,j+1,k-1} \\
e_{i+1,1,k} &= (1/2i\pi c)e_{m+1,1,k-1} \\
e_{i+1,j+1,k} &= (1/2i\pi c)e_{m+1,j+1,k-1} \\
e_{i+1,m+j,k} &= (1/2i\pi c)e_{m+1,m+j,k-1} \\
e_{m+i,1,k} &= (1/2i\pi c)[(-1)^i e_{1,2,k-1} - e_{i+1,1,k-1}] \\
e_{m+i,j+1,k} &= (1/2i\pi c)[(-1)^i e_{1,j+1,k-1} - e_{i+1,j+1,k-1}]
\end{aligned}$$

and

$$e_{m+i,m+j,k} = (1/2i\pi c)[(-1)^i e_{1,m+j,k-1} - e_{i+1,m+j,k-1}]$$

For $k = 2$ and $m = 2$ we have

$$E^2 = (t_f - t_i)^2 \begin{bmatrix} 1/4 - 1/(2\pi^2) & -1/(2\pi^2) & 1/(2\pi) \\ -1/(2\pi)^2 & -1/(2\pi)^2 & 0 \\ -1/(4\pi) & 0 & -3/(2\pi)^2 \end{bmatrix};$$
$$E_2 = (t_f - t_i)^2 \begin{bmatrix} 1/6 & -1/(2\pi^2) & 1/(2\pi) \\ -1/(2\pi)^2 & -1/(2\pi)^2 & 0 \\ -1/(4\pi) & 0 & -3/(2\pi)^2 \end{bmatrix}.$$

Two dimensional square integrable sine-cosine functions

Following the discussion in Section 2, a square integrable function $f(x,t)$ in the region $x_0 \le x \le x_f, t_0 \le t \le t_f$ can be approximated in terms of sine-cosine functions as

$$f(x,t) \approx \hat{\psi}_0(x)\{\hat{f}_{00}\hat{\phi}_0(t) + \sum_{j=1}^{n-1}[\hat{f}_{0j}\hat{\phi}_j(t) + f^*_{0j}\tilde{\phi}_j(t)]\} +$$
$$\sum_{i=1}^{m-1} \hat{\psi}_i(x)\{\hat{f}_{i0}\hat{\phi}_0(t) + \sum_{j=1}^{n-1}[\hat{f}_{ij}\hat{\phi}_j(t) + f^*_{ij}\tilde{\phi}_j(t)]\} +$$

$$\sum_{i=1}^{m-1} \tilde{\psi}_i(x)\{f_{i0}^*\hat{\phi}_0(t) + \sum_{j=1}^{n-1}[f_{ij}\hat{\phi}_j(t) + \tilde{f}_{ij}\tilde{\phi}_j(t)]\}$$
$$\approx \psi^T(x)F\phi(t)$$

where

$$\psi(x) = [\hat{\psi}_0(x), \hat{\psi}_1(x), \cdots, \hat{\psi}_{m-1}(x), \tilde{\psi}_1(x), \cdots, \tilde{\psi}_{m-1}(x)]^T,$$

a $(2m-1)-$ dimensional Fourier basis vector in x

$$\phi(t) = [\hat{\phi}_0(t), \hat{\phi}_1(t), \cdots, \hat{\phi}_{n-1}(t), \tilde{\phi}_1(t), \cdots, \tilde{\phi}_{n-1}(t)]^T,$$

a $(2n-1)-$dimensional Fourier basis vector in t and

$$F = \begin{bmatrix} \hat{f}_{00} & \hat{f}_{01} & \cdots & \hat{f}_{0,n-1} & f_{01}^* & \cdots & f_{0,n-1}^* \\ \hat{f}_{10} & \hat{f}_{11} & \cdots & \hat{f}_{1,n-1} & f_{11}^* & \cdots & f_{1,n-1}^* \\ \vdots & \vdots & & \vdots & \vdots & & \vdots \\ \hat{f}_{m-1,0} & \hat{f}_{m-1,1} & \cdots & \hat{f}_{m-1,n-1} & f_{m-1,1}^* & \cdots & f_{m-1,n-1}^* \\ f_{10}^* & f_{11} & \cdots & f_{1,n-1} & \tilde{f}_{11} & \cdots & \tilde{f}_{1,n-1} \\ \vdots & \vdots & & \vdots & \vdots & & \vdots \\ f_{m-1,0}^* & f_{m-1,1} & \cdots & f_{m-1,n-1} & \tilde{f}_{m-1,1} & \cdots & \tilde{f}_{m-1,n-1} \end{bmatrix}$$

a $(2m-1) \times (2n-1)$ constant matrix whose elements are given by

$$\begin{aligned}
\hat{f}_{00} &= c\int_{t_0}^{t_f}\int_{x_0}^{x_f} f(x,t)dxdt \\
\hat{f}_{0j} &= 2c\int_{t_0}^{t_f}\int_{x_0}^{x_f} f(x,t)\hat{\phi}_j(t)dxdt \\
f_{0j}^* &= 2c\int_{t_0}^{t_f}\int_{x_0}^{x_f} f(x,t)\tilde{\phi}_j(t)dxdt \\
\hat{f}_{i0} &= 2c\int_{t_0}^{t_f}\int_{x_0}^{x_f} f(x,t)\hat{\psi}_i(x)dxdt \\
\hat{f}_{ij} &= 4c\int_{t_0}^{t_f}\int_{x_0}^{x_f} f(x,t)\hat{\psi}_i(x)\hat{\phi}_j(t)dxdt \\
f_{ij}^* &= 4c\int_{t_0}^{t_f}\int_{x_0}^{x_f} f(x,t)\hat{\psi}_i(x)\tilde{\phi}_j(t)dxdt \\
f_{i0}^* &= 2c\int_{t_0}^{t_f}\int_{x_0}^{x_f} f(x,t)\tilde{\psi}_i(t)dxdt
\end{aligned}$$

$$\begin{aligned} f_{ij} &= 4c\int_{t_0}^{t_f}\int_{x_0}^{x_f} f(x,t)\tilde{\psi}_i(x)\hat{\phi}_j(t)dxdt \\ \tilde{f}_{ij} &= 4c\int_{t_0}^{t_f}\int_{x_0}^{x_f} f(x,t)\tilde{\psi}_i(x)\tilde{\phi}_j(t)dxdt \\ c &= \frac{1}{(x_f - x_0)(t_f - t_0)} \end{aligned}$$

5 Lumped parameter system identification

The model of a linear time-invariant single-input single-output, lumped parameter system is described by

$$\begin{aligned} & y^{(n)}(t) + a_{n-1}y^{(n-1)}(t) + \cdots + a_1 y^{(1)}(t) + a_0 y(t) \\ = \; & b_\nu u^{(\nu)}(t) + b_{\nu-1}u^{(\nu-1)}(t) + \cdots + b_1 u^{(1)}(t) + b_0 u(t) \end{aligned} \tag{40}$$

where $u(t)$ and $y(t)$ are respectively the input and output of the system.It is assumed that n and ν with $\nu \leq n$ are known, and the problem is to estimate the parameters $a_{n-1}, a_{n-2}, \cdots, a_0, b_\nu, b_{\nu-1}, \cdots, b_0$ from the measurements of the input $u(t)$ and the output $y(t)$ over an arbitrary but transient state of the system.

The initial conditions required to specify the response of the system (40) are

$$\left.\begin{aligned} \eta_k &= y^{(k)}(t_0), \;\; k = 0,1,2,\cdots,n-1 \\ v_k &= u^{(k)}(t_0), \;\; k = 0,1,2,\cdots,\nu-1 \end{aligned}\right\}. \tag{41}$$

True initial values of these variables are difficult to get by actual measurements because these measurements are likely to be contaminated with noise. Consequently these variables can be looked upon as parameters to be estimated in the identification process. Integrating (40) n times from t_0 to t we have

$$\begin{aligned} & \sum_{k=0}^{n}\left\{I_{n-k}[y(t)] - \sum_{q=0}^{k-1} y^{(q)}(t_0)\frac{t^{n-k+q}}{(n-k+q)!}\right\} \\ = \; & \sum_{k=0}^{\nu} b_k\left\{I_{n-k}[u(t)] - \sum_{q=0}^{k-1} u^{(q)}(t_0)\frac{t^{n-k+q}}{(n-k+q)!}\right\} \end{aligned} \tag{42}$$

where $a_n = 1$ and $(n-k)$ times integration of a function $f(t)$ is denoted by

$$I_{n-k}[f(t)] = \int_{t_0}^{t}\int_{t_0}^{t_1}\cdots\int_{t_0}^{t_{n-k-1}} f(t_{n-k})dt_{n-k}\cdots dt_2 dt_1$$

The coefficients of t^k in the expression (42) are grouped together and are denoted by α_k. So (42) becomes

$$- \sum_{k=0}^{n-1} a_k I_{n-k}[y(t)] + \sum_{k=0}^{\nu} b_k I_{n-k}[u(t)] + \sum_{k=0}^{n-1} \alpha_k t^k = y(t) \tag{43}$$

where

$$\alpha_k = \begin{cases} \sum_{i=n-k}^{n} a_i y^{(k+i-n)}(t_0) \frac{1}{k!}, k = 0, 1, \cdots, n-\nu-1; \\ \left[\sum_{i=n-k}^{n} a_i y^{(k+i-n)}(t_0) - \sum_{i=n-k}^{\nu} b_i u^{(k+i-n)}(t_0)\right] \frac{1}{k!}, \\ k = n-\nu, \cdots, n-1. \end{cases}$$

The input and the output signal and $t^k, k = 0, 1, \ldots, n-1$ are now approximated as a finite $m-$term series of orthogonal functions in the form

$$u(t) \approx \sum_{k=0}^{m-1} u_k \psi_k(t) = \mathbf{u}^T \boldsymbol{\psi}(t) \tag{44}$$

$$y(t) \approx \sum_{k=0}^{m-1} y_k \psi_k(t) = \mathbf{y}^T \boldsymbol{\psi}(t) \tag{45}$$

$$t^k \approx \sum_{j=0}^{m-1} \tau_{j0} \psi_t(t) = \boldsymbol{\tau}_k^T \boldsymbol{\psi}(t) \tag{46}$$

In view of (7), we note that $(n-k)$ times integral of $u(t)$ and $y(t)$ can be written as

$$I_{n-k}[u(t)] = \mathbf{u}^T E_{n-k} \boldsymbol{\psi}(t) \tag{47}$$

$$I_{n-k}[y(t)] = \mathbf{y}^T E_{n-k} \boldsymbol{\psi}(t) \tag{48}$$

Therefore, inserting (44)-(48) in (43) and after simplification we get finally

$$Q\mathbf{x} = \mathbf{y} \tag{49}$$

where

$$\begin{aligned} Q &= [-E_n^T \mathbf{y} \mid -E_{n-1}^T \mathbf{y} \mid \cdots \mid -E_2^T \mathbf{y} \mid -E_1^T \mathbf{y} \mid E_n^T \mathbf{u} \mid \\ &\quad E_{n-1}^T \mathbf{u} \mid \cdots \mid E_{n-\nu}^T \mathbf{u} \mid \tau_0 \mid \tau_1 \mid \cdots \mid \tau_{n-1}] \\ \mathbf{x} &= [a_0, a_1, \cdots, a_{n-1}, b_0, b_1, \cdots, b_\nu, \alpha_0, \alpha_1, \cdots, \alpha_{n-1}]^T, \\ \boldsymbol{\alpha} &= A\boldsymbol{\eta} - B\boldsymbol{v} \\ &= [\alpha_0, \alpha_1, \cdots, \alpha_{n-1}]^T \\ \boldsymbol{\eta} &= [\eta_0, \eta_1, \cdots, \eta_{n-1}]^T \\ \boldsymbol{v} &= [v_0, v_1, \cdots, v_{\nu-1}]^T \end{aligned}$$

$$A = \begin{bmatrix} 1 & 0 & 0 & \cdots & 0 & 0 \\ a_{n-1} & 1 & 0 & \cdots & 0 & 0 \\ \frac{a_{n-2}}{2!} & \frac{a_{n-1}}{2!} & \frac{1}{2!} & \cdots & 0 & 0 \\ \vdots & \vdots & \vdots & \vdots & \vdots & \vdots \\ \frac{a_1}{(n-1)!} & \frac{a_2}{(n-1)!} & \frac{a_3}{(n-1)!} & \cdots & \frac{a_{n-1}}{(n-1)!} & \frac{1}{(n-1)!} \end{bmatrix}$$

$$B = \begin{bmatrix} 0 & 0 & \cdots & 0 & 0 \\ \vdots & \vdots & \vdots & \vdots & \vdots \\ 0 & 0 & \cdots & 0 & 0 \\ \frac{b_\nu}{(n-\nu)!} & 0 & \cdots & 0 & 0 \\ \frac{b_{\nu-1}}{(n-\nu+1)!} & \frac{b_\nu}{(n-\nu+1)!} & \cdots & 0 & 0 \\ \vdots & \vdots & \vdots & \vdots & \vdots \\ \frac{b_1}{(n-1)!} & \frac{b_2}{(n-1)!} & \cdots & \frac{b_{\nu-1}}{(n-1)!} & \frac{b_\nu}{(n-1)!} \end{bmatrix}.$$

The n-square matrix A is always nonsingular and the $n \times \nu$ matrix B does not exist if derivative terms in (40) are absent. Since the matrix Q is of order $m \times (2n+\nu+1)$, for (49) to be solved for $\mathbf{x}$ it is required that

$$m \geq (2n + \nu + 1)$$

Once the vector $\mathbf{x}$ containing the system parameters is computed, the initial conditions of the response $y(t)$ can always be estimated by solving (49), provided B is a null matrix. The continuous time model described by (40) can consequently be identified via any class of orthogonal functions by virtue of the relation (49).

Example:SISO lumped parameter system

The continuous-time model considered to study the identification problem is described by

$$\ddot{y}(t) + a_1\dot{y}(t) + a_0 y(t) = b_0 u(t) \tag{50}$$

The output data in response to a unit ramp input $u(t) = t$ is generated. This gives us the values of $y(t)$ over the interval $1 \leq t \leq 2$ and the task is set to determine the initial conditions $\dot{y}(1)$ and $y(1)$ along with the parameters a_0, a_1, and b_0. The values of these parameters to simulate the stipulated model are $a_0, a_1 = 3$ and $b_0 = 1$ conditions are found to be $y(1) = 0.0840456$ and $\dot{y}(1) = 0.1997882$. It may be seen from the Table 9 that estimated results are in close agreement with the actual values and it is more so as the value of m is increased. Here BPF stands for block pulse functions, TP1 and TP2 Tchebycheff first and second kind polynomials, LeP the Legendre polynomials, LaP the Laguerre polynomials, HeP the Hermite polynomials, and SCF the sine-cosine functions.

Table 9: Estimates of parameters and initial conditions

Approach ↓	m	a_1	a_0	b_0	$y(1)$	$\dot{y}(1)$
actual →		3	2	1	0.0840456	0.1997882
BPF	5	3.0284429	2.0085034	1.0068576	0.0860669	0.1940037
	6	3.0199119	2.0059606	1.0048031	0.0854443	0.1957967
	7	3.0147001	2.0044038	1.0035470	0.0850710	0.1968671
	8	3.0112901	2.0033839	1.0027247	0.0848296	0.1975574
TP1	5	3.0193772	2.0032223	1.0043256	0.0840459	0.1997680
	6	2.9999753	1.9999920	0.9999939	0.0840455	0.1997883
TP2	5	3.0127620	2.0021398	1.0028555	0.0840470	0.1997688
	6	3.0000380	2.0000062	1.0000085	0.0840456	0.1997880
LeP	5	3.0155869	2.0026052	1.0034843	0.0840465	0.1997682
	6	3.0000532	2.0000085	1.0000118	0.0840456	0.1997881
LaP	5	3.0000369	2.0000226	1.0000115	0.0840459	0.1997859
HeP	5	2.9999999	1.9999999	0.9999999	0.0730514	0.2229334
	6	2.9999999	1.9999999	0.9999999	0.0871496	0.1935069
	7	2.9999999	1.9999999	0.9999999	0.0871496	0.1935069
	8	2.9999999	1.9999999	0.9999999	0.0833944	0.2010942
SCF	3	3.0002314	2.0000388	1.0000518	0.1018564	0.1382217
	4	3.0002945	2.0000490	1.0000659	0.0968457	0.1555389
	8	3.0004122	2.0000654	1.0000917	0.0900502	0.1790287
	16	3.0004404	2.0000454	1.0000945	0.0869543	0.1897318
	32	3.0001227	1.9997955	0.9999955	0.0854767	0.1948439

6 Distributed parameter system identification

The model of a linear time-invariant distributed parameter system is described by the following second-order partial differential equation.

$$
\begin{aligned}
a_{tt} + \frac{\partial^2 y(x,t)}{\partial t^2} + a_{xx}\frac{\partial^2 y(x,t)}{\partial x^2} + a_{xt}\frac{\partial^2 y(x,t)}{\partial x \partial t} + \\
a_t \frac{\partial y(x,t)}{\partial t} + a_x \frac{\partial y(x,t)}{\partial x} + a y(x,t) = u(x,t) \qquad (51)
\end{aligned}
$$

Integrating (51) twice with respect to x and twice with respect to t we arrive at the following integral equation after simplification.

$$\begin{aligned}
& a_{tt}\int_{x_0}^{x}\int_{x_0}^{x} y(\chi,t)d\chi^2 + a_{xx}\int_{t_0}^{t}\int_{t_0}^{t} y(x,\tau)d\tau^2 + a_{xt}\int_{t_0}^{t}\int_{x_0}^{x} y(\chi,\tau)d\chi d\tau \\
+ & \ a_t\int_{t_0}^{t}\int_{x_0}^{x}\int_{x_0}^{x} y(\chi,\tau)d\chi^2 d\tau + a_x\int_{t_0}^{t}\int_{t_0}^{t}\int_{x_0}^{x} y(\chi,\tau)d\chi d\tau^2 \\
+ & \ a\int_{t_0}^{t}\int_{t_0}^{t}\int_{x_0}^{x}\int_{x_0}^{x} y(\chi,\tau)d\chi^2 d\tau^2 - a_{tt}\int_{x_0}^{x}\int_{x_0}^{x} f(\chi)d\chi^2 \\
- & \ a_{xx}\int_{t_0}^{t}\int_{t_0}^{t} q(\tau)d\tau^2 - \hat{c}\int_{t_0}^{t}\int_{x_0}^{x} d\chi d\tau - \int_{t_0}^{t}\int_{x_0}^{x}\int_{x_0}^{x} h(\chi)d\chi^2 d\tau \\
- & \int_{t_0}^{t}\int_{t_0}^{t}\int_{x_0}^{x} s(\tau)d\chi d\tau^2 = \int_{t_0}^{t}\int_{t_0}^{t}\int_{x_0}^{x}\int_{x_0}^{x} u(\chi,\tau)d\chi^2 d\tau^2
\end{aligned} \tag{52}$$

where

$$\begin{aligned}
x_0 &= t_0 = 0 \\
f(x) &= y(x,t_0) \\
q(t) &= y(x_0,t) \\
g(x) &= \frac{\partial y(x,t)}{\partial t} \text{ at } t=t_0 \\
\sigma(x) &= a_{tt}g(x) + a_t f(x) && (53) \\
w(x) &= \frac{\partial y(x,t_0)}{\partial x} && (54) \\
r(t) &= \frac{\partial y(x,t)}{\partial x} \text{ at } x=x_0 \\
v(t) &= a_{xx}r(t) + a_x q(t) + a_{xt}\int_{t_0}^{t} q(\tau)d\tau \\
\hat{c} &= a_{xt}y(x_0,t_0) \\
h(x) &= a_{tt}g(x) + a_{xt}\frac{df(x)}{dx} + a_t f(x) && (55) \\
s(t) &= a_{xx}r(t) + a_{xt}\frac{dq(t)}{dt} + a_x q(t) && (56)
\end{aligned}$$

Rearranging various terms containing initial and boundary conditions, approximating all known and unknown functions in x and/or t in terms of a finite set of orthogonal basis functions,introducing the same in (52) and finally simplifying after making use of

$$\int_{t_0}^{t}\int_{t_0}^{t}\int_{x_0}^{x}\int_{x_0}^{x} \psi^T(\chi)F\phi(\tau)d\chi^2 d\tau^2 \approx \psi^T(x)E_{x2}^T F E_{t2}\phi(t)$$

we get a set of simple algebraic equations represented by

$$M\mathbf{p} = \mathbf{v} \tag{57}$$

where

$$\begin{aligned}
M &= [vec(E_{x2}^T Y) \mid vec(Y E_{t2}) \mid vec(E_x^T Y E_t) \mid vec(E_{x2}^T Y E_t) \mid \\
&\quad vec(E_x^T Y E_{t2}) \mid vec(E_{x2}^T Y E_{t2}) \mid -vec(E_{x2}^T \Delta_{11}) \mid \cdots \mid \\
&\quad -vec(E_{x2}^T \Delta_{\alpha 1}) \mid -vec(\Delta_{11} E_{t2}) \mid \cdots \mid -vec(\Delta_{1\beta} E_{t2}) \mid \\
&\quad -vec(E_x^T \Delta_{11} E_t) \mid -vec(E_{x2}^T \Delta_{11} E_t) \mid \cdots \mid -vec(E_{x2}^T \Delta_{\gamma 1} E_t) \mid \\
&\quad -vec(E_x^T \Delta_{11} E_{t2}) \mid \cdots \mid -vec(E_x^T \Delta_{1\delta} E_{t2})] && (58) \\
\mathbf{p} &= [a_{tt}, a_{xx}, a_{xt}, a_t, a_x, a, \hat{f}_0, \cdots, \hat{f}_{\alpha-1}, \hat{q}_0, \cdots, \\
&\quad \hat{q}_{\beta-1}, \hat{c}, h_0, \cdots, h_{\gamma-1}, s_0, \cdots, s_{\delta-1}]^T && (59) \\
\mathbf{v} &= vec(E_{x2}^T U E_{t2}) && (60)
\end{aligned}$$

where Δ_{ij} is an $m \times n$ matrix having (i,j)th element unity and all other elements zero,

$$\begin{aligned}
u(x,t) &\approx \psi^T(x) U \phi(t), \\
y(x,t) &\approx \psi^T(x) Y \phi(t), \\
\hat{f}_i &= a_{tt} f_i, \; i = 0, 1, \cdots, \alpha - 1 && (61) \\
f(x) &\approx \psi^T(x) \sum_{i=0}^{\alpha-1} f_i \Delta_{i+1,1} \phi(t), \alpha \leq m \\
\hat{q}_j &= a_{xx} q_j, \; j = 0, 1, \cdots, \beta - 1 && (62) \\
q_t &\approx \psi^T(x) \sum_{j=0}^{\beta-1} q_j \Delta_{1,j+1} \phi(t), \; \beta \leq n \\
\hat{c} &\approx \psi^T(x) \hat{c} \Delta_{11} \phi(t) && (63)
\end{aligned}$$

$$\begin{aligned}
h(x) &\approx \psi^T(x) \sum_{i=0}^{\gamma-1} h_i \Delta_{i+1,1} \phi(t), \; \gamma \leq m \\
s(t) &\approx \psi^T(x) \sum_{j=0}^{\delta-1} s_j \Delta_{1,j+1} \phi(t), \; \delta \leq n
\end{aligned}$$

and for any matrix A of order $m \times n$ the vector valued function is defined as

$$vec(A)_{(mn\times 1)} = \begin{bmatrix} \mathbf{a}_1 \\ \cdots \\ \mathbf{a}_2 \\ \cdots \\ \vdots \\ \mathbf{a}_n \\ \cdots \end{bmatrix}$$

with $\mathbf{a}_i$ being the ith column of A.

Since M is a $mn \times (7+\alpha+\beta+\gamma+\delta)$ matrix, for the equation (57) to be solved for the parameter vector $\mathbf{p}$ we must have

$$mn \geq (7+\alpha+\beta+\gamma+\delta)$$

This can be arranged by properly selecting the number of polynomials in the basis functions. Once the parameters are obtained, the initial condition $f(x)$ and the boundary condition $q(t)$ can also be obtained from (61) and (62) respectively.To get the other initial condition g(x), we integrate (55) once with respect to x and approximate all functions in x in terms of a finite series of basis functions and simplify to get

$$g(x) \approx \psi^T(x)\{[\mathbf{h} - a_t\mathbf{f}] + a_{xt}(E_x^T)^{-1}[\mathbf{c}\mathbf{e}_x - \mathbf{f}]\}/a_{tt} \tag{64}$$

with

$$\mathbf{e}_x = [1, \underbrace{0, \cdots, 0}_{(m-1)\ elements}]^T \tag{65}$$

In the similar lines the other boundary conditions $r(t)$ in (56) may be obtained from

$$r(t) \approx \{[\mathbf{s}^T - a_x\mathbf{q}^T] + a_{xt}[\mathbf{c}\mathbf{e}_t^T - \mathbf{q}]E_t^{-1}\}\phi(t)/a_{xx} \tag{66}$$

with

$$\mathbf{e}_t = [1, \underbrace{0, \cdots, 0}_{(n-1)\ elements}]^T \tag{67}$$

In order to apply the proposed identification approach via sine-cosine functions,all $m-$ and $n-$ vectors and $m \times m$ and $n \times n$ matrices must be changed to $(2m-1)-$ and $(2n-1)-$ vectors and $(2m-1) \times (2m-1)$ and $(2n-1) \times (2n-1)$ matrices,respectively.

Example: Identification of parameters of a distributed parameter system with noisy data

The following Example shows the numerical strength of the proposed algorithm

Table 10: Distributed parameter system identification with noisy data

Approach ↓ Actual parameters →	NSR	a_t 4	a_x 2	a 1
BPF	0.00	4.0000000	2.0000000	1.0000000
	0.05	4.0096739	1.9980807	1.0104679
	0.10	4.0193903	1.9961488	1.0209641
	0.15	4.0291495	1.9942042	1.0314885
	0.20	4.0389516	1.9922469	1.0420413
	0.25	4.0487969	1.9902769	1.0526223
TP1	0.00	4.0000000	2.0000000	1.0000000
TP2	0.00	3.9835752	1.9917876	1.0524816
	0.05	3.9751867	1.9973049	1.0431734
	0.10	3.9667872	2.0028184	1.0338624
	0.15	3.9583765	2.0083280	1.0245487
	0.20	3.9499549	2.0138337	1.0152323
	0.25	3.9415223	2.0193354	1.0059132
LeP	0.00	4.0000000	1.0000000	1.0000000
	0.05	4.0001419	1.9947943	0.9960242
	0.10	4.0002889	1.9895830	0.9920936
	0.15	4.0004413	1.9843661	0.9882074
	0.20	4.0005998	1.9791438	0.9843644
	0.25	4.0007647	1.9739162	0.9805639
LaP	0.00	2.3388511	0.9079211	8.3603706
	* 0.00	3.9975947	1.9968840	1.0063314
	0.05	4.0015651	1.9985769	1.0043083
	0.10	4.0055413	2.0002716	1.0022811
	0.15	4.0095231	2.0019681	1.0002499
	0.20	4.0023083	1.9923182	1.0044197
	0.25	4.0175040	2.0053665	0.9961753
SCF	0.00	3.9999971	1.9999974	1.0000107
	0.05	3.9998316	1.9998878	0.9879680
	0.10	3.9995152	1.9997253	0.9762400
	0.15	3.9990592	1.9995154	0.9647951
	0.20	3.9984672	1.9992588	0.9536276
	0.25	3.9977429	1.9989562	0.9427323

for the identification of distributed parameter systems via all classes of orthogonal functions. Let the model of a distributed parameter system be given by

$$a_t \frac{\partial y(x,t)}{\partial t} + a_x \frac{\partial y(x,t)}{\partial x} + ay(x,t) = u(x,t) \tag{68}$$

The parameters of the above model will be identified assuming zero initial and boundary conditions. By applying the input $u(x,t) = 4x + 2t + xt$, the output data for $y(x,t)$ is generated.The output and the input data over the interval $x \in [0,1], t \in [0,2]$ are used to estimate the parameters assuming that the output signal $y(x,t)$ is corrupted with a random noise having noise-to-signal ratio (NSR) 0.05, 0.10, 0.15, 0.20 or 0.25.For Laguerre polynomials, the interval selected is $x \in [0,11], t \in [0,12]$. For Hermite polynomials, the estimates are quite inconsistent and therefore are rejected.

There are altogether three parameters to be estimated and we take $m = n = 2$ for signal characterization. The parameter estimates are as shown in Table 10 where BPF stands for block pulse functions, TP1 and TP2 Tchebycheff polynomials of the first and the second kind respectively, LeP the Legendre polynomials, LaP the Laguerre polynomials and SCF the sine-cosine functions.

7 Transfer function matrix identification

Let the MIMO system be described by

$$\sum_{k=0}^{n_i} a_{ik} y_i^{(k)}(t) = \sum_{j=1}^{r} \sum_{k=0}^{n_i-1} b_{ijk} u_j^{(k)}(t), i = 1, 2, \ldots, p \tag{69}$$

where r is the number of inputs and p the number of outputs of the system. In terms of transfer function matrix, the system described by (69) can be writtten as

$$Y(s) = G(s)U(s) \tag{70}$$

where

$$G(s) = \begin{bmatrix} \frac{N_{11}(s)}{D_1(s)} & \frac{N_{12}(s)}{D_1(s)} & \cdots & \frac{N_{1r}(s)}{D_1(s)} \\ \frac{N_{21}(s)}{D_2(s)} & \frac{N_{22}(s)}{D_2(S)} & \cdots & \frac{N_{2r}(s)}{D_2(s)} \\ \vdots & \vdots & \vdots & \vdots \\ \frac{N_{p1}(s)}{D_p(s)} & \frac{N_{p2}(s)}{D_p(s)} & \cdots & \frac{N_{pr}(s)}{D_p(s)} \end{bmatrix} \tag{71}$$

$$D_i(s) = \sum_{k=0}^{n_i} a_{ik} s^k, a_{in_i} = 1, i = 1, 2, \ldots, p \tag{72}$$

$$N_{ij}(s) = \sum_{k=0}^{n_i-1} b_{ijk}s^k, i = 1,2,\ldots,p, j = 1,2,\ldots,r \tag{73}$$

$D_i(s)$ is the least common denominator of the ith row of $G(s)$ having the degree n_i, and the Laplace transform of a time function $f(t)$ is represented by the corresponding capital letter $F(s)$. To get (70) from (69) by taking the Lapalace transform of both sides of (69), we have considered the initial conditions to be zero. In fact, if we consider any record of the output due to a given input, the response should also have components due to these initial conditions. Therefore the system identification requires the determination of the parameters $\{a_{ik}, b_{ijk}\}, i = 1,2,\ldots,p; j = 1,2,\ldots,r; k = 0,1,\ldots,n_i$ and the initial conditions from a transient record of the system inputs and outputs. Integrating (69) n_i times from t_0 to t we have

$$\sum_{k=0}^{n_i} a_{ik}\left\{I_{n_i-k}[y_i(t)] - \sum_{q=0}^{k-1} y_i^{(q)}(t_0)\frac{t^{n_i-k+q}}{(n_i-k+q)!}\right\}$$
$$= \sum_{j=1}^{r}\sum_{k=0}^{n_i-1} b_{ijk}\left\{I_{n_i-k}[u_j(t)] - \sum_{q=0}^{k-1} u_j^{(q)}(t_0)\frac{t^{n_i-k+q}}{(n_i-k+q)!}\right\} \tag{74}$$

where $(n_i - k)$ times integration of a function $f(t)$ is denoted by

$$I_{n_i-k}[f(t)] = \int_{t_0}^{t}\int_{t_0}^{t_1}\cdots\int_{t_0}^{t_{n_i-k-1}} f(t_{n_i-k})dt_{n_i-k}\cdots dt_2dt_1 \tag{75}$$

The coefficients of t^k in the above expression (74) are grouped together and denoted by α_{ik}. So (74) takes the form

$$-\sum_{k=0}^{n_i-1} a_{ik}I_{n_i-k}[y_i(t)] + \sum_{j=1}^{r}\sum_{k=0}^{n_i-1} b_{ijk}I_{n_i-k}[u_j(t)] + \sum_{k=0}^{n_i-1}\alpha_{ik}t^k = y_i(t) \tag{76}$$

where we have used the fact that $a_{in_i} = 1$.

The outputs $y_i(t), i = 1,2,\ldots,p$;the inputs $u_j(t), j = 1,2,\ldots,r$, and $t^k, k = 0,1,\ldots,n_i - 1$ are now expanded into m-term shifted orthogonal series giving us:

$$y_i(t) = \mathbf{y}_i^T\psi(t) = y_{i0}\psi_0(t) + y_{i1}\psi_1(t) + \cdots + y_{i,m-1}\psi_{m-1}(t) \tag{77}$$
$$u_j(t) = \mathbf{u}_j^T\psi(t) = u_{j0}\psi_0(t) + u_{j1}\psi_1(t) + \cdots + u_{j,m-1}\psi_{m-1}(t) \tag{78}$$
$$t^k = \tau_k^T\psi(t) = \tau_{k0}\psi_0(t) + \tau_{k1}\psi_1(t) + \cdots + \tau_{k,m-1}\psi_{m-1}(t) \tag{79}$$

We shall, however, use unshifted polynomials if they are Laguerre and Hermite. In view of (20) we note that $(n_i - k)$-times integral of $y_i(t)$ and $u_j(t)$ can be writte as

$$I_{n_i-k}[y_i(t)] = \mathbf{y}_i^T E_s^{n_i-k} \psi(t) \tag{80}$$
$$I_{n_i-k}[u_j(t)] = \mathbf{u}_j^T E_s^{n_i-k} \psi(t) \tag{81}$$

Substituting (77)-(81) into (76) and equating coefficients of $\psi_i(t), i = 0, 1, \ldots, m-1$ from both sides we get

$$-\sum_{k=0}^{n_i-1} a_{ik}\mathbf{y}_i^T E_s^{n_i-k} + \sum_{j=1}^{r}\sum_{k=0}^{n_i-1} b_{ijk}\mathbf{u}_j^T E_s^{n_i-k} + \\ + \sum_{k=0}^{n_i-1} \alpha_{ik}\tau_k^T = \mathbf{y}_i^T \tag{82}$$

Let us denote the unknown parameter vector of the ith row of $G(s)$ by $\mathbf{x}_i$ then

$$\mathbf{x}_i = [\mathbf{a}_i^T \mathbf{b}_{i1}^T \mathbf{b}_{i2}^T \cdots \mathbf{b}_{ir}^T \boldsymbol{\alpha}_i^T]^T \tag{83}$$

where

$$\mathbf{a}_i = [a_{i0} a_{i1} \cdots a_{in_i-1}]^T \tag{84}$$
$$\mathbf{b}_{ij} = [b_{ij0} b_{ij1} \cdots b_{ijn_i-1}]^T, j = 1, 2, ..., r \tag{85}$$
$$\boldsymbol{\alpha}_i = [\alpha_{i0}\alpha_{i1} \cdots \alpha_{in_i-1}]^T \tag{86}$$

Therefore (82) can be written in the following vector-matrix form:

$$Q_i \mathbf{x}_i = \mathbf{y}_i, i = 1, 2, \ldots, p \tag{87}$$

where

$$Q_i = \big[-(E_s^T)^{n_i}\mathbf{y}_i \mid -(E_s^T)^{n_i-1}\mathbf{y}_i \mid \cdots \mid (-E_s^T)\mathbf{y}_i \mid (E_s^T)^{ni}\mathbf{u}_1 \mid \cdots \mid \\ E_s^T\mathbf{u}_1 \mid \cdots \mid (E_s^T)^{n_i}\mathbf{u}_r \mid \cdots \mid E_s^T\mathbf{u}_r \mid \tau_0 \mid \tau_1 \mid \cdots \mid \tau_{n_i-1}\big] \tag{88}$$

In (87) $\mathbf{x}_i$ is an $(r+2)n_i$-vector, $\mathbf{y}_i$ an m-vector and Q_i an m-vector and Q_i and m by $(r+2)n_i$ matrix. If we denote the set of all unknown parameters by the vector $\mathbf{x}$:

$$\mathbf{x} = [\mathbf{x}_1^T \mathbf{x}_2^T \cdots \mathbf{x}_p^T]^T$$

then (87) can be written in the compact form

$$Qx = y \tag{89}$$

where

$$\begin{aligned} Q &= \text{diag}[Q_1, Q_2, \ldots, Q_p] \\ \mathbf{y} &= [\mathbf{y}_1^T \mathbf{y}_2^T \ldots \mathbf{y}_p^T]^T \end{aligned}$$

Since there are $(r+2)n$ unknown parameters in $\mathbf{x}$ where $n = n_1 + n_2 + \cdots + n_p$, it is necessary to choose m so that

$$mp \geq (r+2)n$$

for a least square solution.

Example:MIMO system identification

We consider the model of a two-input one-output MIMO system described by

$$\ddot{y}(t) + a_{11}\dot{y}(t) + a_{10}y(t) = b_{111}\dot{u}_1(t) + b_{110}u_1(t) + b_{221}\dot{u}(t) + b_{210}u_2(t)$$

With the input functions $u_1(t) = t$ and $u_2(t) = t^3$, the output data are generated with zero initial conditions to estimate the parameters whose true values are

$$b_{111} = 1.0, b_{110} = 0.5, b_{121} = 0, b_{120} = 1.0, a_{11} = 1.5, a_{10} = 0.5$$

The estimates of the parameters and the initial conditions obtained using m=8 are provided in Table 11. The Hermite and Laguerre polynomials are not included, as the estimates offered by them are not satisfactory.

8 Conclusion

An integrated framework is provided in this Chapter to determine the parameters and initial conditons of linear time-invariant continuous time lumped parameter as well as distributed parameter models. Time-invariant multiinput and multioutput lumped parameter systems are treated by extending in a straightforward fashion the techniques for a single-input single-output case. The three-term recurrence relations are very important in this context as they permit us to generate any order of orthogonal polynomials in any class recursively and the differential recurrence re-

Table 11: Patameter estimates of a two-input and one-output system

Parameters	Actuals	Legendre	Tcheby I	Tcheby II	Sine-cosine
a_{10}	0.5	0.50081791	0.50090608	0.50072220	0.46648337
a_{11}	1.5	1.49894343	1.49884961	1.49906578	1.44885305
b_{110}	0.5	0.49894343	0.45686483	0.49905633	0.46648337
b_{111}	1.0	1.00000060	1.02019722	1.00037574	0.96590204
b_{120}	1.0	0.99953655	0.99949612	0.99958958	0.93296674
b_{121}	0.0	0.00033206	0.00102923	0.00029379	0.00000000
$y(0)$	0.0	0.00000000	0.00000000	-0.00000003	0.00207060
$\dot{y}(0)$	0.0	-0.00000002	0.00002079	0.00000021	-0.00037316

lations are convenient to provide a general structure to the integration operational matrix. Although sine-cosine functions are not governed by such relations, in view of the simple structure of these functions and of their derivatives, it is not difficult to construct the integration operational matrix. The shifted orthogonal polynomials are important to study the identification problem over any arbitrary interval. The results provided by infinite-range orthogonal polynomials are not as good as other orthogonal polynomials, and sine-cosine functions give us a good accuracy if the number of basis functions is higher than needed by finite-range orthogonal polynomials.

References

Abramowitz, M. and Stegun, I.A. (1964), *"Handbook of Mathematical Functions with Formulas, Graphs, and Mathematical Tables"*, U.S. Govt. Printing Office, Washington D.C., pp. 771-802.

Bell, E.T. (1945), *"The Development of Mathematics"*, McGraw-Hill, New York.

Chang, R.Y. and Wang, M.L. (1982), "Parameter identification via shifted Legendre polynomials", *Int. J. Syst. Sci.*, Vol. 13, pp. 1125-1135.

Chen, C.F. and Hsiao, C.H. (1975), "Time-domain synthesis via Walsh functions",*Proc. IEE*, Vol. 122, pp. 565-570.

Chen, C.T. (1987), "Techniques for identification of linear time-invariant multivariable systems", *Advances in Control and Dynamical Systems*, (Academic Press), Vol.26, Part 2, pp. 1-34.

Cheng, B., and Hsu, N. S. (1982), " Single input and single output system identification via block pulse functions", *Int. J. Syst. Sci.*, Vol. 13, pp. 697-702.

Chung, H.Y. (1987), "System identification via Fourier series", *Int. J. Syst. Sci.*, Vol. 18, pp. 1191-1194.

Corrington, M.S. (1973), "Solution of differential and integral equations with Walsh functions", *IEEE Trans. Circuit Theory*, Vol. CT-20, pp. 470-476.

Horng, I.R., Chou, J.H. and Tsai, C.H. (1986), "Analysis and identification of linear distributed systems via Chebyshev series", *Int. J. Syst. Sci.*, Vol. 17, pp. 1089-1095.

Hwang, C. and Guo, T.Y. (1984), "Transfer-function matrix identification in MIMO systems via shifted Legendre polynomials", *Int. J. Control*, Vol. 39, pp. 807-814.

Hwang, C. and Shih, Y.P. (1982), "Parameter identification via Laguerre polynomials", *Int. J. Syst. Sci.*, Vol.13, pp.209-217.

Jha, A.N. and Zaman, S. (1985), "Identification of linear distributed systems using Laguerre operational matrices", *Int. J. Control*, Vol.16, pp. 761-767.

Jiang, Z. H. (1987), " Block pulse function approach to the identification of MIMO systems and time delay systems",*Int. J. Syst. Sci.*, Vol.18, pp. 1711-1720.

Kline, M. (1972), *"Mathematical Thought from Ancient to Modern Times"*, Oxford University Press, New York.

Mohan, B.M. and Datta, K.B. (1988). "Lumped and distributed parameter system identification via shifted Legendre polynomials", *J. Dynamic Systems Measurement and Control*, Vol.110, pp. 436-440.

Mohan, B.M., and Datta, K.B. (1989), "Identification via Fourier series for a class of lumped and distributed parameter systems", *IEEE Trans. Circuits and Systems*,Vol. 36, pp. 1454-1458.

Mohan, B.M. and Datta, K.B. (1991a), "Linear time-invariant distributed parameter system identification via orthogonal functions", *Automatica*, Vol. 27, March 1991.

Mohan, B.M. and Datta, K.B. (1991b), "Identification of linear time-invariant single-input single-output continuous-time lumped parameter systems via orthogonal functions",*Control - Theory and Advanced Technology*, to be published.

Palanisamy, K.R. and Bhattacharyya, D.K. (1981), " System identification via block pulse functions",*Int. J. Syst. Sci.*, Vol. 12, pp. 643-647.

Paraskevopoulos, P.N. (1978), "Transfer-function matrix identification via Walsh functions", Proc. MECO '78, (Athens), Vol.1. pp. 194-197.

Paraskevopoulos, P.N. (1983), "Chebyshev approach to system identification, analysis and optimal control", *J. Franklin Institute*, Vol. 316, pp. 135-157.

Paraskevopoulos, P.N. and Bounas, A.C. (1978), "Distributed parameter system identification via Walsh functions", *Int. J. Syst. Sci.*, Vol. 9, pp. 75-83.

Paraskevopoulos, P.N. and Kekkeris, G. Th. (1983), "Identification of time-invariant and time-varying distributed parameter systems using Chebyshev functions", Proc. Int. ASME Conf. on Modeling and Simulation, (Nice), Vol. 1, pp. 51-69.

Paraskevopoulos, P.N. and Kekkeris, G.Th. (1983), "Hermite series apporoach to system identification, analysis and optimal control", Proc. Sixth Int. Symp. MECO'83 - Measurement and Control, (Athens), Vol.1, pp. 146-149.

Ranganathan, V., Jha, A.N. and Rajamani, V.S. (1984), "Identification of linear distributed systems via Laguerre polynomials", *Int. J. Syst. Sci.*, Vol. 15, pp. 1101-1106.

Rao, G.P. and Sivakumar, L., "System identification via Walsh functions ", *Proc. IEE*, Vol.122, pp.1160-1161.

Rao, G.P. and Sivakumar, L. (1981), "Transfer-function matrix identification in MIMO systems via Walsh functions", *Proc. IEEE*, Vol. 69, pp. 465-466.

Rao, G.P. and Palanisamy, K.R. (1983), "Improved algorithms for parameter identification in continuous systems via Walsh functions", *IEE Proc.*, Vol. 130, Pt.D, pp. 9-16.

Rao, G.P. (1983), *"Piecewise Constant Orthogonal Functions and their Application to Systems and Control"*, Springer-Verlag, Berlin.

Sansone, G. (1959), *"Orthogonal Functions"*,Interscience Publishers, New York.

Szegö, G. (1959), *"Orthogonal Polynomials"*, American Mathematical Society, New York.

Use of numerical integration methods

H. Dai and N.K. Sinha
Department of Electrical & Computer Engineering
McMaster University
Hamilton, Canada L8S 4L7

Abstract

Identification of the continuous-time system model from samples of input-output data can be carried out directly by integrating the system dynamic equations over various sampling intervals and determining the model parameters which provide the best fit to the data in some sense. Using numerical integration methods, three methods are proposed for identification of continuous-time systems which are either single-input single-output systems or multi-input multi-output systems. Several methods of numerical integration are examined for their suitability, especially in the presence of noise. Robustness of the presented identification algorithms is also investigated.

1. Introduction

As pointed out in earlier chapters, it is often more desirable to estimate the parameters of the model of a continuous-time system directly from the samples of the input-output observations. The main idea here is to integrate the system differential equation between the sampling instants to obtain an expression relating the samples of the input and output to the parameters of the system model. For example, consider the following state-space model.

$$\begin{aligned} \dot{\boldsymbol{x}} &= \boldsymbol{A}\boldsymbol{x} + \boldsymbol{B}\boldsymbol{u} \\ \boldsymbol{y} &= \boldsymbol{C}\boldsymbol{x} + \boldsymbol{D}\boldsymbol{u} \end{aligned} \tag{1}$$

where $\boldsymbol{x} \in \mathrm{R}^n$, $\boldsymbol{y} \in \mathrm{R}^p$, and $\boldsymbol{u} \in \mathrm{R}^m$. $\boldsymbol{A}$, $\boldsymbol{B}$, $\boldsymbol{C}$, and $\boldsymbol{D}$ are parameter matrices.

Our problem is to determine the elements of the matrices $\boldsymbol{A}$, $\boldsymbol{B}$, $\boldsymbol{C}$, and $\boldsymbol{D}$ from observations of the input, $\boldsymbol{u}(t)$, and output, $\boldsymbol{y}(t)$, at the sampling instants, $t = kT$, where k is an integer in the range 0 to N. It is assumed that the sampling interval, T, has been selected carefully, so that there is negligible loss of information produced by the sampling process. Complications are introduced by the fact that, in practice, the data are contaminated with noise.

If we integrate the differential equation in (1) between the limits $(k\text{-}1)T$ and kT, we obtain

N. K. Sinha and G. P. Rao (eds.), Identification of Continuous-Time Systems, 259–289.

$$
\begin{aligned}
x(k) - x(k \;\; 1) &= A \int_{(k-1)T}^{kT} x(t)\,dt + B \int_{(k-1)T}^{kT} u(t)\,dt \\
y(k) &= Cx(t) + Du(k)
\end{aligned}
\tag{2}
$$

where, for convenience in notation, $\boldsymbol{x}(k)$ represents $\boldsymbol{x}(kT)$, $\boldsymbol{u}(k)$ represents $\boldsymbol{u}(kT)$ and $\boldsymbol{y}(k)$ represents $\boldsymbol{y}(kT)$. It is evident that it would be possible to estimate the values of the elements of the matrices $\boldsymbol{A}$, $\boldsymbol{B}$, $\boldsymbol{C}$, and $\boldsymbol{D}$ from the records of the $\boldsymbol{u}(k)$ and $\boldsymbol{y}(k)$ if we could evaluate the integrals in equation (2) for various values of k, as we would now get a set of linear equations in terms of the unknowns.

In practice, we cannot actually perform the integrations indicated in equation (2), since we do not have exact expressions for $\boldsymbol{x}(t)$ and $\boldsymbol{u}(t)$. In fact, we only have the values of $\boldsymbol{y}(k)$ and $\boldsymbol{u}(k)$ to which some noise has been added. Thus our main problem is to obtain approximate expressions for the integrals in terms of $\boldsymbol{y}(k)$ and $\boldsymbol{u}(k)$ by assuming a structure for the system model.

Using numerical integration approaches, several methods have been proposed for identification of continuous-time systems (Hung, Liu, and Chan 1980; Sinha and Lastman 1982; Zhou and Sinha 1982; Prasad and Sinha 1983; and Bingulac and Sinha 1989). However, some of the approaches are suitable for only single-input single-output systems. In this chapter, we shall first compare these methods and then extend them to multi-input multi-output systems.

It is convenient to assume that these matrices are a minimal representation of the system, i.e., the resulting model is completely controllable and observable. Furthermore, it is desirable to consider certain canonical forms which simplify the identification problem. In this chapter, three different model structures are employed to perform identification of continuous-time systems

To simplify matters, we shall first consider single-input single-output systems for the noise-free case and then extend the procedures to multivariable systems. Finally, we shall examine the effect of noise on parameter estimates given by these methods.

2. Identification of single-input full state output systems

Single-input full state output systems

In this case, we assume that the $u(t)$ is scalar and the number of outputs equals the number of states of the systems. It follows that $\boldsymbol{B}$ is a column vector of dimension n and $\boldsymbol{C}$ is an $n \times n$ non-singular and known matrix. Without losing generality, we assume that the matrix $\boldsymbol{D}$ is zero so that equation (1) can be rewritten as

$$\dot{x} = Ax + Bu \tag{3a}$$

$$y = Cx + E \tag{3b}$$

where C is a known matrix, and the matrix A and vector B are given by

$$A = \begin{bmatrix} a_{11} & \cdots & a_{1n} \\ \vdots & & \vdots \\ a_{n1} & \cdots & a_{nn} \end{bmatrix} \quad and \quad B = \begin{bmatrix} b_1 \\ \vdots \\ b_n \end{bmatrix} \tag{4}$$

where n is the model order of the system. In equation (3), E is the noise vector contaminating the output, that is

$$E^T = [e_1(t) \cdots e_n(t)] \tag{5}$$

where the elements $\{e_i(t)\ i=1, \ldots, n\}$ of E are supposed to be white noise.

Review of some numerical integration methods

Numerical integration has been extensively discussed in the literature and various methods have been described in many references (Burden and Faires 1981, Lastman and Sinha 1989). Some commonly used methods are the Mid-point rule, the Trapezoidal rule, Simpson's rule, and the Newton-Cotes formulas. The basic idea behind these methods is same, that is to approximate the integral, the area surrounded by the integrand and the upper and lower limits of integration on the axis, by a finite sum of panels. Here, the kth panel is the region between the graph of $f(t)$ and the subinterval $[(k\text{-}1)T,\ kT]$ on the t-axis.

In the Mid-point rule, we choose the intermediate points at the base-midpoint of each panel. It leads to the following approximation formula

$$I(k) = Tf\left(\frac{(k-1)T+kT}{2}\right) \tag{6}$$

where $k=1,\ldots,N$, and T is the sampling interval and given as

$$T = \frac{t_f}{N}. \tag{7}$$

Here we assume that the integration is performed over the interval $[0, t_f]$. Fig. 1 shows a typical panel for the Mid-point rule. Equation (6) must be modified since we do not know the values of $f(t)$ except at the sampling instants. The resulting equation is

$$I(k+1) = 2Tf(kT) \tag{8}$$

where $I(k+1)$ denotes the integral of $f(t)$ on the subinterval $[(k\text{-}1)T, (k+1)T]$.

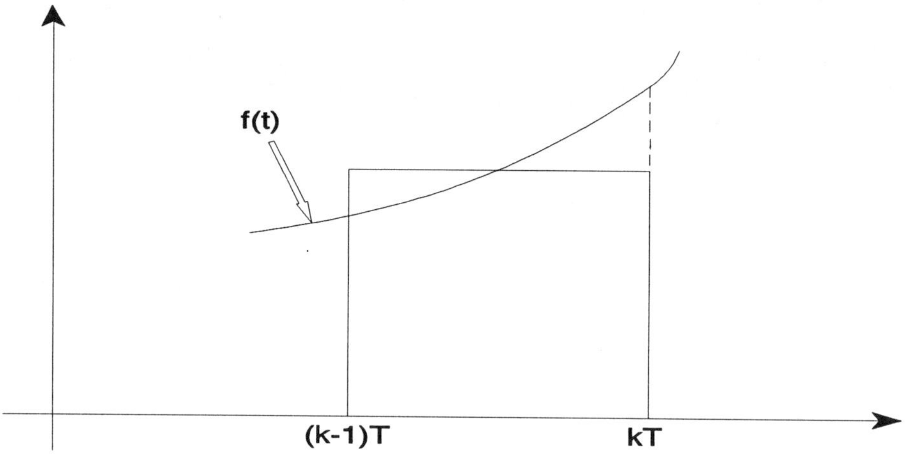

Figure 1. Typical panel of mid-point rule

Evidently, the approximation of numerical integration using the Mid-point rule is not precise and other kinds of panels are needed for determining the area more accurately.

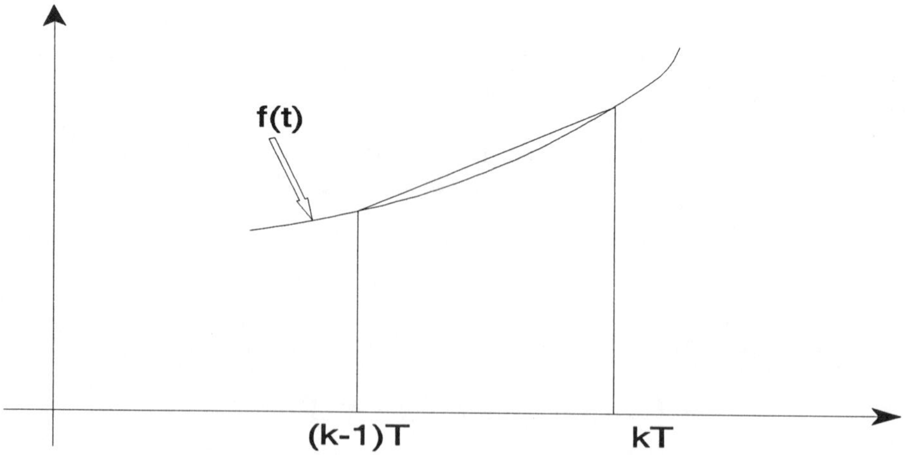

Figure 2. Typical panel of trapezoidal rule

If we choose a trapezoid to approximate the area of a typical panel, then we have the Trapezoidal rule, that is

$$I(k) = T\frac{f((k-1)T) + f(kT)}{2} \tag{9}$$

for $k=1, 2, 3, \ldots, N$. As shown in Fig. 2, the area of a typical panel is approximated by a trapezoid and the accuracy of numerical integration is improved.

Instead of using either a rectangle or a trapezoid for approximating the panel area, we can utilize a more complicated panel. Introducing the panel like the one shown in Fig. 3 gives Simpson's rule, described as

$$I(k) = \frac{T}{3}[f((k-2)T) + 4f((k-1)T) + f(kT)] \tag{10}$$

for $k=2, 4, 6, \ldots$. From Fig. 3, we can clearly see that the panel of Simpson's rule provides a closer approximation to the integral than obtained with the Mid-point and the Trapezoidal rules.

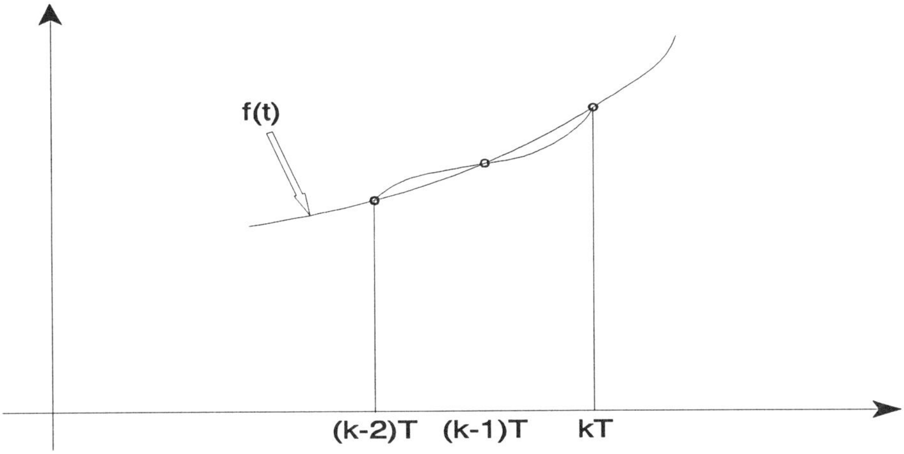

Figure 3. Typical panel of simpson's rule

More general numerical integration methods are based on the Newton-Cotes quadrature method. In the Newton-Cotes family of methods, there are two sub-families. One utilizes a closed formula while other uses an open formula. Here, we choose a closed integration formula, given by the following equation

$$I(k) = \frac{3}{8}T[f((k-3)T) + 3f((k-2)T) + 3f((k-1)T) + f(kT)]. \tag{11}$$

It should be noted that the Mid-point rule, the Trapezoidal rule, and the Simpson's rule are special examples of the Newton-Cotes family of methods.

It should be pointed out that there are many other methods available for numerical integration, such as the open Newton-Cotes formulas, the other closed Newton-Cotes formulas, Gaussian quadrature, Romberg quadrature, and adaptive quadrature (Lastman and Sinha 1989). In general, there is no fixed rule for deciding which numerical integration method is the best for a particular integration problem. It depends mainly upon the integrand f(t) and the individual problem itself.

Another issue associated with numerical integration is the selection of the size of the sampling interval, as well as whether equal spacing or variable spacing should be used. This issue relies not only on the given integrand $f(t)$ and the particular integration problem but also the method that has been selected for numerical integration. As is usual with most problems in system identification, the sampling interval is a constant, as described by equation (7). Consequently, equal spacing is utilized for numerical integration throughout this chapter.

In the rest of this section, we shall apply the numerical integration methods described above to estimation of system parameters.

Identification using numerical integration methods

As described earlier, equation (3a) can be rewritten as

$$\boldsymbol{x}(k) - \boldsymbol{x}(k-1) = \boldsymbol{A}\boldsymbol{I}_x(k) + \boldsymbol{B}I_u(k) \tag{12}$$

by integrating both sides of (3a) from $(k\text{-}1)T$ to kT. Here the $\boldsymbol{I}_x(k)$ and $I_u(k)$ are the numerical integrals of $\boldsymbol{x}(t)$ and $u(t)$, respectively. The former is a vector and the latter is a scalar. The vector $\boldsymbol{I}_x(k)$ is defined as

$$\boldsymbol{I}_x^T(k) = [I_{x_1}(k), \cdots, I_{x_n}(k)]. \tag{13}$$

Substituting (13) into (12) and considering the structure of $\boldsymbol{A}$ and $\boldsymbol{B}$ in (4), we then have

$$\begin{bmatrix} x_1(k)-x_1(k-1) \\ \vdots \\ x_n(k)-x_n(k-1) \end{bmatrix} = \begin{bmatrix} a_{11} & \cdots & a_{1n} \\ \vdots & & \vdots \\ a_{n1} & \cdots & a_{nn} \end{bmatrix} \begin{bmatrix} I_{x_1}(k) \\ \vdots \\ I_{x_n}(k) \end{bmatrix} + \begin{bmatrix} b_1 \\ \vdots \\ b_n \end{bmatrix} I_u(k). \tag{14}$$

Clearly, if we have all the state measurements and the input samples, the parameters a_{ij} and b_i can be easily estimated by applying the least squares approach to every individual equation in (14). For example, the least-squares estimates of a_{ij} and b_i can be obtained by solving the following equation.

$$\boldsymbol{\theta}_i = [\Phi^T\Phi]^{-1}\Phi^T\Delta\boldsymbol{X}_i \tag{15}$$

where the matrix $\boldsymbol{\Phi}$ and vectors $\boldsymbol{\theta}_i$ and $\Delta\boldsymbol{X}_i$ are defined as

$$\Phi = \begin{bmatrix} I_{x_1}(1) & I_{x_2}(1) & \cdots & I_{x_n}(1) & I_u(1) \\ \vdots & \vdots & & \vdots & \vdots \\ I_{x_1}(N) & I_{x_2}(N) & \cdots & I_{x_n}(N) & I_u(N) \end{bmatrix} \tag{16a}$$

$$\theta_i^T = [\, a_{i1} \quad a_{in} \; b_i \,] \tag{16b}$$

$$\Delta X_i^T = [\, x_i(1) \;\; x_i(0) \qquad x_i(N) \;\; x_i(N \;\; 1) \,] \tag{16c}$$

for $i=1, \ldots, n$. Here we assume the integral I_{x_i} is given by either equation (8) or equation (9). If other numerical integration methods are used, for example, equations (10) and (11), the indices in equations (16b) and (16c) should be changed accordingly.

It is interesting to note that the identification procedures using the Trapezoidal rule and the Block-pulse function coincide. It is worth emphasizing that all the other numerical integration methods should be, in principle, available for the proposed identification procedure. The only modification needed is calculating the integrals of the states and the input by using the selected numerical integration method.

Examples

To illustrate the identification procedure discussed above, we shall consider the following two examples. In these examples, it is assumed that the output data are not contaminated with any noise or measurement errors. The effect of noise on the estimated parameters will be considered in section 5.

Example 1. Let the system is described by an observable canonical form. The matrix $\boldsymbol{A}$ and vector $\boldsymbol{B}$ are given as

$$\boldsymbol{A} = \begin{bmatrix} -0.7 & 1.0 \\ -1.4 & 0.0 \end{bmatrix}, \quad \boldsymbol{B} = \begin{bmatrix} 1.0 \\ 2.0 \end{bmatrix}, \quad \text{and} \quad \boldsymbol{C} = \boldsymbol{I} \tag{17}$$

where $\boldsymbol{I}$ denotes the identity matrix. The sampling interval is taken as 0.01 while the lower and upper integration limits are 0.0 and 4.0, respectively. The initial values of states are zero and the input signal $u(t)$ is chosen as an exponential function, i.e.

$$u(t) = e^{-0.5t}. \tag{18}$$

The parameters of the system matrices were estimated by using the four different numerical integration methods described earlier. The resulting values, corresponding to the system dscribed by equation (17) are listed in Table 1. It may be noted that all methods give the exact values for this example.

Table 1. Parameter estimates of the system in Example 1

	Mid-point rule	Trapezoidal rule	Simpson's rule	Newton-Cotes rule
a_{11}	-0.7000	-0.7000	-0.7000	-0.7000
a_{12}	1.0000	1.0000	1.0000	1.0000
a_{21}	-1.4000	-1.4000	-1.4000	-1.4000
a_{22}	-0.0000	-0.0000	-0.0000	-0.0000
b_1	0.9999	1.0000	1.0000	1.0000
b_2	2.0000	2.0000	2.0000	2.0001

Example 2. In this example, a second-order observability canonical form is considered, which is

$$A = \begin{bmatrix} 0.0 & 1.0 \\ -1.4 & -0.7 \end{bmatrix}, \quad B = \begin{bmatrix} 1.0 \\ 2.0 \end{bmatrix}, \quad \text{and} \quad C = I. \tag{19}$$

The input signal, the initial states, the sampling interval, and the lower and upper integration limits are the same as in Example 1. Also, the same four numerical integration methods are applied to identify the system parameters. The estimated parameters are given in Table 2.

Table 2. Parameter estimates of the system in Example 2

	Mid-point rule	Trapezoidal rule	Simpson's rule	Newton-Cotes rule
a_{11}	0.0000	-0.0000	-0.0000	0.0000
a_{12}	1.0000	1.0000	1.0000	1.0000
a_{21}	-1.4000	-1.4000	-1.4000	-1.4000
a_{22}	-0.7000	-0.7000	-0.7000	-0.7000
b_1	1.0000	1.0001	1.0000	1.0000
b_2	2.0001	2.0000	2.0001	2.0000

Again, we can see that the estimated parameters are identical although different numerical integration methods were employed. However, these are ideal cases, with no noise. If we take into account noise or measurement errors included in the input-output data, the parameter estimates will certainly deviate from their true values. This is the subject of section 5 in this chapter.

Identification of multi-input multi-output systems

In the last subsection, we discussed the identification of single-input and full state output systems using numerical integration methods. Now we shall extend the procedure to multi-input multi-output systems.

Let the multi-input multi-output system be described by equation (3). For the sake of simplicity, we assume that the matrix $\boldsymbol{D}$ is zero. In fact, it is very straightforward to modify the proposed procedure if $\boldsymbol{D}$ is not a null matrix. Following the reasoning given earlier, we get the integrated equation

$$\boldsymbol{x}(k) - \boldsymbol{x}(k-1) = \boldsymbol{A}\boldsymbol{I}_x(k) + \boldsymbol{B}\boldsymbol{I}_u(k) \tag{20}$$

where $\boldsymbol{I}_x(k)$ is given in (13) and $\boldsymbol{I}_u(k)$ is defined by

$$\boldsymbol{I}_u^T(k) = [I_{u_1}(k), \cdots, I_{u_m}(k)]. \tag{21}$$

Here, I_{u_i} is the numerical integral of the ith input signal. Similar to equation (14), the following equation can be obtained.

$$\begin{bmatrix} x_1(k)-x_1(k-1) \\ \vdots \\ x_n(k)-x_n(k-1) \end{bmatrix} = \begin{bmatrix} a_{11} & \cdots & a_{1n} \\ \vdots & & \vdots \\ a_{n1} & \cdots & a_{nn} \end{bmatrix} \begin{bmatrix} I_{x_1}(k) \\ \vdots \\ I_{x_n}(k) \end{bmatrix} + \begin{bmatrix} b_{11} & \cdots & b_{1m} \\ \vdots & & \vdots \\ b_{n1} & \cdots & b_{nm} \end{bmatrix} \begin{bmatrix} I_{u_1}(k) \\ \vdots \\ I_{u_m}(k) \end{bmatrix}. \tag{22}$$

The parameters a_{ij} and b_{il}, $i,j = 1, \ldots, n$ and $l = 1, \ldots, m$, can be directly estimated using the least squares approach provided that all the states and inputs of the system are available. Note that equation (15) can still be used to solve for the ith set of parameters a_{ij} and b_{il}, but the matrix $\boldsymbol{\Phi}$ and vectors $\boldsymbol{\theta}_i$ and $\Delta\boldsymbol{X}_i$ should be defined as

$$\boldsymbol{\Phi} = \begin{bmatrix} I_{x_1}(1) & \cdots & I_{x_n}(1) & I_{u_1}(1) & \cdots & I_{u_m}(1) \\ \vdots & & \vdots & \vdots & & \vdots \\ I_{x_1}(N) & \cdots & I_{x_n}(N) & I_{u_1}(N) & \cdots & I_{u_m}(N) \end{bmatrix} \tag{23a}$$

$$\boldsymbol{\theta}_i^T = [\, a_{i1} \ \cdots \ a_{in} \ b_i \ \cdots \ b_{im}] \tag{23b}$$

$$\Delta\boldsymbol{X}_i^T = [\, x_i(1)-x_i(0) \ \cdots \ x_i(N)-x_i(N-1)\,] \tag{23c}$$

for $i=1, \ldots, n$. The indices in equations (23b) and (23c) should be correspondingly changed if another numerical integration method is selected. Notice that equation (23) is almost the same as equation (16). The only difference is in the dimensions of the matrix $\boldsymbol{\Phi}$ and vector $\boldsymbol{\theta}_i$ due to the extra input signals introduced.

Example 3. Consider a system described by equation (3), with the matrices $\boldsymbol{A}$, $\boldsymbol{B}$, and $\boldsymbol{C}$ chosen as

$$A = \begin{bmatrix} -2.0 & 1.0 & 1.3 & 0.0 \\ -3.0 & 0.0 & 2.0 & 0.0 \\ -3.0 & 0.0 & -1.4 & 1.0 \\ 2.0 & 0.0 & -2.7 & 0.0 \end{bmatrix}, \quad B = \begin{bmatrix} 1.0 & 1.0 \\ 0.0 & 0.0 \\ -1.0 & 1.0 \\ 1.0 & 0.0 \end{bmatrix}, \quad \text{and} \quad C = I. \tag{24}$$

The input signals are selected as

$$\begin{aligned} u_1(t) &= e^{-0.5t} \\ u_2(t) &= e^{-0.1t}. \end{aligned} \tag{25}$$

The sampling interval is taken 0.01 and the lower and upper integration limits are chosen as 0.0 and 4.0, respectively. The initial values of the states are taken as zero.

Four numerical integration methods are applied to the identification procedure in order to estimate the system parameters. The results are listed in Table 3 and show that all methods give good estimates of parameters in the absence of noise.

Table 3. Parameter estimates of the system in Example 3

	Mid-point rule	Trapezoidal rule	Simpson's rule	Newton-Cotes rule
a_{11}	-1.9994	-1.9999	-1.9997	-2.0000
a_{12}	0.9995	0.9996	0.9996	1.0000
a_{13}	1.3000	1.3002	1.3001	1.3000
a_{14}	-0.0003	-0.0002	-0.0002	-0.0000
a_{21}	-2.9994	-3.0003	-3.0000	-2.9995
a_{22}	-0.0000	0.0000	-0.0000	-0.0008
a_{23}	2.0000	2.0000	2.0000	2.0003
a_{24}	-0.0001	0.0001	-0.0000	-0.0004
a_{31}	-3.0000	-3.0001	-3.0000	-3.0002
a_{32}	0.0001	0.0001	-0.0001	0.0004
a_{33}	-1.3996	-1.4002	-1.4000	-1.4001
a_{34}	0.9998	1.0002	1.0000	1.0002
a_{41}	1.9996	2.0001	2.0002	2.0001
a_{42}	-0.0003	0.0003	-0.0003	-0.0002
a_{43}	-2.6998	-2.7002	-2.6999	-2.6999
a_{44}	-0.0001	0.0001	-0.0002	-0.0001
b_{11}	1.0008	1.0008	1.0007	1.0000
b_{12}	0.9991	0.9992	0.9993	1.0000
b_{21}	0.0003	-0.0002	0.0001	0.0014
b_{22}	-0.0002	0.0001	-0.0001	-0.0014
b_{31}	-0.9995	-1.0005	-0.9999	-1.0007
b_{32}	0.9999	1.0003	0.9999	1.0007
b_{41}	1.0005	0.9995	1.0006	1.0003
b_{42}	-0.0004	0.0004	-0.0006	-0.0003

3. Identification of continuous-time system in diagonal form

As pointed out earlier, the identification method described in the last section requires the observation of all state variables and inputs in order to obtain the state-space model. Obviously, this is not a reasonable assumption in practice since in many situations, we can only measure the output instead of all states.

To identify parameters of continuous-time systems from samples of only input-output data, some special technique is needed. Zhou and Sinha (1982) have developed an approach to identify a state-space model of single-input single-output continuous-time systems from given input-output samples. The key issue of the approach is to utilize a special form of state-space model, i.e. the diagonal form. Identification can be performed after integrating the state-space model of the diagonal form using the Trapezoidal rule.

In this section, we shall first utilize this approach to the identification of single-input single-output systems, and then extend it to multi-input multi-output systems.

Diagonal form of single-input single-output systems

It is well known that any linear lumped continuous-time system, that has distinct eigenvalues, can be always described by a diagonal form. That is,

$$\dot{x} = Ax + Bu \tag{26a}$$

$$y = Cx + E \tag{26b}$$

where the diagonal matrix A and vectors B and C are given as

$$A = diag\,[\lambda_1\ \lambda_2\ \cdots\ \lambda_n] \tag{27a}$$

$$B^T = [\bar{b}_1\ \bar{b}_2\ \cdots\ \bar{b}_n] \tag{27b}$$

$$C^T = [1\ 1\ \cdots\ 1]. \tag{27c}$$

The non-zero elements of the diagonal matrix A are the distinct real eigenvalues of the system. In the case of conjugate complex eigenvalues, the above equation is slightly modified for the matrix A. For example, if a second order system has a pair of conjugate complex eigenvalues $\alpha \pm j\beta$, then the corresponding matrix A is

$$A = \begin{bmatrix} \alpha & \beta \\ -\beta & \alpha \end{bmatrix}. \tag{28}$$

If some eigenvalues of the system are identical, the Jordan form can be used to replace equation (27a). For instance, the matrix $\boldsymbol{A}$ of a second order system should be

$$\boldsymbol{A} = \begin{bmatrix} \lambda & 1 \\ 0 & \lambda \end{bmatrix} \tag{29}$$

where λ is the eigenvalue of the system.

Since it is always possible to use a diagonal or Jordan form to represent a linear system, we shall develop the identification procedure only for the case of distinct real eigenvalues in the following part of this section. However, an example will be provided to illustrate the approach for a case of complex conjugate eigenvalues.

Identification procedure for diagonal form single-input single-output systems

As in the previous section, the following equation can be obtained by integrating both sides of equation (26)

$$\boldsymbol{x}(k) - \boldsymbol{x}(k-1) = \boldsymbol{A}\boldsymbol{I}_x(k) + \boldsymbol{B}\boldsymbol{I}_u(k). \tag{30}$$

Integrating the above equation applying the Trapezoidal rule, we get

$$(\boldsymbol{I} - \frac{\boldsymbol{A}T}{2})\boldsymbol{x}(k) = (\boldsymbol{I} + \frac{\boldsymbol{A}T}{2})\boldsymbol{x}(k-1) + \frac{\boldsymbol{B}T}{2}[u(k) + u(k-1)]. \tag{31}$$

Since $\boldsymbol{A}$ is a diagonal matrix, the above equation can be rewritten as

$$diag[(1 - \frac{\lambda_1 T}{2}) \;\; (1 - \frac{\lambda_2 T}{2}) \;\; \cdots \;\; (1 - \frac{\lambda_n T}{2})]\boldsymbol{x}(k)$$

$$= diag[(1 + \frac{\lambda_1 T}{2}) \;\; (1 + \frac{\lambda_2 T}{2}) \;\; \cdots \;\; (1 + \frac{\lambda_n T}{2})]\boldsymbol{x}(k-1) + \frac{T}{2}\boldsymbol{B}[u(k) + u(k-1)]. \tag{32}$$

Evidently, $\boldsymbol{x}(k)$ can be obtained by solving (32), that is,

$$\boldsymbol{x}(k) = diag\,[f_1 \;\; f_2 \;\; \cdots \;\; f_n]\boldsymbol{x}(k-1) + \boldsymbol{G}^T[u(k) + u(k-1)] \tag{33}$$

where

$$f_i = (1 - \frac{\lambda_i T}{2})^{-1}(1 + \frac{\lambda_i T}{2}) \tag{34a}$$

$$\boldsymbol{G}^T = [g_1 \; g_2 \; \cdots \; g_n] \tag{34b}$$

with

$$g_i = \frac{T}{2}(1 - \frac{\lambda_i T}{2})^{-1}\bar{b}_i. \tag{34c}$$

Considering the output equation (26b), we can rewrite equation (33) in terms of output $y(k)$, that is

$$y(k) = \sum_{i=1}^{n} C(1-f_i z^{-1})^{-1} G^T w(k) \tag{35}$$

where z^{-1} is the backward shift operator and

$$w(k) = u(k) + u(k-1). \tag{36}$$

If the all eigenvalues of the system are real and distinct, the above equation is reduced to the following,

$$y(k) = \sum_{i=1}^{n} \frac{g_i}{1-f_i z^{-1}} w(k). \tag{37}$$

Note that if we define an "auxiliary" linear discrete-time system with input $w(k)=u(k)+u(k\text{-}1)$ and output $y(k)$, we can obtain the following transfer function

$$H(z^{-1}) = \sum_{i=1}^{n} \frac{g_i}{1-f_i z^{-1}} \tag{38}$$

which can be further simplified as

$$H(z^{-1}) = \frac{b_0 + b_1 z^{-1} + \cdots + b_{n-1} z^{-n+1}}{1 + a_1 z^{-1} + \cdots + a_n z^{-n}}. \tag{39}$$

Using equation (39), identification of continuous-time systems can be reduced to the standard identification problem of a discrete-time system. That is

$$y(k) + a_1 y(k-1) + \cdots + a_n y(k-n) = b_0 w(k) + b_1 w(k-1) + \cdots + b_{n-1} w(k-n+1). \tag{40}$$

Estimates of a_i and b_i can be obtained by applying the least squares method

$$\theta = [\Phi^T \Phi]^{-1} \Phi^T Y \tag{41}$$

where

$$\theta^T = [a_1 \cdots a_n \;\; b_0 \cdots b_{n-1}] \tag{42a}$$

$$\Phi = \begin{bmatrix} y(n) & \cdots & y(1) & w(n+1) & \cdots & w(2) \\ \vdots & & \vdots & \vdots & & \vdots \\ y(N-1) & \cdots & y(N-n) & w(N) & \cdots & w(N-n+1) \end{bmatrix} \tag{42b}$$

$$Y^T = [y(n+1) \;\; \cdots \;\; y(N)]. \tag{42c}$$

After obtaining the estimates of a_i and b_i using any standard identification method, for example the least squares method, some calculations are still needed to determine the estimates of the parameters $\lambda_1, \ldots, \lambda_n$ and $\bar{b}_1, \ldots, \bar{b}_n$ of the continuous-time system. Clearly, it is a straightforward calculation of several sets of equations.

For further clarification, the detailed procedure is given below:

1. From the given input-output data, obtain the least squares estimates of parameters of a_i and b_i using equation (41);
2. Calculate f_i and g_i from a_i and b_i by performing the partial fraction expansion of $H(z^{-1})$;
3. From f_i and g_i, solve equation (34) for the parameters λ_i and $\bar{b}_i$ of the continuous-time system (26).

It should be pointed out that in the case of complex conjugate eigenvalues, the corresponding procedure used to estimate parameters of continuous-time systems is slightly more complicated. This will be illustrated in Example 5.

Examples

Example 4. Consider a second order system which has two distinct real eigenvalues as follows:

$$A = \begin{bmatrix} -3 & 0 \\ 0 & -7 \end{bmatrix}, \quad B = \begin{bmatrix} 1 \\ 15 \end{bmatrix}, \quad and \quad C = [\,1 \;\, 1\,]. \tag{43}$$

The initial values of states are zero. The sampling interval T is 0.01 and the lower and upper integration limits are 0.0 and 2.0, respectively. The input signal is generated by the following equation

$$u(t) = \frac{2t}{1+2t} + \sum_{i=1}^{10} \cos(it). \tag{44}$$

Using the suggested identification procedure, we obtain the following parameter estimates.

Table 4. Parameter estimates of the system in Example 4

Parameter	Estimated value	True value
λ_1	-2.955037	-3.0
λ_2	-7.002294	-7.0
$\bar{b}_1$	0.985777	1.0
$\bar{b}_2$	15.024349	15.0

From the results, we can see clearly that the method can give very good estimates when the input-output data do not contain any noise or measurement errors.

Example 5. Here we show how the approach works in the case of conjugate complex eigenvalues. Without losing generality, suppose $\boldsymbol{A}$, $\boldsymbol{B}$ and $\boldsymbol{C}$ are chosen as

$$\boldsymbol{A} = \begin{bmatrix} -1 & 2 \\ -2 & -1 \end{bmatrix}, \quad \boldsymbol{B} = \begin{bmatrix} 2 \\ 2.3 \end{bmatrix}, \quad \text{and} \quad \boldsymbol{C} = [1 \ 1]. \tag{45}$$

Again, the initial values of states and sampling interval are chosen as 0.0 and 0.01, respectively. The integration is performed on [0, 2]. The input variable is

$$u(t) = \frac{2t}{1+2t}. \tag{46}$$

After obtaining the estimates $\hat{a}_1$, $\hat{a}_2$, $\hat{b}_0$, and $\hat{b}_1$, we can follow the procedure given below for calculating the parameters of the continuous-time system.

1. Given $\hat{a}_1$, $\hat{a}_2$, $\hat{b}_0$, and $\hat{b}_1$, obtain f_{12} which is defined as

$$f_{12} = \begin{bmatrix} \sigma & \gamma \\ -\gamma & \sigma \end{bmatrix}, \quad \sigma = -\frac{\hat{a}_1}{2}, \quad \text{and} \quad \gamma = \sqrt{\hat{a}_2 - \sigma^2}. \tag{47}$$

2. From (47), calculate the conjugate complex eigenvalues $\alpha \pm j\beta$ by

$$\alpha_T = \frac{\gamma^2 + \sigma^2 - 1}{(1+\sigma)^2 + \gamma^2} \qquad \text{where} \quad \alpha_T = \alpha\frac{T}{2}$$

$$\beta_T = \frac{2\gamma}{(1+\sigma)^2 + \gamma^2} \qquad \text{where} \quad \beta_T = \beta\frac{T}{2}. \tag{48}$$

3. Based on (48), calculate g_1 and g_2 using the following equation

$$g_1 = \frac{\hat{b}_0\gamma - \hat{b}_1 - \sigma\hat{b}_0}{2\gamma}\frac{2}{T}$$

$$g_2 = \frac{\hat{b}_1 + \sigma\hat{b}_0 + \gamma\hat{b}_0}{2\gamma}\frac{2}{T}. \tag{49}$$

4. Solve for the parameters of the continuous-time system by substituting equation (49) into the following equation

$$\begin{bmatrix} \bar{b}_1 \\ \bar{b}_2 \end{bmatrix} = \begin{bmatrix} 1-\alpha_T & -\beta_T \\ \beta_T & 1-\alpha_T \end{bmatrix} \begin{bmatrix} g_1 \\ g_2 \end{bmatrix} . \tag{50}$$

The final estimates of the parameters of the continuous-time system are listed in Table 5.

Table 5. Parameter estimates of the system in Example 5

Parameter	Estimated value	True value
α	-1.0000	-1.0
β	2.0000	2.0
$\bar{b}_1$	1.9998	2.0
$\bar{b}_2$	2.2999	2.3

Again the results confirm that the suggested approach works very well for the noise-free case. The effect of noise will be investigated in section 5.

Identification of multi-input multi-output systems in diagonal form

We shall assume that the diagonal form of a multi-input multi-output continuous-time system can be described by equation (26) except for difference in the number of inputs and outputs. Consequently, the matrices $\boldsymbol{B}$ and $\boldsymbol{C}$ should be

$$\boldsymbol{B} = \begin{bmatrix} \bar{b}_{11} & \cdots & \bar{b}_{1m} \\ \vdots & & \vdots \\ \bar{b}_{n1} & \cdots & \bar{b}_{nm} \end{bmatrix}; \quad \boldsymbol{C} = \begin{bmatrix} c_{11} & \cdots & c_{1n} \\ \vdots & & \vdots \\ c_{p1} & \cdots & c_{pn} \end{bmatrix} . \tag{51}$$

Since we usually assume that the matrix $\boldsymbol{C}$ is known, our problem then is to estimate the parameters of the matrices $\boldsymbol{A}$ and $\boldsymbol{B}$ from samples of input-output data of the system. Following the same procedure as described earlier, we may come to the following equation

$$\boldsymbol{x}(k) = diag[f_1 \;\; f_2 \;\; \cdots \;\; f_n]\boldsymbol{x}(k-1) + \boldsymbol{G}^T[\boldsymbol{u}(k) + \boldsymbol{u}(k-1)] \tag{52}$$

which is almost the same as equation (33) with the only difference that $\boldsymbol{u}(k)$ is now an m-dimensional vector and $\boldsymbol{G}$ is an $n \times m$ matrix.

$$\boldsymbol{G} = \begin{bmatrix} g_{11} & \cdots & g_{1m} \\ \vdots & & \vdots \\ g_{n1} & \cdots & g_{nm} \end{bmatrix} \tag{53a}$$

where

$$g_{ij} = \frac{T}{2}(1 - \frac{\lambda_i T}{2})^{-1}\bar{b}_{ij} \tag{53b}$$

for $i=1, \ldots, n$ and $j=1, \ldots, m$.

Combining equations (26b) and (51) with equation (52) yields a multi-input multi-output transfer function matrix, i.e.

$$\boldsymbol{y}(k) = \sum_{i=1}^{n} \boldsymbol{C}(1 - f_i z^{-1})^{-1}\boldsymbol{G}^T\boldsymbol{w}(k) \tag{54}$$

with

$$\boldsymbol{y}^T(k) = \begin{bmatrix} y_1(k) & \cdots & y_p(k) \end{bmatrix} \tag{55a}$$

$$\boldsymbol{w}(k) = \boldsymbol{u}(k) + \boldsymbol{u}(k-1) = \begin{bmatrix} u_1(k) + u_1(k-1) \\ \vdots \\ u_m(k) + u_m(k-1) \end{bmatrix}. \tag{55b}$$

From equation (54), we can have the following equation

$$\begin{aligned} & y_i(k) + a_1 y_i(k-1) + \cdots + a_n y_i(k-n) \\ & = b_{01}w_1(k) + b_{11}w_1(k-1) + \cdots + b_{n-11}w_1(k-n+1) \\ & \quad + \quad \cdots \\ & \quad + b_{0m}w_m(k) + b_{1m}w_m(k-1) \cdots + b_{n-1m}w_m(k-n+1) . \end{aligned} \tag{56}$$

Therefore, applying the standard least squares method leads to the following least squares estimates of (56)

$$\theta = [\Phi_i^T\Phi_i]^{-1}\Phi_i^T Y_i \qquad i = 1, \cdots, n \tag{57}$$

where

$$\theta^T = [a_1 \cdots a_n \;\; b_{01} \cdots b_{n-11} \;\; \cdots \;\; b_{0m} \cdots b_{n-1m}] \tag{58a}$$

$$\Phi_i = \begin{bmatrix} y_i(n) & y_i(1) & w_1(n+1) & & w_1(2) & & w_m(n+1) & & w_m(2) \\ \\ y_i(N-1) & y_i(N-n) & w_1(N) & \cdots & w_1(N-n+1) & \cdots & w_m(N) & \cdots & w_m(N-n+1) \end{bmatrix}$$

(58b)

$$Y_i^T = [\, y_i(n+1) \quad \cdots \quad y_i(N) \,]. \tag{58c}$$

Once we obtain the parameter estimate $\hat{\theta}$, the parameters of the continuous-time system can be obtained directly by following the procedure developed in the earlier. It should be noted that although the system may have p output and m input sequences, we only need one output and all input sequences in order to estimate all the parameters of the continuous-time system. The following example will illustrate the procedure for identification of multi-input multi-output systems.

Example 6. Consider a one-output two-input second order system which is described by equation (26). The parameter matrices are given as

$$A = \begin{bmatrix} -1 & 0 \\ 0 & -4 \end{bmatrix}, \quad B = \begin{bmatrix} 2.0 & 1.0 \\ 3.5 & -2.0 \end{bmatrix}, \quad and \quad C = [\,1 \;\; 1\,]. \tag{59}$$

The sampling interval and the lower and upper integration limits are 0.01, and 0.0 and 2.0, respectively. The initial states are zero and the input variables are

$$u_1(t) = \frac{2t}{1+2t} + \sum_{i=1}^{5} \cos(2it) \tag{60a}$$

$$u_2(t) = \frac{2t}{1+2t} + \sum_{i=1}^{5} \sin(2it). \tag{60b}$$

Calculating the least squares estimates obtained by solving equation (57) and following the suggested procedure give the parameter estimates of the continuous-time system. The parameter estimates are listed in Table 6.

Table 6. Parameter estimates of the system in Example 6

Parameter	Estimated value	True value
λ_1	-1.0028	-1.0
λ_2	-4.0067	-4.0
b_{11}	2.0145	2.0
b_{21}	3.4912	3.5
b_{12}	1.0027	1.0
b_{22}	-2.0066	-2.0

It is clear that the proposed identification approach can give accurate estimates of the parameters of multi-input multi-output continuous-time systems when the input-output data are noise-free. The results have also confirmed the advantage claimed earlier, that is, only one output sequence, together with all input sequences, is required in order to identify all parameters of the system.

4. Identification using Trapezoidal Pulse Functions

It is well known that there are mainly two ways to describe dynamics of a linear continuous-time system. One is the so-called "interior approach", which uses a state-space model to describe the dynamic behaviour of the system. Another is to express the input-output relation of the system by a differential equation in time-domain, or an algebraic equation of degree n in s-domain, i.e. the transfer function.

In the previous two sections, we have discussed two methods for identification of continuous-time systems using numerical integration methods. They are quite different from the point of view of input-output data required for performing identification. The method developed in section 2 requires that all states and inputs should be measurable. On the contrary, the second method described in section 3 requires only one output sequence together with all input sequences to estimate parameters of the continuous-time systems. On the other hand, however, these two methods are similar since both of them utilize numerical integration techniques for identification of the system.

In this section, we shall give another approach for identification of input-output models for continuous-time systems from samples of input-output data. Basically, this approach follows the same line as those of the previous two methods, i.e. applying integration technique to a differential equation. The integration approach used in this method is nothing more than the Trapezoidal rule. This method has been proposed for identification of single-input single-output continuous-time systems by Prasad and Sinha (1983). In this section, we shall extend this method to identification of multi-input multi-output continuous-time systems.

Review of Trapezoidal Pulse Functions

The Trapezoidal rule is one of the commonly used integration approaches and has been discussed in section 2. From equation (9) and Figure 2, we can see clearly that the basic idea of the rule is to approximate the integrand $y(t)$ by a linear function over the interval $(k\text{-}1)T \leq t \leq kT$. In the interval, the linear function, i.e.,the Trapezoidal Pulse Functions, can be expressed as

$$I_t(y) = \frac{1}{T}\{ (kT-t)y(k-1) + [t-(k-1)T]y(k) \} \tag{61}$$

for $(k\text{-}1)T \leq t \leq kT$. Here $y(k)$ represents $y(kT)$.

During the development of this identification method, it is required to have the values of successive integrals of $y(t)$ for estimating the parameters of a dynamic model for the continuous-time system. Therefore, we shall give a set of approximate formulas to express these integrals. The detailed derivation of these formulas can be found in the paper by Prasad and Sinha (1983).

The expression for the first integration of $y(t)$ is given as

$$I_{1,t}(y) = \int_0^t y(\tau)\,d\tau = I_{1,k-1}(y) + \int_{(k-1)T}^{t} y(\tau)\,d\tau \tag{62}$$

where

$$I_{1,k-1}(y) = \int_0^{(k-1)T} y(\tau)\,d\tau. \tag{63}$$

In equations (62) and (63), the first subscript of $I_{.,.}$ represents the number of integration and the second one indicates the time instant of the integration. Substituting equation (61) into (62) leads to

$$I_{1,t}(y) = I_{1,k-1}(y) + \frac{1}{2T}\left[T^2-(kT-t)^2\right]y(k-1) + \frac{1}{2T}\left[t-(k-1)T\right]y(k). \tag{64}$$

When $t=kT$, the above equation gives the standard integral of $y(t)$, same as that obtained by using the Trapezoidal rule. That is

$$I_{1,k}(y) = I_{1,k-1}(y) + \frac{T}{2}\left[y(k-1)+y(k)\right]. \tag{65}$$

Similarly, the second integral of $y(t)$ can be expressed as

$$\begin{aligned} I_{2,t}(y) &= I_{2,k-1}(y) + \int_{(k-1)T}^{t} I_{1,\tau}(y)d\tau \\ &= I_{2,k-1}(y) + \{t-(k-1)T\}I_{1,k-1}(y) \\ &+ \{\frac{T}{2}\left[t-(kT-t)T\right]-\frac{1}{3!T}\left[T^3-(kT-t)^3\right]\}y(k-1) \\ &+ \frac{1}{3!T}\{t-(k-1)T\}^3 y(k). \end{aligned} \tag{66}$$

At the instant $t=kT$, we have the second integral as

$$I_{2,k}(y) = I_{2,k-1}(y) + T I_{1,k-1}(y) + \frac{T}{3!}\{2y(k-1) + y(k)\}. \tag{67}$$

In general, we can express the nth integral of $y(t)$ by the following equation

$$I_{n,t}(y) = \sum_{i=1}^{n} \frac{1}{(i-1)!}\left[t-(k-1)T\right]^{i-1} I_{n-i+1,k-1}$$

$$+ \{\sum_{i=1}^{n-1} \frac{(-1)^{i+1}T^i}{(i+1)!\,(n-i)!}\left[t-(k-1)T\right]^{n-i} + \frac{(-1)^{n+1}}{(n+1)!}\left[T^{n+1}-(kT-t)^{n+1}\right]\}\, y(k-1)$$

$$+ \frac{1}{(n+1)!\,T}\{t-(k-1)T\}^{n+1} y(k). \tag{68}$$

Again, for $t=kT$, the integration $I_{n,t}(y)$ has the following form

$$I_{n,k}(y) = \sum_{i=1}^{n} \frac{T^{n-i}}{(i-1)!} I_{n-i+1,k-1}(y) + \frac{nT^n}{(n+1)!} y(k-1) + \frac{T^n}{(n+1)!} y(k). \tag{69}$$

Now, the identification problem can be easily solved by simple applying equations (65), (67), and (69) to the differential equation of continuous-time systems.

Input-output model of continuous-time systems

As an important way to represent a linear continuous-time system, the transfer function has been widely utilized in many engineering areas. The transfer function of an nth order linear continuous-time system may be written as

$$H(s) = \frac{y(s)}{u(s)} = \frac{b_0 s^{n-1} + b_1 s^{n-2} + \cdots + b_{n-1}}{s^n + a_1 s^{n-1} + \cdots + a_n}. \tag{70}$$

Its counterpart in the time domain is an nth order differential equation of the input $u(t)$ and output $y(t)$, that is

$$\frac{d^n y}{dt^n} + a_1 \frac{d^{n-1}y}{dt^{n-1}} + \cdots + a_n y = b_0 \frac{d^{n-1}u}{dt^{n-1}} + b_1 \frac{d^{n-2}u}{dt^{n-2}} + \cdots + b_{n-1} u. \tag{71}$$

For a multi-input and multi-output system, the input-output relations can be expressed by a transfer function matrix. For instance, the transfer function matrix of an m-input and p-output continuous-time system is

$$\boldsymbol{H}(s) = \frac{\boldsymbol{Y}(s)}{\boldsymbol{U}(s)} = \begin{bmatrix} H_{11}(s) & \cdots & H_{1m}(s) \\ \vdots & & \vdots \\ H_{p1}(s) & \cdots & H_{pm}(s) \end{bmatrix} \tag{72}$$

where $Y(s)$ and $U(s)$ are output and input vectors, respectively and $H_{ij}(s)$ is the transfer function which relates the ith output to the jth input. In fact, the ith output can be given by a combination of all the inputs, that is

$$y_i(s) = H_{i1}(s)\,u_1(s) + \cdots + H_{im}(s)\,u_m(s). \tag{73}$$

Equivalently, in the time-domain, the above equation can be written as an nth order differential equation if the order of systems is n, i.e.

$$\begin{aligned}\frac{d^n y_i}{dt^n} + a_1\frac{d^{n-1}y_i}{dt^{n-1}} + \cdots + a_n y_i = {} & b_{01}\frac{d^{n-1}u_1}{dt^{n-1}} + b_{11}\frac{d^{n-2}u_1}{dt^{n-2}} + \cdots + b_{n-11}u_1 + \cdots \\ & + b_{0m}\frac{d^{n-1}u_m}{dt^{n-1}} + b_{1m}\frac{d^{n-2}u_m}{dt^{n-2}} + \cdots + b_{n-1m}u_m.\end{aligned} \tag{74}$$

Identification of single-input single-output systems

Consider a single-input single-output continuous-time system described by equation (71), the identification problem now is to estimate the parameters a_i and b_{i-1} for $i=1,\ldots,n$ in (71).

Integrating both sides of equation (71) n times with respect to t over the subinterval $(k\text{-}1)T \le t \le kT$, we obtain

$$\Delta y(k) + \sum_{i=1}^{n} a_i \Delta I_{i,k}(y) = \sum_{i=1}^{n} b_{i-1}\Delta I_{i,k}(u) \tag{75}$$

where $\Delta y(k)$, $\Delta I_{i,k}(y)$, and $\Delta I_{i,k}(u)$ are defined as

$$\begin{aligned}\Delta y(k) &= y(k) - y(k-1)\\ \Delta I_{i,k}(y) &= I_{i,k}(y) - I_{i,k-1}(y)\\ \Delta I_{i,k}(u) &= I_{i,k}(u) - I_{i,k-1}(u).\end{aligned} \tag{76}$$

The quantities $I_{i,k}(y)$ and $I_{i,k}(u)$ are the ith integrals of y and u at the kth sampling instant, respectively.

With the above equation we can directly use the least squares method to obtain parameter estimates a_i and b_{i-1} for $i=1,\ldots,n$, as shown below.

$$\theta = [\Phi^T\Phi]^{-1}\Phi^T\Delta Y \tag{77}$$

where

$$\theta^T = [a_1 \ a_2 \ \cdots \ a_n \ \ b_0 \ b_1 \ \cdots \ b_{n-1}] \tag{78a}$$

$$\Phi = \begin{bmatrix} \Delta I_{1,2}(y) & \cdots & -\Delta I_{n,2}(y) & \Delta I_{1,2}(u) & \cdots & \Delta I_{n,2}(u) \\ \vdots & & \vdots & \vdots & & \vdots \\ -\Delta I_{1,N}(y) & \cdots & -\Delta I_{n,N}(y) & \Delta I_{1,N}(u) & \cdots & \Delta I_{n,N}(u) \end{bmatrix} \tag{78b}$$

$$\Delta Y^T = [\,\Delta y(2) \ \cdots \ \Delta y(N)\,]. \tag{78c}$$

To demonstrate the validity of the suggested method for identification of single-input single-output systems, we provide two examples. In example 7, a second-order system will be considered. Since one of the potential problems in using nth integration for system identification is the equation stiffness, we shall give an example, Example 8 where parameters of a high order system will be estimated, to investigate this important issue. It will be proved that the equation stiffness does not occur in parameter identification of continuous-time systems when using this suggested approach.

Example 7. The system considered is the same as one in Example 1 except that the output matrix $\boldsymbol{C}$ is selected as

$$\boldsymbol{C} = \begin{bmatrix} 1 & 0 \end{bmatrix}. \tag{79}$$

The sampling interval is taken as 0.01 and the lower and upper integration limits are 0.0 and 2.0, respectively. The initial values of the states are zero and the input signal is generated by equation (18). The estimates of the parameters are listed in Table 7.

Table 7. Parameter estimates of the system in Example 7

Parameter	Estimated value	True value
a_1	0.7000	0.7
a_2	1.4000	1.4
b_0	1.0000	1.0
b_1	2.0000	2.0

Clearly, the method can give fairly accurate estimates when the input-output data is noise-free.

Example 8. In this example, a fourth order system is selected, which is given by

$$\frac{d^4y}{dt^4}+\frac{d^3y}{dt^3}+0.6\frac{d^2y}{dt^2}+1.5\frac{dy}{dt}+2.7y = \frac{d^3u}{dt^3}+1.3\frac{d^2u}{dt^2}+0.7\frac{du}{dt}+2.1u. \quad (80)$$

Again, the sampling interval is 0.01, the initial values of states are zero and the input variable is taken by equation (18). Two different integration limits are selected, one is from 0.0 to 2.0 and another is from 0.0 to 8.5. It implies the number of sampling points are different. The results are listed in Table 8.

Table 8. Parameter estimates of the system in Example 8

Parameters	Estimated values ($0 \leq t \leq 2.0$)	Estimated values ($0 \leq t \leq 8.5$)	True parameters
a_1	0.9888	0.9999	1.0
a_2	0.6322	0.5999	0.6
a_3	1.4567	1.5000	1.5
a_4	2.7072	2.6999	2.7
b_0	1.0000	1.0000	1.0
b_1	1.2888	1.2999	1.3
b_2	0.7293	0.7000	0.7
b_3	2.0649	2.0999	2.1

From the results in Examples 7 and 8, we can observe that this method can provide fairly accurate parameter estimates when the input-output data are not contaminated with noise. In other words, this method does not suffer the problem of equation stiffness which may occur when using *n*th integration for system identification. Secondly, under the same conditions, the parameter estimates may be affected by the length of input-output data. It follows that the accuracy of the estimates can be significantly improved by simply increasing the number of sampling points. Furthermore, it can be seen that if we have the same number of data for two systems, the method can provide more accurate parameter estimates for the system which has lower order. It simply means that the system of high order usually requires more sampling points if the suggested method is used to estimate parameters.

Identification of multi-input multi-output systems

Extension of the method discussed above to identification of multi-input multi-output continuous-time systems is very straightforward. The whole procedure remains essentially the same. The only change is in the use of a different matrix $\boldsymbol{\Phi}$ and vectors $\boldsymbol{\theta}$ and $\Delta \boldsymbol{Y}$.

In this case, we can form the regression equation (77) by integrating both sides of equation (74) n times. The resulting equation is

$$\Delta y_l(k) + \sum_{i=1}^{n} a_i \Delta I_{i,k}(y_l) = \sum_{j=1}^{m} \{ \sum_{i=1}^{n} b_{i-1,j}(l) \Delta I_{i,k}(u_j) \} \qquad for\ l = 1, \cdots, p. \tag{81}$$

The regression equation corresponding to (77) is then

$$\theta_l = [\Phi_l^T \Phi_l]^{-1} \Phi_l^T \Delta Y_l \tag{82}$$

where

$$\theta_l^T = [\, a_1 \ \cdots \ a_n \ \ b_{01}(l) \ \cdots \ b_{n-11}(l) \ \ \cdots \ \ b_{0m}(l) \ \cdots \ b_{n-1m}(l)] \tag{83a}$$

$$\Phi_l = \begin{bmatrix} -\Delta I_{1,2}(y_l) & \cdots & -\Delta I_{n,2}(y_l) & \Delta I_{1,2}(u_1) & \cdots & \Delta I_{n,2}(u_1) & \cdots & \Delta I_{1,2}(u_m) & \cdots & \Delta I_{n,2}(u_m) \\ \vdots & & \vdots & \vdots & & \vdots & & \vdots & \vdots & \\ -\Delta I_{1,N}(y_l) & \cdots & -\Delta I_{n,N}(y_l) & \Delta I_{1,N}(u) & \cdots & \Delta I_{n,N}(u) & \cdots & \Delta I_{1,N}(u_m) & \cdots & \Delta I_{n,N}(u_m) \end{bmatrix} \tag{83b}$$

$$\Delta Y_l^T = [\, \Delta y_l(2) \quad \cdots \quad \Delta y_l(N) \,]. \tag{83c}$$

A simulation example is provided in order to illustrate the proposed method.

Example 9. Consider a continuous-time system with two inputs and three outputs. The transfer function matrix is

$$H(s) = \begin{bmatrix} \dfrac{3}{s+4} & \dfrac{2.7(s+5)}{(s+4)(s+0.6)} \\ \dfrac{4}{(s+4)(s+0.6)} & \dfrac{1.2}{(s+0.6)} \\ \dfrac{3.5(s\quad 1)}{(s+4)(s+0.6)} & \dfrac{0.5(s+10)}{(s+4)(s+0.6)} \end{bmatrix}. \tag{84}$$

The initial states are taken as zero and the input variables are selected as

$$\begin{aligned} u_1(t) &= e^{-0.5t} \\ u_2(t) &= \frac{2t}{1+2t}. \end{aligned} \tag{85}$$

The sampling interval is chosen as 0.01 and the lower and upper integration limits are 0.0 and 6.0, respectively. The parameter estimates are listed in Tables 9 to 11.

Table 9. Parameter estimates of the first output of the system in Example 9

Parameters	Estimated values	True values
a_1	4.5997	4.6
a_2	2.3999	2.4
$b_{01}(1)$	2.9998	3.0
$b_{11}(1)$	1.7997	1.8
$b_{02}(1)$	2.7004	2.7
$b_{12}(1)$	13.4992	13.5

Table 10. Parameter estimates of the second output of the system in Example 9

Parameters	Estimated values	True values
a_1	4.5990	4.6
a_2	2.3994	2.4
$b_{01}(2)$	0.0003	0.0
$b_{11}(2)$	3.9990	4.0
$b_{02}(2)$	1.2000	1.2
$b_{12}(2)$	4.7987	4.8

Table 11. Parameter estimates of the third output of the system in Example 9

Parameters	Estimated values	True values
a_1	4.5994	4.6
a_2	2.3997	2.4
$b_{01}(3)$	3.4995	3.5
$b_{11}(3)$	3.4993	3.5
$b_{02}(3)$	0.5002	0.5
$b_{12}(3)$	4.9993	5.0

From the estimates in Tables 9 to 11, it is confirmed that the proposed method can be used to obtain accurate estimates of parameters of multi-input multi-output systems if the input-output data do not contain noise or measurement errors.

5. Effect of noise in the input-output data

It is common knowledge that, in real life situations, the input-output data of a system almost always contain certain kinds of noise, or measurement errors. These may arise from various sources, such as the limited accuracy of sensors, complicated industrial environment, data transmission, transducers and A/D converters, and sometimes even from some inherent properties of the system itself. The parameter estimates from most identification methods obtained using these contaminated data may be highly inaccurate. Consequently, investigation of this aspect is a very important issue in developing any method for system identification.

To study this aspect for the identification procedure developed in section 2, we generated three sets of noise sequences and added them to the states of the system. The system parameters are estimated based on these contaminated data.

Example 10. Suppose the system is as described in Example 1. The output $y(t)$ of

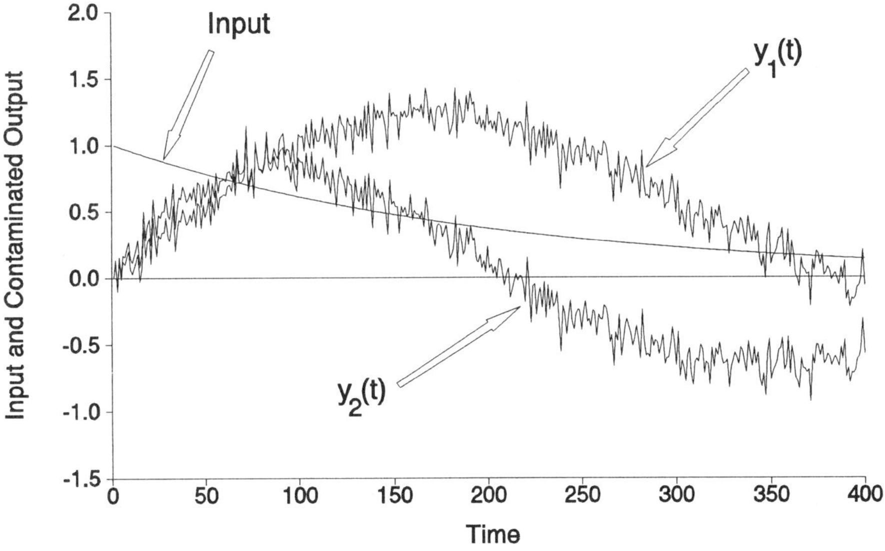

Figure 4. Input-output signals (noise variance is 0.1)

the system follows the equation (3b) with $\boldsymbol{C}=\boldsymbol{I}$. Replacing $x_i(t)$ in equations (15) and (16) by $y_i(t)$ leads to the following equation.

$$\boldsymbol{\theta}_i = [\boldsymbol{\Phi}^T\boldsymbol{\Phi}]^{-1}\boldsymbol{\Phi}^T\Delta \boldsymbol{Y}_i \tag{86}$$

where the matrix $\boldsymbol{\Phi}$ and vectors $\boldsymbol{\theta}_i$ and $\Delta\boldsymbol{Y}_i$ are defined as

$$\boldsymbol{\theta}_i^T = \left[a_{i1} \cdots a_{in} \; b_i\right] \tag{87a}$$

$$\Phi = \begin{bmatrix} I_{y_1}(1) & \cdots & I_{y_n}(1) & I_u(1) \\ \vdots & & \vdots & \vdots \\ I_{y_1}(N) & \cdots & I_{y_n}(N) & I_u(N) \end{bmatrix} \tag{87b}$$

$$\Delta Y_i^T = [\, y_i(1) - y_i(0) \quad \cdots \quad y_i(N) - y_i(N \;\; 1) \,] \tag{87c}$$

for $i=1, \ldots, n$. Again, for different numerical integration methods the indices in equation (23b) and (23c) should be correspondingly modified.

The noise sequences added to the states are white with zero mean and variances 0.01,0.05,and 0.1,respectively. Fig. 4, shows the input signal $u(t)$ and the contaminated output $\{y_i(t)\ i=1,2\}$ when the noise variance is equal to 0.1.

Based on the contaminated data, all four numerical integration methods mentioned above, are applied to the estimation of the system parameters. The results are listed in Tables 12, 13, and 14, respectively.

Table 12. Parameter estimates of the system (noise variance is 0.01)

	Mid-point rule	Trapezoidal rule	Simpson's rule	Newton-Cotes rule
a_{11}	-0.7018	-0.7001	-0.7028	-0.6985
a_{12}	0.9856	0.9991	0.9845	0.9990
a_{21}	-1.4037	-1.4023	-1.4050	-1.4005
a_{22}	-0.0138	-0.0008	-0.0153	-0.0006
b_1	1.0131	1.0003	1.0151	0.9976
b_2	2.0159	2.0030	2.0181	2.0000

Table 13. Parameter estimates of the system (noise variance is 0.05)

	Mid-point rule	Trapezoidal rule	Simpson's rule	Newton-Cotes rule
a_{11}	-0.6409	-0.7014	-0.6664	-0.6908
a_{12}	1.0059	0.9942	0.9769	0.9967
a_{21}	-1.3398	-1.4064	-1.3721	-1.3967
a_{22}	0.0209	0.0020	-0.0157	0.0041
b_1	0.9355	1.0033	0.9847	0.9845
b_2	1.9289	2.0049	1.9886	1.9882

Table 14. Parameter estimates of the system (noise variance is 0.1)

	Mid-point rule	Trapezoidal rule	Simpson's rule	Newton-Cotes rule
a_{11}	-0.4178	-0.7055	-0.5131	-0.6771
a_{12}	1.1939	0.9871	1.0875	0.9983
a_{21}	-1.0894	-1.3996	-1.2105	-1.3784
a_{22}	0.2530	0.0189	0.1178	0.0248
b_1	0.5594	1.0114	0.7432	0.9607
b_2	1.4964	1.9846	1.7235	1.9482

A couple of points can be observed from the results. The first is that all the numerical integration methods work well when the noise variance is small. This can be verified from the results in Tables 12 to 14. Another point drawn from the results is that the Trapezoidal rule (9) and the Newton-Cotes rule (11) seem to work better than the other two. Especially, the Trapezoidal rule was found to give the most accurate estimates.

It should be mentioned that the other two methods developed in sections 3 and 4 did not give estimates as good as the approach presented in section 2. In fact, they are quite sensitive to noise or measurement error which may be included in the input-output data. One possible reason may the complexity of the algorithms themselves. Improvement in the robustness of these two methods is a topic for future research.

6. Conclusions

Utilizing numerical integration methods, three identification procedures have been proposed for parameter estimation of both single-input single-output and multi-input multi-output continuous-time systems from samples of input-output data.

Four different numerical integration methods have been applied to perform the identification procedure developed in section 2. They are the Mid-point rule, the Trapezoidal rule, Simpson's rule, and Newton-Cotes formula. All these methods can provide very good estimates for parameters when the input-output data are not contaminated with noise. In the presence of small amount white noise, these methods can still give good parameter estimates. When the noise becomes larger, only the Trapezoidal rule and the Newton-Cotes formula can produce reasonable estimates for the system parameters. In this chapter, the identification procedure has been extended to multi-input multi-output systems. It has been confirmed by a simulation example that the identification procedure works very well for the multivariable system.

It should be noted that the identification procedure discussed in section 2 does not have restrictions on the numerical integration method to be used. In other words, any other methods for numerical integration can be, in principle, used to perform the procedure. Of course, the accuracy of parameter estimates will depend upon the use of particular numerical integration method selected. The only restriction of the approach is that all the states should be measurable.

In sections 3 and 4, we have discussed another two methods for identification of continuous-time systems including not only single-input single-output but also multi-input multi-output systems. Both methods demand only samples of input-output data rather than the all state outputs. A distinctive advantage of the method developed in section 3 is that only one output sequence, together with all input samples, is needed in order to estimate all parameters of multi-input multi-output systems. One of the disadvantage is that the method needs some extra calculation compared with other methods.

A good feature of the third identification method discussed in section 4 is that it can give more accurate estimates than the method described in section 3. This can be easily verified by simply checking the estimates in the Tables. A reasonable explanation is that some extra computation errors may be introduced in the second method. The third method can provide very accurate parameter estimates because it utilizes the Trapezoidal Pulse Functions which are more accurate than the commonly used Block Pulse Functions. It has been verified that the parameter estimates given by the third method are not affected by the problem of equation stiffness which often occurs when the *n*th integration is used in the identification procedure. It has been also illustrated by an example that more input-output data is usually required as the order of the system is increased.

Unfortunately, from the robustness point of view, both the second and third methods have the drawback that they are quite sensitive to noise or measurement errors contained in the input-output data. The improvement of robustness of those two methods is a topic for future research.

References

Bingulac, S. and Sinha, N.K. (1989), "On the identification of continuous-time multivariable systems from samples of input-output data", Proc. Seventh Int. Conf. on Mathematical and Computer Modelling, (Chicago, Ill, August 1989), pp.231-239.

Burden, R.L. and Faircs, J.D. (1981), "*Numerical Analysis*", PWS Publishers, Boston, Massacusetts.

Hung, J.C., Liu, C.C., and Chan, P.Y. (1980), "An algebraic method for system parameter identification", Proc. 14th Asilomac Conf. on Circuits, Systems and Computers, (Pacific Grove, CA, USA).

Lastman, G.J. and Sinha, N.K. (1989), "*Microcomputer-based numerical methods for science and engineering*", Saunders College Publishing, New York.

Prasad, T. and Sinha, N.K. (1983), "Modelling of continuous-time systems from sampled data using trapezoidal pulse functions", Proc. International Conference on Systems, Man and Cybernetics (Bombay, India, Dec. 1983), pp. 427-430.

Sinha, N.K. and Lastman, G.J. (1982), "Identification of continuous-time multivariable systems from sampled data", *Int. J. Contr.*, vol. 35, pp.117-126.

Zhou, Q.-J. and Sinha, N.K. (1982), "Identification of continuous-time state-space model from samples of input-output data", *Electronics Letters*, vol.18, pp.50-51.

Application of digital filtering techniques

S. Sagara and Z.Y. Zhao
Department of Electrical Engineering
Kyushu University
Hokozaki, Fukuoka 812, Japan

Abstract

Continuous-time system identification usually consists of two main parts: signal processing (or pre-filtering) and parameter estimation. Both analog and digital pre-filters for signal processing can be used, where analog pre-filters are implemented in a digital computer by using such techniques as the numerical integration and the bilinear transformation. As for parameter estimation, an emphasis is put on on-line identification algorithms. Using the pre-filters of digital form, a discrete-time identification model which retains the continuous-time model parameters is derived. Some fundamental identification methods such as the least squares method, bias-compensating methods and instrumental variable methods are reviewed. Finally the choice of the input signal is discussed with simulation experiments.

1. Introduction

For the identification of continuous-time systems, signal processing is required before applying an identification algorithm like the least squares (LS) algorithm because the time derivatives of the input-output signals must be determined from their measurements that always contain noise. In practice, the principal way is to use the filtered signals, which are generated by passing the measurements through a pre-filter, rather than the actual signals themselves (Unbehauen and Rao 1987; 1990). In the early period of continuous-time system identification, the analog pre-filters such as the state variable filter were well used (Young 1981). Owing to the rapid development of digital computers, however, it requires that the identification of continuous-time systems be carried out in a complete digital way. Thus digital filtering techniques became necessary for signal processing based on discrete-time measurements. In recent years, the techniques such as the numerical integration, the bilinear transformation and the orthogonal functions have been used to implement the analog pre-filters in

N. K. Sinha and G. P. Rao (eds.), Identification of Continuous-Time Systems, 291–325.

digital computers. In most cases, the implementations of analog pre-filters have a convenient form of finite or infinite impulse response filters. By using analog pre-filters discretized in digital form, a discrete-time model which retains the parameters of a differential equation (d.e.) model is derived. Based on the derived model, the d.e. model parameters are estimated using a direct or modified one of the identification algorithms that are fairly well developed in discrete-time model identification (Eykhoff 1974; Ljung and Söderström 1983). It is also possible to directly use a digital pre-filter in the same context as the analog pre-filters. But the d.e. model should first be discretized. In order to have the discretized model retain the d.e. model parameters, the bilinear transformation can be used.

To improve the statistical efficiency or the accuracy of parameter estimates in system identification, it is important to design a pre-filter properly. Although the design of analog pre-filters and the design of digital pre-filters are done differently, all pre-filters should have some specified frequency response characteristics to filter as much as possible the unwanted frequency components in system signals. It is shown that if the pre-filter can be designed in such a way that it has similar or close frequency response characteristics to the system under study, even the LS algorithm can yield quite good results in the case of a low level of noise on signal ($< 10\%$).

When the noise level is high, the LS algorithm is no longer suitable for consistent parameter estimation because the asymptotic bias on the LS estimates becomes remarkable. Several bias-eliminating identification algorithms such as the bias-compensating methods and the instrumental variable methods have been proposed using the state variable filter or the linear integral filter (Young 1970; Young and Jakeman 1980; Sagara and Zhao 1989; 1990; Zhao, Sagara and Wada 1990). Here we examine how these methods can be extended to the general case of using both discretized analog pre-filters and digital pre-filters.

2. Statement of the problem

System model descriptions

Consider an nth-order linear SISO continuous-time system modeled by the following differential equation with $x^{(j)}(t) \stackrel{\Delta}{=} d^j x(t)/dt^j$

$$x^{(n)}(t) + \sum_{i=1}^{n} a_i x^{(n-i)}(t) = \sum_{i=1}^{n} b_i u^{(n-i)}(t) \tag{1}$$

subject to an arbitrary initial condition of the system states

$$\boldsymbol{u}_0 = (u(0), \cdots, u^{(n-2)}(0))^T, \qquad \boldsymbol{x}_0 = (x(0), \cdots, x^{(n-1)}(0))^T \tag{2}$$

where the superscript T denotes the transpose. Here $u(t)$ is the input signal and $x(t)$ is the output response to $u(t)$.

For convenience, denote p as a differential operator, i.e. $px(t) = dx(t)/dt$ with $p^0 = 1$. Then eqn.(1) can be rewritten as

$$A(p)x(t) = B(p)u(t) \quad \text{subject to } (\boldsymbol{u}_0, \boldsymbol{x}_0) \tag{3a}$$

where A and B are polynomials in p, given by

$$\begin{aligned} A(p) &= p^n + a_1 p^{n-1} + \cdots + a_n \\ B(p) &= b_1 p^{n-1} + b_2 p^{n-2} + \cdots + b_n. \end{aligned} \tag{3b}$$

In the frequency domain, the system model (1) or (3) is described by

$$A(s)X(s) = B(s)U(s) + C(s) \tag{4}$$

where s represents the Laplace operator; $X(s)$ and $U(s)$ are the Laplace transforms of $x(t)$ and $u(t)$, respectively; and $C(s)$ is a polynomial of order $n-1$ in s, given by

$$\begin{aligned} &C(s) = c_0 s^{n-1} + c_1 s^{n-2} + \cdots + c_{n-1} \\ &c_0 = x(0), \; c_i = \left[\sum_{j=0}^{i} a_{i-j} x^{(j)}(0) - \sum_{j=0}^{i-1} b_{i-j} u^{(j)}(0) \right] \quad i = 1, 2, \cdots, n-1. \end{aligned} \tag{5}$$

If the system is under zero initial condition, i.e. $\boldsymbol{u}_0 = 0$, $\boldsymbol{x}_0 = 0$, the polynomial $C(s)$ will disappear, then eqn.(4) can also be described by the following transfer function model

$$\frac{X(s)}{U(s)} = G(s) = \frac{B(s)}{A(s)}. \tag{6}$$

In practice, the initial condition of the system under study is usually nonzero. Therefore, a system model like (3) or (4) should be considered for use.

Parameter identification problem

It is assumed that the system signals are sampled with a sampling interval T. $u(kT)$, $x(kT)$ represent the sampled data of the input and the output at sampling instants respectively. For convenience, notations like $u(k)$, $x(k)$ will be used instead of $u(kT)$, $x(kT)$ in the sequel. The observation equation is given as

$$y(k) = x(k) + v(k) \tag{7}$$

where $\{v(k)\}$ is a stationary noise sequence.

It is assumed that the system under study is asymptotically stable, i.e. $A(s)$ has all zeros in the left-hand side of the s plane. The case of the open-loop operation of the system i.e. the noise $v(k)$ being independent of the input $u(k)$ will be considered.

Assume that the structure of the system (i.e. the order of the system) is known. The problem treated here is to identify system parameters $\{a_j, b_j\}, j = 1, 2, \cdots, n$ from sampled data of input-output measurements $\{u(k), y(k)\}$.

3. Signal processing

Signal processing (or pre-filtering) is required both for handling time derivative terms and for improving the statistical efficiency or the accuracy of parameter estimates. For handling derivative terms, the difficulty encountered is that the derivatives of the input-output signals, $u^{(1)}, \cdots, u^{(n-1)}$, $y^{(1)}$, $\cdots$, $y^{(n)}$ must be determined from their sampled data $u(k), y(k)$ that usually contain noise. The use of "measures" is a principal way to overcome the difficulty. The measures are generated by passing the sampled data through a pre-filtering scheme that usually has low-pass filtering function. Both analog and digital filters can be used as the filtering scheme. For computer use, the filtering scheme in digital form is required. Thus the analog filters are implemented by using the techniques such as the numerical integration and the bilinear transformation.

Analog pre-filters

Assume that a general analog pre-filter has a Laplace transform $\mathcal{F}(s)$. Performing the filter $\mathcal{F}(s)$ on both sides of eqn.(4) yields

$$\mathcal{F}(s)A(s)X(s) = \mathcal{F}(s)B(s)U(s) + \mathcal{F}(s)C(s)$$

or

$$X_n(s) + \sum_{i=1}^{n} a_i X_{n-i}(s) = \sum_{i=1}^{n} b_i U_{n-i}(s) + \mathcal{F}(s)C(s) \tag{8}$$

where

$$X_i(s) = s^i \mathcal{F}(s) X(s), \quad U_i(s) = s^i \mathcal{F}(s) U(s).$$

Denoting the inverse Laplace transforms as

$$x_i(t) = \mathcal{L}^{-1}[X_i(s)], \quad u_i(t) = \mathcal{L}^{-1}[U_i(s)], \quad c(t) = \mathcal{L}^{-1}[\mathcal{F}(s)C(s)] \tag{9}$$

we have

$$x_n(t) + \sum_{i=1}^{n} a_i x_{n-i}(t) = \sum_{i=1}^{n} b_i u_{n-i}(t) + c(t). \tag{10}$$

At the sampling instant, $t = kT$, eqn.(10) is written as

$$x_n(k) + \sum_{i=1}^{n} a_i x_{n-i}(k) = \sum_{i=1}^{n} b_i u_{n-i}(k) + c(k). \tag{11}$$

Considering the noise contained in the output, we get

$$y_n(k) + \sum_{i=1}^{n} a_i y_{n-i}(k) = \sum_{i=1}^{n} b_i u_{n-i}(k) + c(k) + e(k) \tag{12}$$

where the equation error $e(k)$ is given by

$$e(k) = v_n(k) + \sum_{i=1}^{n} a_i v_{n-i}(k).$$

Note that in eqn.(12) there is a term $c(k)$ which resulted from the initial condition. Whether it requires estimation or it can be neglected depends upon the selected analog filter $\mathcal{F}(s)$. There are many possibilities for the choice of $\mathcal{F}(s)$. Three typical analog filters are as follows:

$$\mathcal{F}_1(s) = \frac{1}{s^n}; \quad \mathcal{F}_2(s) = \frac{(1 - e^{-lTs})^n}{s^n}; \quad \mathcal{F}_3(s) = \frac{1}{Q(s)} \tag{13}$$

where $Q(s)$ is an nth-order stable polynomial in s. Here $\mathcal{F}_1(s)$ represents the usual multiple integral operation; $\mathcal{F}_2(s)$ is referred to a linear integral filter (Sagara and Zhao 1990); and $\mathcal{F}_3(s)$ is known as the state variable filter (Young 1970).

Multiple integral operation

The description of the multiple integral operation $\mathcal{F}_1(s)$ in the time domain is given by

$$\mathcal{L}^{-1}[\mathcal{F}_1(s)] = I_n(\cdot)$$

or

$$\mathcal{L}^{-1}[\mathcal{F}_1(s)X(s)] = I_n x(t) = \int_0^t \int_0^{t_1} \cdots \int_0^{t_{n-1}} x(t_n)dt_n dt_{n-1} \cdots dt_1 \tag{14}$$

with $I_0 x(t) \triangleq x(t)$. Then we have

$$x_i(t) = \mathcal{L}^{-1}[s^i \mathcal{F}_1(s)X(s)] = I_{n-i}x(t) \tag{15}$$

and

$$c(t) = \sum_{j=0}^{n-1} c_j(t^j/j!).$$

To calculate $I_i x(k)$ from sampled data we have several approaches e.g. via the numerical integration, the bilinear transformation and the orthogonal functions.

Numerical integration approach. The integral of a continuous-time signal $x(t)$ over a fixed time interval $[t - lT, t)$ can be approximately calculated by

$$\partial I_1 x(t) \triangleq \int_{t-lT}^{t} x(\tau)d\tau \approx f_0 x(t) + f_1 x(t-T) + \cdots + f_l x(t-lT) \tag{16}$$

where T is the step size (here also taken the same as the sampling interval for convenience), and l is a positive integer. The coefficients f_i are determined by formulae of numerical integration. For example, when the trapezoidal rule and Simpson's rule are taken, they are given respectively as follows:

$$\left.\begin{aligned} &f_0 = f_l = T/2; \\ &f_i = T \quad (i = 1, 2, \ldots, l-1) \end{aligned}\right\} \text{ for the trapezoidal rule,} \tag{17}$$

and

$$\left.\begin{aligned} &f_0 = f_l = T/3; \\ &f_i = \begin{cases} 2T/3 & (i = 2, 4, \ldots, l-2); \\ 4T/3 & (i = 1, 3, \ldots, l-1) \end{cases} \end{aligned}\right\} \text{ for Simpson's rule.} \tag{18}$$

Introducing a unit-delay operator q^{-1}, i.e. $q^{-1}x(t) = x(t-T)$, we can rewrite eqn.(16) in the form

$$\partial I_1 x(t) \approx \sum_{i=0}^{l} f_i q^{-i} x(t) \triangleq \nabla_l (q^{-1}) x(t)$$

where

$$\nabla_l(q^{-1}) = f_0 + f_1 q^{-1} + \cdots + f_l q^{-l}.$$

By definition, we have

$$I_j x(t) = I_j x(t - lT) + \partial I_j x(t) \tag{19}$$

or

$$I_j x(t) = \frac{1}{1-q^{-l}} \partial I_j x(t) \approx \frac{\nabla_l(q^{-1})}{1-q^{-l}} I_{j-1} x(t). \tag{20}$$

Therefore we can calculate the multiple integrals $I_j x(k)$ in a recursive way, as follows:

$$\begin{aligned} I_0 x(k) &= x(k) \\ I_j x(k) &\approx I_j x(k-l) + \nabla_l(q^{-1}) I_{j-1} x(k) \quad j = 1, 2, \cdots, n. \end{aligned} \tag{21}$$

The initial values of $I_j x(i), i = 0, 1, \cdots, l$ are taken as

$$\left.\begin{aligned} I_j x(0) &= 0 \\ I_j x(1) &= (T/2)^j (x(0) + x(1)) \\ I_j x(i) &= \nabla_i(q^{-1}) I_{j-1} x(i) \quad i = 2, 3, \cdots, l \end{aligned}\right\} \quad j = 1, 2, \cdots, n. \tag{22}$$

There may be another way to calculate $I_j x(k)$ by using $x(k)$ directly. By eqn.(20), we get

$$I_j x(k) \approx \frac{\nabla_l(q^{-1})}{1-q^{-l}} I_{j-1} x(k) \approx \frac{\nabla_l(q^{-1})^j}{(1-q^{-l})^j} x(k) \tag{23a}$$

or

$$I_j x(k) \approx [1 - (1-q^{-l})^j] I_j x(k) + \nabla_l(q^{-1})^j x(k) \quad j = 1, 2, \cdots, n. \tag{23b}$$

The initial conditions are taken as

$$\left.\begin{aligned} I_j x(0) &= 0 \\ I_j x(1) &= (T/2)^j (x(0) + x(1)) \\ I_j x(i) &= \nabla_i(q^{-1})^j x(i) \quad i = 2, 3, \cdots, nl \end{aligned}\right\} \quad j = 1, 2, \cdots, n. \tag{24}$$

Clearly the way of calculating the multiple integrals via eqn.(21) is much easier than the way of eqn.(23). It is found that the approximation in the numerical calculation of multiple integrals depends upon the step size or the sampling interval T.

Bilinear transformation approach. The use of the bilinear transformation is a very simple way of calculating $I_j x(k)$. It is done as follows:

$$I_j x(k) = x_{n-j}(k) \approx s^{n-j} \mathcal{F}_1(s)\Big|_{s=s_0} x(k) = \frac{[\frac{T}{2}(1+q^{-1})]^j}{(1-q^{-1})^j} x(k) \tag{25}$$

where s_0 is the operator of the bilinear transformation, given by

$$s_0 = \frac{2}{T}\frac{1-q^{-1}}{1+q^{-1}}. \tag{26}$$

It is found that the bilinear transformation method is identical with the trapezoidal integrating rule of numerical integration approach because eqn.(25) accords with eqn.(23a) if the trapezoidal rule with $l = 1$ is taken.

Orthogonal function (OF) approach. The OF approach is based on the so-called integration operational matrix which is developed from orthogonal functions such as block-pulse functions, Walsh functions, Jacobi polynomials, Chebyshev polynomials of the first or second kind, Legendre polynomials and Laguerre polynomials (Rao 1983). It is known that a function or signal, $x(t)$, which is square-integrable in $t \in [0, T_0]$ can be approximated as

$$x(t) \approx \sum_{i=1}^{m} x_i \phi_i(t) = \boldsymbol{x}_m^T \Phi_{(m)}(t).$$

Here $\boldsymbol{x}_m$ is the spectrum of $x(t)$ and $\Phi_{(m)}(t)$ is the OF basis vector, respectively, defined as

$$\boldsymbol{x}_m^T = (x_1 \ x_2 \ \cdots \ x_m)$$

$$\Phi_{(m)}(t)^T = (\phi_1(t) \ \phi_2(t) \ \cdots \ \phi_m(t)).$$

The spectral coefficients x_i is the inner product

$$x_i = (x(t), \phi_i(t)) = c \int_0^{T_0} w(t) x(t) \phi_i(t) dt \tag{27}$$

with c being the constant and $w(t)$ being the weight function of the OF system. Since

$$I_1 x(t) \approx \boldsymbol{x}_m^T \int_0^t \Phi_{(m)}(\tau) d\tau \approx \boldsymbol{x}_m^T M \Phi_{(m)}(t)$$

we have

$$I_j x(t) \approx \boldsymbol{x}_m^T M^j \Phi_{(m)}(t) \triangleq \boldsymbol{x}_{mj}^T \Phi_{(m)}(t) \quad t \in [0, T_0]$$

or

$$I_j x(k) \approx \boldsymbol{x}_{mj}^T \Phi_{(m)}(t)|_{t=kT} \tag{28}$$

where the matrix M is the integration operational matrix; and

$$\boldsymbol{x}_{mj}^T = \boldsymbol{x}_m^T M^j = (x_1^j \ x_2^j \ \cdots \ x_m^j) = \boldsymbol{x}_{m(j-1)}^T M.$$

Note that to calculate $I_j x(t)$, the spectrum coefficients x_j must be obtained using $x(t)$ in the whole identification interval $[0, T_0]$ via eqn.(27), and thus generally $I_j x(k)$ can not be calculated in a time recursive way. Clearly the approximation by the OF approach depends on m, the number of the orthogonal functions used. In the sampling case, it is required to recover $x(t)$ from sampled data $\{x(k)\}$. But it should be pointed out that the block pulse functions (BPF) method is an exception. With gradually incremented time, the related BPF spectral coefficients can be recursively obtained. In fact, the BPF method is a special case of the numerical integration approach wherein the trapezoidal integrating rule is used (Sagara and Zhao 1987).

For the OF approach, there is another way used to estimate the parameters. Instead of using the equations generated by eqn.(12) for different sampling points, the other method is based on the m equations which are obtained by equating the components of m orthogonal functions (Rao 1983).

Although the use of the multiple integral operation is a simple way of handling time derivatives and can be implemented in a recursive way via the numerical integration, the block-pulse functions and the bilinear transformation, the term due to the unknown initial condition, $c(k)$, has to be estimated together with the system parameters. However, Whitfield and Messali (1987) and Sagara and Zhao (1988) have found that the term under estimation causes many difficulties for on-line parameter identification. For example, the number of the parameters to be estimated increases, and thus a heavier burden of computation occurs especially for multivariable systems. Care must be taken in the choice of input signals in order to get a unique solution to parameter estimation problems. Some action must be taken to limit the magnitude of multiple integrals for on-line identification.

Linear integral filter

The description of the linear integral filter $\mathcal{F}_2(s)$ in time domain is given by

$$\mathcal{L}^{-1}[\mathcal{F}_2(s)] \triangleq \delta I_n(\cdot)$$

or

$$\mathcal{L}^{-1}[\mathcal{F}_2(s)X(s)] = \delta I_n x(t) = \int_{t-lT}^{t} \int_{t_1-lT}^{t_1} \cdots \int_{t_{n-1}-lT}^{t_{n-1}} x(t_n) dt_n dt_{n-1} \cdots dt_1. \qquad (29)$$

Lemma 1. The calculation of $x_j(t) = \mathcal{L}^{-1}[s^j \mathcal{F}_2(s)X(s)]$ is carried out as follows:

$$x_j(t) = \delta I_n x^{(j)}(t) \approx \mathcal{J}_j x(t) \triangleq \sum_{i=0}^{nl} p_i^j q^{-i} x(t)$$

or
$$x_j(k) \approx \mathcal{J}_j x(k) = \sum_{i=0}^{nl} p_i^j q^{-i} x(k). \tag{30}$$

Here the polynomial $\mathcal{J}_j$ is given by

$$\mathcal{J}_j = (1 - q^{-l})^j (f_0 + f_1 q^{-1} + \cdots + f_l q^{-l})^{n-j} \quad j = 0, 1, \cdots, n \tag{31}$$

where the coefficients f_i are determined by formulae of numerical integration like the trapezoidal rule and Simpson's rule (see eqns.(17), (18)).

Proof. The proof of the lemma can be found in Sagara and Zhao (1990).

It is interesting to find out that

$$s^j \mathcal{F}_2(s)\Big|_{s=s_0} \approx \frac{(1 - q^{-l})^n}{s^{n-j}\big|_{s=s_0}} = T(q^{-1})\bar{\mathcal{J}}_j$$

where

$$T(q^{-1}) = (1 + q^{-1} + \cdots + q^{-l+1})^n$$
$$\bar{\mathcal{J}}_j = (1 - q^{-1})^j [\frac{T}{2}(1 + q^{-1})]^{n-j}. \tag{32}$$

Here $T(q^{-1})\bar{\mathcal{J}}_j$ is the same as $\mathcal{J}_j$ where the trapezoidal rule is used.

Clearly, the linear integral filter is an operation of numerical integration over a fixed time interval. The approximation of eqn.(30) depends upon the step size or the sampling interval T. The calculation of $\mathcal{J}_j x(k)$ is easily done by calculating an inner product of a coefficient vector $\boldsymbol{p}_j$ and a data vector $\boldsymbol{x}(k)$, i.e.

$$\mathcal{J}_j x(k) = \boldsymbol{p}_j^T \boldsymbol{x}(k) \qquad j = 0, 1, \cdots, n$$

where

$$\boldsymbol{p}_j = (p_0^j \; p_1^j \; \cdots \; p_{nl}^j)^T$$
$$\boldsymbol{x}(k) = (x(k) \; x(k-1) \; \cdots \; x(k - nl))^T.$$

The linear integral filter can be used not only to handle the time derivatives but also to eliminate completely the effect of the initial condition problem. This is seen from the following lemma.

Lemma 2.

$$\mathcal{L}^{-1}[\mathcal{F}_2(s)C(s)] = (1-q^{-l})^n \sum_{i=0}^{n-1} c_i(t^i/i!) = 0. \tag{33}$$

Proof. It is sufficient to show

$$(1-q^{-l})^{j+1}t^j = 0 \quad j \geq 0.$$

By induction on index j, we prove the above equation as follows.

(1) It is evident that $(1-q^{-l})t^0 = (1-q^{-l})1 = 0$ when $j = 0$.

(2) Suppose that the equation holds when j equals k (integer), i.e.

$$(1-q^{-l})^{k+1}t^k = 0 \quad k \geq 0$$

or

$$(1-q^{-l})^{k+1}t^i = 0 \quad 0 \leq i \leq k,\ k \geq 0.$$

For convenience, we define

$$\sum_{i=0}^{k} t^i(t-lT)^{k-i} = \sum_{i=0}^{k} c_i' t^i.$$

Thus we obtain

$$\begin{aligned}
(1-q^{-l})^{k+2}t^{k+1} &= (1-q^{-l})^{k+1}[(1-q^{-l})t^{k+1}] \\
&= (1-q^{-l})^{k+1}[t^{k+1} - (t-lT)^{k+1}] \\
&= (1-q^{-l})^{k+1} lT \sum_{i=0}^{k} t^i(t-lT)^{k-i} \\
&= lT \sum_{i=0}^{k} c_i'(1-q^{-l})^{k+1}t^i \\
&= 0
\end{aligned}$$

which completes the proof.

By Lemma 2, the term due to the initial condition, $c(k)$, no longer appears in eqn.(12). Thus only system parameters requires estimation. It is also find that the linear integral filter has a finite impulse response (FIR), so it belongs to the kind of FIR filters.

State variable filter

The state variable filter, which appeared in the early 1960's, has been widely used in continuous-time system identification and adaptive control (see Kaya and Yamamura 1962 and Young 1965). The state variable filter functions as a pre-filter. How to choose the polynomial $Q(s)$ is a main problem. At least it is required that $Q(s)$ is stable, i.e.

$$\text{Real}(s_i) < 0 \text{ for all } s_i \text{ such that } Q(s_i) = s^n + \sum_{i=1}^{n} q_i s^{n-i} = 0. \tag{34}$$

Let $C_3(s) = C(s)/Q(s)$ and $c_3(t) = \mathcal{L}^{-1}[C_3(s)]$. Since $Q(s)$ is stable, term $c_3(t)$, due to the initial condition, will decrease to zero exponentially as time increases. Therefore, to remove the effect of the initial condition as soon as possible, the state variable filter is required to have well damped transient response characteristics. It is worth noting that $c_3(t)$ will take longer to disappear when $C(s)/Q(s)$ has a non-minimum-phase character.

To implement the state variable filter via a computer, a way is directly to use the Runge-Kutta method to solve the following differential equations

$$y_i(t) = \frac{p^i}{Q(p)} y(t), \quad u_i(t) = \frac{p^i}{Q(p)} u(t) \tag{35}$$

from zero initial states. Another way is to solve a set of difference equations that are obtained by using the bilinear transformation. That is,

$$\begin{aligned} \left.\frac{s^i}{Q(s)}\right|_{s=s_0} &\triangleq \frac{\beta_i(q^{-1})}{\alpha(q^{-1})} \\ y_i(k) = \frac{\beta_i(q^{-1})}{\alpha(q^{-1})} y(k), &\quad u_i(k) = \frac{\beta_i(q^{-1})}{\alpha(q^{-1})} u(k) \end{aligned} \tag{36}$$

where

$$\begin{aligned} \alpha(q^{-1}) &= \bar{\mathcal{J}}_n + \sum_{i=1}^{n} q_i \bar{\mathcal{J}}_{n-i} \triangleq \alpha_0 + \alpha_1 q^{-1} + \cdots + \alpha_n q^{-n} \quad (\alpha_0 \neq 0) \\ \beta_i(q^{-1}) &= \bar{\mathcal{J}}_i \triangleq \beta_{i0} + \beta_{i1} q^{-1} + \cdots + \beta_{in} q^{-n}. \end{aligned} \tag{37}$$

Here $\bar{\mathcal{J}}_i$ is given by eqn.(32).

Note that the implementation via the bilinear transformation is simpler than that of Runge-Kutta method, and $\alpha(q^{-1})$ obtained from a stable $Q(s)$ via the bilinear transformation is also stable. Clearly the state variable filter has an infinite impulse response (IIR), and thus belongs to the kind of IIR filters.

Digital pre-filters

As shown above, analog pre-filters have been implemented in digital forms in order to make it convenient for computers to handle them. The digital implementations lead to a form of FIR filters or IIR filters. On the other hand, it is also possible to use a digital pre-filter $D(q^{-1})$ directly for signal processing. In this case, the d.e. model should first be discretized. The z transformation is perhaps most used, and it gives a precise discrete-time model without any approximation when the input signal remains constant between sampling instants. But to obtain the d.e. model parameters from the identified discrete-time model is not with ease (Sinha 1972). Moreover, the computational burden is usually heavy, thus such an indirect approach is not suitable for some on-line applications. Another simple way is to use the bilinear transformation. Although this way results in an approximated model, the model keeps the structure of the d.e. model, i.e. it has the same parameters as the d.e. model. Thus the continuous system parameters can be directly estimated.

Discretize the d.e. model (3) by the bilinear transformation to get

$$A(p)|_{p=s_0} x(k) = B(p)|_{p=s_0} u(k)$$

or

$$\bar{\mathcal{J}}_n x(k) + \sum_{i=1}^{n} a_i \bar{\mathcal{J}}_{n-i} x(k) = \sum_{i=1}^{n} b_i \bar{\mathcal{J}}_{n-i} u(k) \tag{38}$$

where s_0 is the operator of the bilinear transformation, and $\bar{\mathcal{J}}_j$ is given by eqn.(32).

Performing the digital pre-filter $D(q^{-1})$ on both sides of eqn.(38), we have

$$D(q^{-1})\bar{\mathcal{J}}_n x(k) + \sum_{i=1}^{n} a_i D(q^{-1})\bar{\mathcal{J}}_{n-i} x(k) = \sum_{i=1}^{n} b_i D(q^{-1})\bar{\mathcal{J}}_{n-i} u(k).$$

Substituting eqn.(7) into the above equation gives

$$D(q^{-1})\bar{\mathcal{J}}_n y(k) + \sum_{i=1}^{n} a_i D(q^{-1})\bar{\mathcal{J}}_{n-i} y(k) = \sum_{i=1}^{n} b_i D(q^{-1})\bar{\mathcal{J}}_{n-i} u(k) + e(k) \tag{39}$$

where the equation error is as

$$e(k) = D(q^{-1})\bar{\mathcal{J}}_n v(k) + \sum_{i=1}^{n} a_i D(q^{-1})\bar{\mathcal{J}}_{n-i} v(k).$$

Denoting

$$y_i(k) = D(q^{-1})\bar{\mathcal{J}}_i y(k), \quad u_i(k) = D(q^{-1})\bar{\mathcal{J}}_i u(k)$$

then eqn.(39) can be rewritten by

$$y_n(k) + \sum_{i=1}^{n} a_i y_{n-i}(k) = \sum_{i=1}^{n} b_i u_{n-i}(k) + e(k). \tag{40}$$

It is interesting to note that if $D(q^{-1})$ is taken as $T(q^{-1})$ of eqn.(32) the identification model (40) is identical with the model obtained via the linear integral filter of the trapezoidal rule. The model (40) is also the same as the model obtained via the state variable filter when $D(q^{-1})$ has the form of $1/\alpha(q^{-1})$.

According to its impulse response characteristic, $D(q^{-1})$ can be classified as an FIR filter or an IIR filter.

Design of pre-filters

Pre-filtering signals is a very useful and important way to improve the statistical efficiency or the accuracy of parameter estimates in system identification. The unwanted frequency components in signals should be filtered off as much as possible. For this purpose, the pre-filters should have some specified frequency response characteristics. From frequency response point of view, the pre-filter $\mathcal{F}(s)$ should be designed in such a way that the pre-filter has similar or close frequency response characteristics to the system under study, as shown in Fig.1. Thus the free parameters of a pre-filter, if any, should be chosen in such a way.

In case of the state variable filter, it is often suggested that the optimal choice is to use the exact system polynomial $A(s)$, i.e. $Q(s) = A(s)$ in the white noise case (Young and Jakeman 1980). In this case, the equation error $e(k)$ is identical with $v(k)$. In the white noise case, therefore, even the LS method can yield satisfactory parameter estimates. Another form, $Q(s) = (s + \tau_0)^n$ $(\tau_0 > 0)$ is also frequently used, where τ_0 is a design parameter. In the design of the state variable filter, as stated before, the initial condition effect should also be considered. For the linear

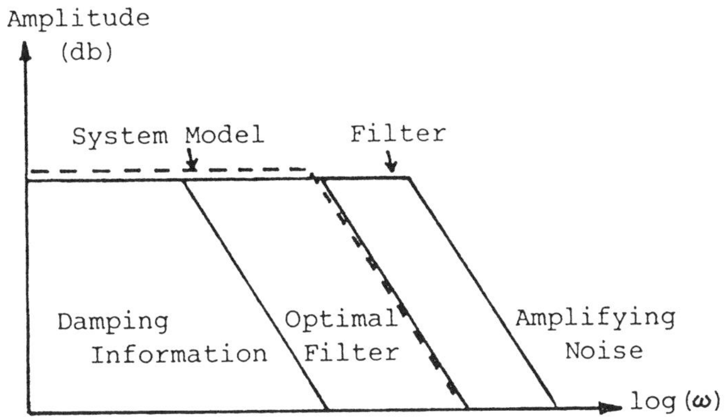

Fig.1 Comparison of frequency responses

integral filter, the free parameter l is a design parameter. As for $\mathcal{F}_1(s)$, it has no free parameter, and thus it can not be designed.

In a similar way to the design of analog pre-filters, the digital pre-filters should also have similar or close frequency response characteristics to the system under study. There are many approaches available to the designer of digital filters (see monographs e.g. Oppenheim and Schafer 1975; Mitani 1988). Whether to use an FIR filter or an IIR filter depends on the practical situation. The IIR filters are in a recursive form and thus can produce a desired frequency characteristic with fewer coefficients than the FIR filters. On the contrary, the FIR filters have a simpler form which makes it possible to employ the bias-compensating techniques for consistent parameter identification.

4. Parameter estimation

In this section, we focus only on the on-line identification problem, thus the usual multiple integral operation $\mathcal{F}_1(s)$ is not used, and only system parameters $\{a_i,\ b_i\}$ are under estimation.

Identification model

As shown in Section 3, using the linear integral filter, the state variable filter or the digital pre-filters, the d.e. model can be put in the following form

$$y_n(k) + \sum_{i=1}^{n} a_i y_{n-i}(k) = \sum_{i=1}^{n} b_i u_{n-i}(k) + e(k) \tag{41}$$

where

$$\begin{aligned} y_i(k) &= F(i, q^{-1})y(k), \quad u_i(k) = F(i, q^{-1})u(k) \\ e(k) &= F(n, q^{-1})v(k) + \sum_{i=1}^{n} a_i F(n-i, q^{-1})v(k). \end{aligned} \tag{42}$$

For analog pre-filters, $F(i, q^{-1}) = \mathcal{J}_i,\ \beta_i(q^{-1})/\alpha(q^{-1})$ are the digital implementations of the linear integral filter and the state variable filter, respectively; while $F(i, q^{-1}) = D(q^{-1})\bar{\mathcal{J}}_i$ is the case of digital pre-filters.

$F(i, q^{-1})$ can have a finite impulse response

$$F_F(i, q^{-1}) = n_{i0} + n_{i1}q^{-1} + \cdots + n_{im}q^{-m} \triangleq N_i(q^{-1}) \tag{43}$$

or an infinite impulse response

$$F_I(i, q^{-1}) = \frac{n_{i0} + n_{i1}q^{-1} + \cdots + n_{im}q^{-m}}{d_0 + d_1 q^{-1} + \cdots + d_m q^{-m}} \triangleq \frac{N_i(q^{-1})}{D_I(q^{-1})}. \tag{44}$$

In case of an FIR filter, using eqn.(43) the equation error $e(k)$ can be written as

$$e_F(k) = \sum_{i=0}^{n} a_i N_{n-i}(q^{-1})v(k) \triangleq H(q^{-1})v(k) \qquad (a_0 = 1) \tag{45}$$

where

$$\begin{aligned} H(q^{-1}) &= h_0 + h_1 q^{-1} + \cdots + h_m q^{-m} \\ h_j &= \sum_{i=0}^{n} a_i n_{(n-i)j} \quad (j = 0, 1, \cdots, m). \end{aligned} \tag{46}$$

In case of an FIR filter, similarly we get from eqn.(44)

$$e_I(k) = \sum_{i=0}^{n} a_i \frac{N_{n-i}(q^{-1})}{D_I(q^{-1})} v(k) = \frac{H(q^{-1})}{D_I(q^{-1})} v(k). \tag{47}$$

For convenience, we introduce the following notations

$$\begin{aligned}
\theta^* &= (a_1 \ \cdots \ a_n \ b_1 \ \cdots \ b_n)^T \\
\hat{\theta} &= (\hat{a}_1 \ \cdots \ \hat{a}_n \ \hat{b}_1 \ \cdots \ \hat{b}_n)^T \\
\varphi(u,k) &= (u_{n-1}(k) \ u_{n-2}(k) \ \cdots \ u_0(k))^T \\
\varphi(y,k) &= (y_{n-1}(k) \ y_{n-2}(k) \ \cdots \ y_0(k))^T \\
\varphi(k) &= (-\varphi(y,k)^T \ \varphi(u,k)^T)^T
\end{aligned} \tag{48}$$

where θ^* and $\hat{\theta}$ denote the vector of true parameters and the vector of estimated parameters respectively. Then eqn.(41) can be rewritten as

$$y_n(k) = \varphi(k)^T \theta^* + e(k). \tag{49}$$

which is referred to the identification model.

Least squares method

Based on the identification model (49), the least squares estimate that minimizes the sum of the squared equation errors is given by

$$\hat{\theta}_{LS}(N) = [\sum_{k=1}^{N} \varphi(k)\varphi(k)^T]^{-1} \sum_{k=1}^{N} \varphi(k) y_n(k) \tag{50}$$

provided that the inverse exists. Here N represents the number of data. The above off-line LS estimate can also be calculated in the following recursive way

$$\begin{aligned}
\hat{\theta}_{LS}(k) &= \hat{\theta}_{LS}(k-1) + L(k)\epsilon(k) \\
\epsilon(k) &= y_n(k) - \varphi(k)^T \hat{\theta}_{LS}(k-1) \\
L(k) &= \frac{P(k-1)\varphi(k)}{1 + \varphi(k)^T P(k-1)\varphi(k)} \\
P(k) &= [I - L(k)\varphi(k)^T]P(k-1).
\end{aligned} \tag{51}$$

Assume that the noise is stationary and ergodic. As the data number N tends to infinity, eqn.(50) can be rewritten as

$$\begin{aligned}
&E[\varphi(k)\varphi(k)^T] = R, \quad \text{nonsingular}, \\
&\hat{\theta}_{LS} = \lim_{N \to \infty} \hat{\theta}_{LS}(N) = R^{-1} E[\varphi(k) y_n(k)]
\end{aligned} \tag{52}$$

and

$$\hat{\theta}_{LS} - \theta^* = R^{-1}E[\varphi(k)e(k)] \tag{53}$$

where E denotes the expectation.

When $\{v(k)\}$ is a white noise sequence with zero-mean and variance σ_v^2, it can be shown that

$$\begin{aligned} E[\varphi_F(k)e_F(k)] &= (-E[\varphi_F(y,k)^T e_F(k)] \;\; E[\varphi_F(u,k)^T e_F(k)])^T \\ &= -(E[\varphi_F(v,k)^T H(q^{-1})v(k)] \;\; 0)^T \\ &= -\mathcal{G}(\theta^*)\sigma_v^2 \neq 0 \end{aligned} \tag{54}$$

where the subscript F indicates that $F(i, q^{-1})$ has a finite impulse response, and

$$\mathcal{G}(\theta^*) = (g_{n-1} \;\; \cdots \;\; g_0 \;\; \underbrace{0 \; \cdots \; 0}_{n})^T$$

$$g_i = \sum_{j=0}^{m} n_{ij}h_j \qquad (i = 0, 1, \cdots, n-1). \tag{55}$$

Equation (54) implies that the LS estimate has an asymptotic bias even in the case of white noise.

Let $w(k) = v(k)/D_I(q^{-1})$. When $F(i, q^{-1})$ has an infinite impulse response, we have

$$E[\varphi_I(k)e_I(k)] = -(E[\varphi_F(w,k)^T H(q^{-1})w(k)] \;\; 0)^T \neq 0 \tag{56}$$

because $\{w(k)\}$ is a colored noise sequence.

On the LS estimator, therefore, the following conclusion can be made.

Theorem 1. In general, the LS estimator (50) is asymptotically biased.

On the above result, several remarks are in order here.

Remark 1. We may conclude that the statement in the theorem is still valid in general cases of measurement noise. Therefore, applying the LS method directly to the identification model (49) cannot result in consistent estimators.

Remark 2. By eqn.(54), it is evident that the bias of the LS estimator depends upon the noise variance σ_v^2, i.e. the level of noise on signal. As shown later by simulation experiments, when the level of noise on output signals is low ($\leq$ 10%), the above LS estimator does not introduce noticeable bias because of the filtering features of the pre-filter. However, when the level of noise to signal is high, the bias becomes intolerable, and thus bias-eliminating methods are required.

Bias-compensating methods

In case of an FIR pre-filter, i.e. $F(i, q^{-1})$ having a finite impulse response, rewrite eqn.(53) to get

$$\theta^* = \lim_{N\to\infty} \hat{\theta}_{LS}(N) - \{-R^{-1}\mathcal{G}(\theta^*)\sigma_v^2\} \tag{57}$$

which implies that a consistent estimate of the unknown parameters θ^* can be obtained by subtracting an estimate of the bias term from the LS estimate. Based on such an observation, a bias-compensating LS algorithm is given as (Zhao, Sagara and Wada 1990)

$$\hat{\theta}^*(k) = \hat{\theta}_{LS}(k) + k\sigma_v^2 P(k)\mathcal{G}(\hat{\theta}^*(k-1)) \tag{58}$$

for a known variance σ_v^2, and

$$\hat{\theta}^*(k) = \hat{\theta}_{LS}(k) + k\hat{\sigma}_v^2(k) P(k)\mathcal{G}(\hat{\theta}^*(k-1)) \tag{59}$$

for an unknown variance σ_v^2. The estimate of σ_v^2 is obtained by

$$\begin{aligned}
\hat{\sigma}_v^2(k) &= \frac{1}{k}\frac{Q(k)}{d(k)} \\
Q(k) &= Q(k-1) + \frac{1}{1+\varphi_F(k)^T P(k-1)\varphi_F(k)}\epsilon(k)^2 \\
d(k) &= \sum_{j=0}^{m} \hat{h}_j^2 + \mathcal{G}(\hat{\theta}^*(k-1))^T[\hat{\theta}_{LS}(k) - \hat{\theta}^*(k-1)].
\end{aligned} \tag{60}$$

Another similar bias-compensating method is based on Newton's stochastic method. It is easy to show that eqn.(54) can also be rewritten as

$$E[\varphi_F(k)e_F(k)] = \omega + \Omega\theta \triangleq -\frac{d}{d\theta}J(\theta) \tag{61}$$

where the criterion is defined as $J(\theta) = E[\frac{1}{2}e_F(k)^2]$; and the $2n$ vector ω and the $2n \times 2n$ matrix Ω are given by

$$\omega = \begin{pmatrix} -\mathcal{N}_{n-1}\boldsymbol{n}_n \\ 0 \end{pmatrix}\sigma_v^2, \quad \Omega = \begin{pmatrix} -\mathcal{N}_{n-1}\mathcal{N}_{n-1}^T & 0 \\ 0 & 0 \end{pmatrix}\sigma_v^2$$

$$\mathcal{N}_{n-1} = \begin{pmatrix} n_{(n-1)0} & n_{(n-1)1} & \cdots & n_{(n-1)m} \\ n_{(n-2)0} & n_{(n-2)1} & \cdots & n_{(n-2)m} \\ \vdots & \vdots & & \vdots \\ n_{00} & n_{01} & \cdots & n_{0m} \end{pmatrix}, \quad \boldsymbol{n}_n = \begin{pmatrix} n_{n0} \\ n_{n1} \\ \vdots \\ n_{nm} \end{pmatrix}. \tag{62}$$

Introduce an additive term into the gradient of the criterion function to get

$$\frac{d}{d\theta}J(\theta^*) + \omega + \Omega\theta^* = -[E\varphi_F(k)e(k) - \omega - \Omega\theta^*] = 0. \tag{63}$$

Then a recursive algorithm of the bias-compensating method is of the form

$$\begin{aligned} \hat{\theta}^*(k) &= \hat{\theta}^*(k-1) + P(k)[(\varphi_F(k)y_n(k) - \omega) - (\varphi_F(k)\varphi_F(k)^T + \Omega)\hat{\theta}^*(k-1)] \\ P(k) &= P(k-1) - \frac{P(k-1)\varphi_F(k)\varphi_F(k)^T P(k-1)}{1 + \varphi_F(k)^T P(k-1)\varphi_F(k)}. \end{aligned} \tag{64}$$

Note that in case of IIR pre-filters, the above bias-compensating methods are not available because of the difficulty in the calculation of the bias term.

Instrumental variable methods

Another effective way to remove the bias is to use the IV methods which need little information on the noise statistics. The key point is how to choose the instrumental variables. There are many choices. For example, the following two choices are very useful (Sagara and Zhao 1989; 1990)

$$\zeta(k) = (-\varphi(u, k-k_0)^T \;\; \varphi(u,k)^T)^T \tag{65a}$$

and

$$\zeta(k) = (-\varphi(\hat{x}, k)^T \;\; \varphi(u,k)^T)^T. \tag{65b}$$

Here k_0 is a delay parameter; $\hat{x}(k)$ is generated using an auxiliary model which is obtained by discretizing the transfer function $B(s)/A(s)$ via the bilinear transformation, as follows:

$$\hat{x}(k) = \left.\frac{\hat{B}(s)}{\hat{A}(s)}\right|_{s=s_0} u(k) \tag{66}$$

where $\hat{A}(s)$, $\hat{B}(s)$ are constructed using the previous estimates of a_i, b_i.

The recursive IV algorithm is then given as follows:

$$\begin{aligned} \hat{\theta}_{IV}(k) &= \hat{\theta}_{IV}(k-1) + L(k)\epsilon(k) \\ \epsilon(k) &= \mathcal{J}_n y(k) - \varphi(k)^T\hat{\theta}_{IV}(k-1) \\ L(k) &= \frac{P(k-1)\zeta(k)}{1 + \varphi(k)^T P(k-1)\zeta(k)} \\ P(k) &= [I - L(k)\varphi(k)^T]P(k-1). \end{aligned} \tag{67}$$

Note that the above IV methods are valid both in the case of FIR and IIR filters and in the case of colored noise.

Simulation experiments

Numerical simulations are presented to illustrate the above identification algorithms. In the following simulations, the linear integral filter using the trapezoidal integrating rule is employed for signal processing. The system simulated is as follows:

$$(p^2 + a_1 p + a_2)x(t) = (b_1 p + b_2)u(t)$$
$$y(k) = x(k) + v(k)$$

with $a_1 = 2.8, a_2 = 4.0, b_1 = 0.0, b_2 = 5.0$ and subject to initial conditions $x(0) = 1.0$, $x^{(1)}(0) = 5.0$. The input signal is chosen as the following combination of three sinusoidal signals

$$u(t) = \sin 0.714t + \sin 1.428t + \sin 2.142t.$$

The noise $\{v(k)\}$ is a zero-mean identically and independently distributed (i.i.d.) Gaussian sequence. Its variance is adjusted to obtain the desired ratio of noise to signal (N/S) defined by

$$N/S = \frac{\text{SD of } v(k)}{\text{SD of } x(k)}$$

where SD represents standard deviation.

Table 1 shows the LS estimates obtained using the on-line LS algorithm (51) from a Monte-Carlo simulation of 20 experiments for three different values of l in a low level of N/S ratio, 10%. In the table, the computed mean $\hat{\theta}_{mean}$ and standard deviation σ_i of each estimate are included. The mean normalized error (MNE) and average standard deviation (ASD) are also included for quick comparison. They are defined as

$$\text{MNE} = \frac{||\hat{\theta}_{mean} - \theta^*||^2}{||\theta^*||^2}, \qquad \text{ASD} = \frac{\sum_{i=1}^{2n} \sigma_i}{2n}.$$

Figure 2 plots the frequency responses of the given system and the linear integral filter with different values of l. All amplitudes are normalized for comparison.

It is found that the frequency bandwidth of the linear integral filter matches most closely the frequency band of the given system when $l = 20$ for $T = 0.05$. It can be observed from Table 1 that in the case of "optimal" choice of l (i.e. $l = 20$ for $T = 0.05$) convergence seems to take the least samples or time and the best parameter estimates are obtained for any given sample size. When l is taken less than 20, the noise effect increases and the LS estimates become worse. When l is taken larger than 20, ASD has a large value for the small sample size $N = 100$, but both ASD and MNE are improved greatly when N increases. As known, the increase of N implies that more information on the system enters the on-line LS estimator, thus the parameter estimates are improved.

It can also be observed from Table 1 that when l is chosen properly, the bias in the LS estimates is not so noticeable and may be tolerable. Clearly this is due to the filtering feature of the linear integral filter.

When the ratio of noise to signal is taken as 50%, which corresponds to $\sigma_v^2 = 0.341$, the LS estimates of one experiment are plotted in Fig.3. As can be seen from the figure, the bias on the LS estimates becomes remarkable and can no longer be tolerable.

Figure 4 shows the results obtained by using the bias-compensating LS algorithm (59). The unknown noise variance is estimated together with the system parameters via the algorithm (60), as shown in Fig.5. When the noise variance is known, the bias-compensating algorithm (64) can also be used. By using algorithm (64), the results are shown in Fig.6.

When the on-line IV algorithm (67) with the second instrumental variables (65b) is used, the results from 20 Monte-Carlo experiments in the case of white noise with $N/S = 20\%$ are presented in Table 2. The results shown in Table 3 are obtained when $\{v(k)\}$ is generated from an i.i.d. white noise sequence $\{w(k)\}$ via a second-order autoregressive moving average (ARMA) model

$$v(k) = \frac{1 - 1.5q^{-1} + 0.5725q^{-2}}{1 - 1.6q^{-1} + 0.68q^{-2}} w(k)$$

where the ratio of noise to signal is taken $N/S = 20\%$.

Table 1 LS estimates; $N/S = 10\%$, $T = 0.05$

l	N	a_1 (2.8)	a_2 (4.0)	b_1 (0.0)	b_2 (5.0)	MNE (10^{-2})	ASD (10^{-1})
	100	2.251±.505	3.284±.585	0.229±.202	4.028±.793	3.71	5.21
10	500	2.646±.089	3.808±.119	0.070±.048	4.724±.170	0.29	1.06
	1000	2.602±.055	3.745±.086	0.098±.038	4.634±.115	0.51	0.74
	100	2.618±.359	3.783±.471	0.066±.150	4.689±.641	0.37	4.05
20	500	2.761±.086	3.955±.130	0.010±.053	4.937±.176	0.02	1.11
	1000	2.751±.071	3.919±.107	0.033±.044	4.902±.146	0.04	0.92
	100	2.621±1.14	3.760±1.02	0.074±.277	4.667±1.58	0.42	10.0
30	500	2.757±.094	3.934±.136	0.020±.056	4.916±.190	0.03	1.19
	1000	2.754±.074	3.912±.109	0.037±.043	4.897±.151	0.04	0.94

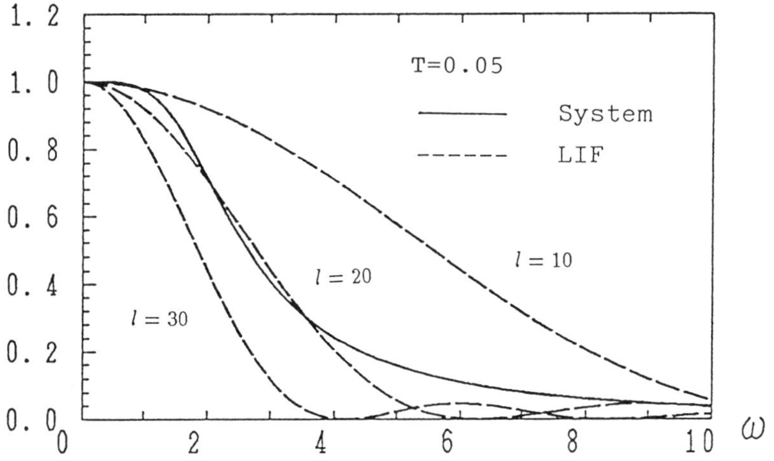

Fig.2 Frequency response of the LIF and the system

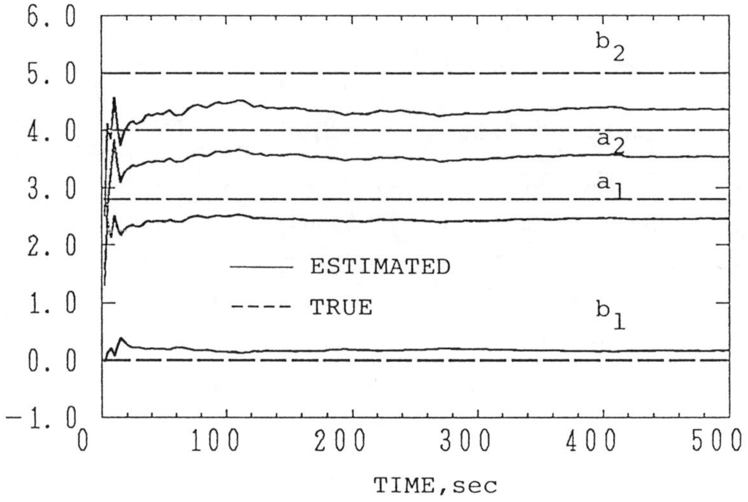

Fig.3 LS estimates; $N/S = 50\%$, $T = 0.05$

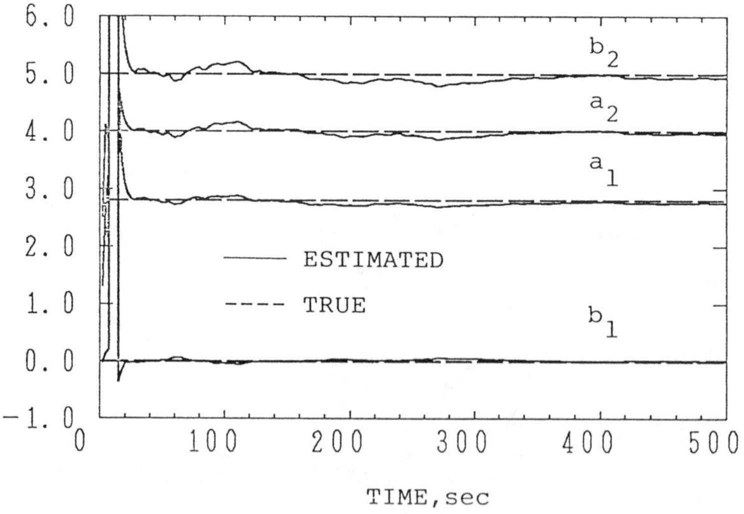

Fig.4 Bias-compensating LS estimates; $N/S = 50\%$, $T = 0.05$

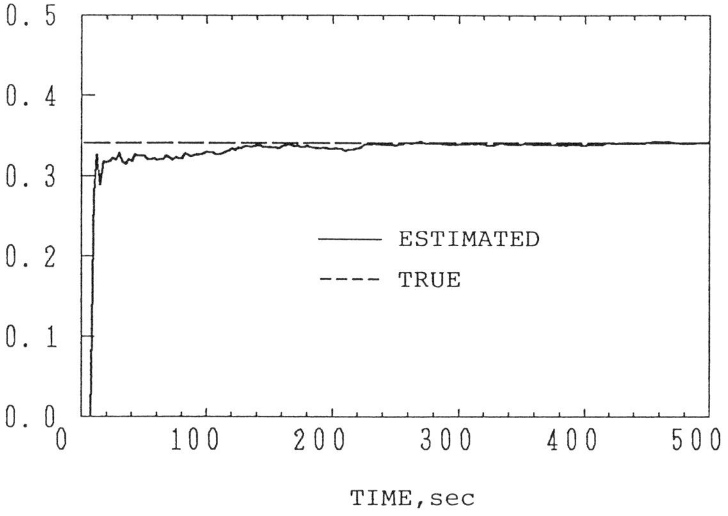

Fig.5 Estimates of σ_v^2

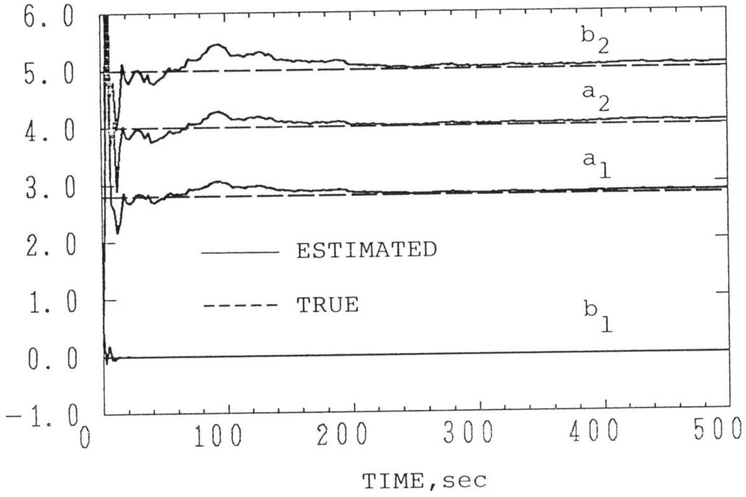

Fig.6 Bias-compensating estimates with known σ_v^2;
$N/S = 50\%$, $T = 0.05$

Table 2 IV estimates in case of white noise; $N/S = 20\%$, $T = 0.05$

	True value	Time(sec) 50.0 (N=2500)	100.0 (N=5000)	200.0 (N=10000)	500.0 (N=25000)
a_1	2.8	2.768±.094	2.780±.069	2.808±.052	2.808±.030
a_2	4.0	3.940±.153	3.971±.100	4.006±.078	4.002±.050
b_1	0.0	0.037±.071	0.020±.044	0.003±.037	0.004±.023
b_2	5.0	4.922±.199	4.956±.135	5.008±.103	5.006±.060
MNE($\times 10^{-4}$)		2.46	0.75	0.03	0.02
ASD($\times 10^{-1}$)		1.29	0.87	0.68	0.41

Table 3 IV estimates in case of ARMA process noise; $T = 0.05$

	True value	Time(sec) 50.0 (N=2500)	100.0 (N=5000)	200.0 (N=10000)	500.0 (N=25000)
a_1	2.8	2.915±.190	2.869±.152	2.853±.075	2.828±.056
a_2	4.0	3.943±.229	3.946±.139	3.982±.074	3.986±.049
b_1	0.0	-0.010±.060	-0.001±.048	-0.003±.024	0.001±.019
b_2	5.0	5.188±.309	5.128±.230	5.095±.117	5.055±.086
MNE($\times 10^{-3}$)		1.06	0.49	0.25	0.08
ASD($\times 10^{-1}$)		1.97	1.42	0.72	0.52

5. On the choice of input signal

To ensure that the system is identifiable, the minimum requirement of the input signal is the persistently exciting (p.e.) conditions, i.e. the system dynamics have to be excited persistently by the input signal over the measuring time (Åström and Bohlin 1965). For discrete-time model identification, the results of p.e. conditions on the input signal have been well developed (see e.g. Söderström and Stoica 1989). But general theoretical results are not available for the continuous-time model identification based on discrete-time measurements. One major reason is that the p.e. conditions depend upon the digital implementations of the analog pre-filters or the digital pre-filters and thus it becomes more difficult and complicated to derive the p.e. conditions explicitly. Here an attempt is made to examine the problem, i.e. under what condition of the input the matrix R defined in eqn.(52) is nonsingular. This is required for the LS algorithm and its alternatives like the bias-compensating algorithms and the bootstrap IV algorithm.

Note that the matrix R can be expressed as

$$R = E[\varphi(k)\varphi(k)^T] = R' + R_v \tag{68}$$

where

$$\begin{aligned} R' &= E[\bar{\varphi}(k)\bar{\varphi}(k)^T] \\ R_v &= E[\varphi_v(k)\varphi_v(k)^T] \end{aligned} \tag{69}$$

and

$$\begin{aligned} \varphi(k) &= (-y_{n-1}(k) \ \cdots \ -y_0(k) \ u_{n-1}(k) \ \cdots \ u_0(k))^T \\ \bar{\varphi}(k) &= (-x_{n-1}(k) \ \cdots \ -x_0(k) \ u_{n-1}(k) \ \cdots \ u_0(k))^T \\ \varphi_v(k) &= (-v_{n-1}(k) \ \cdots \ -v_0(k) \ 0 \ \cdots \ 0)^T. \end{aligned}$$

It is clear that $R_v \geq 0$, i.e. $\boldsymbol{x}^T R_v \boldsymbol{x} \geq 0$ for any non-zero vector $\boldsymbol{x}$. Thus if the symmetric matrix R' is nonsingular, i.e. $R' > 0$, the matrix R is also nonsingular.

When the trapezoidal integrating rule is used in the linear integral filter, we have

$$\mathcal{J}_j = T(q^{-1})\bar{\mathcal{J}}_j \tag{70}$$

where $T(q^{-1})$ and $\bar{\mathcal{J}}_j$ are given by eqn.(32). It is assumed that the digital pre-filter $D(q^{-1})$ takes the form of

$$D(q^{-1}) = \frac{N_D(q^{-1})}{D_D(q^{-1})}$$

where $D_D(q^{-1})$ is stable. Then $F(i, q^{-1})$ can be put in a general form as follows:

$$F(i, q^{-1}) = \frac{N_F(q^{-1})}{D_F(q^{-1})}\bar{\mathcal{J}}_i \tag{71}$$

where

(1) the linear integral filter : $N_F(q^{-1}) = T(q^{-1}),\ D_F(q^{-1}) = 1$;

(2) the state variable filter : $N_F(q^{-1}) = 1,\ D_F(q^{-1}) = \alpha(q^{-1})$;

(3) the digital pre-filters : $N_F(q^{-1}) = N_D(q^{-1}),\ D_F(q^{-1}) = D_D(q^{-1})$.

Rewriting eqn.(38) gives

$$x(k) = \frac{\beta_0(q^{-1})}{\alpha_0(q^{-1})}u(k) \tag{72}$$

where

$$\alpha_0(q^{-1}) = \bar{\mathcal{J}}_n + \sum_{i=1}^{n} a_i \bar{\mathcal{J}}_{n-i} = \sum_{i=0}^{n} a_i \bar{\mathcal{J}}_{n-i} \quad (a_0 = 1)$$

$$\beta_0(q^{-1}) = \sum_{i=1}^{n} b_i \bar{\mathcal{J}}_{n-i}. \tag{73}$$

Let us now make some definitions and present preliminary results which are used in the later analysis of the p.e. conditions for consistency.

Definition 1. The concept of *persistently exciting* related to the input signals will be used. Its formal definition is as follows (cf. Åström and Bohlin 1965).

Let $\{u(k)\}$ be such that the limits

$$\lim_{N\to\infty} \frac{1}{N} \sum_{k=1}^{N} u(k)u(k+\tau) \triangleq r_u(\tau)$$

exist for all $0 \le \tau \le m$. Define

$$R_m = \begin{pmatrix} r_u(0) & \cdots & r_u(m-1) \\ \vdots & \cdots & \vdots \\ r_u(m-1) & \cdots & r_u(0) \end{pmatrix}_{(m \times m)}$$

Then the sequence $\{u(k)\}$ is said to be persistently exciting of order m, if R_m is positive and definite.

An intuitive understanding of this concept in the frequency domain is given by Ljung (1971).

Definition 2. Define a new polynomial $\mathcal{J}_j^*$ as

$$\begin{aligned} \mathcal{J}_j^* &= (1-q^{-1})^j[\frac{T}{2}(1+q^{-1})]^{2n-j} \\ &\triangleq \sum_{i=0}^{2n} \bar{p}_{ij} q^{-i}, \quad j = 0, 1, \cdots, 2n. \end{aligned} \tag{74}$$

Then it is easy to verify that

$$\bar{\mathcal{J}}_i \bar{\mathcal{J}}_j = \mathcal{J}_{i+j}^*, \quad i, j = 0, 1, \cdots, n. \tag{75}$$

Lemma 3. Consider a $2n \times (2n+1)$ matrix $\bar{P}$ defined by

$$\bar{P} = \begin{pmatrix} \bar{p}_{0(2n-1)} & \bar{p}_{1(2n-1)} & \cdots & \bar{p}_{2n(2n-1)} \\ \bar{p}_{0(2n-2)} & \bar{p}_{1(2n-2)} & \cdots & \bar{p}_{2n(2n-2)} \\ \vdots & \vdots & & \vdots \\ \bar{p}_{00} & \bar{p}_{10} & \cdots & \bar{p}_{2n0} \end{pmatrix} \tag{76}$$

where the coefficients $\bar{p}_{ij}$ are given by eqn.(74). Then the matrix $\bar{P}$ has full rank, i.e. $\text{rank}\bar{P} = 2n$.

Proof. See Sagara and Zhao (1990).

Lemma 4. Consider two polynomials in s

$$\begin{aligned} A(s) &= s^n + a_1 s^{n-1} + \cdots + a_n \\ B(s) &= b_1 s^{n-1} + \cdots + b_n \end{aligned}$$

and a $2n \times 2n$ matrix $\Re(-B, A)$

$$\Re(-B, A) = \left(\begin{array}{cccc|cccc} 0 & & & 0 & 1 & & & 0 \\ -b_1 & \ddots & & \vdots & a_1 & \ddots & & \\ \vdots & & \ddots & 0 & \vdots & & \ddots & 1 \\ -b_n & & & -b_1 & a_n & & & a_1 \\ & \ddots & & \vdots & & \ddots & & \vdots \\ 0 & & & -b_n & 0 & & & a_n \end{array}\right). \tag{77}$$

The matrix $\Re(-B, A)$ is nonsingular if and only if $A(s)$, $B(s)$ are relatively prime, i.e. $A(s)$, $B(s)$ have no common zeros.

Proof. The determinant of $\Re(-B, A)$ is known as a resultant. The lemma has been shown by e.g. Lancaster and Tismenetsky (1985).

Lemma 5. Let $u(k)$ be persistently exciting of order m. Assume that $F(q^{-1})$ is an asymptotically stable filter with m_0 zeros on the unit circle. Then the filtered signal $\xi(k) = F(q^{-1})u(k)$ is persistently exciting of order m' with $m - m_0 \leq m' \leq m$.

Proof. See Söderström and Stoica (1989, page123).

Now we are ready to analyze the matrix R' defined by eqn.(69).

By eqns.(71) through (73) and *Definition 2*, we obtain

$$\begin{aligned} x_j(k) &= F(j, q^{-1})x(k) \\ &= \frac{N_F(q^{-1})}{D_F(q^{-1})}\bar{\mathcal{J}}_j x(k) \\ &= \frac{N_F(q^{-1})}{D_F(q^{-1})\alpha_0(q^{-1})}\bar{\mathcal{J}}_j \beta_0(q^{-1})u(k) \\ &= \frac{N_F(q^{-1})}{D_F(q^{-1})\alpha_0(q^{-1})}\sum_{i=1}^{n} b_i \mathcal{J}^*_{n-i+j} u(k) \quad (0 \leq j < n). \end{aligned} \tag{78}$$

In a similar way,

$$u_j(k) = \frac{N_F(q^{-1})}{D_F(q^{-1})\alpha_0(q^{-1})}\sum_{i=0}^{n} a_i \mathcal{J}^*_{n-i+j} u(k) \quad (0 \leq j < n). \tag{79}$$

Then some straight-forward calculating gives

$$\begin{aligned} R' &= E\begin{pmatrix} -x_{n-1}(k) \\ \vdots \\ -x_0(k) \\ u_{n-1}(k) \\ \vdots \\ u_0(k) \end{pmatrix} \begin{pmatrix} -x_{n-1}(k) & \cdots & -x_0(k) & u_{n-1}(k) & \cdots & u_0(k) \end{pmatrix} \\ &= \Re(-B, A)^T \bar{P} \wp(G_0, u) \bar{P}^T \Re(-B, A) \end{aligned} \tag{80}$$

where the $2n \times 2n$ matrix $\Re(-B, A)$ is given by eqn.(77) and

$$G_0(q^{-1}) = \frac{N_F(q^{-1})}{D_F(q^{-1})\alpha_0(q^{-1})}$$

$$\wp(G_0, u) = EG_0(q^{-1})\begin{pmatrix} u(k) \\ \vdots \\ u(k-2n) \end{pmatrix} G_0(q^{-1}) \begin{pmatrix} u(k) & \cdots & u(k-2n) \end{pmatrix}. \tag{81}$$

Using the above lemmas, it is easily seen that R' is nonsingular if $\wp(G_0, u)$ has full rank, and that $\wp(G_0, u)$ must have rank of at least $2n$ if R' has full rank. Therefore, we have the following results.

Lemma 6. Assume that $N_F(q^{-1})$ has m' zeros on the unit circle. For a stable $A(s)$, $G_0(q^{-1})$ defined in eqn.(81) is an asymptotically stable filter with m' zeros on the unit circle. Thus $\wp(G_0, u)$ has full rank if $u(k)$ is persistently exciting of order $2n+1+m'$.

Clearly $T(q^{-1})$ of the linear integral filter has $n(l-1)$ zeros on the unit circle, while the state variable filter does not have any zero on the unit circle.

Theorem 2. Assume that $A(s)$ and $B(s)$ are coprime.
(1) [Sufficient condition on $u(k)$] the matrix R is nonsingular if $u(k)$ is persistently exciting of order $2n+1+m'$.
(2) [Necessary condition on $u(k)$] $u(k)$ has to be persistently exciting of order at least $2n$ in order to assure R to be nonsingular.

Proof. *Lemma 6* and *Theorem 2* can easily be verified by using *Lemma 3* through *Lemma 5*.

Remark 1. There are some types of input signals which are often used in practice. A sum of sinusoids, steps with variable height and a pseudorandom binary sequence (PRBS) are good examples. For an nth-order system, a sum of sinusoids with $n+1$ distinct frequencies in $(0, \pi)$ may be used as the input to meet the condition of persistent excitation if the frequencies of sinusoids do not coincide with the zeros of $N_F(q^{-1})$ of the filter $F(q^{-1})$.

Remark 2. As described earlier, approximation techniques like the numerical integration are required to calculate the "measures" of derivative terms from sampled data. Sometimes the technique is based on approximating the derivative terms by calculating the differences of the sampled values (Neumann, Isermann and Nold 1988). Thus if there are many fast changes in the input signal e.g. a PRBS or a white noise, the analog pre-filters are not able to follow the signal dynamics and huge errors will occur. This problem can be overcome by using the input signal generated from a low-order process, named an input mode, which is driven by a PRBS or a white noise. To obtain good accuracy of parameter estimates, the input model should have a high frequency bandwidth to encompass the frequency bandwidth of the d.e. model of the system. This is illustrated by the following experiment.

Consider a second-order system described by

$$x(t) = \frac{K\omega_n^2}{s^2 + 2\zeta\omega_n s + \omega_n^2} u(t) = \frac{b_1 s + b_2}{s^2 + a_1 s + a_2} u(t)$$
$$y(k) = x(k) + v(k)$$

with $K = 1.25$, $\omega_n = 2$, $\zeta = 0.7$ and $u(t)$ is generated from a second-order process

$$u(t) = \frac{K_u \omega_u^2}{s^2 + 2\zeta\omega_u s + \omega_u^2} w_u(t), \quad K_u = 5.0, \ \zeta = 0.7$$

driven by a zero-mean white noise $\{w_u(t)\}$ that is independent of the noise contained in the output.

The parameters $\theta = (a_1, a_2, b_1, b_2)^T$ are estimated by the LS algorithm (21). The sampling interval $T = 0.02$, the noise-to-signal ratio $N/S = 10\%$ are taken.

Figure 7 plots the frequency responses of the system, the linear integral filter with $l = 20, 50$ and the input model with $\omega_u = 1, 4$. All amplitudes are normalized for comparison. Table 4 shows the LS estimates obtained from Monte-Carlo simulation of 20 experiments. It can be observed from the table that when the frequency bandwidth of the input model encompasses that of the system, i.e. $\omega_u = 4$, the LS estimates are much better than those of the case when $\omega_u = 1$.

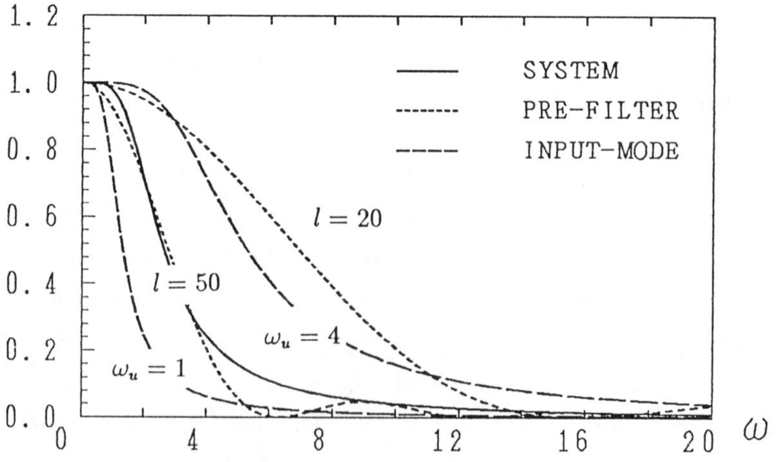

Fig.7 Frequency responses

Table 4 LS estimates; $N/S = 10\%$, $T = 0.02$, $N = 5,000$

ω_u	l	a_1 (2.8)	a_2 (4.0)	b_1 (0.0)	b_2 (5.0)	MNE ($\times 10^{-3}$)	ASD ($\times 10^{-1}$)
1	20	2.569±0.063	3.698±0.120	0.111±0.051	4.583±0.150	6.762	0.961
	50	2.708±0.048	3.830±0.081	0.096±0.033	4.779±0.105	1.953	0.666
4	20	2.770±0.028	3.997±0.033	0.039±0.015	4.972±0.047	0.065	0.304
	50	2.790±0.028	3.992±0.039	0.041±0.012	4.984±0.049	0.043	0.322

6. Conclusions

Digital filtering techniques have been applied to the identification problem of continuous-time systems based on discrete-time measurements. The importance of signal processing has been emphasized. Three typical analog pre-filters have been implemented in convenient digital forms of FIR or IIR filters by using such techniques as the numerical integration in the time domain and the bilinear transformation in the frequency domain. It has been confirmed by simulation experiments that to improve the accuracy of the parameter estimates the pre-filter should be designed in such a way that the frequency bandwidth of the pre-filter matches as closely as possible the frequency band of the system under identification. Since the system is yet unknown at the beginning, several experiments should be performed to update the pre-filter. It is also possible to adapt the pre-filter to an on-line algorithm.

Using the pre-filters in digital form, a discrete-time identification model which retains the continuous-time model parameters has been derived. It has been shown that if the pre-filters are properly designed even the LS algorithm yields satisfactory results in the case of a low level of noise on signal. When the noise level is high, effective bias-eliminating algorithms such as the bias-compensating methods and the IV methods have been reviewed. All the algorithms can be applied to the on-line identification.

Persistently exciting conditions on the input signals for consistency have been investigated. Both the sufficient condition and the necessary condition are explicitly derived . Simulation experiments have shown that the input model should have a high frequency bandwidth to encompass the frequency bandwidth of the system model.

References

Åström, K.J. and Bohlin, T. (1965), "Numerical identification of linear dynamic systems from normal operating records", IFAC Symp. on Theory of Adaptive Control Systems, Plenum Press, pp.96-111.

Eykhoff, P. (1974), *"System Identification"*, Wiley, New York.

Kaya, Y. and Yamamura, S. (1962), "A self adaptive system with a PID controller", AIEE Trans. App. Ind., vol.80, 378.

Lancaster, P. and Tismenetsky, M. (1985), *"The Theory of Matrices - Second Edition with Applications"*, Academic Press, Orlando.

Ljung, L. (1971), "Characterization of the concept of 'persistently exciting' in the frequency domain", Report 7119, Division of Automatic Control, Lund Institute of Technology, Sweden.

Ljung, L. and Söderström, T. (1983), *"Theory and Practice of Recursive Identification"*, The MIT Press, Cambridge, Mass.

Mitani, M. (1988), *"Design of Digital Filters"*, Shoukoudou, Tokyo (in Japanese).

Neumann, D., Isermann, R. and Nold, S. (1988), "Comparison of some parameter estimation methods for continuous-time models", Preprints of 8th IFAC/IFORS Symp. on Identification and System Parameter Estimation, (Beijing, August 1988), pp.1171-1176.

Oppenheim, A.V. and Schafer, R.W. (1975), *"Digital Signal Processing"*, Prentice-Hall, New Jersey.

Rao, G.P. (1983), *"Piecewise Constant Orthogonal Functions and Their Applications to Systems and Control"*, Springer, Berlin.

Sagara, S. and Zhao, Z.Y. (1987), "Parameter identification in continuous systems via numerical integration", Technology Reports of the Kyushu University, vol.60, pp.443-449 (in Japanese).

Sagara, S. and Zhao, Z.Y. (1988), "On-line identification of continuous systems using the operation of numerical integration", Trans. IEE of Japan, vol.108-C, pp.603-610 (in Japanese).

Sagara, S. and Zhao, Z.Y. (1989), "Recursive identification of transfer function matrix in continuous systems via linear integral filter", Int. J. Control, vol.50, pp.457-477.

Sagara, S. and Zhao, Z.Y. (1990), "Numerical integration approach to on-line identification of continuous time systems", Automatica, vol.26, pp.63-74.

Sinha, N.K. (1972), "Estimation of transfer function of continuous systems from sampled data", IEE Proc., vol.119, pp.612-614.

Söderström, T. and Stoica, P. (1989), *"System Identification"*, Prentice Hall, London.

Unbehauen, H. and Rao, G.P. (1987), *"Identification of Continuous Systems"*, North-Holland, Amsterdam.

Unbehauen, H. and Rao, G.P. (1990), "Continuous-time approaches to system identification -a survey", Automatica, vol.26, pp.23-35.

Whitfield, A.H. and Messali, N. (1987), "Integral-equation approach to system identification", Int. J. Control, vol.45, pp.1431-1445.

Young, P.C. (1965), "Process parameter estimation and adaptive control", In P.H. Hammond (Ed.), *"Theory of Self Adaptive Control Systems"*, Plemum Press, New York.

Young, P.C. (1970), "An instrumental variable method for real-time identification of a noisy process", Automatica, vol.6, pp.271-287.

Young, P.C. (1981), "Parameter estimation for continuous-time models - a survey", Automatica, vol.17, pp.23-39.

Young, P.C. and Jakeman, A. (1980), "Refined instrumental variable methods of recursive time-series analysis. Part III. Extensions", Int. J. Control, vol.31, pp.741-764.

Zhao, Z.Y., Sagara, S. and Wada, K. (1990), "Bias-compensating least squares method for identification of continuous-time systems from sampled data", Int. J. Control (to appear).

The Poisson moment functional technique – Some new results

D.C.Saha
Department of Electrical Engineering
Indian Institute of Technology
Kharagpur, India 721 302

V.N.Bapat[1]
Department of Electrical Engineering
Walchand College of Engineering
Sangli, India 416 415
and
B.K.Roy
Department of Electrical Engineering
Regional Engineering College
Silchar, India 788 010

Abstract

We present some generalisations of the classical (ordinary) Poisson moment functionals (PMF) approach towards improving the quality of estimation in continuous-time models of lumped linear single input single output(SISO) dynamical systems. Some results of an investigation on the influence of Poisson filter constant on the quality of estimation and useful guidelines for the proper choice of filter constant are discussed. This generalised PMF approach has been used for combined parameter and state estimation in a linear time-invariant system based on recursive least squares algorithm. This algorithm can be extended for identification of time-varying systems. A recursive instrumental variable (IV) algorithm based on PMFs of the process data is developed to reduce the bias in the estimation. Finally, an attempt has been made to apply the algorithm for the identification of parameters of a system operating under closed loop.

1. Introduction

Considerable work has been reported so far on continuous-time (CT) model identification (Saha and Rao 1983, Unbehauen and Rao 1987, Sagara and Zhao 1988, Unbehauen and Rao 1990). The Poisson moment functional (PMF) technique is known to be one of the effective methods for the identification of the systems

[1]Presently working for Ph.D. at Electrical Engineering Department, I.I.T., Kharagpur.

N. K. Sinha and G. P. Rao (eds.), Identification of Continuous-Time Systems, 327–361.

in CT domain. This technique has been applied to identify time-varying, nonlinear, multivariable and distributed parameter systems using the data en-block. Recently, a recursive algorithm has been derived in the least squares sense using this technique for time-invariant and time-varying systems, making it suitable for real-time applications (Saha and Mandal 1990). The PMFs of the signals are generated by passing them through Poisson filter chain (PFC). In case of classical PFC, each element of the PFC has a transfer function of the form $1/(s+\lambda)$, where λ is termed as a Poisson filter constant. PFC in this form has some limitations and so an improved version of PFC, known as normalised PFC (NPFC) has been suggested recently. Each element of the normalised PFC has a transfer function of the form $\lambda/(s+\lambda)$. To increase the flexibility of the PMF approach, a generalised PFC (GPFC) has been proposed here which may be used to generate either classical or normalised PMFs. Hence each element of the GPFC has a transfer function $\beta/(s+\lambda)$. Clearly, $\beta = 1$ and $\beta = \lambda$ correspond to classical and normalised PFC respectively. It may be noted that $\beta = 1$ and $\lambda = 0$ give rise to the chain of pure integrators. It has been found that parameter estimation in practice depends to some extent on the Poisson filter constant. An attempt has been made here to investigate the influence of the Poisson filter constant, based on some numerical results, on parameter estimation and to suggest some useful guidelines for proper choice of the filter constant in practical computational environment.

An algorithm for parameter estimation using generalised PMF (GPMF) approach has been developed. The parameter vector has been augmented by n (order of the system) number of terms related to initial conditions. For combined parameter and state estimation, observable phase variable form is used for state space representation of the system. The advantage of the observable phase variable form of representation is that the terms relating to initial conditions in the parameter vector are the initial states. Thus the initial states are estimated along with the parameters of the system. The state vector at subsequent time instants is then estimated simultaneously with the parameters in a recursive manner using the estimated initial states (Rao and Patra 1988).

Algorithms presented so far are based on the simple least squares (LS) estimation, which are adequate upto medium levels of noise. This is in view of the inherent noise immunity character of all these algorithms. In the event of higher noise levels, the bias in the LS estimates becomes considerable. This necessitates the use of suitable bias removing strategies (Young and Jakeman 1980, Söderström and Stoica 1983). Here a recursive instrumental variable (IV) algorithm based on PMFs of the process data is also presented.

Estimation of the parameters of a system operating in closed loop is of practical interest since many systems operate under some form of feedback. Not much work has been reported so far on parameter identification of closed loop systems (Graupe

1975, Gevers 1978, Wellstead 1978, Gustavson, Ljung and Söderström 1978) in general and for continuous-time models in particular (Kwong and Chen 1981, Kung and Lee 1982). Based on continuous-time model, C.P.Kwong and C.F.Chen (1981) have suggested a procedure for identifying linear feedback systems via block pulse functions in a deterministic situation. Fan-Chu Kung and Hua Lee (1982) have presented a method for estimation in a closed loop system using Laguerre polynomial expansion, considering only noise-free data. An attempt has been made here to estimate the parameters of both forward and feedback path transfer functions using noisy data in a recursive manner (Roy and Saha 1989). The method proposed here is close in spirit to the indirect identification method in which the overall closed loop system is first estimated from which the plant is separated.

The organisation of the chapter is as follows: Section 2 reviews briefly the generalised Poisson moment functionals (GPMF) definition and basic GPMF relations for signals and their derivatives. Section 3 describes the algorithm for combined parameter and state estimation and presents some results of investigation showing the influence of the Poisson filter constant, λ, in parameter estimation. The IV algorithm based on GPMFs of process data is illustrated in Section 4. Finally, the algorithm for estimating transfer function parameters in systems operating in closed loop is presented in Section 5.

2. Generalised Poisson moment functionals

The GPMF transformation converts a signal $f(t)$ over $(0, t_0)$ into a set of real numbers corresponding to t_0,

$$\left\{ M_{k_g}\left[f(t)\right]_{t_0} \stackrel{\text{def}}{=} f^0_{k_g} = \beta^{k+1} \int_0^{t_0} f(t) p_{k_g}(t_0 - t) dt \right\}, k = 0, 1, 2 \ldots \tag{1}$$

where $p_{k_g}(t)$, the generalised Poisson pulse function (GPPF), is defined as

$$p_{k_g}(t) = \beta^{k+1} t^k exp(-\lambda t)/k! \tag{2}$$

and $\lambda(real) > 0$.

Extensive studies have already been done on parameter estimation using classical PMFs (Saha and Rao 1983, Unbehauen and Rao 1987). NPMFs and GPMFs can be obtained through minor modifications of the classical PMFs. The state-space description of GPFC is

$$(\dot{\mathrm{f}}_0)_g = D_g(\mathrm{f}_0)_g + \mathrm{q}_g f(t) \tag{3}$$

where

$$(\mathbf{f}_0)_g = (f^0_{0_g}\ f^0_{1_g} \dots f^0_{k_g})^T$$

$$D_g = \begin{bmatrix} -\lambda & 0 & 0 & \dots & 0 \\ \beta & -\lambda & 0 & \dots & 0 \\ 0 & \beta & -\lambda & \dots & 0 \\ \cdot & \cdot & \cdot & & \cdot \\ \cdot & \cdot & \cdot & & \cdot \\ \cdot & \cdot & \cdot & & \cdot \\ 0 & 0 & \dots & \beta & -\lambda \end{bmatrix}$$

and

$$\mathbf{q}_g = (\beta\ 0\ 0\ \dots\ 0)^T$$

The GPMFs of signals can be realised by using (3).

The advantage of PMF approach lies in the fact that the GPMFs of derivatives of signals can be obtained from the PMF measures of the signals. GPMFs of signal derivatives of different orders about time $t = t_j$ can be derived as

$$M_{k_g}[\frac{d^n f(t)}{dt^n}]_{t_j} \stackrel{\text{def}}{=} f^{(n)j}_{k_g} = \sum_{i=0}^{n}(-1)^i\ {}^nc_i\beta^{n-i}\lambda^i f^j_{(k-n+i)_g}$$
$$-\sum_{l=1}^{n} f^{(l-1)}(0)\sum_{i=0}^{n-l}(-1)^i\ {}^{n-l}c_i\beta^{n-l-i}\lambda^i p^j_{(k-n+l+i)_g} \tag{4}$$

3. Combined parameter and state estimation

Parameter estimation equations

Consider an $n-th$ order linear time-invariant, single-input, single-output system modelled as

$$\sum_{i=0}^{n} a_i\frac{d^{n-i}y(t)}{dt^{n-i}} = \sum_{i=0}^{m} b_i\frac{d^{m-i}u(t)}{dt^{m-i}}, \quad m \le n. \tag{5}$$

The parameter estimation equations for (5) assuming $a_0 = 1$, may be written as

$$(\phi_y^T \ \phi_u^T \ \phi_i^T)\mathbf{p} = c \tag{6}$$

where

$$\phi_y^T = -(\mathbf{y}^j_{(k-n+1)_g})^T [\gamma_{n-1}]_g^T,$$

$$\phi_u^T = (\mathbf{u}^j_{(k-m)_g})^T [\gamma_m]_g^T,$$

$$\phi_i^T = (p^j_{(k-n+1)_g})^T [\gamma_{n-1}]_g^T,$$

$$\mathbf{p} = (a_1 \dots a_n, b_0 \dots b_m, \theta_1 \dots \theta_n)^T,$$

$c = [\mathbf{y}^j_{(k-n)_y}]^T [\gamma 1_n]_g^T$, a known term related to output,

$$[\gamma_n]_g = \begin{bmatrix} (-1)^0\, {}^n c_0\, \beta^n \lambda^0 & (-1)^1\, {}^n c_1\, \beta^{n-1}\lambda^1 & \dots & (-1)^n\, {}^n c_n\, \beta^0 \lambda^n \\ 0 & (-1)^0\, {}^{n-1}c_0\, \beta^{n-1}\lambda^0 & \dots & (-1)^{n-1}\, {}^{n-1}c_{n-1}\, \beta^0\lambda^{n-1} \\ \cdot & \cdot & \cdot & \cdot \\ \cdot & \cdot & \cdot & \cdot \\ \cdot & \cdot & \cdot & \cdot \\ 0 & 0 & \cdot & (-1)^0\, {}^0 c_0\, \beta^0 \lambda^0 \end{bmatrix}$$

$[\gamma 1_n]_g$ = the first row of $[\gamma_n]_g$ matrix,

$$\mathbf{y}^j_{i_g} = (y^j_{i_g} \ y^j_{(i+1)_g} \ y^j_{(i+2)_g} \cdots)^T$$

Similarly $\mathbf{u}^j_{i_g}$ and $p^j_{i_g}$ are defined.

The initial condition terms (assuming $m = n$) are related by

$$\begin{bmatrix} \theta_1 \\ \theta_2 \\ \theta_3 \\ \cdot \\ \cdot \\ \cdot \\ \theta_n \end{bmatrix} = \begin{bmatrix} 1 & 0 & \dots & 0 \\ a_1 & 1 & \dots & 0 \\ a_2 & a_1 & \dots & 0 \\ \cdot & \cdot & \cdot & \cdot \\ \cdot & \cdot & \cdot & \cdot \\ a_{n-1} & a_{n-2} & \dots & 1 \end{bmatrix} \begin{bmatrix} y^{(0)}(0) \\ y^{(1)}(0) \\ y^{(2)}(0) \\ \cdot \\ \cdot \\ \cdot \\ y^{(n-1)}(0) \end{bmatrix} + \text{(to next page\dots)}$$

$$(\text{from previous page}\ldots) + \begin{bmatrix} -b_0 & 0 & \ldots & 0 \\ -b_1 & -b_0 & \ldots & 0 \\ -b_2 & -b_1 & \ldots & 0 \\ . & . & . & . \\ . & . & . & . \\ . & . & . & . \\ -b_{n-1} & -b_{n-2} & \ldots & -b_0 \end{bmatrix} \begin{bmatrix} u^{(0)}(0) \\ u^{(1)}(0) \\ u^{(2)}(0) \\ . \\ . \\ . \\ u^{(n-1)}(0) \end{bmatrix}$$

or,

$$0 = A(\mathbf{y}0) + B(\mathbf{u}0) \tag{7}$$

where

$$(\mathbf{y}0) = (y^{(0)}(0)\ y^{(1)}(0)\ldots y^{(n-1)}(0))^T,$$

$$(\mathbf{u}0) = (u^{(0)}(0)\ u^{(1)}(0)\ldots u^{(n-1)}(0))^T,$$

$$y^{(i)}(0) = \frac{d^i y(t)}{dt^i}|_{t=0},$$

$$u^{(i)}(0) = \frac{d^i u(t)}{dt^i}|_{t=0}$$

$$A(i,j) = \begin{cases} 0 & , \ \forall\, j > i \\ 1 & , \ \forall\, j = i \\ a_{i-j} & , \ \forall\, j < i \end{cases}$$

$$B(i,j) = \begin{cases} 0 & , \ \forall\, j > i \\ -b_{i-j} & , \ \forall\, j \leq i \end{cases}$$

Off-line solution of the estimation equation (6) requires "x" number of linearly independent equations for estimating the parameters, where "x" is greater than or equal to the length of the parameter vector. These equations can be generated by

Method I: keeping t_j fixed and by varying k, and also by

Method II: keeping k fixed and by varying t_j.

The linearly independent equations can be written in matrix form as

$$\mathbf{\Phi p} = \mathbf{c} \tag{8}$$

where

$\mathbf{\Phi}$ is the measurement matrix in terms of GPMFs of process data, $\mathbf{p}$ is the unknown parameter vector and $\mathbf{c}$ is a vector of known GPMF measures of output signal.

From (8) we estimate,

$$\hat{\mathbf{p}} = (\mathbf{\Phi}^T\mathbf{\Phi})^{-1}\mathbf{\Phi}^T\mathbf{c} \tag{9}$$

For real-time applications the estimation of parameters has to be done in a recursive manner in time (Saha and Mandal 1990). For the recursive algorithm, only Method II can be used. The value of k is set equal to the order of the system. Then j is varied to correspond to different time-instants. Using least squares error criterion the usual recursive relations with GPMF measures of process data may be developed as follows:

$$\left.\begin{array}{rcl} \hat{\mathbf{p}}(j+1) & = & \hat{\mathbf{p}}(j) + \hat{e}(j+1)q(j+1) \\ q(j+1) & = & P(j)\phi(j+1)[1+\phi^T(j+1)P(j)\phi(j+1)]^{-1} \\ P(j+1) & = & P(j) - q(j+1)\phi^T(j+1)P(j) \\ \hat{e}(j+1) & = & c(j+1) - \phi^T(j+1)\hat{\mathbf{p}}(j) \end{array}\right\} \tag{10}$$

where

$\hat{\mathbf{p}}$ - Estimated parameter vector

q - Kalman gain vector

P - Error covariance matrix

$\hat{e}$ - Prediction error

ϕ - Measurement vector comprised of GPMF measures of process data.

Effect of λ in parameter estimation

The definition of GPMF does not impose any other constraint in theory except that λ is real and greater than zero. However, it has been found in practice that accuracy and convergence of parameter estimates depend significantly on λ.

To study these effects, consider a second order model

$$\frac{d^2y(t)}{dt^2} + a_1\frac{dy(t)}{dt} + a_2y(t) = b_1u(t),$$

for three systems of different bandwidths and with different initial conditions

Example I) $a_1 = 4, a_2 = 3, b_1 = 1.5$ and $\theta_1 = \theta_2 = 0$.

Example II) $a_1 = 0.6, a_2 = 0.05, b_1 = 1$ and $\theta_1 = \theta_2 = 0$.

Example III) $a_1 = 11, a_2 = 10, b_1 = 1.5$ and $\theta_1 = 0.2, \theta_2 = 2$.

The parameters of this model are estimated taking various values of λ fron 0.1 to 10. Both classical and normalised PMFs (corresponding to $\beta\ = 1$ and respectively) are used.

The value of λ should be such that the PFC, (i) passes the signals within th bandwidth of the system and (ii) attenuates high frequency components such a measurement noise. The error norm and the noise to signal ratio (NSR) are define as

$$\text{error norm} \stackrel{\text{def}}{=} \frac{\|\text{ (True parameter vector} - \text{Estimated parameter vector) }\|}{\|\text{ True parameter vector }\|}$$

$$NSR \stackrel{\text{def}}{=} \frac{\text{Standard deviation of noise}}{\text{Standard deviation of signal}}$$

Numerical experiments have been carried out for the above three cases wit classical and normalised PMFs using noise-free and noisy data for various values o $\lambda(0.1 \leq \lambda \leq 10)$. These results are presented in the form of several tables and plot whose details are given below for easy reference.

	Description	Examples I	II	III
A.	Estimated parameters with Classical and Normalised PMFs for			
	i)$NSR = 0\%$, in Tables	1,2	5,6	9,10
	ii)$NSR = 20\%$, in Tables	3,4	7,8	11,12
B.	Parameter convergence with recursions for noisy data ($NSR = 20\%$) in Figures	1–3	7–9	13–15
C.	Variation of error norm with λ for			
	i)$NSR = 0\%$, in Figures	4	10	16
	ii)$NSR = 20\%$, in Figures	4	10	16
D.	Decay of error norm with recursions for different values of λ			
	i)$NSR = 0\%$, in Figures	5	11	17
	ii)$NSR = 20\%$, in Figures	6	12	18

Table 1: Estimated parameters in Example I using classical PMF with no noise

λ	$\hat{a}_1$	$\hat{a}_2$	$\hat{b}_1$	$\hat{\theta}_1$	$\hat{\theta}_2$	error norm
0.1	4.0347	3.0345	1.5178	-0.001249	0.000274	0.0010
0.5	4.0340	3.0338	1.5174	-0.001326	0.000275	0.0098
1.0	4.0220	3.0244	1.5244	-0.001620	-0.000056	0.0067
2.0	3.9205	2.9584	1.4729	-0.002347	-0.00289	0.0167
5.0	2.2426	1.9362	0.8498	-0.001534	-0.05577	0.4129
10.0	0.2003	0.2701	0.2846	0.002861	-0.09220	0.9262

Table 2: Estimated parameters in Example I using NPMF with no noise

λ	$\hat{a}_1$	$\hat{a}_2$	$\hat{b}_1$	$\hat{\theta}_1$	$\hat{\theta}_2$	error norm
0.1	0.0618	-0.0424	-0.0530	-0.08163	-0.00737	0.9988
0.5	3.4463	2.5696	1.2845	-0.01762	-0.00093	0.1406
1.0	4.0222	3.0244	1.5244	-0.00162	-0.00006	0.0067
2.0	4.0583	3.0478	1.5233	-0.00081	-0.00007	0.0151
5.0	4.0800	3.0497	1.5226	-0.00038	-0.00257	0.0186
10.0	4.1135	3.0459	1.5208	-0.00018	-0.00389	0.0238

Table 3: Estimated parameters in Example I using classical PMF with NSR = 20%

λ	$\hat{a}_1$	$\hat{a}_2$	$\hat{b}_1$	$\hat{\theta}_1$	$\hat{\theta}_2$	error norm
0.1	3.4851	2.6155	1.3101	-0.007952	-0.000144	0.1285
0.5	3.5261	2.6453	1.3261	-0.007881	0.000227	0.1182
1.0	3.6446	2.7316	1.3724	-0.006427	0.001132	0.0889
2.0	3.7209	2.7739	1.3928	-0.004452	-0.00209	0.0718
5.0	2.2055	1.8787	0.8309	-0.001264	-0.05687	0.4253
10.0	0.2012	0.2695	0.2758	0.004101	-0.10090	0.9266

Table 4: Estimated parameters in Example I using NPMF with NSR = 20%

λ	$\hat{a}_1$	$\hat{a}_2$	$\hat{b}_1$	$\hat{\theta}_1$	$\hat{\theta}_2$	error norm
0.1	0.0624	-0.041	-0.053	-0.08180	-0.007609	0.9986
0.5	3.1308	2.3321	1.1685	-0.01966	0.000965	0.2194
1.0	3.6446	2.7316	1.3724	-0.00643	0.001130	0.0889
2.0	3.8333	2.8496	1.4361	-0.00313	-0.000160	0.0408
5.0	3.9156	2.8901	1.4469	0.00025	-0.008690	0.0285
10.0	3.9357	2.9183	1.4430	0.002246	-0.01824	0.0230

Table 5: Estimated parameters in Example II using classical PMF with no noise

λ	$\hat{a}_1$	$\hat{a}_2$	$\hat{b}_1$	$\hat{\theta}_1$	$\hat{\theta}_2$	error norm
0.1	0.5999	0.0499	1.0000	0.000000	0.000000	0.0000
0.5	0.5999	0.0499	0.9999	0.000000	0.000000	0.0000
1.0	0.5999	0.0499	0.9998	0.000000	-0.000020	0.0001
2.0	0.5986	0.0500	0.9996	0.000090	-0.000660	0.0026
5.0	0.4985	0.0533	0.8612	0.003870	-0.030030	0.1496
10.0	0.0176	0.0567	0.2407	0.01063	-0.083110	0.8125

Table 6: Estimated parameters in Example II using NPMF with no noise

λ	$\hat{a}_1$	$\hat{a}_2$	$\hat{b}_1$	$\hat{\theta}_1$	$\hat{\theta}_2$	error norm
0.1	0.3144	0.0523	0.1928	-0.01994	0.22915	0.7595
0.5	0.5996	0.0500	0.9991	0.00001	-0.00025	0.0008
1.0	0.5999	0.0499	0.9998	0.00000	-0.00002	0.0001
2.0	0.6006	0.0499	0.9994	0.00001	0.00005	0.0001
5.0	0.5999	0.0499	0.9998	0.00001	0.00000	0.0007
10.0	0.6065	0.0499	0.9977	0.00001	0.00017	0.0059

Table 7: Estimated parameters in Example II using classical PMF with NSR = 20%

λ	$\hat{a}_1$	$\hat{a}_2$	$\hat{b}_1$	$\hat{\theta}_1$	$\hat{\theta}_2$	error norm
0.1	0.5981	0.0499	0.9865	-0.00627	-0.00498	0.0136
0.5	0.5994	0.0498	0.9948	-0.00799	-0.00165	0.0083
1.0	0.6023	0.0497	1.0054	-0.00829	0.00092	0.0087
2.0	0.6174	0.0499	1.0250	-0.00217	-0.00109	0.0262
5.0	0.5295	0.0545	0.8971	0.00944	-0.03956	0.1125
10.0	0.0395	0.0618	0.2565	0.01213	-0.09038	0.8010

Table 8: Estimated parameters in Example II using NPMF with NSR = 20%

λ	$\hat{a}_1$	$\hat{a}_2$	$\hat{b}_1$	$\hat{\theta}_1$	$\hat{\theta}_2$	error norm
0.1	0.3143	0.0521	0.1915	-0.02518	-0.22900	0.7607
0.5	0.5992	0.0498	0.9939	-0.00798	-0.00191	0.0088
1.0	0.6023	0.0497	1.0054	-0.00829	-0.00092	0.0087
2.0	0.6187	0.0499	1.0276	-0.00225	-0.00044	0.0287
5.0	0.6367	0.0588	1.0426	0.00558	-0.00824	0.0489
10.0	0.6006	0.0553	1.0490	0.00108	-0.00282	0.0423

Table 9: Estimated parameters in Example III using classical PMF with no noise

λ	$\hat{a}_1$	$\hat{a}_2$	$\hat{b}_1$	$\hat{\theta}_1$	$\hat{\theta}_2$	error norm
0.1	10.736	9.7660	1.4644	0.2067	2.3431	0.0238
0.5	9.747	8.8011	1.3147	0.2008	2.1035	0.1169
1.0	7.098	6.0324	0.8836	0.2082	1.4104	0.3757
2.0	2.758	1.7912	0.2066	0.2161	0.3201	0.7854
5.0	0.421	0.0209	0.0845	0.2196	-0.2062	0.9818
10.0	0.059	0.0069	0.0144	0.2126	-0.1575	0.9982

Table 10: Estimated parameters in Example III using NPMF with no noise

λ	$\hat{a}_1$	$\hat{a}_2$	$\hat{b}_1$	$\hat{\theta}_1$	$\hat{\theta}_2$	error norm
0.1	-0.003	0.0124	-0.0185	0.0327	-0.0273	0.9999
0.5	1.333	0.7964	0.0802	0.1187	0.1280	0.8996
1.0	7.098	6.0324	0.8836	0.2082	1.4104	0.3757
2.0	10.470	9.4460	1.4136	0.2138	2.2599	0.0518
5.0	10.950	9.8586	1.4872	0.2135	2.3698	0.0102
10.0	10.849	9.5716	1.4726	0.2129	2.3150	0.0306

Table 11: Estimated parameters in Ex.III using classical PMF with NSR = 20%

λ	$\hat{a}_1$	$\hat{a}_2$	$\hat{b}_1$	$\hat{\theta}_1$	$\hat{\theta}_2$	error norm
0.1	4.229	3.2814	0.4439	0.1965	1.0847	0.6439
0.5	6.213	5.0126	0.7215	0.2186	1.1491	0.4643
1.0	5.746	4.6323	0.6518	0.2205	1.0358	0.5075
2.0	2.549	1.6249	0.1685	0.2189	0.2613	0.8036
5.0	0.420	0.0348	-0.0910	0.2215	-0.2126	0.9814
10.0	0.060	0.0085	0.0458	0.2129	-0.1592	0.9981

Table 12: Estimated parameters in Example III using NPMF with NSR = 20%

λ	$\hat{a}_1$	$\hat{a}_2$	$\hat{b}_1$	$\hat{\theta}_1$	$\hat{\theta}_2$	error norm
0.1	-0.003	0.0120	-0.0180	0.0318	-0.0267	0.9999
0.5	1.647	1.0415	0.1149	0.1284	0.1826	0.8730
1.0	5.746	4.6323	0.6518	0.2205	1.0358	0.5075
2.0	9.175	8.4387	1.2152	0.2186	1.9435	0.1626
5.0	9.305	8.8471	1.2554	0.2128	2.0035	0.1389
10.0	8.795	8.2378	1.2061	0.2122	1.8542	0.1910

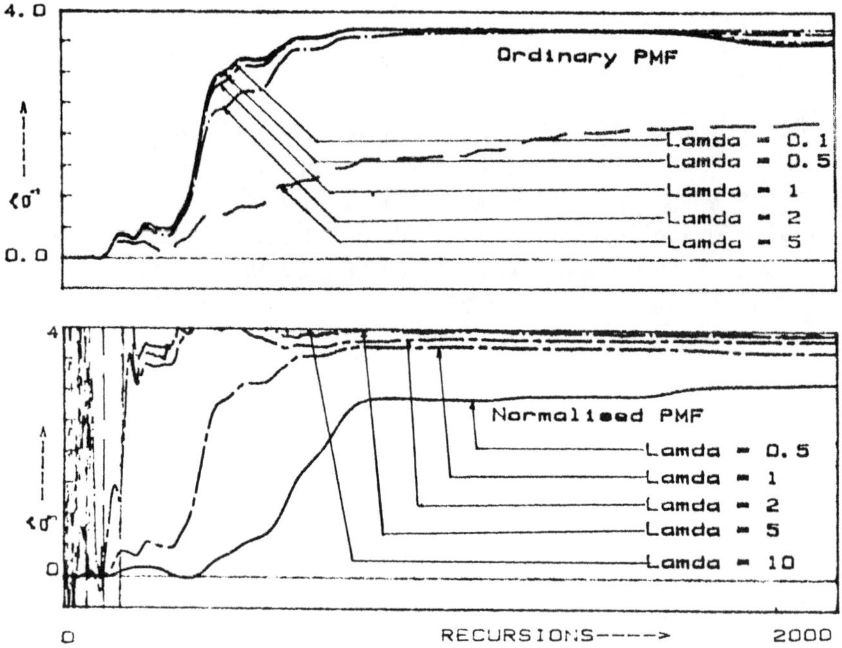

Fig.1. Convergence of $\hat{a}_1$ in Example I with NSR = 20%

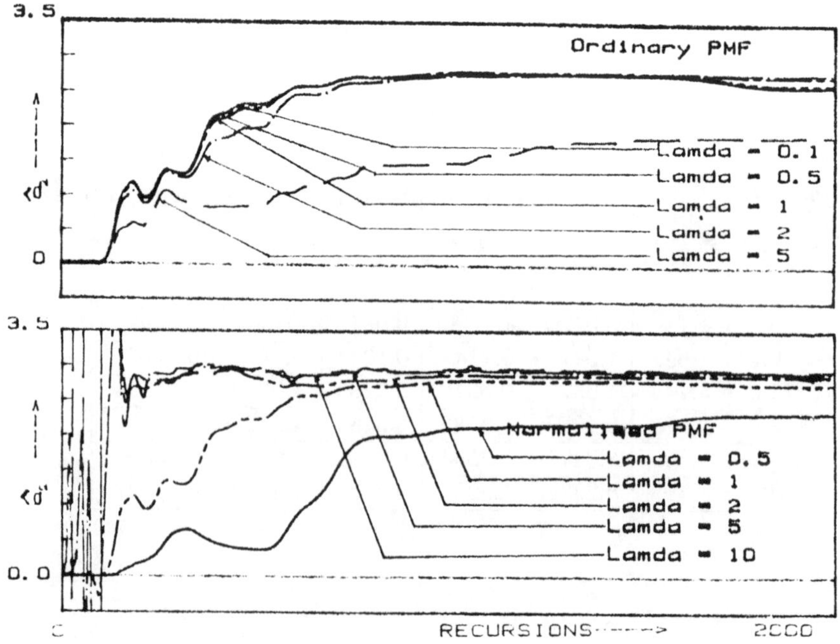

Fig.2. Convergence of $\hat{a}_2$ in Example I with NSR = 20%

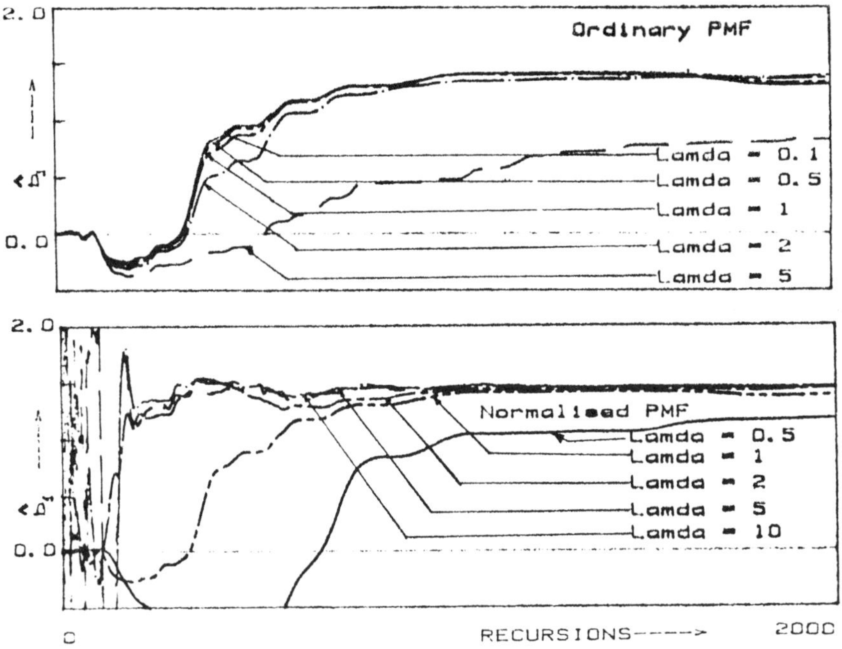

Fig.3. Convergence of $\hat{b}_1$ in Example I with NSR = 20%

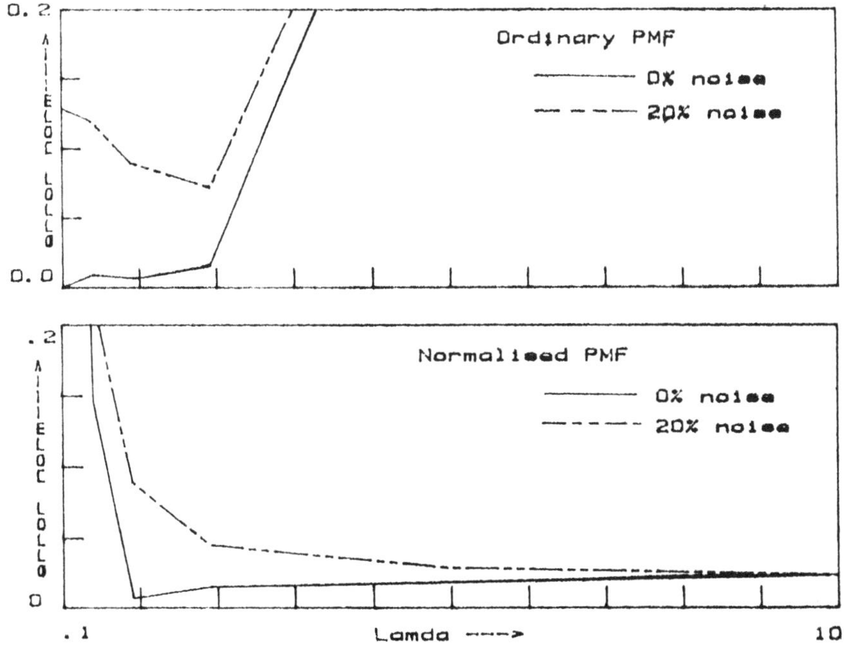

Fig.4. Variation of error norm with λ for Example I

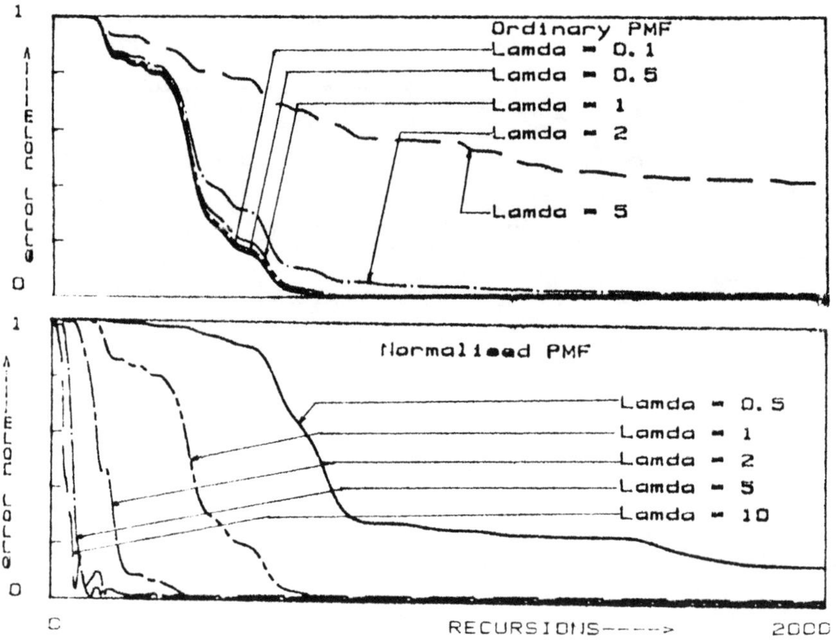

Fig.5. Decay of error norm for Example I with no noise

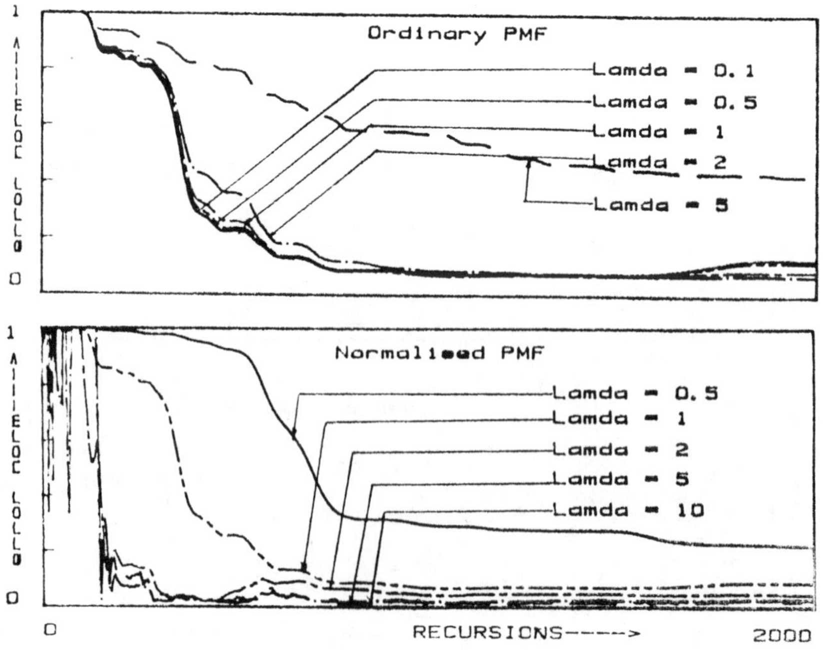

Fig.6. Decay of error norm for Example I with NSR = 20%

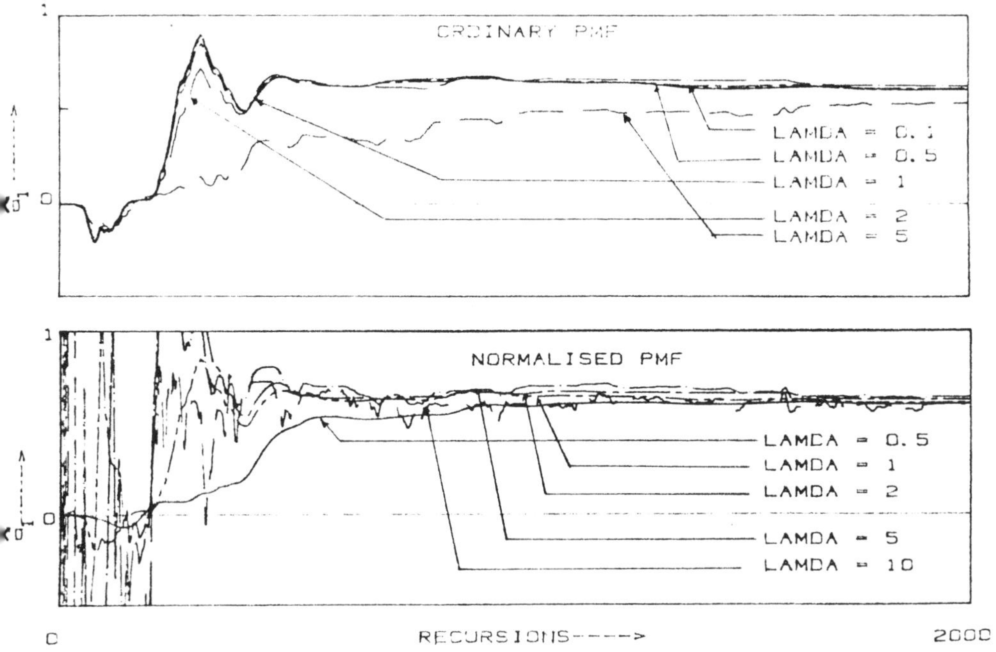

Fig.7. Convergence of $\hat{a}_1$ in Example II with NSR = 20%

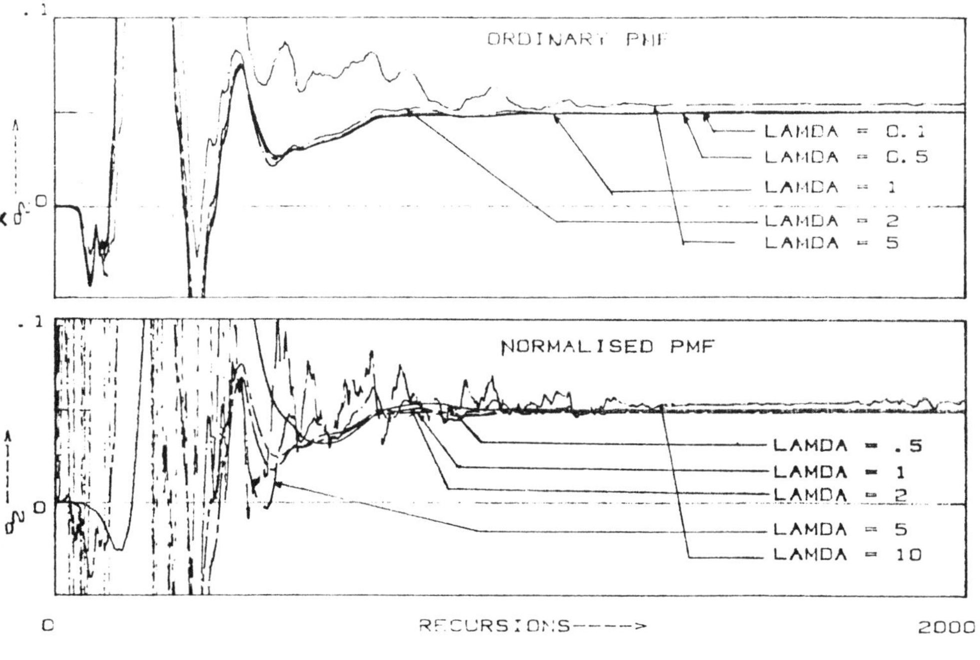

Fig.8. Convergence of $\hat{a}_2$ in Example II with NSR = 20%

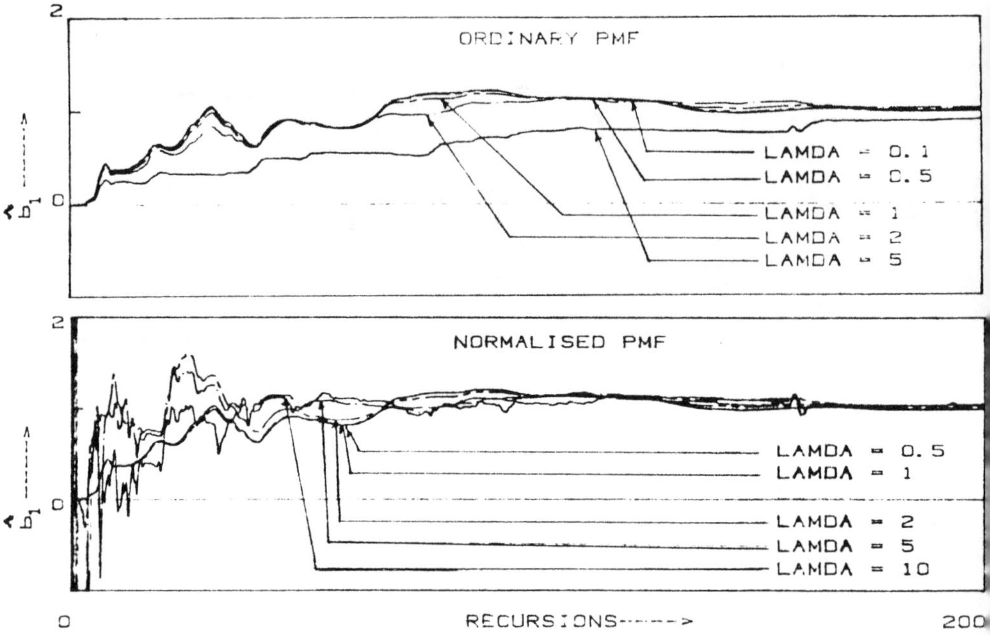

Fig.9. Convergence of $\hat{b}_1$ in Example II with NSR = 20%

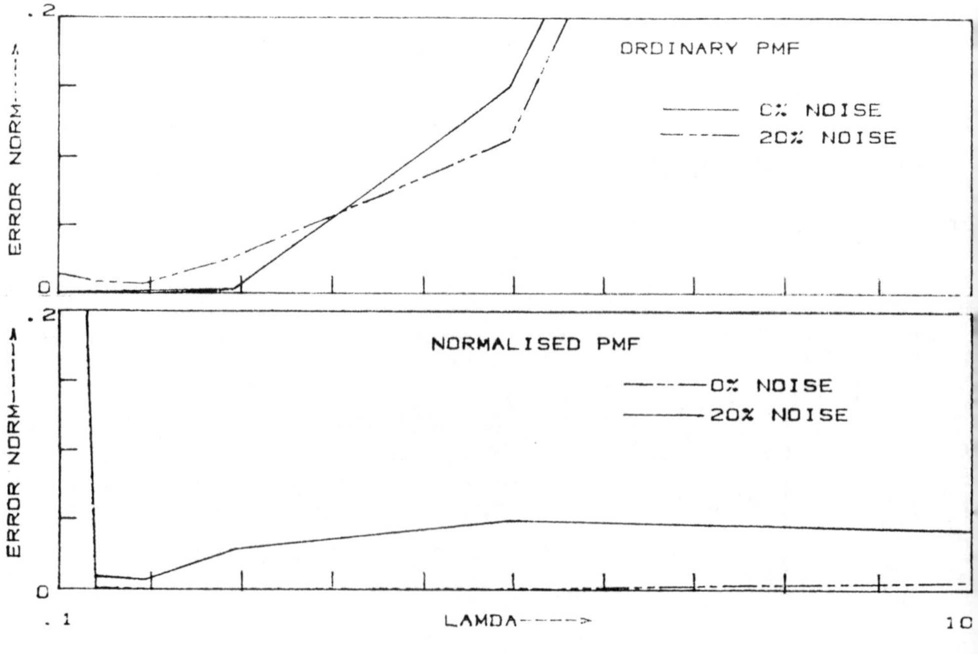

Fig.10. Variation of error norm with λ for Example II

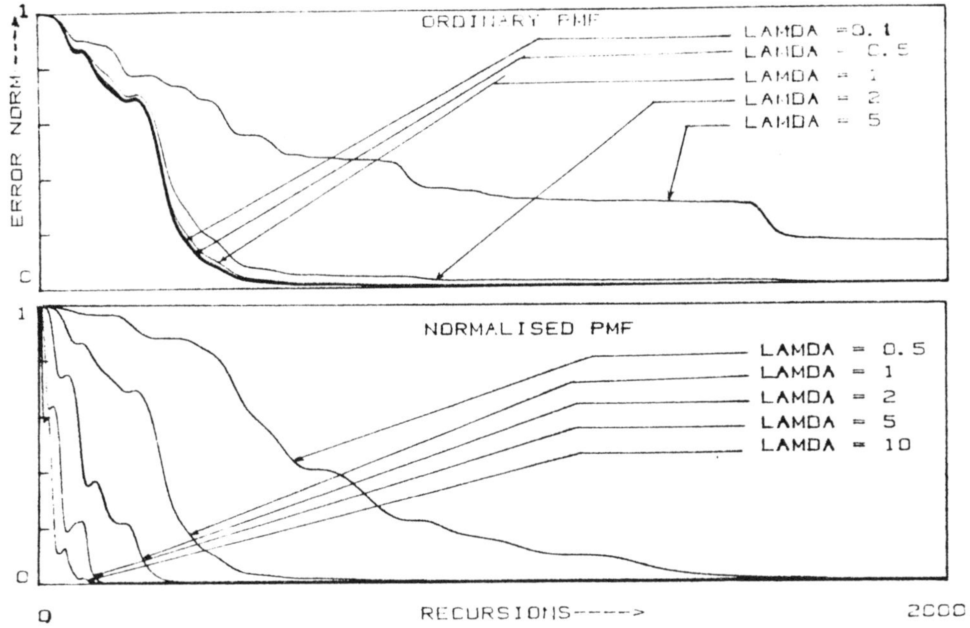

Fig.11. Decay of error norm for Example II with no noise

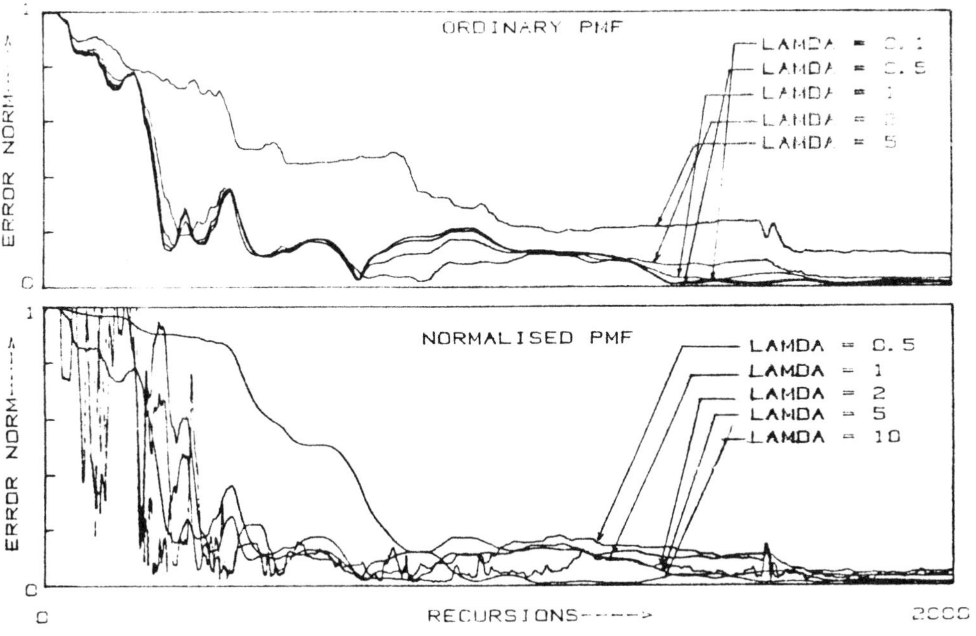

Fig.12. Decay of error norm for Example II with NSR = 20%

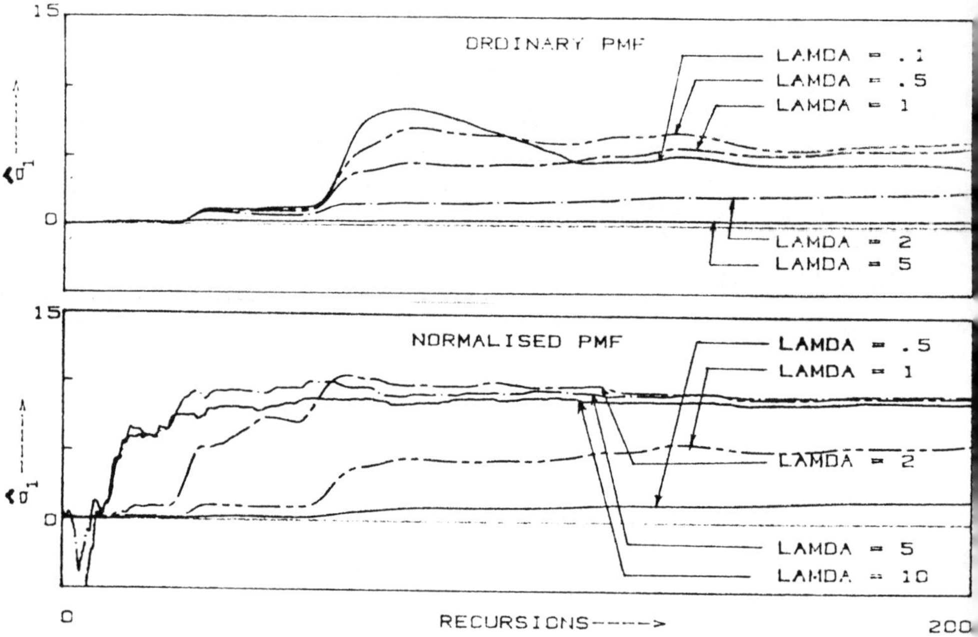

Fig.13. Convergence of $\hat{a}_1$ in Example III with NSR = 20%

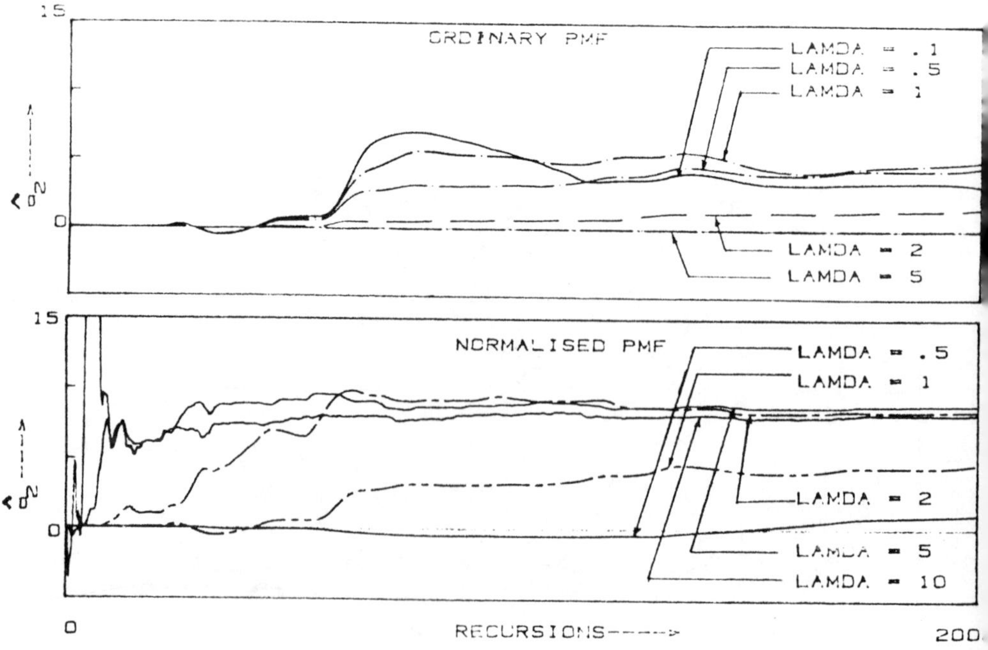

Fig.14. Convergence of $\hat{a}_2$ in Example III with NSR = 20%

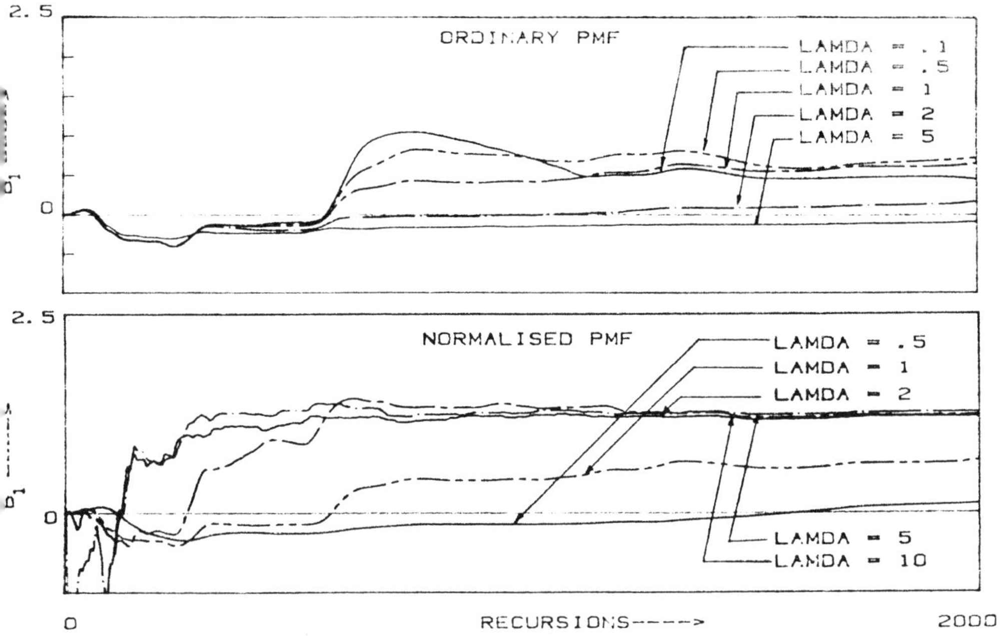

Fig.15. Convergence of $\hat{b}_1$ in Example III with NSR = 20%

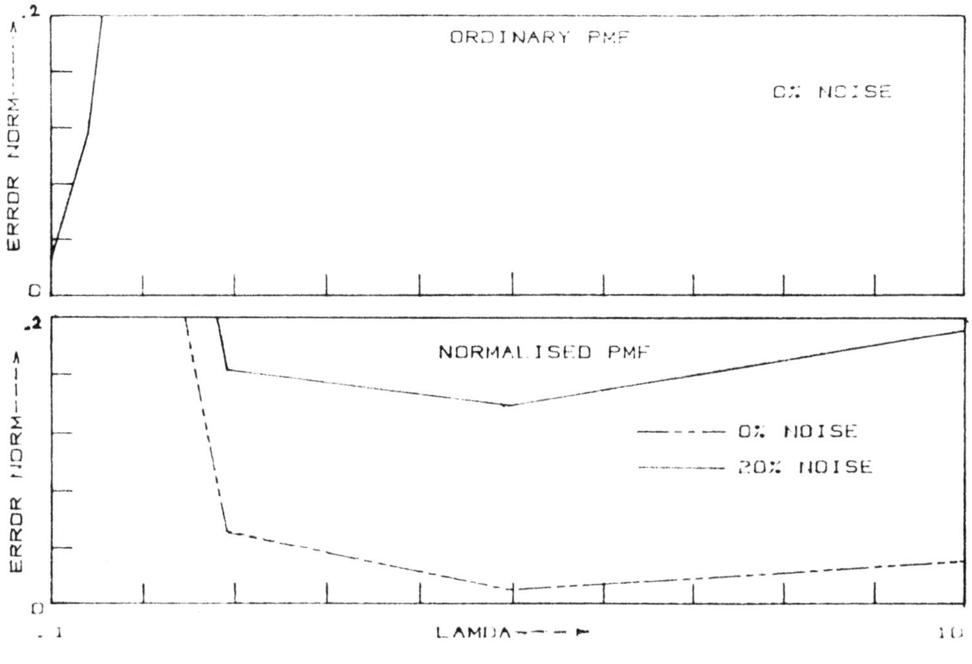

Fig.16. Variation of error norm with λ for Example III

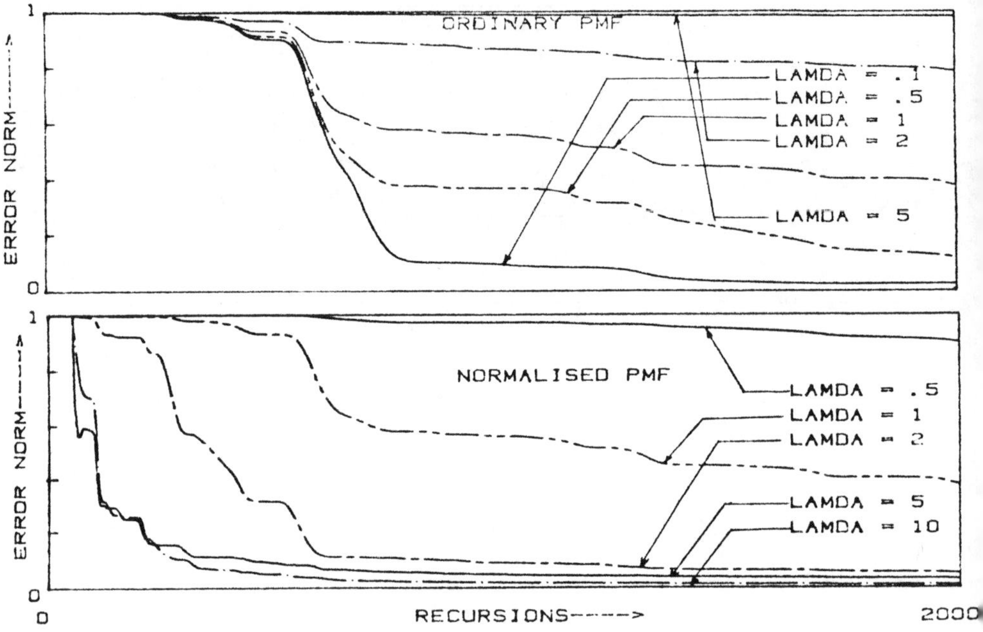

Fig.17. Decay of error norm for Example III with no noise

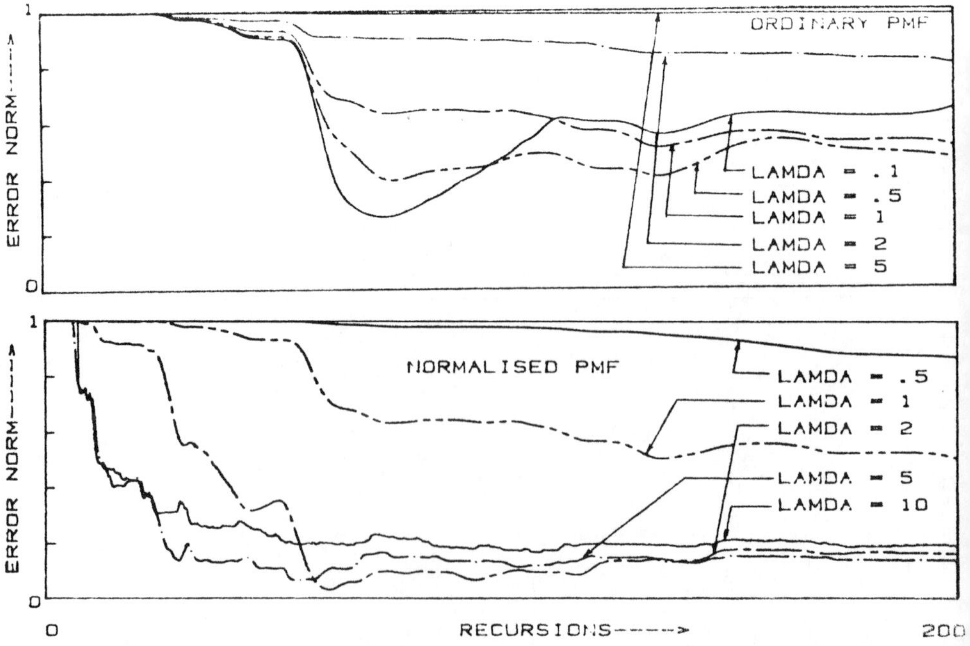

Fig.18. Decay of error norm for Example III with NSR = 20%

The following observations may be made from the results

i)In the case of classical PMFs, both for noise-free and noisy data, the parameter error norm remains within a certain acceptable limit (about 20%) for only a small range of λ in all the examples. When λ exceeds a certain value the estimation of the parameters becomes unsatisfactory. This is evident from Figures 4 and 10. In example III, the error norm exceeds 20% for all values of λ (0.1 to 10) with noisy data (NSR=20%) which may be seen from Table 11.

ii) The normalised PMFs give satisfactory estimates of the parameters for a wider range of λ. In this case, λ should be chosen higher than the smallest corner frequency of the system for better performance. It is also observed that if λ is less than this quantity (such as $\lambda < 1.0$ in example I) the estimates become inaccurate.

iii) It may be clearly seen from the plots of the decay of the error norm versus recursions for various values of λ that the parameter error norm decays at a much faster rate for normalised PMFs in both noise-free and noisy situations.

State estimation equations

State space formulation of (5) for $m = n$ is

$$\begin{bmatrix} \dot{x}_1 \\ \dot{x}_2 \\ \cdot \\ \cdot \\ \cdot \\ \dot{x}_{n-1} \\ \dot{x}_n \end{bmatrix} = \begin{bmatrix} -a_1 & 1 & 0 & 0 & \dots & 0 \\ -a_2 & 0 & 1 & 0 & \dots & 0 \\ \cdot & \cdot & \cdot & \cdot & \dots & \cdot \\ \cdot & \cdot & \cdot & \cdot & \dots & \cdot \\ \cdot & \cdot & \cdot & \cdot & \dots & \cdot \\ -a_{n-1} & 0 & 0 & 0 & \dots & 1 \\ -a_n & 0 & 0 & 0 & \dots & 0 \end{bmatrix} \begin{bmatrix} x_1 \\ x_2 \\ \cdot \\ \cdot \\ \cdot \\ x_{n-1} \\ x_n \end{bmatrix} + \begin{bmatrix} b_1 - a_1 b_0 \\ b_2 - a_2 b_0 \\ \cdot \\ \cdot \\ \cdot \\ b_{n-1} - a_{n-1} b_0 \\ b_n - a_n b_0 \end{bmatrix} u \quad (11)$$

$$y = x_1 + b_0 u \quad (12)$$

The initial state is

$$\begin{bmatrix} x_1(0) \\ x_2(0) \\ \cdot \\ \cdot \\ \cdot \\ x_n(0) \end{bmatrix} = \begin{bmatrix} 1 & 0 & 0 & \dots & 0 \\ a_1 & 1 & 0 & \dots & 0 \\ \cdot & \cdot & \cdot & \dots & \cdot \\ \cdot & \cdot & \cdot & \dots & \cdot \\ \cdot & \cdot & \cdot & \dots & \cdot \\ a_{n-1} & a_{n-2} & a_{n-3} & \dots & 1 \end{bmatrix} \begin{bmatrix} y^{(0)}(0) \\ y^{(1)}(0) \\ \cdot \\ \cdot \\ \cdot \\ y^{(n-1)}(0) \end{bmatrix}$$

+ (to next page...)

$$\text{(from previous page...)} + \begin{bmatrix} -b_0 & 0 & 0 & \dots & 0 \\ -b_1 & -b_0 & 0 & \dots & 0 \\ \cdot & \cdot & \cdot & \cdots & \cdot \\ \cdot & \cdot & \cdot & \cdots & \cdot \\ \cdot & \cdot & \cdot & \cdots & \cdot \\ -b_{n-1} & -b_{n-2} & -b_{n-3} & \dots & -b_0 \end{bmatrix} \begin{bmatrix} u^{(0)}(0) \\ u^{(1)}(0) \\ \cdot \\ \cdot \\ \cdot \\ u^{(n-1)}(0) \end{bmatrix}$$

$$or, \mathbf{x}(0) = \mathbf{A}(\mathbf{y0}) + \mathbf{B}(\mathbf{u0}) \tag{13}$$

Note that the expression for the parameter related to initial condition θ_i's in equation (7) is similar to that of the $x_i(0)$'s in equation (13). Thus along with the parameter, the initial state is also estimated. If the initial state of the system is first estimated, the state at any other instant of time can be recursively estimated as

$$x_i(t) = \sum_{j=1}^{n+1-i} \frac{t^{j-1}}{(j-1)!} x_{i+j-1}(0) - a_j y^j_{(j-1)_l} + (b_j - a_j b_0) u^j_{(j-1)_l} \tag{14}$$

where $u^j_{l_l}$ and $y^j_{l_l}$ are the l-th GPMF with $\beta = 1$ and $\lambda = 0$.

Illustrative example IV

Consider the following second order system

$$\frac{d^2y(t)}{dt^2} + 2\frac{dy(t)}{dt} + y(t) = u(t)$$

Here $u(t)$ is a PRBS signal of period 1023 and amplitude ± 1. The system is simulated using (11) and (12). Different levels of zero mean Gaussian noise are added in $y(t)$. The estimated parameters are given in Table 13. Estimated state is compared with the actual for 0% and 20% noise levels and the comparison is shown in Figures 19 and 20.

The actual values are

$$a_1 = 2.0, a_2 = 1.0, b = 1.0, x_1(0) = 0.2 \text{ and} x_2(0) = 0.6$$

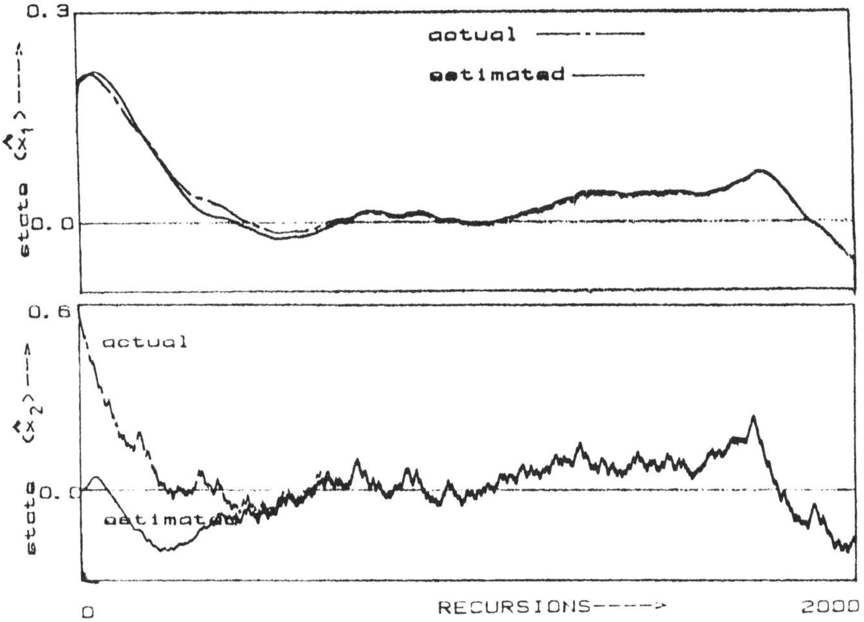

Fig.19. Comparison of the estimated states $\hat{x}_1$ and $\hat{x}_2$ with their true values for no noise

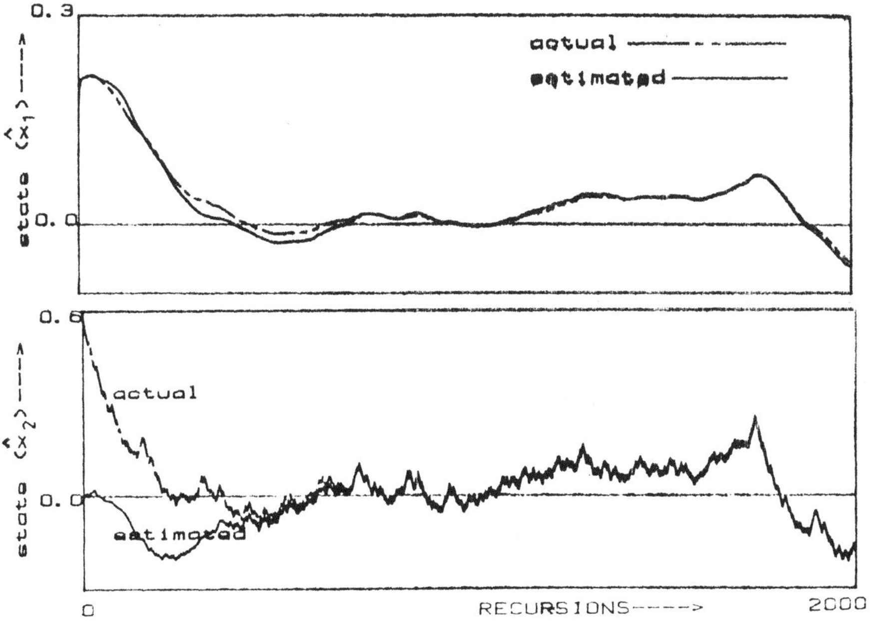

Fig.20. Comparison of the estimated states $\hat{x}_1$ and $\hat{x}_2$ with their true values for NSR = 20%

4. GPMF based IV algorithm

Recursive parameter estimation equations

The GPMF based recursive IV algorithm for continuous-time models may be developed following the model shown in Fig.21.

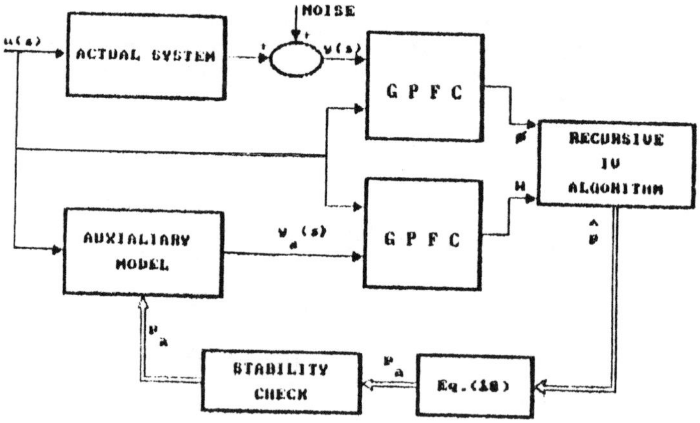

Fig.21. Model of recursive scheme using GPMFs of process data

The recursive IV estimation equations are as follows

$$\left.\begin{aligned} \hat{\mathbf{p}}(j+1) &= \hat{\mathbf{p}}(j) + \hat{e}(j+1)q(j+1) \\ q(j+1) &= P(j)\mathbf{w}(j+1)[1+\phi^T(j+1)P(j)\mathbf{w}(j+1)]^{-1} \\ P(j+1) &= P(j) - q(j+1)\phi^T(j+1)P(j) \\ \hat{e}(j+1) &= c(j+1) - \phi^T(j+1)\hat{\mathbf{p}}(j) \end{aligned}\right\} \tag{15}$$

where $\mathbf{w}$ is the instrumental variable (IV) vector given by

$$\begin{aligned} \mathbf{w}^T &= \left[\phi_{y_a}^T \; \phi_u^T \; \phi_i^T\right] \\ &= \left[-[y^0_{a_{(k-n+1)}}]^T[\gamma_{n-1}]^T \; [u^0_{(k-m)}]^T[\gamma_m]^T \; [p^0_{k-n+1}]^T[\gamma_{n-1}]^T\right] \end{aligned} \tag{16}$$

and y_a is the output of the auxiliary model. Vector $\mathbf{w}$ must be such that it satisfies

$$\left.\begin{array}{lll} prob\ \lim_{N\to\infty} \frac{1}{N}\sum_{i=1}^{N} \mathbf{w}(i)\phi^T(i) & & \text{is positive definite} \\ prob\ \lim_{N\to\infty} \frac{1}{N}\sum_{i=1}^{N} \mathbf{w}(i)\hat{e}(i) & = & 0 \end{array}\right\} \quad (17)$$

Auxiliary parameter update

The conditions in (17) require that the output y_a of the auxiliary model should be strongly correlated with the noise-free output of the system but be uncorrelated with the noise. Proper choice of the parameters of the auxiliary model is necessary to satisfy this. A discrete low pass filter suggested in (Sagara and Zhao 1988) is being used here to update the auxiliary parameter vector $\mathbf{p}_a$. The filter equation is

$$\mathbf{p}_a(k) = (1 - \alpha)\mathbf{p}_a(k-1) + \alpha\hat{\mathbf{p}}(k-d) \quad (18)$$

where d is the dead time between the current parameter estimates and the estimates being used to form the vector $\mathbf{p}_a$. Selection of d is such that $e(k+d)$ is independent of $e(k)$. α is the weight attached to the past parameter estimates in forming the auxiliary parameters.

Stability check

To certain extent, auxiliary parameters are selected arbitrarily. Hence there is a chance that at some stage during estimation, they give rise to an unstable auxiliary model leading to convergence problems. To avoid this, a stability check has been introduced in the algorithm. This checks the stability of the auxiliary model and if it is found to be unstable at any step, it holds on to the previous stable set of auxiliary model parameters untill the appearance of the next stable set of parameters.

Adaptive switchover from LS to IV algorithm

To ensure good initialization of the auxiliary model parameters, the IV procedure is started only after processing certain number of data off-line or on-line by LS method. The instant of this switchover is of importance. An adaptive switchover strategy has been employed here. When the estimates of all the parameters, over certain number of recursions settle to within a specified tolerance limit, switchover will take place.

Illustrative Example V

A second order model of a system having transfer function $1/s(s+1)$ is considered for estimation. The time-domain description of the above system is

$$\frac{d^2y(t)}{dt^2} + \frac{dy(t)}{dt} = u(t)$$

With reference to the general description of the linear time-invariant system given in Eq.(5), different parameters of the above system are

$$a_1 = 1.0,\ a_2 = 0.0,\ b_0 = 1.0$$

Initial conditions θ_1 and θ_2 are assumed zero and are also estimated alongwith a_1 , a_2 and b_0 .

These parameters are estimated using both LS and IV algorithms. Both the algorithms use GPMF with $\beta = \lambda$. Figures 22 and 23 show the convergence plots for the estimates of the system parameters a_1, a_2 and b_0 and also the variation of error norm, by both LS and IV methods with 50% noise for $\lambda = 5$ and 10 respectively. Table 14 presents the estimated parameters and error norm for LS and IV algorithms.

5. Closed loop system identification

The proposed method for the identification of the closed loop system is close in sprit to the indirect identification method. The method used here for the identification of both forward and feedback path transfer functions without the knowledge of feedback path transfer function. The stucture of the closed-loop system is as shown in Fig. 24 where

$G_m(s)$ and $G_c(s)$: forward and feed-back path transfer functions respectively,

$R_1(s)$: disturbance/noise signal,

$R_2(s)$: test signal and

$W(s)$: reference signal, which is usually assumed to be zero.

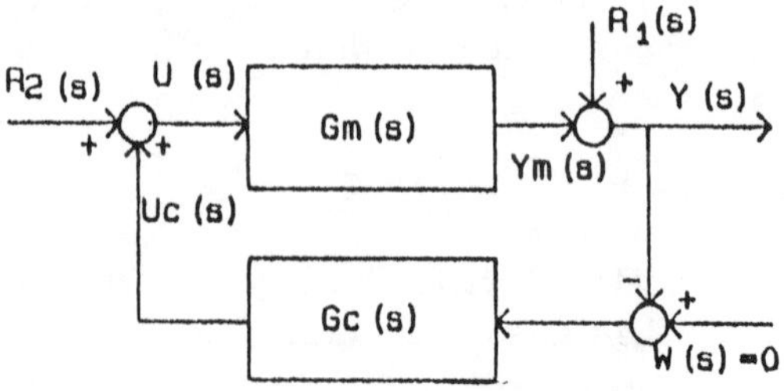

Fig.24. Closed loop system

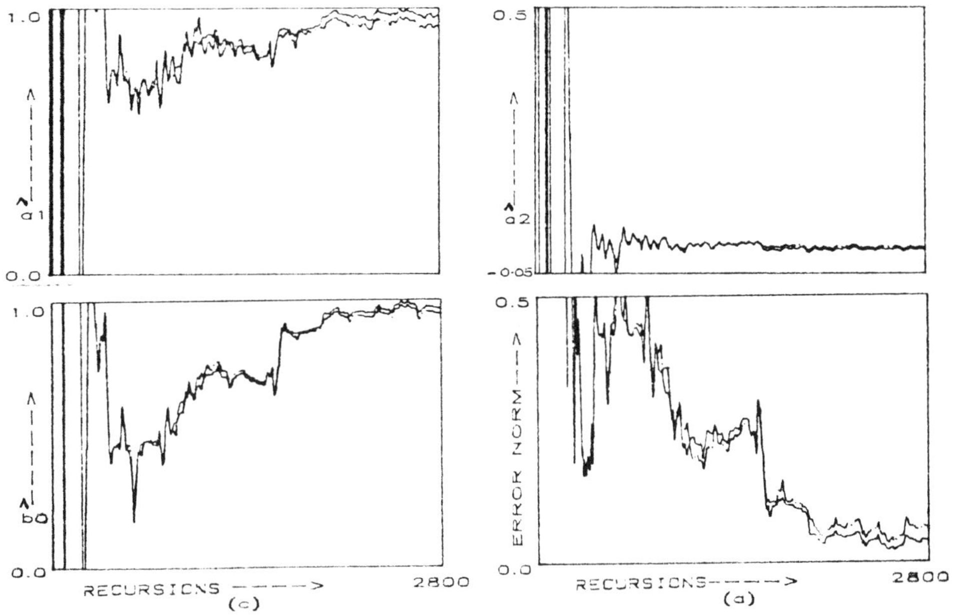

Fig.22. Convergence of parameters for example V
$\lambda = 5, NSR = 2$ (IV —, LS -.-)

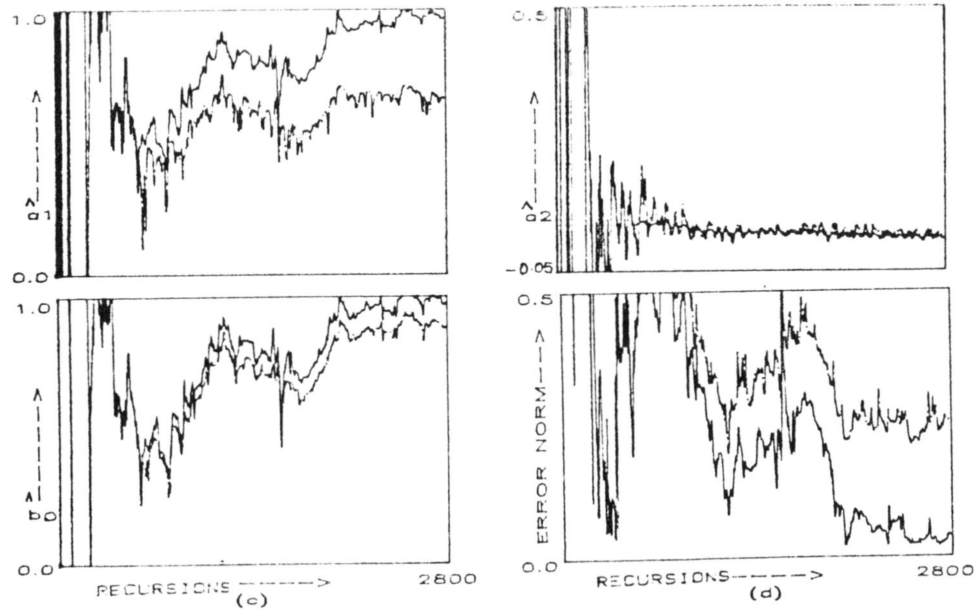

Fig.23. Convergence of parameters for example V
$\lambda = 10, NSR = 2$ (IV —, LS -.-)

The measurable test signal $r_2(t)$ is stationary and not correlated with the disturbance signal $r_1(t)$, then $r_2(t)$ can be interpreted as an independent variable of the overall system. Thus

$$Y(s) = \frac{G_m(s)}{1 + G_m(s)G_c(s)} R_2(s) + \frac{1}{1 + G_m(s)G_c(s)} R_1(s) \tag{19}$$

$$U(s) = \frac{1}{1 + G_m(s)G_c(s)} R_2(s) - \frac{G_C(S)}{1 + G_m(s)G_c(s)} R_1(s) \tag{20}$$

Eqns.(19) and (20) describe the open-loop systems according to Figs. 25 and 26.

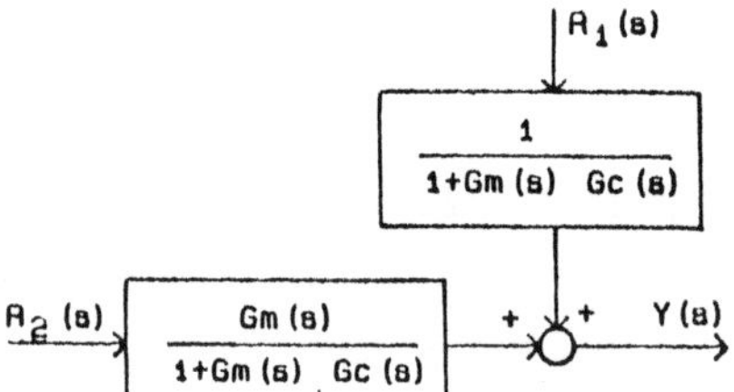

Fig.25. Transform representation of the losed loop system for $Y(s)$ and $R_2(s)$

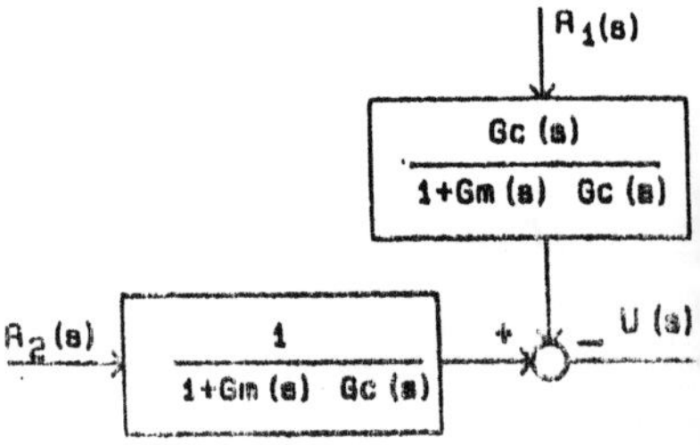

Fig.26. Transform representation of the clo loop system for $U(s)$ and $R_2(s)$

This transformation suggests estimation of G_1 and G_2 where

$$G_1(s) = \frac{G_m(s)}{1 + G_m(s)G_c(s)} \tag{21}$$

$$G_2(s) = \frac{1}{1 + G_m(s)G_c(s)} \tag{22}$$

which represents the transfer function $Y(s)/R_2(s)$ and $U(s)/R_2(s)$ respectively.

Denoting $G_m(s) = \frac{N_m(s)}{D_m(s)}$ and $G_c(s) = \frac{N_c(s)}{D_c(s)}$ the above equation reduces to

$$\{D_m(s)D_c(s) + N_m(s)N_c(s)\}Y(s) = D_c(s)N_m(s)R_2(s)$$

or,

$$H(s)Y(s) = H_1(s)R_2(s) \tag{23}$$

and

$$\{D_m(s)D_c(s) + N_m(s)N_c(s)\}U(s) = D_m(s)D_c(s)R_2(s)$$

or,

$$H(s)U(s) = H_2(s)R_2(s) \tag{24}$$

where

$$\begin{aligned} H(s) &= D_m(s)D_c(s) + N_m(s)N_c(s) \\ H_1(s) &= D_c(s)N_m(s) \\ H_2(s) &= D_m(s)D_c(s) \end{aligned}$$

Eqns.(23) and (24) are to be used in estimating the parameters of the transfer functions using GPMF algorithm derived in Section 3.

Illustrative Example VI

Consider the closed-loop system with

$$G_m(s) = \frac{1}{s^2 + s + 2} \text{ and } G_c(s) = 1 + \frac{1}{s}$$

The measurable test signal $r_2(t)$ used here is a PRBS signal of period 1023 and amplitude ± 1. $r_1(t)$, a zero mean gaussian noise signal with various noise to signal ratio (NSR) is added to $y(t)$. The sampling time is taken as 0.01 for both system discretisation and estimation. The estimated parameters and transfer functions for various noise levels are presented in Tables 15 and 16.

In this case Eqns. (23) and (24) take the form

$$(s^3 + a_{12}s^2 + a_{11}s + a_{10})Y(s) = (b_{11}s + b_{10})R_2(s)$$

and

$$(s^3 + a_{12}s^2 + a_{11}s + a_{10})Y(s) = (b_{23}s^3 + b_{22}s^2 + b_{21}s + b_{20})R_2(s)$$

The true values are

$$a_{12} = 1,\ a_{11} = 3,\ a_{10} = 1, b_{11} = 1,\ b_{10} = 0$$

$$b_{23} = 1,\ b_{22} = 1,\ b_{21} = 2,\ b_{20} = 0$$

Table 13: Estimated parameters and initial states for Example IV.

NSR	parameters			initial states	
	$\hat{a}_1$	$\hat{a}_2$	$\hat{b}$	$\hat{x}_1(0)$	$\hat{x}_2(0)$
0%	2.0057	1.0013	1.0015	0.1998	0.5999
20%	1.9972	1.0163	1.0200	0.1965	0.6074

Table 14: Estimates of parameters of Example V with NSR = 50%

algorithm		parameters					error norm
	λ	$\hat{a}_1$	$\hat{a}_2$	$\hat{b}_0$	$\hat{\theta}_1$	$\hat{\theta}_2$	
LS	5	0.9328	0.0029	0.9471	0.0085	-0.0363	0.0660
	10	0.6579	0.0200	0.8846	0.0163	-0.0328	0.2570
IV	5	0.9714	-0.0012	0.9662	0.0086	-0.0343	0.0400
	10	0.9656	0.0146	0.9744	0.0163	-0.0253	0.0384

Table 15: Estimated parameters of Example VI

NSR	parameters								
	$\hat{a}_{12}$	$\hat{a}_{11}$	$\hat{a}_{10}$	$\hat{b}_{11}$	$\hat{b}_{10}$	$\hat{b}_{23}$	$\hat{b}_{22}$	$\hat{b}_{21}$	$\hat{b}_{20}$
0%	0.99	3.01	1.01	1.01	0.00	1.00	0.99	1.99	0.00
10%	0.98	2.99	0.98	0.99	0.00	1.00	0.98	2.02	0.00
20%	0.97	2.97	0.96	0.99	0.00	1.00	0.98	2.03	-0.01

Table 16: Estimated transfer functions of Example VI

NSR	$G_m(s)$	$G_c(s)$
0%	$\frac{1.005}{s^2 + 0.99s + 1.99}$	$\frac{1.01s + 1.01}{s + 0}$
10%	$\frac{0.996}{s^2 + 0.98s + 2.02}$	$\frac{0.98s + 0.99}{s + 0}$
20%	$\frac{0.986}{s^2 + 0.98s + 2.03}$	$\frac{0.96s + 0.99}{s - 0.01}$

6. Discussion and conclusion

The classical PMF approach is generalised and a recursive LS algorithm is presented. The ordinary PMF with $\lambda \gg 1$ has a small steady state gain. It attenuates the low frequency part of the input-output signal spectra which may be vital in the process of estimation. On the other hand $\lambda \ll 1$, the low frequency components of process data are amplified. This may be advantageous only if such parts of the signal spectra are vital and if there is no significant noise in this band to accentuate noise. The bandwidth of the unknown system under identification actually determines the vital sections of the signal spectra during the process of estimation. In the case of NPMF, the steady state gain is independent of λ. For very small λ, the NPFC attenuates the essential signals uniformly over almost the entire frequency band. Consequently the NPMFs with $\lambda \ll 1$ are not recommended for parameter estimation. On the other hand, it may be said that the GPMFs have the desired flexibility, due to facility to choose β and λ independently of each other, to conserve the information content of the process data during identification. The actual values of β and λ may be chosen to match the requirements of the situation. Thus the concept of GPMFs introduced here is likely to give the degree of flexibility to considerably enhance accuracy and rate of convergence of parameters.

The algorithm developed with GPMFs is used for combined parameter and state estimation. Normalised PMF ($\beta = \lambda$) is used for parameter and initial state estimation and the state at subsequent instants of time is estimated using repeated integration technique which is a special case of GPMF ($\beta = 1$ and $\lambda = 0$).

To reduce bias in the estimates of the parameters, a recursive IV algorithm is formulated based on the GPMF measures of the process signals. It is found that at high noise levels, for example 50%, the IV algorithm gives considerably less bias than the LS algorithm. It is also observed that as λ is increased, LS estimates are degraded considerably while the quality of the IV estimates remains unaffected. In the case of systems with larger bandwidth, where it becomes necessary to have λ sufficiently large so as to accomdate all the significant frequencies in the process signals, IV algorithm offers advantage over LS justifying the additional computational burden in IV implementation.

Finally, a simple and straight forward method for the estimation of parameters of the forward and feedback path transfer functions of the system operating in closed loop is presented using GPMFs of the noisy process data. The method has the generality that it does not assume the knowledge of feedback path transfer function.

Acknowledgement

The authors wish to thank Professor G.P. Rao, Dr. A. Patra and Mr. S. Mukhopadhyay of Electrical Engineering Department, I.I.T., Kharagpur for their useful discussion during the preparation of the manuscript.

References

Gevers, M.R. (1978), "On the Identification of Feedback Systems", in *Identification and System Parameter Estimation*, Ed. Rajbman, N.S., North Holland, Amsterdam, Part 3, pp. 1621–1630.

Graupe, D. (1975), "On Identifying Stochastic Closed-loop Systems", *IEEE Trans. Automatic Control*, vol. 20, pp. 553–555.

Gustavsson, I., Ljung, L. and Söderström, T. (1978),; "Identification of Processes in Closed loop - Identifiability and Accuracy Aspects",in *Identification and System Parameter Estimation*, Ed. Rajbman, N.S., North Holland, Amsterdam, Part 1, pp. 39–77.

Kung, F.C. and Lee, H. (1982), "Solution of Linear State Space Equations and Parameter Estimation in Feedback Systems using Lagurre Polynomial Expansion", *Journal of Franklin Institute*, vol.314, no.6, pp.393–403.

Kwong, C.P. and Chen C.F. (1981); "Linear Feedback System Identification via Block Pulse Functions",*Int. Journal of Systems Science*, vol.12, no.5, pp.635–642.

Rao, G.P. and Patra, A. (1988),"Continuous-time Approaches to Combined State and Parameter Estimation in Linear Continuous Systems", *Proc. 8th IFAC/IFORS Symposium on Identification and System Parameter Estimation*, (Beijing, P.R.China, August 1988), pp.1287–1291.

Roy, B.K. and Saha, D.C. (1989),"Parameter Estimation in Closed-loop Continuous-time Models - A Recursive Least Squares Poisson Moment Functional Approach",*Proc. 13th National Systems Conference*,(I.I.T., Kharagpur, India, December 1989), pp.67–70.

Sagara, S. and Zhao, Zhen-Yu (1988),"Numerical Integration Approach to On-line Identification of Continuous Systems in the presence of Measurement Noise",*Proc. 8th IFAC/ IFORS Symposium on Identification and System Parameter Estimation*, (Beijing, P.R.China, August 1988),pp.41–46.

Saha, D.C. and Rao, G.P. (1983),"*Identification of Continuous Dynamical Systems - The Poisson Moment Functional (PMF) Approach*", LNCIS, vol.56, Springer-Verlag, Berlin.

Saha, D.C. and Mandal, S.K. (1990),"Recursive Least Squares Parameter Estimation in SISO Systems via Poisson Moment Functionals-Part I. Open Loop Systems ", *Int. Journal of Systems Science*, vol.21,pp.1205–1216.

Söderström, T. and Stoica, P.G. (1983), "*Instrumental Variable Methods for System Identification*", LNCIS, vol.57, Springer-Verlag, Berlin.

Unbehauen, H. and Rao, G.P. (1987), "*Identification of Continuous Systems*", North Holland, Amsterdam.

Unbehauen, H. and Rao, G.P. (1990), "Continuous-time Approaches to System Identification – A Survey",*Automatica*, vol.26, no.1, pp.23–35.

Wellstead, P.E. (1978),"The Identification and Parameter Estimation in Feedback Systems", in *Identification and System Parameter Estimation*, Ed. Rajbman, N.S., North Holland, Amsterdam, part 3, pp.1593–1600.

Young, P.C. and Jakeman, A. (1980), "Refined Instrumental Variable Methods of Recursive Time-Series Analysis - Part III", *Int. Journal of Control*, vol.31, no.4, pp.741–764.

Identification, Estimation and Control of Continuous-Time Systems Described by Delta Operator Models

Peter C. Young
Arun Chotai and Wlodek Tych
Centre for Research on Environmental Systems
Institute of Environmental and Biological Sciences
University of Lancaster, Lancaster, LA1 4YQ
United Kingdom

Abstract

This Chapter outlines a unified approach to the identification, estimation and control of linear, continuous-time, stochastic, dynamic systems which can be described by delta (δ) operator models with constant or time-variable parameters. It shows how recursive refined instrumental variable estimation algorithms can prove effective both in off-line model identification and estimation, and in the implementation of self-tuning or self-adaptive True Digital Control (*TDC*) systems which exploit a special Non-Minimum State Space (*NMSS)* formulation of the δ operator models.

1. Introduction

When serious research on data-based methods of continuous-time system identification and parameter estimation began in the control and systems community over thirty years ago, it was natural to think directly in terms of models characterised by continuous-time, differential equations; or, equivalently, transfer functions in the Laplace operator. And even the methods used for estimating the parameters in such models were often based on continuous-time analog computer techniques (see e.g. Young, 1965 and the references therein). However, the demise of the analog computer and the rapid rise to prominence of digital computation, coupled with innovatory research on statistical methods of discrete-time model identification and estimation by research workers such as Astrom and Bohlin (1966), proved very influential during the next quarter century. This spawned an enormous literature on alternative, data-based methods of discrete-time system modelling which, apparently, owed little to their continuous-time progenitors.

Surprisingly, however, these developments in discrete-time, sampled data modelling failed to make too many inroads into the underlying faith of many traditional control systems designers in continuous-time models and design methods. As a result, although adaptive control systems based directly on discrete-time models are now becoming relatively common, the majority of practical control systems still rely on the ubiquitous, three-term, Proportional-Integral-Derivative (*PID*) controller, with its predominantly continuous-time heritage. And when such systems, or their more complex relatives, are designed off-line (rather than simply "tuned" on-line), the design procedure is normally based on traditional continuous-time concepts, with the resultant design then

N. K. Sinha and G. P. Rao (eds.), Identification of Continuous-Time Systems, 363–418.

being, rather artificially, "digitised" prior to implemention in digital form.

But does this 'hybrid' approach to control system design really make sense? Would it not be both more intellectually satisfying and practically advantageous to evolve a unified, truly digital approach which could satisfy the predilection of the traditional control systems designer for continuous-time models and methods whilst, at the same time, allowing for the full exploitation of discrete-time theory and digital implementation? In a number of previous publications (Young et al, 1987a,b, 1988, 1991a; Young, 1989), we have promoted one such alternative philosophy based on the idea of "True Digital Control" (*TDC*). This rejects the idea that a digital control system should be initially designed in continuous-time terms. Rather it suggests that the designer should consider the design from a digital, sampled-data standpoint, even when rapidly sampled, near continuous-time operation is required. In this chapter, we concentrate on one particular aspect of the general *TDC* approach; namely the recursive identification and optimal estimation of discrete-time models for rapidly sampled, continuous-time systems and their use in adaptive control system design.

Central to development of these recursive methods is the discrete differential or "finite-difference" operator. Although the idea of such an operator is not new, its relevance in control system terms was not fully recognised until Goodwin (1985, 1988) and Goodwin et al. (1988) renamed it the "delta" (δ) operator and exposed its many attractive qualities. More recently still, Middleton and Goodwin (1990) have produced an excellent, definitive text on a unified approach to digital control and estimation which demonstrates how all of the traditional concepts of continuous-time modelling and control have their closely related equivalents in the δ operator domain. Here, we explore this new world of discrete-time modelling; show how the parameter estimation procedures currently being proposed for δ operator models are closely related to continuous-time modelling concepts and methods evolved by the first author in the 1960's; and develop improved recursive instrumental variable (*IV*) methods for δ operator model estimation that are direct developments of these earlier algorithms.

2. The Discrete Differential (δ) Operator *TF* Model

In all model-based control system design procedures, the form of the models and the associated theoretical background is of paramount importance. Prior to 1960, most control systems were based on analog methodology. Where sampled data implementation was required, it was normally obtained by the digitisation of continuous-time analog designs using *z* transform theory and sample-hold circuitry. As a result, when the first author began his research on dynamic system identification and estimation in the early nineteen sixties, continuous-time differential equation models were *de rigueur*: because the conrol systems designer utilised analytical procedures based on such models, it seemed quite natural to investigate estimation methods based on the analysis of continuous-time, analog signals.

Within this continuous-time context, the most common model is the ordinary differential equation or its equivalent, the transfer function model in terms of the time derivative operator (which we denote here by $s=d/dt$, because of its close relationship with the Laplace operator) i.e.,

$$\frac{d^n x(t)}{dt^n} + a_1 \frac{d^{n-1} x(t)}{dt^{n-1}} + ... + a_n x(t) = b_0 \frac{d^m u(t)}{dt^m} + ... + b_m u(t)$$

or,

$$x(t) = \frac{b_0 s^m + ... + b_m}{s^n + a_1 s^{n-1} + ... + a_n} u(t)$$

where $x(t)$ and $u(t)$ denote, respectively, the output and input signals of the system; while a_i, $i=1, 2, ... ,n$ and b_j, $j=0, 1, ... , m$ are the $m+n+1$ model coefficients or parameters. Although this continuous-time formulation of the model was natural at a time where control systems were implemented in largely analog terms, its virtues became less obvious during the nineteen sixties as the improvements in digital computer technology made it increasingly clear that the days of the analog computer and analog circuitry were numbered. It was not surprising, therefore, that the control world began to consider alternative discrete-time, sampled data model forms from the late nineteen fifties; and papers on the estimation of parameters in these discrete-time models began to appear (e.g. Kalman, 1958; Joseph et al, 1961), culminating in the seminal work of Astrom and Bohlin (1966).

The two major forms of the transfer function model in the discrete-time domain are the well known backward shift operator form; and the alternative, but less well known, transfer function in the discrete differential operator. Although the latter has closer connections with the continuous-time models that found favour with the majority of control systems analysts, the advantages of the former in *estimation* terms were immediately clear. And so there has been little, if any, discussion of the discrete differential operator model until quite recently, when it was revived under the title "δ operator" by Goodwin (1985, 1988) and Goodwin et al.(1988). Rather, those research workers who insisted on a completely continuous-time formulation went in one direction, retaining their dedication to differential equations and continuous-time, despite the digital computer revolution (e.g. see references to "model reference methods" in Young, 1979, 1981). A few, such as the first author (e.g. Young, 1965, 1966, 1969b, 1970), took a "hybrid" path, retaining interest in the differential equation model but posing the estimation problem in discrete-time terms. The majority, however, followed the lead of Astrom and Bohlin, and later Box and Jenkins (1970), preferring to address the problems of model identification and parameter estimation associated directly with the conventional discrete-time model, or its equivalent, the z^{-1} operator transfer function.

This apparent disregard of the discrete differential operator is rather misleading. As we shall see, those research workers who took the "hybrid" path rarely had access to a true hybrid (analog-digital) computer and so, accepting the virtues of the digital computer, they were forced to implement the inherently continuous-time aspects of their algorithm (such as the data pre-filters[1] required to avoid direct differentiation of noisy signals) in discrete-time terms using discrete approximations to the differential equations. And, of course, the simplest such approximations are equivalent to those used to define the discrete differential operator. In this Chapter, however, this relationship is overtly

1. these prefilters are an inherent part of the most successful continuous-time estimation algorithms and were originally and, as we shall see, rather appropriately termed "state variable filters"; see e.g. Young, 1964, 1965, 1979, 1981.

acknowledged and we consider the direct identification and estimation of the discrete differential operator model, which can be written in the following form,

$$y(k) = \frac{B(\delta)}{A(\delta)} u(k-\tau)$$

where, for generality, we have introduced a pure time delay τ on the input signal. In this model, $A(\delta)$ and $B(\delta)$ are polynomials in the δ operator , defined as follows,

$$A(\delta) = \delta^n + a_1 \delta^{n-1} + \ldots\ldots + a_n$$

$$B(\delta) = b_0 \delta^m + b_1 \delta^{m-1} + \ldots\ldots + b_m$$

in which the numerator order m will normally be less than the denominator order n. Here, the δ operator, for the sampling interval Δt, is normally defined in terms of the backward shift operator as,

$$\delta = \frac{1 - z^{-1}}{z^{-1} \Delta t}$$

or, alternatively, in terms of the forward shift operator z,

$$\delta = \frac{z-1}{\Delta t} \;\; ; \; i.e. \;\; \delta x(k) = \frac{x(k+1) - x(k)}{\Delta t}$$

which is more convenient for the current analysis.

In some cases, it may be convenient to accommodate the pure time delay directly in the denominator polynomial $A(\delta)$: this is obtained by substituting for $z^{-\tau}$ in terms of δ and the resulting *TF* equation takes the form,

$$y(k) = \frac{B(\delta)}{A(\delta)} u(k)$$

where now,

$$A(\delta) = (\delta^n + a_1 \delta^{n-1} + \ldots\ldots + a_n)(1 + \Delta t \, \delta)^{\tau}$$

$$B(\delta) = b_0 \delta^m + b_1 \delta^{m-1} + \ldots\ldots + b_m$$

Remarks

(1) The above definition of the δ operator is based on the forward difference but there may be some advantages in considering the alternative backward difference definition (see Appendix 1, comment 1).
(2) In the limit as $\Delta t \rightarrow 0$, the δ operator reduces to the derivative operator $(s = d/dt)$ in continuous time (i.e $\delta \rightarrow s$).
(3) Given a polynomial of any order p in the z operator, this will be exactly equivalent to some polynomial in δ, also of order p.
(4) The δ operator model coefficients are related to forward z operator coefficients by simple vector matrix equations (Chotai et al., 1990a; Young et al., 1991a).
(5) Subsequently, it is convenient to refer to (3) as an (n, m, τ) model.

One attraction of the δ operator model to those designers who prefer to think in continuous-time terms is that it can be considered as a direct approximation to a continuous-time system, with the accuracy increasing as $\Delta t \rightarrow 0$. For example, it is easy to see that the unit circle in the complex z plane maps to a circle with centre $-1/\Delta t$ and radius $1/\Delta t$ in the complex δ plane; so that, as $\Delta t \rightarrow 0$, this circular stability region is transformed to the left half of the complex s plane. For rapidly sampled systems, therefore, the δ operator model can be considered in almost continuous-time terms, with the pole positions in the δ plane close to those of the 'equivalent' continuous-time system in the s plane; and with the *TF* parameters directly yielding information on factors such as the approximate natural frequency and damping ratio. Those readers who are not aquainted with these and many other attractive aspects of the δ operator are strongly recommended to consult the recent book on the subject by Middleton and Goodwin (1990).

3. Recursive Identification and Parameter Estimation

Recursive identification and estimation tools are particularly useful in control and systems analysis since they facilitate the modelling of *Time Variable Parameter* (*TVP*) systems and the development of self-adaptive control systems. Many recursive estimation procedures have become available in the past few years but we believe that the Instrumental Variable (*IV*) approach is probably the most flexible and easy to use in practical applications since, unlike all other commonly used procedures (such as recursive approximate maximum likelihood (*AML*), prediction error recursion (*PER*), and generalised least squares (*GLS*) methods; see e.g Ljung and Soderstrom, 1983; Norton, 1986; Soderstrom and Stoica, 1989), it does not demand concurrent estimation of parameters in a model for any stochastic disturbances that may affect the system. *IV* methods of model order identification and parameter estimation are well known in the literature on time-series analysis and, for a general introduction to *IV* estimation, the reader should consult Young (1984); Young and Armitage (1991); and Soderstrom and Stoica (1983).

When considered in δ operator terms, these *IV* methods are applied to the following stochastic version of the *TF* model (omitting the pure time delay τ for convenience),

$$y(k) = \frac{B(\delta)}{A(\delta)} u(k) + \xi(k) \qquad (1)$$

where $\xi(k)$ is a general disturbance term introduced to account for all stochastic or non-measureable deterministic inputs to the system. This model can be written in the alternative vector form,

$$\delta^n y(k) = z(k)^T \boldsymbol{a} + \eta(k) \qquad (2)$$

where,

$$z(k) = [-\delta^{n-1}y(k), -\delta^{n-2}y(k), \ldots\ldots, -y(k), \delta^n u(k), \delta^{m-1}u(k), \ldots, u(k)]^T$$

$$\boldsymbol{a} = [a_1, a_2, \ldots\ldots, a_n, b_0, b_1, \ldots\ldots, b_m]^T$$

and $\eta(k)$ is defined accordingly as a function of $\xi(k)$. Note that the specification of the highest numerical derivative on the left hand side of this equation is arbitrary and is selected here to be consistent with the later control systems analysis. Past research on continuous-time systems (Young, 1969a), and our current experience, both suggest that there may be some practical advantages with this specification. However, other model forms are available and may be preferable, as discussed in comment 2. of Appendix 1.

The linear-in-the-parameters model (2) could be used as the basis for ordinary *IV* identification and estimation, but this would imply direct computation of the numerical derivatives of the input and output signals, a task which is not to be recommended if the data are noisy. However, as in the equivalent continuous-time case, this does not prove necessary since all *TF* estimation problems of this type, when considered from a "refined" or optimum standpoint, require all of the data to be prefiltered in a special manner, as discussed in the next sub-section.

The Recursive Refined (or quasi-Optimal) *IV* Algorithm

The refined or optimal *IV* (*RIV*) algorithm for z^{-1} operator models (Young 1976, 1984; Young and Jakeman 1979; Soderstrom and Stoica 1983[2]) involves special adaptive prefiltering of the data used in the *IV* algorithm in order to induce asymptotic optimality in a minimum variance sense. This same approach can be used for δ operator models and can be justified most simply by considering the following special case of equation (1), where $\xi(k)$ is assumed to be a zero mean, stochastic disturbance with rational spectral density, which can be described by the following *TF* model,

$$\xi(k) = \frac{D(\delta)}{C(\delta)}\, e(k) \tag{3}$$

where,

$$C(\delta) = \delta^{p} + c_1 \delta^{p-1} + ... + c_p \quad ; \quad D(\delta) = \delta^{q} + d_1 \delta^{q-1} + ... + d_q$$

while $e(k)$ is a zero mean, serially uncorrelated sequence of random variables (white noise) with variance σ^2; and the *TF* is assumed to be stable, i.e. the roots of the characteristic equation $A(\delta)=0$ all lie within the stability circle of the complex δ plane (see above). This model is the δ operator equivalent of the well known *ARMA* process. As we discuss later, however, it is may often be sufficient to consider only the simpler equivalent of the *AR* process, i.e. when $D(\delta)=1$.

The complete model obtained in this manner can be written in the form,

$$y(k) = \frac{B(\delta)}{A(\delta)}\, u(k) + \frac{D(\delta)}{C(\delta)}\, e(k) \tag{4}$$

---

2. Soderstrom and Stoica (1983) refer to their z^{-1} operator implementation of the refined IV algorithm (first suggested by Young, 1976; see also Young and Jakeman, 1979) as *"IV-4"*, a non-recursive version of which is available in the Matlab "Identification Toolbox". The simpler recursive *SRIV* algorithm, which is available in the *micro*CAPTAIN and *TDC* program packages developed at Lancaster has, we believe, some advantages in practical application. The alternative δ operator versions of SRIV discussed in this Chapter will be available soon in *micro*CAPTAIN and *TDC*, the latter written within Matlab.

Perhaps the simplest way of illustrating the need for prefiltering is to consider the problem of estimating the model parameters in terms of the following least squares cost function,

$$J = \sum_{k=1}^{k=N} \hat{e}(k)^2$$

where $\hat{e}(k)$ is the following error function obtained directly by inspection of the model (4),

$$\hat{e}(k) = \frac{\hat{C}(\delta)}{\hat{D}(\delta)} \left\{ y(k) - \frac{\hat{B}(\delta)}{\hat{A}(\delta)} u(k) \right\} \tag{5a}$$

while N is the total sample size and the 'hat' indicates estimated values.

This error function is clearly nonlinear in the parameters of the unknown polynomials. However, it can be written alternatively as,

$$\hat{e}(k) = \frac{\hat{C}(\delta)}{\hat{D}(\delta)\hat{A}(\delta)} \left\{ \hat{A}(\delta)\, y(k) - \hat{B}(\delta)\, u(k) \right\} \tag{5b}$$

or,

$$\hat{e}(k) = \hat{A}(\delta)\, y^*(k) - \hat{B}(\delta)\, u^*(k) \tag{5c}$$

where $y^*(k)$ *and* $u^*(k)$ are "prefiltered" variables defined as follows,

$$y^*(k) = \frac{\hat{C}(\delta)}{\hat{D}(\delta)\hat{A}(\delta)} y(k) \quad ; \quad u^*(k) = \frac{\hat{C}(\delta)}{\hat{D}(\delta)\hat{A}(\delta)} u(k) \tag{6}$$

Equation (5c) is now linear-in-the-parameters of the *TF* model, so that normal *IV* methods could be used to estimate the parameters *if it were possible to perform the prefiltering operations in (6).* Moreover, the inclusion of such prefilters would nicely solve the problem of derivative measurement by providing the physically realisable filtered derivatives as by-products of the prefiltering operation.

In practice, of course, the parameters of the $A(\delta)$, $C(\delta)$ and $D(\delta)$ polynomials are unknown *a priori* and so this prefiltering operation must be made adaptive, with the algorithm "learning" the parameters of the polynomials on a recursive or recursive-iterative basis (Young, 1976, 1984; Young and Jakeman, 1980). A block diagram of the prefilter for the case where $C(\delta)=D(\delta)=1.0$ is shown in Fig.1.

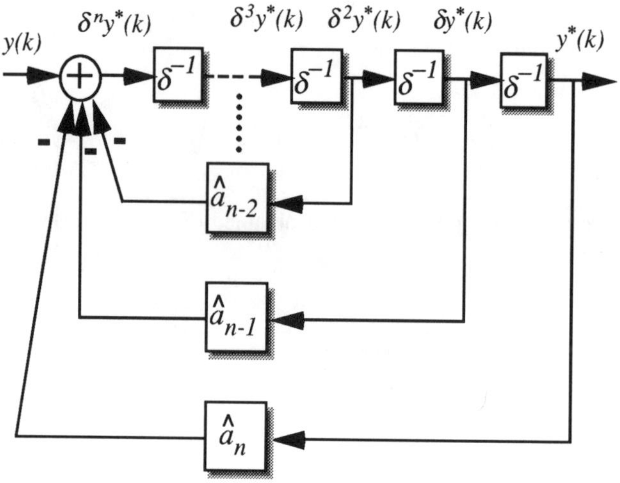

Fig.1 Block diagram of the prefilter in the SRIV recursive algorithm (shown operating on the output y(k) signal)

The resultant recursive Refined Instrumental Variable (*RIV*) algorithm for estimating the parameter vector $\boldsymbol{a}$ in equation (2) takes the form,

$$\hat{\boldsymbol{a}}(k) = \hat{\boldsymbol{a}}(k-1) + \boldsymbol{P}(k-1)\hat{\boldsymbol{x}}^*(k)\ \varepsilon(k/k-1) \qquad \text{(i)}$$

where,

$$\varepsilon(k/k-1) = \frac{\delta^n y^*(k) - z^{*T}(k)\hat{\boldsymbol{a}}(k-1)}{1 + z^*(k)^T \boldsymbol{P}(k-1)\hat{\boldsymbol{x}}^*(k)} \qquad \text{(ii)} \qquad (7)$$

and,

$$\boldsymbol{P}(k) = \boldsymbol{P}(k-1) - \frac{\boldsymbol{P}(k-1)\hat{\boldsymbol{x}}^*(k) z^*(k)^T \boldsymbol{P}(k-1)}{1 + z^*(k)^T \boldsymbol{P}(k-1)\hat{\boldsymbol{x}}^*(k)} \qquad \text{(iii)}$$

In this algorithm,

$$z^*(k)=[-\delta^{n-1}y^*(k), -\delta^{n-2}y^*(k), ..., -y^*(k), \delta^n u^*(k), \delta^{n-1}u^*(k), ..., u^*(k)]^T$$

$$\hat{\boldsymbol{x}}^*(k)=[-\delta^{n-1}\hat{x}^*(k), -\delta^{n-2}\hat{x}^*(k), ..., -\hat{x}^*(k), \delta^n u^*(k), \delta^{n-1}u^*(k), ..., u^*(k)]^T$$

where $\hat{x}^*(k)$, is the "instrumental variable", and the star superscript indicates that the associated variables are adaptively prefiltered in the manner described above. The matrix $\boldsymbol{P}(k)$, which is the recursively computed inverse of the *IV* cross product matrix, i.e.,

$$\boldsymbol{P}(k) = \left[\sum_{i=1}^{i=k} \hat{\boldsymbol{x}}^*(k) z^*(k)^T\right]^{-1} \qquad (8)$$

is related to the error covariance matrix $\boldsymbol{P}^*(k)$ of the estimated parameter vector $\hat{\boldsymbol{a}}(k)$ by the equation,

$$\boldsymbol{P}^*(k) = \sigma^2 \boldsymbol{P}(k) \tag{9}$$

and an estimate $\hat{\sigma}^2(k)$ of the variance σ^2 can be obtained from an additional recursive least squares estimation equation (see Young (1984), p.100). In the important special case (see below) where $C(\delta)=D(\delta)=1.0$, this equation takes the form,

$$\hat{\sigma}^2(k) = \hat{\sigma}^2(k-1) + g(k)\{\varepsilon^2(k/k-1) - \hat{\sigma}^2(k-1)\} \tag{10}$$

where the scalar gain $g(k)$ is defined by,

$$g(k) = \frac{p(k-1)+q}{1+p(k-1)+q}$$

and $p(k)$ is generated by the recursion,

$$p(k) = p(k-1)+q - g(k)\{p(k-1)+q\}$$

Here q is the *noise variance ratio* (*NVR*) parameter which, for assumed homoscedastic (constant variance) residuals, will be equal to zero.

The complete recursive *RIV* algorithm obtained in the above manner is simply a modification of the well known Recursive Least Squares (*RLS*) algorithm, with the data vector $z^*(k)$ replaced alternately by $\hat{x}^*(k)$. In the off-line case, a recursive-iterative approach is utilised where, at the *jth* iteration, the prefiltered instrumental variable $\hat{x}^*(k)$ required in the definition of $\hat{x}^*(k)$ is generated by adaptively prefiltering the output of an "auxiliary model" defined in the following manner,

$$\hat{x}^*(k) = \frac{\hat{C}_{j-1}(\delta)}{\hat{D}_{j-1}(\delta)\hat{A}_{j-1}(\delta)}\hat{x}(k) \;\; ; \;\; \hat{x}(k) = \frac{\hat{B}_{j-1}(\delta)}{\hat{A}_{j-1}(\delta)}u(k) \tag{11}$$

adaptive prefilter *adaptive auxiliary model*

where $\hat{A}_{j-1}(\delta)$ etc. are estimates of the polynomials obtained by reference to the estimates obtained by the algorithm at the end of the *previous (j-1)*th iteration. The prefiltered variables $y^*(k)$ and $u^*(k)$ are obtained in a similar fashion. Such a *recursive-iterative* or *relaxation* approach normally requires only two to four iterations to converge on a reasonable set of model parameters (see later; comment (vi)).

The recursive equations of the *RIV* estimation algorithm described in this Section are, of course, dynamic systems in their own right. Given the present δ operator context, therefore, it is tempting, finally, to present the algorithm in its alternative δ operator form. In order to conserve space, however, we will leave this as an exercise for the reader, who can consult Middleton and Goodwin (1990) for assistance, if this proves necessary.

The Simplified Refined Instrumental Variable (*SRIV*) Algorithm

The full *RIV* algorithm described above is quite complicated. Ideally, the adaptive prefiltering in (11) requires identification and estimation of the noise model (3), which we discuss in the next sub-Section. However, the first author (1985) has demonstrated the value of a special version of the refined *IV* algorithm, which he termed the *Simplified Refined Instrumental Variable (SRIV)* algorithm. This is obtained if we assume that the stochastic disturbance $\xi(k)$ is a white noise process, so that $C(\delta)=D(\delta)=1$. The *SRIV* algorithm is then straightforward to implement: it is identical to that given in equations (7), (9) and (10), but with the adaptive prefiltering operation in (11) simplified by setting $\hat{C}_{j-1}(\delta)=\hat{D}_{j-1}(\delta)=1$. This algorithm is not only easy to implement, since it does not require concurrent estimation of a disturbance model *TF*[3], but it also provides very good estimates in many practical situations. For example, it is particularly effective in estimating the *TF* model parameters when the input-output data have been obtained from single impulse response testing; and it provides an excellent approach to model order reduction (see example 4 in Section 4., and Tych and Young, 1990). It is this *SRIV* algorithm, therefore, which we propose as the normal basis for *TF* model identification and estimation (note, however, the later caveat in comment (ii) below).

Noise Model Estimation

Although we favour the *SRIV* algorithm because of its simplicity, it is possible to implement the more complex *RIV* algorithm if this is desired. Moreover, the recursive analysis procedure required to estimate the noise model in such an *RIV* algorithm is useful in its own right for univariate time series analysis.

In order to develop the recursive algorithm for noise model estimation, let us first consider again the δ operator *ARMA* model (3). It is not so easy to develop a recursive estimation algorithm for this model since the introduction of adaptive prefiltering, so essential for continuous-time or δ operator model estimation, is not so obvious as in the *TF* model case. It is possible to conceive of an iterative procedure in which the parameters of the $C(\delta)$ polynomial are estimated, with the adaptive prefiltering based on the $D(\delta)$ polynomial. However, this involves various difficulties and it is more straightforward to restrict consideration to the simpler *AR* model estimation problem.

The *AR* model can be written as,

$$\xi(k) = \frac{1}{C(\delta)}\, e(k) = \frac{1}{\delta^{p} + c_1\delta^{p-1} + c_2\delta^{p-2} + \dots + c_p}\, e(k)$$

i.e,

$$\delta^{p}\xi(k) = -c_1\delta^{p-1}\xi(k) - c_2\delta^{p-2}\xi(k) - \dots - c_p\,\xi(k) + e(k)$$

or (cf equation (2),

3. and is, therefore, independent of any stochastic assumptions about the noise process, which can take on any form provided it is statistically independent of the the input signal $u(k)$

$$\delta^p \xi(k) = \boldsymbol{n}(k)^T \boldsymbol{c} + e(k)$$

with obvious definitions for the data vector $\boldsymbol{n}$ and the parameter vector $\boldsymbol{c}$. Most forms of prefiltering would clearly impair the estimation of this model since they would destroy the white noise properties of the residual $e(k)$. However, the selection of the special prefilter $F(\delta)$, where,

$$F(\delta) = \frac{1}{(\delta + \frac{1}{\Delta t})^p}$$

obviates this problem since, from the definition of δ, we see that $F(\delta)$ is simply a scaled *pth* order backward shift operation $(\Delta t)^p z^{-p}$, which can be implemented in the same manner as the prefilters used in the *TF* model estimation algorithms, but will avoid the frequency dependent filtering problems associated with most other parameterisations of the prefilters.

The estimation of the *AR* parameter vector $\boldsymbol{c}$ now follows straightforwardly using a single pass of the recursive least squares algorithm; i.e. (cf equations (7)-(9)),

$$\hat{c}(k) = \hat{c}(k-1) + \boldsymbol{P}(k-1)\,\boldsymbol{n}^*(k)\;\varepsilon(k/k-1)$$

where,

$$\varepsilon(k/k-1) = \frac{\delta^p \xi^*(k) - \boldsymbol{n}^{*T}(k)\,\hat{c}(k-1)}{1 + \boldsymbol{n}^*(k)^T \boldsymbol{P}(k-1)\,\boldsymbol{n}^*(k)}$$

and,

$$\boldsymbol{P}(k) = \boldsymbol{P}(k-1) - \frac{\boldsymbol{P}(k-1)\,\boldsymbol{n}^*(k)\,\boldsymbol{n}^*(k)^T\,\boldsymbol{P}(k-1)}{1 + \boldsymbol{n}^*(k)^T \boldsymbol{P}(k-1)\,\boldsymbol{n}^*(k)}$$

while,

$$\boldsymbol{n}^*(k) = [-\delta^{p-1}\xi^*(k), -\delta^{p-2}\xi^*(k), \ldots, -\xi^*(k)]^T$$

and the star superscript now indicates *fixed* prefiltering by $F(\delta)$ defined above.

Although this *AR* estimation algorithm appears to work satisfactorily for low dimensional models (see example 3 in Section 4), further research is clearly required on this topic. For example, the generation of the higher order derivatives required by higher dimensional models may be deleteriously affected by the absence of low-pass filtering in the $F(\delta)$ prefilter. And the alternative approach of estimating the z^{-1} operator *AR* model and then converting this to the δ operator form could prove sensitive to uncertainty on the estimated parameters at high sampling rates.

The above *AR* estimation algorithm can be used for δ operator modelling of univariate time-series or for noise model estimation in full *RIV* estimation. In the latter application, the *RIV* algorithm is given by equations (7) to (11) except that the adaptive prefilter is simplified by setting $\hat{D}_{j-1}(\delta)=1.0$ in equation (11).

Model Order Identification

Model order identification is extremely important in time series modelling. In the case of the system *TF* polynomials $A(\delta)$ and $B(\delta)$, the approach proposed in this paper is based on the following identification statistic (see e.g. Young, 1989),

$$YIC = \log_e\left\{\frac{\hat{\sigma}^2}{\sigma_y^2}\right\} + \log_e\{NEVN\} \tag{12}$$

where,

$\hat{\sigma}^2$ is the sample variance of the model residuals $e(k)$
σ_y^2 is the sample variance of the measured system output $y(k)$ about its mean value.

while *NEVN* is a "*Normalised Error Variance Norm*" (Young et al, 1980) defined as,

$$NEVN = \frac{1}{\pi}\sum_{i=1}^{i=\pi}\frac{\hat{\sigma}^2 p_{ii}}{\hat{a}_i^2}$$

Here, in relation to the *TF* model (1), $\pi=m+n+1$ is the total number of parameters estimated; $\hat{a}_i^2$ the estimate of the ith parameter in the parameter vector $\boldsymbol{a}$; while p_{ii} is the ith diagonal element of the $\boldsymbol{P}(N)$ matrix (so that $\hat{\sigma}^2 p_{ii}$ is an estimate of the error variance associated with the ith parameter estimate after N samples). Examination of this criterion shows that the model which minimises the *YIC* provides a good compromise between goodness of fit and parametric efficiency: as the model order is increased, so the first term tends always to decrease; while the second term tends to decrease at first and then to increase quite markedly when the model becomes over-parameterised and the standard error on its parameter estimates becomes large in relation to the estimated values.

Noise model identification is less critical for most applications and, in the case of the δ operator *AR* or *ARMA* models, the most obvious approach is to utilise the Akaike Information Criterion (*AIC*; Akaike, 1974), which was developed for the more conventional backward shift models but is applicable within the present δ operator context. In relation to the pth order *AR(p)* model, this requires the minimisation of the following criterion,

$$AIC(p) = \log_e \hat{\sigma}^2(p) + \alpha\frac{p}{N}$$

where $\hat{\sigma}^2(p)$ is the variance of the *AR(p)* model residuals, N is the sample size and, for the *AIC*, $\alpha=2$. In other words, the optimal order $\hat{p}$, in this sense, is given by

$$AIC(\hat{p}) = \min\{AIC(p);\, p=1,2,\dots,M\}$$

where M is chosen sufficiently large to encompass all reasonable model orders. However, note that, for control applications, the *AIC* may tend to over identify the model order.

Some Remarks on *RIV/SRIV* Identification and Estimation

(i) The z^{-1} operator versions of the refined *IV* and *SRIV* algorithms are obtained straightforwardly by redefining the model (2) as,

$$y(k) = z(k)^T \mathbf{a} + \eta(k) \tag{13}$$

where now[4],

$$z(k) = [-y(k-1), -y(k-2), \ldots\ldots, -y(k-n), u(k), \ldots\ldots, u(k-m)]^T$$

$$\mathbf{a} = [a_1, a_2, \ldots\ldots, a_n\ b_0\ b_1\ b_2, \ldots., b_m]^T$$

with the IV vector defined by,

$$\hat{\mathbf{x}}(k) = [-\hat{x}(k-1), -\hat{x}(k-2), \ldots\ldots, -\hat{x}(k-n), u(k), \ldots\ldots, u(k-m)]^T$$

and then replacing the various δ operator *TF* definitions by the appropriate z^{-1} equivalents, i.e., $A(z^{-1})$, $B(z^{-1})$, $C(z^{-1})$ and $D(z^{-1})$. The *SRIV* algorithm is then based on these definitions with the appropriate prefiltering introduced on the variables (Young, 1985).

ii) By analogy with the equivalent z^{-1} operator methods (see e.g. Ljung and Soderstrom, 1983; Young, 1976, 1984), it is clear from equation (5a) that, in relation to the full stochastic model (4), the *RIV* algorithm provides an *indirect* approach to "prediction error" minimisation (or Maximum Likelihood estimation if $e(k)$ has Gaussian amplitude distribution). The simplified *SRIV* and *IV* algorithms are, of course, utilised to avoid these more complicated *RIV* procedures. In such situations, the *SRIV* algorithm can be associated with the following "response error" function,

$$\hat{e}(k) = y(k) - \frac{\hat{B}(\delta)}{\hat{A}(\delta)} u(k)$$

which nominally assumes that $\xi(k)$ is white noise (i.e. $C(\delta)=D(\delta)=1$); whereas the basic IV algorithm, without prefiltering, relates to the "equation error" function,

$$\hat{e}(k) = \hat{A}(\delta)\, y(k) - \hat{B}(\delta)\, u(k)$$

which is strictly applicable only in the special situation where $C(\delta) \equiv A(\delta)$ *and* $D(\delta)=1$. Fortunately, of course, the *IV* modification to the *RLS* algorithm ensures that both the *IV* and *SRIV* methods are robust to the violation of such assumptions. However, bearing the above observations in mind, the *SRIV* algorithm should be preferred to the IV method in all situations, except where the passband of the stochastic disturbances is similar to that of the system itself ($C(\delta) \equiv A(\delta)$ *and* $D(\delta)=1$; or $C(z^{-1}) \equiv A(z^{-1})$ and $D(z^{-1})=1$ in the z^{-1} operator case). In these latter, classic, equation error circumstances, we see from equation

4. for convenience, we utilise the same nomenclature for the parameters here as in the δ operator model; the parameters will, of course be different in each model and the relationship between them is given in Chotai et al (1990a)

(6) that no prefiltering is necessary, so that the basic IV algorithm is most appropriate and the *SRIV* may yield inferior results.

(iii) A more detailed discussion on the prefilters in the z^{-1} case is given in Young et al. (1988), including their use in overcoming low frequency noise effects such as drift, bias or load disturbances. Similar arguments apply in the δ operator case.

(iv) The prefiltering approach proposed here is very similar to the "state variable filtering" method proposed previously (Young, 1979, 1981; Young and Jakeman, 1980) for continuous-time systems; which was itself an optimal generalisation of a previous, more heuristic, procedure where the prefilters were designed simply as band-pass filters chosen to span the passband of the system under study (Young, 1964, 1965, 1966, 1969a; see also Gawthrop (1988); Goodwin (1988). Moreover, simple discretisation of these earlier continuous-time designs (as required for their implementation in a digital computer) yields very similar implementations to the present δ operator algorithms. *Nevertheless, the δ operator formulation is more satisfying from a TDC standpoint.*

(v) Within the δ operator model context, the *SRIV* prefilters can also be considered optimal equivalents of the more arbitrary and heuristically defined prefilters proposed by Goodwin (1988) and Middleton and Goodwin (1990). However, the performance of the *SRIV* algorithm is not particularly sensitive to the choice of prefilters and the *SRIV* estimates remain reasonably efficient (see below: Table 1, example 1, in Section 4) even if the filters are not optimal in a theoretical sense. And, as we shall see later, sub-optimal prefilters designed on a more heuristic basis may prove essential in those practical situations where the theoretical assumptions cannot be satisfied; e.g. when the open loop system to be controlled is unstable or marginally stable.

(vi) Research is proceeding on both the initialisation and the convergence of the δ operator IV estimation algorithms. In general, they appear to be not quite so robust as their z^{-1} operator equivalents and the algorithmic performance depends to a greater extent on factors such as the nature of initial prefilter parameters, the initial conditions on the auxiliary model used to generate the instrumental variables, and the number of iterations required for convergence. In the *RIV and IV* algorithms, for example, no more than 3 iterations should be used or some divergence may occur.

(vii) In both the $z^{-1}$and δ operator model cases, it will be noted that all the variables required for each recursive RIV or *SRIV* parameter estimation update are available from the prefilters at any *single* sampling instant. This means that the recursion does not need to be carried out at the same sampling interval as that used by the control algorithm; nor, indeed, does it even need to be at uniformly spaced intervals of time. This attractive feature, which was pointed out by the first author (1966, 1969a), and recently emphasised by Gawthrop (1988), has important practical implications, particularly in self adaptive and self-tuning system design: for example, it can help overcome any problems associated with the computational time required for implementing the adaptive algorithms, and can also allow for easier allocation of time in multi-tasking operations. Moreover, by employing multiple sets of heuristically designed prefilters, each with different bandpass characteristics, it is possible to extract the information required for multiple *SRIV* parameter estimation updates at a single time instant (see e.g. Young, 1964, 1966, 1969a).

Time Variable Parameter (*TVP*) Estimation

The recursive formulation of the *SRIV* algorithm allows immediately for Time Variable Parameter (*TVP*) estimation. There are two basic and interrelated approaches to *TVP* estimation: an "on-line" or "filtering" form involving simple modifications to the recursive estimation algorithm (7); and an "off-line" or "smoothing" form, which exploits fixed-interval recursive smoothing and involves backwards recursion. The on-line algorithms are, of course, important in the implementation of self-adaptive control based on "identification and synthesis", where they provide the sequential updates of the model parameter estimates required to adapt the control system gains. As we shall see, the off-line algorithms are of primary use in the prior analysis of data, where they allow the analyst to investigate nonstationary and/or nonlinear aspects of the system dynamics.

On-line *TVP* versions of the *SRIV* algorithm can be developed in a variety of ways. The two best known approaches (see e.g. Young, 1984) are: (a) "exponential forgetting"; and (b) "modelling the parameter variations". The former is probably the most popular and simply involves the introduction of an exponential data weighting factor which defines the memory of the estimator. However, we prefer the latter, which is a little more complex but adds much greater flexibility. At its most general, the approach assumes that the model parameter vector $\boldsymbol{a} = \boldsymbol{a}(k)$ is changing over time and that its evolution can be described by a stochastic Gauss-Markov (*GM*) process of the form,

$$\boldsymbol{a}(k) = \Phi(k-1)\,\boldsymbol{a}(k-1) + \Gamma(k-1)\,\mu(k-1)$$

where $\Phi(k-1)$ and $\Gamma(k-1)$ are, respectively, state transition and input matrices that are assumed known (and possibly time variable); while $\mu(k)$ is a vector of zero mean, serially uncorrelated random variables with covariance matrix $\boldsymbol{Q}(k)$ introduced to add a stochastic degree of freedom to the equation and allow for the specification of random variations in the parameters that characterise the parameter vector $\boldsymbol{a}(k)$. Normally, $\boldsymbol{Q}(k)$ is chosen as a constant coefficient, diagonal matrix with elements q_{ii}, $i=1,2, \ldots , m+n+1$. When applied within the context of the present recursive algorithms, these elements can be considered as *Noise Variance Ratios* (*NVR*'s) relating the variance of the white noise elements in $\mu(k)$ to the variance of $e(k)$; see e.g references cited below in relation to nonstationary univariate time-series estimation. These *NVR* values define the parameter tracking characteristics of the algorithm: if $\Phi(k)=\Gamma(k)=\boldsymbol{I}$ for all k , where $\boldsymbol{I}$ is the identity matrix (see below), then zero values inform the algorithm that the associated parameters are assumed to be time-invariant; small values (e.g. 10^{-6}) indicate that only slow parameter variations are expected; while larger values allow for more rapid variations but progressively allow the estimates to be contaminated by the effects of the measurement noise as the *NVR* value is increased.

In practice, of course, the specification of the Φ and Γ matrices is difficult since, more often than not, little is known *a priori* about the potential variation in these *TF* model parameters. For this reason, it is common practice in such circumstances to restrict the model to a class of vector random walk models, of which the *Random Walk* (*RW*) and *Integrated Random Walk* (*IRW*) are the best known and used (see e.g. Norton, 1975, 1986; Young, 1988; Young et al, 1989, 1991b; Ng and Young, 1990).

With this *GM* assumption on the nature of the parameter variations, the recursive *SRIV* algorithm in the *TVP* case takes the following prediction-correction form (cf equation (7)),

Prediction:

$$\hat{\boldsymbol{a}}(k/k\text{-}1) = \boldsymbol{\Phi}(k\text{-}1)\,\hat{\boldsymbol{a}}(k\text{-}1) \qquad \text{(i)}$$

$$\boldsymbol{P}(k/k\text{-}1) = \boldsymbol{\Phi}(k\text{-}1)\,\boldsymbol{P}(k\text{-}1)\,\boldsymbol{\Phi}(k\text{-}1)^T + \boldsymbol{\Gamma}(k\text{-}1)\,\boldsymbol{Q}(k\text{-}1)\,\boldsymbol{\Gamma}(k\text{-}1)^T \qquad \text{(ii)}$$

Correction: (14)

$$\hat{\boldsymbol{a}}(k) = \hat{\boldsymbol{a}}(k/k\text{-}1) + \boldsymbol{P}(k/k\text{-}1)\,\hat{\boldsymbol{x}}^*(k)\,\varepsilon(k/k\text{-}1) \qquad \text{(iii)}$$

$$\boldsymbol{P}(k) = \boldsymbol{P}(k/k\text{-}1) - \frac{\boldsymbol{P}(k/k\text{-}1)\,\hat{\boldsymbol{x}}^*(k)\,\boldsymbol{z}^*(k)^T\,\boldsymbol{P}(k/k\text{-}1)}{1 + \boldsymbol{z}^*(k)^T\boldsymbol{P}(k/k\text{-}1)\,\hat{\boldsymbol{x}}^*(k)} \qquad \text{(iv)}$$

where $\varepsilon(k/k\text{-}1)$ is the normalised recursive residual defined as in equation (7)(ii) but with $\hat{\boldsymbol{a}}(k\text{-}1)$ and $\boldsymbol{P}(k\text{-}1)$ replaced by their *a priori* predictions; namely $\hat{\boldsymbol{a}}(k/k\text{-}1)$ and $\boldsymbol{P}(k/k\text{-}1)$ obtained from (i) and (ii).

The off-line fixed interval smoothing algorithm utilises equations (14) in a forward pass through the data to obtain the filtered estimate of the parameter vector at each sample k conditioned on the data up to sample k; in other words, $\hat{\boldsymbol{a}}(k)$ is interpreted explicitly as the estimate at the *kth* sample based on all data up to the *kth* sampling instant, i.e., $\hat{\boldsymbol{a}}(k/k)$ [5]. The smoothed recursive estimate $\hat{\boldsymbol{a}}(k/N)$ (i.e. the estimate at sample k conditioned on *all* N samples of the data) is then obtained from the following *backward* recursions,

$$\hat{\boldsymbol{a}}(k/N) = \hat{\boldsymbol{a}}(k) - \hat{\boldsymbol{P}}(k)\,\boldsymbol{\Phi}(k)^T\,\lambda(k) \qquad \text{(i)}$$

or alternatively, using the more memory efficient form,

$$\hat{\boldsymbol{a}}(k/N) = \boldsymbol{\Phi}(k)^{-1}\left[\hat{\boldsymbol{a}}(k+1/N) + \boldsymbol{\Gamma}(k)\,\boldsymbol{Q}(k)\,\boldsymbol{\Gamma}(k)^T\lambda(k\text{-}1)\right] \qquad \text{(ii)} \qquad (15)$$

where,

$$\lambda(k\text{-}1) = [\boldsymbol{I} - \hat{\boldsymbol{P}}(k)\,\hat{\boldsymbol{x}}^*(k)\,\hat{\boldsymbol{x}}^*(k)^T]^T\,\boldsymbol{r}(k) \qquad \text{(iii)}$$

is a Lagrange multiplier vector with $\boldsymbol{r}(k)$ defined as,

$$\boldsymbol{r}(k) = \boldsymbol{\Phi}(k)^T\lambda(k) - \hat{\boldsymbol{x}}^*(k)\{\hat{\boldsymbol{x}}^*(k) - \hat{\boldsymbol{x}}^*(k)^T\,\boldsymbol{\Phi}(k\text{-}1)\,\hat{\boldsymbol{a}}(k\text{-}1)\}$$

and the backwards recursion initiated at $\lambda(N)=\boldsymbol{0}$.

Although the algorithm (15) appears to be of the conventional fixed interval form

5. note that in the following smoothing equations, we retain the previous, less rigorous nomenclature in order to conserve space in the equations.

(see e.g. Bryson and Ho, 1969; Gelb et al., 1974; Norton, 1975, 1986), it is, in fact, an *IV* modification of the algorithm. In particular, it uses the IV vector $\hat{x}^*(k)$ rather than the data vector $z^*(k)$ and introduces a symmetric matrix $\hat{P}(k)$ to replace the $P(k)$ matrix in (14). This symmetric matrix is obtained using the following procedure:

(1) Apply the *constant parameter SRIV* algorithm (7) to the data.
(2) Define the $Q(k)$ matrix in (14)(ii) as a constant diagonal matrix Q with its *NVR* elements proportional to the diagonal elements of the $P(N)$ matrix obtained at the convergence of the *SRIV* algorithm, i.e.,

$$Q = \gamma \, diag[P(N)]$$

Now repeat step (1) *using the TVP algorithm* (14) with $Q(k)$ set to Q defined above and with γ set initially to unity.
(3) On convergence of step (2), perform an additional iteration of the equation (14) but with $z^*(k)$ replaced everywhere by $\hat{x}^*(k)$, to yield a symmetric $\hat{P}(k)$.
(4) Apply the fixed interval smoothing algorithm (15) to yield the smoothed *TVP* estimates $\hat{a}(k/N)$.
(5) Repeat steps (1) to (4) for different values of γ in order to investigate the nature of the parameter variation for different levels of smoothing.

This rather complicated multi-step *IV* approach proves necessary because the conventional fixed interval recursive smoothing algorithm is defined for normal linear least squares problems and is not entirely appropriate within the present *IV* context[6]; in particular, the $P(k)$ matrix from algorithm (14) is not symmetric, as it would be in the least squares case, and it is necessary to replace it by the symmetric matrix $\hat{P}(k)$ obtained in the manner discussed above. The modification can be justified theoretically by noting (see Pierce 1972; Young 1984, p 184) that, within the context of maximum likelihood, estimation the theoretically optimum $P(k)$ matrix is defined by,

$$P^0(k) = \left[\sum_{i=1}^{i=k} \hat{x}^*(k) \hat{x}^*(k)^T \right]^{-1}$$

where $\hat{x}^*(k)$ is defined as in the *SRIV* algorithm above but with the parameters in the auxiliary model and prefilters set to their *actual* rather than estimated values. It is clear by reference to equation (8), therefore, that the $\hat{P}(k)$ matrix obtained from the additional iteration of the *SRIV* algorithm is an estimate of this $P^0(k)$ matrix, based on the converged *SRIV* parameter estimates.

In practice, the general *TVP* estimation procedure outlined above can be simplified in various ways. First, as we have pointed out, the *GM* models for the parameter variations can normally be restricted to *RW* and *IRW* processes, i.e. the $\Phi(k)=\Phi$ *and* $\Gamma(k)=\Gamma$ matrices are defined as follows[7],

6. it will be noted that the IV algorithm is not obtained directly by least squares minimisation but by modification to the normal least squares algorithm to remove the noise induced bias produced by that algorithm when it is applied to structural time-series models (see Young, 1984)

7. sometimes prior information may be available that suggests more complex GM models for the parameter variations (see e.g Young, 1984, p.84 *et seq*; Chotai et al, 1991 for examples associated with the use of the on-line TVP algorithms in adaptive control applications)

$$\text{RW}: \ \Phi = 1 \ ; \ \Gamma = 1$$

$$\text{IRW}: \ \Phi = \begin{bmatrix} 1 & 1 \\ 0 & 1 \end{bmatrix}; \ \Gamma = \begin{bmatrix} 0 \\ 1 \end{bmatrix}$$

Second, since the *SRIV* algorithm is relatively insensitive to the prefilter characteristics, it is best to use the *SRIV* algorithm with fixed rather than adaptive prefilters. Thirdly, in the case of z^{-1} operator *TF* models, the procedure can be further simplified by omitting the prefilters altogether and replacing step (1) by a single least squares iteration. This is the procedure used in the phloem transport example discussed in Example 5 of Section 4.

Finally, it should be noted that, although in this paper we concentrate on the use of fixed interval smoothing algorithms in *TF* modelling, similar algorithms can be developed for the extrapolation, interpolation and seasonal adjustment of nonstationary univariate data, as discussed in a series of recent papers (e.g. Young 1988, 1989; Young and Ng 1989; Young et al. 1989, 1991b; Ng and Young 1990; Young and Runkle, 1989). These papers also discuss various topics which are relevant to the present paper: these include the interpretation of the fixed interval smoothing algorithms in frequency domain terms; the choice of *NVR* values in *TVP* estimation and fixed interval smoothing; and the use of a technique termed "variance intervention". In this latter case, points of severe change (or even discontinuity) in the parameters and/or their rates of change are identified, either subjectively or objectively, and the relative variance of the white noise input in the appropriate random walk model (as defined by the *NVR*) is increased transiently to allow for the detected change.

4. Simulation and Practical Examples of *SRIV* Modelling

The ultimate test of any identification and estimation procedure is that it works and is easy to use in practice. In this Section, we consider a number of examples based on both real and simulated data. The real data examples demonstrate the practical utility of the general approach and its ability to yield models that are meaningful and useful in scientific terms. The simulation examples, which are considered first, help to evaluate of the statistical performance of the estimation algorithms. They are all based on single stochastic realisations since it has not yet been possible to carry out a full stochastic (Monte Carlo) simulation study. However, as pointed out previously, the δ operator refined IV algorithms described here are directly related to the equivalent z^{-1} operator algorithms and these have been evaluated comprehensively in this manner (see Young and Jakeman, 1979,1980; Young, 1984).

Example 1: A Simulated, Noisy, First Order System

In their recent book, Middleton and Goodwin (*MG*; 1990) discuss δ operator model parameter estimation in some detail[8] and use the following, first order, simulated system

8. surprisingly, *MG* fail completely to place their approach within the context of previous research on continuous-time parameter estimation and provide no references to this extensive literature, which is so germane to their methodlogy: indeed, their approach is simply a particular, discrete-time implementation of continuous-time methods proposed in the early nineteen sixties; and their prefilters are examples of the "state variable filters" proposed at that time (see for example the many references in the present author's review of continuous-time identification and estimation (Young, 1979, 1981)).

as a basic illustration of their ideas,

$$\delta x(t) = -0.1x(t) + 0.1u(t)$$

$$y(t) = x(t) + \xi(t)$$

where $u(t)$ and $\xi(t)$ are two independent white noise sequences, uniformly distributed between ±0.5. As in most previous proposals for continuous time estimation over the past thirty years, *MG* utilise prefilters to avoid direct numerical differentiation. In contrast to the unified approach described in the present chapter, however, they employ fixed parameter prefilters with heuristically chosen characteristics.

MG compare the results obtained from the analysis of *2000* input-output samples using *least squares* estimation algorithms: (a) with a first order model and fixed parameter prefilter polynomial $\hat{A}(\delta)=\delta+1$; (b) with a first order model and $\hat{A}(\delta)=\delta+0.2$; and (c) with a *second* order model and $\hat{A}(\delta)=(\delta+0.2)^2$. Since in this case $\Delta t=1$, the first choice of prefilter is simply equivalent to the forward shift and the *MG* rightly counsel against its use in the context of δ operator *TF* model estimation. The second is heuristically chosen, presumably on the well known basis that it spans the bandwidth of the system, so passing information which is useful to the estimation of the system parameters, whilst attenuating high frequency noise effects on the data. Of these, *MG* recommend the third choice, which is based on the idea of an "extended" model and is probably motivated by the concept of "repeated least squares" analysis, as suggested by Astrom and Eykhoff (1971; see also Young 1984, p.126).

Fig.2 is a typical *200* sample segment of the simulated data, with the noise free signal $x(t)$ shown as a bold line. The *2000* sample set used by *MG* seems a rather long for this level of noise (see below), and so our own results have been obtained with only 1000 samples.

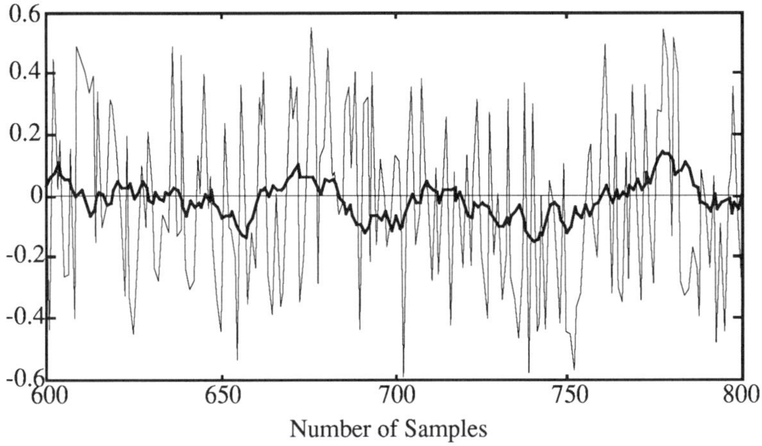

Fig.2 Middleton/Goodwin (*MG*) Example: comparison of noise free output (full) with noisy output (light).

Table 1 : Comparison of *SRIV* and *MG* Estimation Results

Method	$\hat{a}_1$	$\hat{b}_0$	R_T^2	*YIC*
True	*0.1*	*0.1*		
SRIV	*0.0948 (0.0217)*	*0.1019 (0.0189)*	*0.045*	*-3.182*
MG	*0.2008 (0.0182)*	*0.1170 (0.0203)*	*0.034*	*-3.995*
SRIV (FP)	*0.0622 (0.0405)*	*0.0776 (0.0236)*	*0.042*	*-1.397*

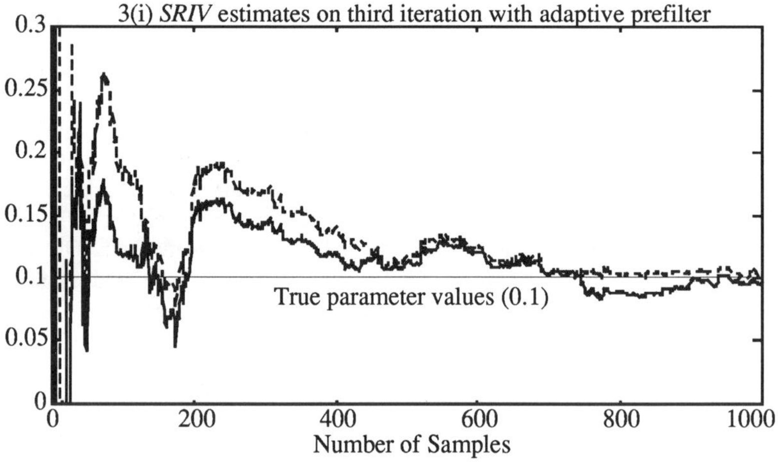

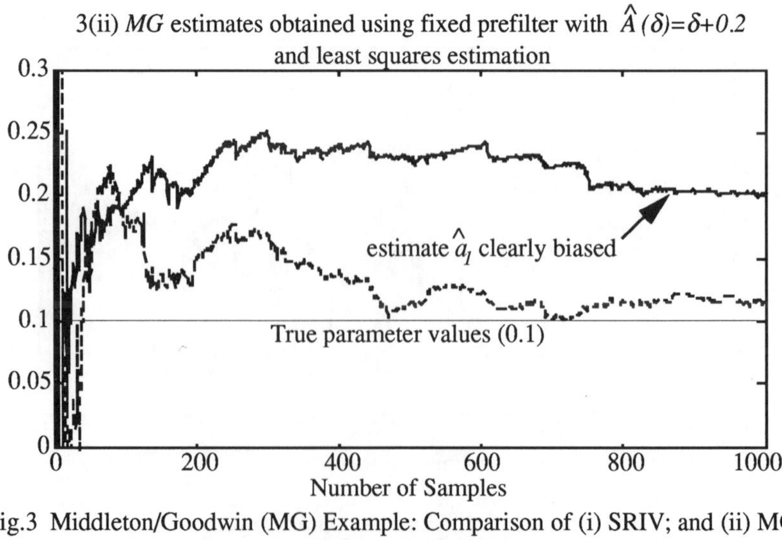

Fig.3 Middleton/Goodwin (MG) Example: Comparison of (i) SRIV; and (ii) MG recursive parameter estimates; $\hat{a}_1$ (full), $\hat{b}_0$ (dashed); fig. cont. over page

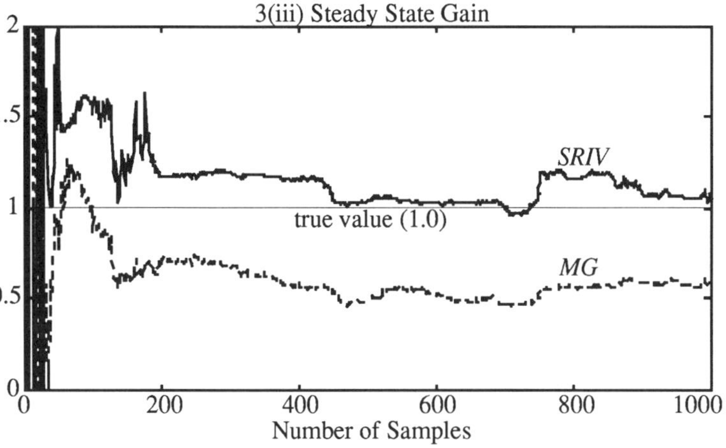

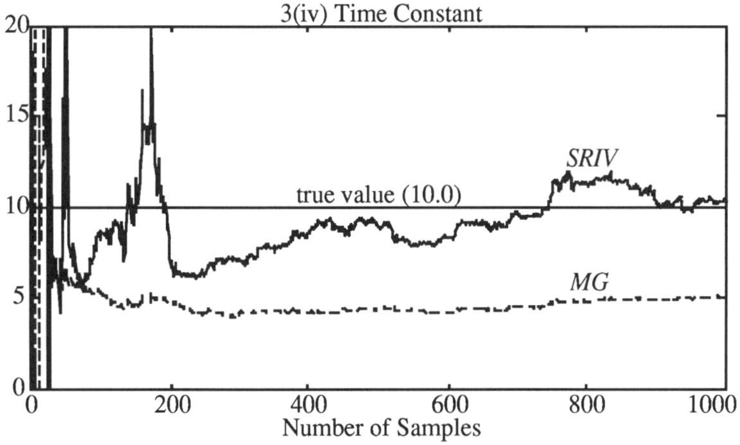

Fig.3 cont. Middleton/Goodwin Example: comparison of *SRIV* and *MG* recursive estimates of (iii) steady state gain; and (iv) time constant

The *YIC* clearly identifies a first order model: the final adaptive prefilter *SRIV* parameter estimates after *1000* samples, together with their standard errors (shown in parentheses) are compared above in Table 1 with the estimates obtained using the *MG* method (b); and the *SRIV(FP)* results obtained with a *fixed* prefilter set to $\hat{A}(\delta)=\delta+0.2$: The low R_T^2 values in all cases are, of course, a direct consequence of the low signal/noise ratio in thesesimulation experiments.

Two points are worth noting from the results shown in Table 1. First, as usual in cases where least squares estimation is applied incorrectly in structural model or "errors-in-variables" situations (see Young, 1984), the *MG* estimates are not only asymptotically biased (see below) but the computed standard errors indicate rather erroneously high accuracy. Second, the fixed prefilter *SRIV(FP)* estimates appear asymptotically unbiased, i.e. the errors in the estimates are ecompassed by their computed standard errors; and these standard errors are noticeably larger for $\hat{a}_1$ estimate than those obtained

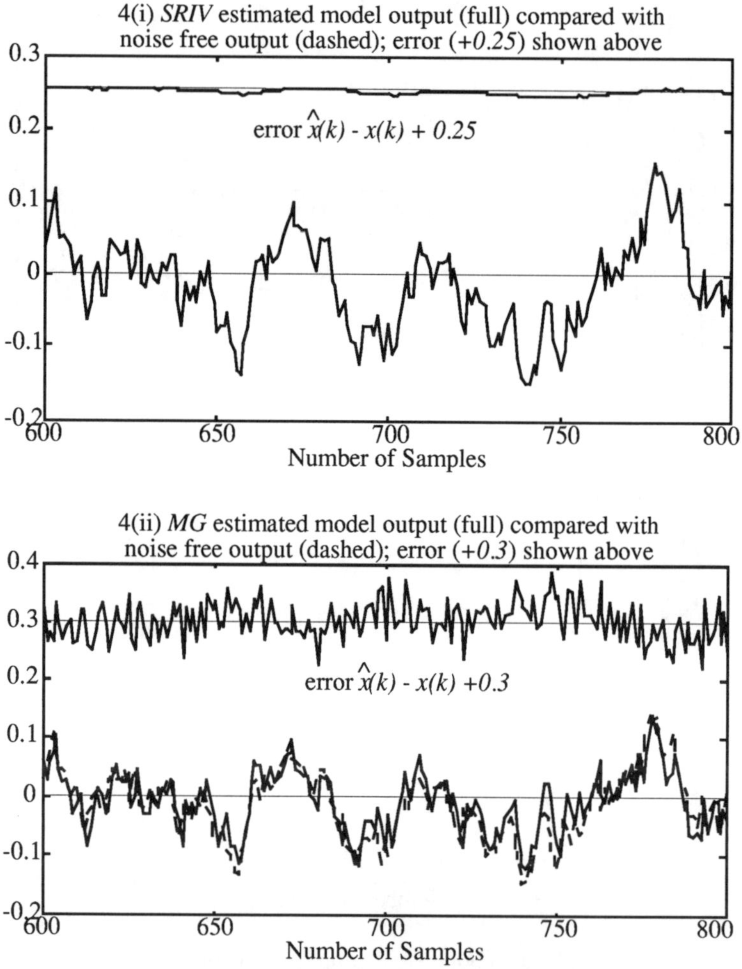

Fig.4. Middleton/Goodwin Example: comparison of (i) *SRIV* and (ii) *MG* recursive estimation results.

in the *MG* case (this accounts for the substantially less negative *YIC* value of *-1.397* in this case). However, the results are still rather poor in comparison with those obtained using the adaptive prefilter *SRIV* method and demonstrate the general superiority of this more sophisticated approach.

Figs.3 and 4 examine the estimation results more closely. First, Fig.3 compares the recursive estimates obtained during the 3rd iteration of the *SRIV* algorithm in 3(i) with those obtained using the *MG* procedure (b) in 3(ii). This confirms that the *MG* estimates are rather poor, with clear asymptotic bias on the $\hat{a}_1$ parameter estimate. In contrast, the *SRIV* estimates, obtained after 3 adaptive iterations *starting* with the same prefilter (i.e. $\hat{A}(\delta)=\delta+0.2$), are unbiased and have converged quite reasonably to the region of the *0.1* true values after about *700* samples. Indeed, if we consider the steady state gain and time

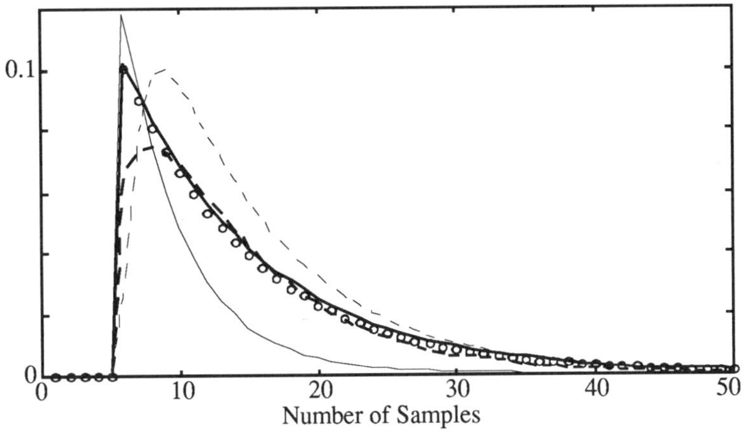

Fig.5 Middleton/Goodwin Example: comparison of various model impulse responses: (a) actual (circles); (b) *SRIV* model (full); (c) *MG* second order model after *2000* samples (dashed); (d) *MG* second order model after *1000* samples (light dashed, poor fit); (e) *MG* first order model (light, poor fit)

constant estimates obtained from these parameter estimates, as shown in Figs.3(iii) and (iv), then satisfactory convergence has occurred in about 400 samples, which has implications for self-tuning and self-adaptive control applications.

This superiority of the *SRIV* estimation is further illustrated in Figure 4, where the model outputs $\hat{x}(t)$, based on the final parameter estimates at $k=1000$, are compared with the noise free outputs $x(t)$: the error between them is shown above (raised by adding 0.25 for clarity). Here, the *SRIV* estimated model in 4(i) is clearly much superior, with the model output (shown dashed) hardly visible and the error very small over the whole observation interval. In contrast, the *MG* model in 4(ii) has quite large residuals with considerable autocorrrelation.

Finally, in Fig.5, the impulse responses obtained from various models are compared with the the actual impulse response (shown as circles in plot (5(a)). Here, the first order *MG* estimated model discussed above and shown in plot 5(e) is particularly poor, mainly as the result of the asymptotic bias on the estimates. The *MG* estimated second order model response (plot 5(c)) is much better and compares reasonably, in the exponential recession part of the response, with the *SRIV* model response (plot 5(b)), although the *SRIV* model would clearly be preferred since it is minimally parameterised and, therefore has better defined estimates[9]. Also, it is interesting to note that the *MG* second order model response in 5(c) was obtained after *2000* samples: as we see in plot 5(d), the results are not nearly so good after *1000* samples, and this is probably why *MG* used the larger sample size.

9. *MG* suggest that they would utilise a first order approximant to this second order model which is then similar to the estimated SRIV first order model: however, it is unclear to the present author exactly how *MG* obtained this approximant or how they would map the statistical uncertainty in the second order model to the first order model

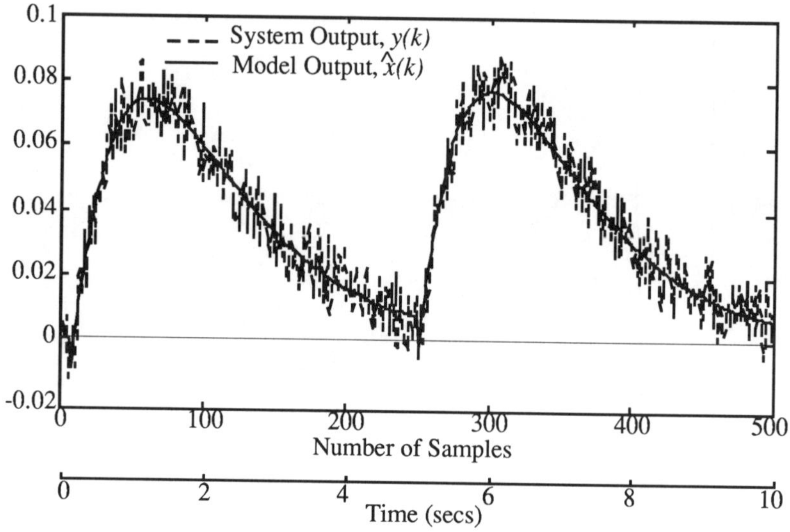

Fig. 6 Comparison of *SRIV* estimated model output $\hat{x}(k)$ with measured output series $y(k)$.

Table 2 : *SRIV* Identification Results

MODEL	*YIC*	R_T^2
1,1,0	*-4.83*	*0.341*
1,2,0	*0.025*	*0.341*
2,1,0	*-11.05*[1]	*0.919*
2,2,0	*-7.04*[2]	*0.920*
2,3,0	*-3.01*	*0.921*
3,2,0	*-5.76*	*0.920*
3,3,0	*-4.98*	*0.922*
4,4,0	*-4.84*	*0.922*

[1] denotes best identified model
[2] denotes 2nd best identified model

Example 2: Simulated, Noisy, Second Order System

As a second simulation example, let us consider the identification and estimation of the following non-minimum phase, second order, system,

$$y(k) = \frac{-0.05\delta + 1}{\delta^2 + 2\delta + 1}\, u(k)$$

with a sampling interval $\Delta t=0.02$ *secs*. Fig.6 shows the output $y(k)$, $k=1,2, \ldots ,500$ (i.e. *10* seconds in real time) from a simulated identification experiment in which the input $u(k)$ consists of two impulse signals of magnitude *10* applied at $k=7$ and $k=250$, respectively. Here, the additive noise $\xi(k)$ is in the form of zero mean, white noise with the noise/signal ratio set to about 10%. The rather poor excitation signal in this case is selected to show how quite reasonable results can be obtained, even when experimental restrictions limit the nature of the planned identification experiments. Indeed, satisfactory results can be obtained in this example even when only a single impulsive input is utilised.

Table 2 shows the results obtained from the *SRIV* identification analysis: The best identified model in *YIC* terms is actually the *(2,1,0)* transfer function ($YIC=-11.05$; $R_T^2=0.919$), with the correct order *(2,2,0)* model displaying a marginally better fit ($R_T^2=0.922$) but a considerably less negative $YIC=-7.04$. These slightly ambiguous results are easily explained by the nature of the system and the experimental data in this example. In particular, the input excitation is rather poor as regards the identification of the non-minimum phase (*NMP*) characteristics, which are relatively insignificant in relation to the noise effects. Not surprisingly, therefore, the second order model, without these *NMP* characteristics, is preferred by the *YIC*, since it provides almost as good an explanation of the data, but with better defined parameter estimates (largely because the poorly estimate of the *NMP* parameter b_0 is absent). In this regard, it is interesting to note that, with the b_0 parameter changed to *-0.3*, the position is reversed with the *(2,2,0)* model now clearly the best identified both in terms of *YIC* and R_T^2.

The parameter estimates and their standard errors for the two best identified models are given below, together with the associated, δ operator defined, steady state gain (*SSG*; actual value *1.0*); natural period (P_n; actual value *6.28* secs./cycle); and Damping Ratio (ζ; actual value *1.0*):

(2,1,0) Model

$$\hat{a}_1 = 1.841(0.029)\ ;\ \hat{a}_2 = 0.937\ (0.011)\ ;\ \hat{b}_0 = 0.929\ (0.013)$$

$$SSG=0.991;\ P_n = 6.49;\ \zeta = 0.951$$

(2,2,0) Model

$$\hat{a}_1 = 2.027(0.049)\ ;\ \hat{a}_2 = 1.018\ (0.020)\ ;\ \hat{b}_0 = -0.044\ (0.009)\ ;\ \hat{b}_1 = 1.011\ (0.021)$$

$$SSG=0.992;\ P_n = 6.23;\ \zeta = 1.004$$

The full estimation results (i.e. the estimated parameter vector and associated covariance matrix for the *(2,2,0)* model) are given below:

$$\hat{\mathbf{a}} = \begin{bmatrix} 2.027 \\ 1.018 \\ -0.044 \\ 1.011 \end{bmatrix} \quad \mathbf{P}^* = 10^{-3} \begin{bmatrix} 2.3875 & 0.8694 & -0.3521 & 0.9754 \\ 0.8694 & 0.4106 & -0.1536 & 0.3946 \\ -0.3521 & 0.1536 & 0.0820 & -0.1550 \\ 0.9754 & 0.3946 & -0.1550 & 0.4450 \end{bmatrix}$$

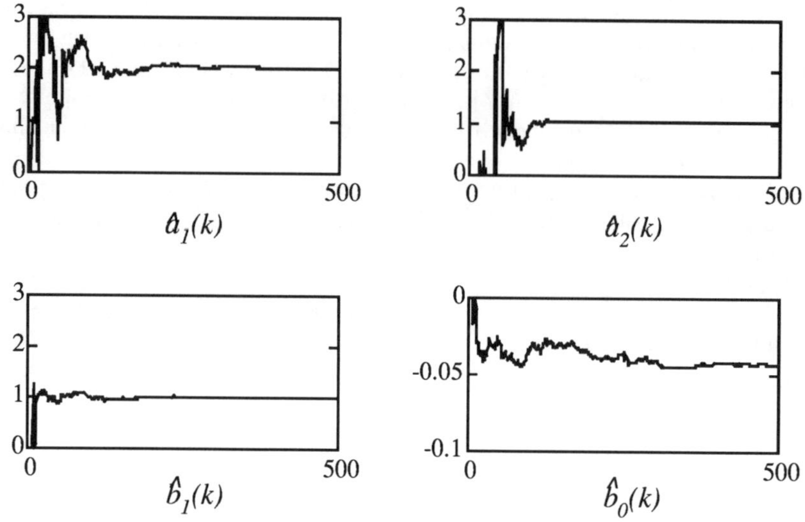

Fig. 7 Recursive estimates of parameters in *(2,2,0)* δ operator transfer function model

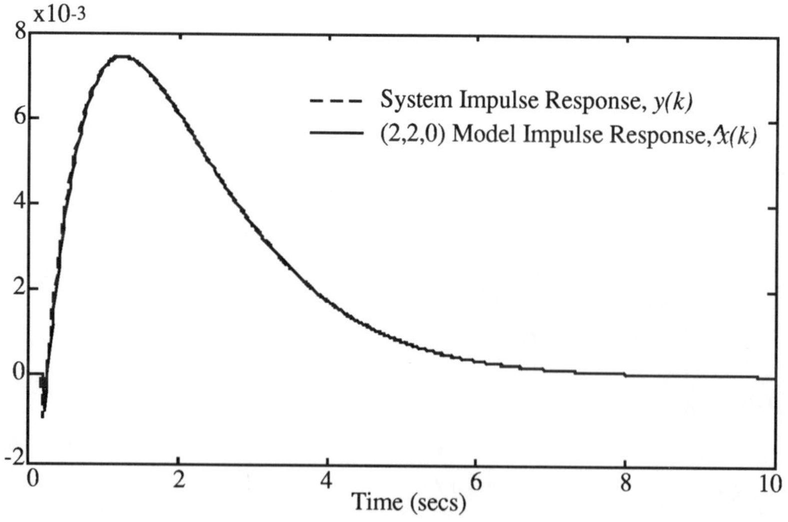

Fig. 8 Comparison of *SRIV* estimated model unit impulse response $\hat{x}(k)$ with the actual impulse response $y(k)$

These results were obtained after only 1 iteration (i.e the first, least squares iteration) of the *SRIV* algorithm with the initial estimates of the model parameters (and those of the pre-filter) all set to unity. The recursive estimates of the parameters are shown in Fig.7, where we see that, except for the estimate of the non-minimum phase parameter b_0, the estimates converge fairly quickly and are very well defined. For illustration, the *(2,2,0)* model ouput is compared with the measured ouput in Fig.6, while the impulse response of this model is compared with the actual impulse response in Fig.8. These results demonstrate that, in contrast to the previous high noise example, simple least squares with a fixed parameter, sub-optimal prefilter can yield quite acceptable results *if the noise level is relatively low.* Again, this has important implications if these algorithms are used for self-tuning and self-adaptive control system design.

Very similar results, but naturally with larger standard errors, are obtained with the estimation sampling interval set larger than the sampling interval Δt: e.g with a recursion interval of $3\Delta t$ (i.e.recursive updates every *0.06* secs.), the estimates for the *(2,2,0)* model are as follows,

***(2,2,0)* Model (sampling interval $3\Delta t$)**

$$\hat{a}_1 = 2.023(0.098)\ ;\ \hat{a}_2 = 1.041\ (0.041)\ ;\ \hat{b}_0 = -0.065(0.0180)\ ;\ \hat{b}_1 = 1.017\ (0.043)$$

$$SSG=0.977;\ P_n = 6.159;\ \zeta = 0.991$$

In this case, the model does not explain the data quite so well, with $R_T^2=0.889$ and $YIC=-6.085$.

Finally, it is not surpring to find that the combination the high sampling rate of 0.02 secs and measurement noise make identification and estimation of an alternative z^{-1} operator model rather difficult: LS estimation yields a very poor model; while ordinary IV estimation fails completely to converge on sensible estimates. However, the adaptive prefiltering in the z^{-1} *SRIV* algorithm cures these problems and yields a quite reasonable model, with $R_T^2=0.947$ and

$$\mathbf{a}=[1,\ -1.95932,\ 0.95972,\ -0.00136,\ 0.00176]^T$$

However, because of the high sampling frequency, the roots are very near the unit circle in the complex z plane: consequently, the model response is quite sensitive to the modelling uncertainty and the model will not be particularly good for certain applications such as control system design (see Tych et al, 1991, where this model is compared with the δ operator model for control system design).

Example 3: Univariate Noise Modelling

Fig.9 is a plot of a simulated noise series generated by the following δ operator *AR(2)* model at a sampling interval $\Delta t=0.1$ time units,

$$\xi(k) = \frac{1}{\delta^2 + \delta + 1}\, e(k)$$

where $e(k)$ is a zero mean, white noise sequence with unity variance. Also shown on the plot are the one-step ahead prediction errors obtained from the *SRIV* estimated model after 50 time units (500 samples). The recursive estimates of the model parameters are given in Fig.10; and Fig.11 compares the impulse response of the estimated model with the actual impulse response.

These results, which were obtained using the recursive least squares algorithm described in Section 3, seem quite satisfactory and compare well with the estimates

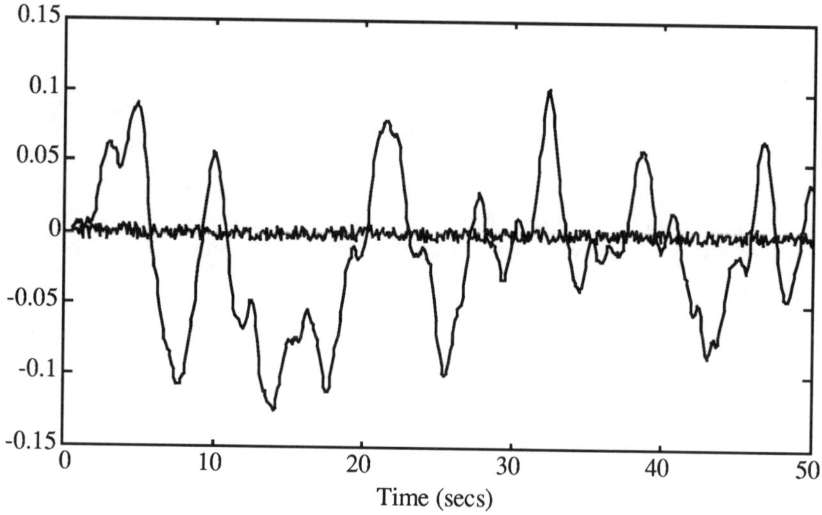

Fig.9 Delta Operator *AR* modelling Example: data and one step ahead prediction errors for *AR(2)* mode.

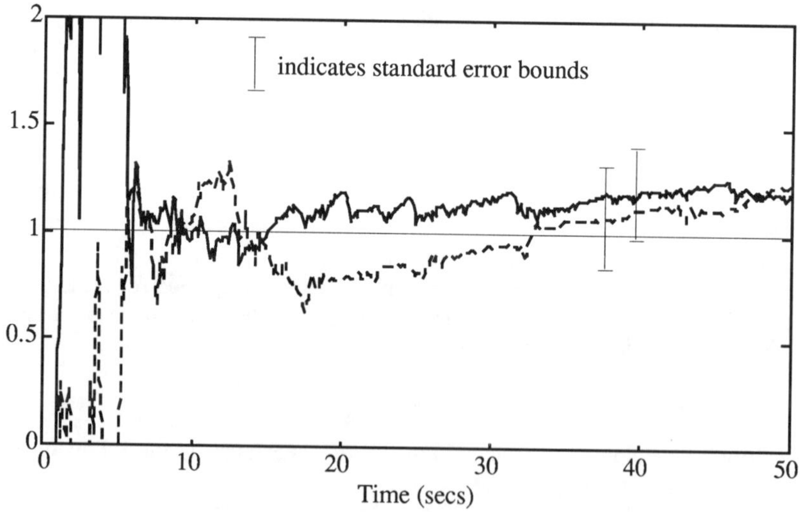

Fig.10 Delta Operator *AR* modelling Example: recursive estimates of *AR(2)* parameters with standard error bounds (a) $\hat{c}_1$ (full); $\hat{c}_2$ (dashed).

obtained by first estimating the z^{-1} operator *AR(2)* model parameters and then converting the model to δ operator form. The results of this exercise are as follows

(i) z^{-1} operator *AR(2)* model: $C(z^{-1}) = 1 - 1.8794z^{-1} + 0.8919z^{-2}$

(ii) δ operator form of (i): $C(\delta) = \delta^2 + 1.2063\,\delta + 1.2490$

(ii) estimated δ operator model: $C(\delta) = \delta^2 + 1.1976\,\delta + 1.2441$

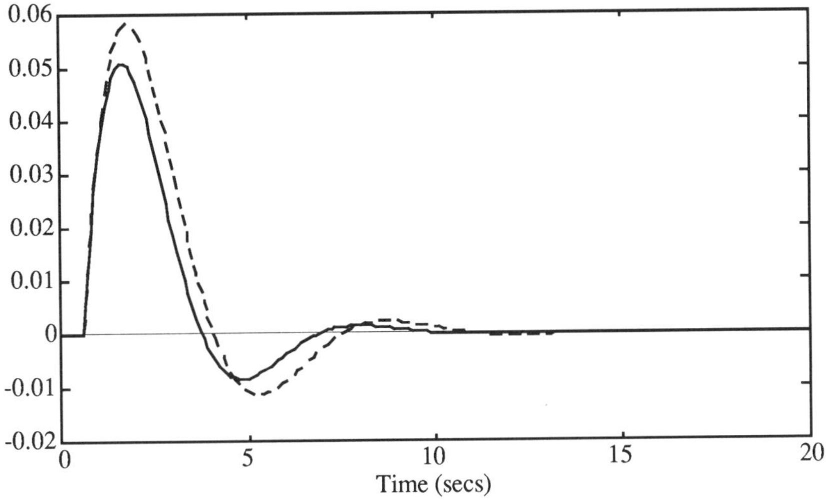

Fig.11 Delta Operator *AR* modelling Example: true impulse response (dashed) compared with estimated *AR(2)* impulse response (full).

Both models (i) and (iii) explain the data well with coefficient of determination $R_T^2=0.997$, although this is, of course, biased towards unity because of the high sampling rate.

Example 4: The *SRIV* algorithm and Model Reduction

In a previous research study (Tych and Young, 1990), we have demonstrated how the z^{-1} operator *SRIV* algorithm can be very useful as a tool in data-based model reduction. Fig.12(i) shows the step response of an eighth order, z^{-1} operator transfer function model (see Shieh, 1975). The *TF* numerator and denominator polynomials in this model are as follows,

$$A(z^{-1}) = 1 - 0.63075z^{-1} - 0.4185z^{-2} + 0.07875z^{-3} - 0.057z^{-4} + 0.1935z^{-5} + 0.09825z^{-6} - 0.0165z^{-7} + 0.00225z^{-8}$$

$$B(z^{-1}) = 0.1625z^{-1} + 0.125z^{-2} - 0.0025z^{-3} + 0.00525z^{-4} - 0.022625z^{-5} - 0.000875z^{-6} + 0.003z^{-7} - 0.0004125z^{-8}$$

The reduced order z^{-1} polynomials obtained by Liaw et al (1986) are given by,

$$A(z^{-1}) = 1 - 1.7598z^{-1} + 0.8341z^{-2} \quad ; \quad B(z^{-1}) = 0.1625z^{-1} - 0.0825z^{-2}$$

whilst those obtained using the z^{-1} operator *SRIV* approach are,

$$A(z^{-1}) = 1 - 1.7568z^{-1} + 0.8309z^{-2} \quad ; \quad B(z^{-1}) = 0.1669z^{-1} - 0.0872z^{-2}$$

Here, for comparison, we simply apply the δ operator *SRIV* algorithm to the step response data set and allow the *YIC* criterion to identify the best low order model. This confirms that the second order model is best and yields the following estimates of the δ

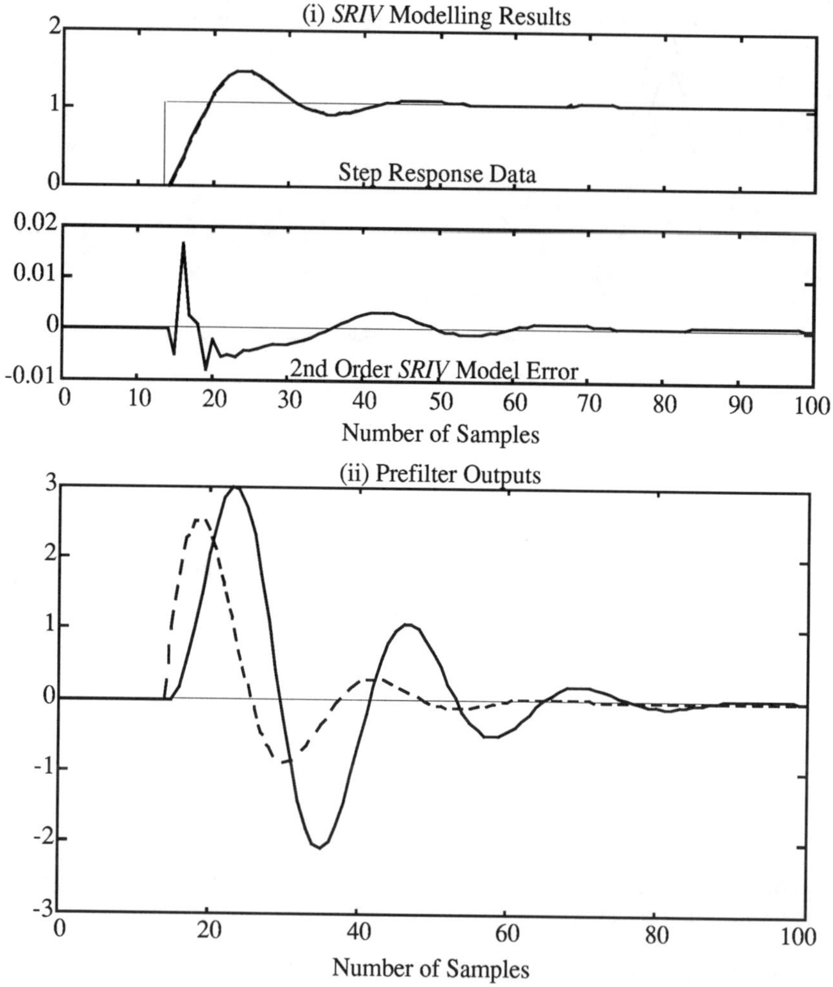

Fig. 12 Model Reduction Example: (i) *SRIV* model output (full), 8th order Shieh model response (dashed, but hardly visible), and 2nd order *SRIV* model error; (ii) prefiltered first derivatives of the output (full) and input (dashed) signals.

polynomials,

$$A(\delta) = \delta^2 + 0.2414\ \delta + 0.07356 \quad ; \qquad B(\delta) = 0.1673\ \delta + 0.0792$$

which, when converted to z^{-1} operator form for comparison with the above results, yields,

$$A(z^{-1}) = 1 - 1.7586z^{-1} + 0.8322z^{-2} \quad ; \qquad B(z^{-1}) = 0.1673z^{-1} - 0.0881z^{-2}$$

This model has a coefficient of determination R_T^2 very close to unity and the very small residual error shown in the lower plot of Fig.12(i). In the z domain, its poles are at

0.8793±0.24296j, which can be compared with *0.8784±0.24354j* for the z^{-1} operator *SRIV* model; and *0.8799±0.2470j* for the Liaw method. A graph of the the adaptively prefiltered first derivatives of the input and output signals obtained in the final, third iteration of the *SRIV* algorithm is given in Fig.12(ii).

Example 5: Environmental and Ecological Modelling

In this Section, we consider a number of practical examples which illustrate the value of *SRIV* identification and estimation in a variety of different applications. In particular, they illustrate how objective, data-based, time-series analysis, involving transfer function models, can reveal important physical aspects of natural systems that are not immediately obvious from the original measured data. The first two examples, which concern the modelling of solute transport and dispersion processes, have been described elsewhere and so are discussed here only briefly. The third example is considered in more detail : it shows how the *SRIV* algorithm has been particularly successful in revealing the parallel nature of the relationship between rainfall and flow in a typical river catchment.

ADZ Modelling of Pollution Transport and Dispersion in Rivers and Soils

The *SRIV* algorithm has proven extremely useful in a number of hydrological applications (Young and Wallis 1985; Young 1986). For example, it provided the main modelling tool in the development of the "Aggregated Dead Zone" (*ADZ*) model for the transport and dispersion of pollutants in river systems (e.g. Young and Wallis 1986; Wallis et al. 1989). The basic element of this model is a simple first order *TF* characterised by a time constant (or "residence time") T and a pure "advective" time delay τ. The first comprehensive evaluation of the model was carried out on four rivers in North West England and was supported by the U.K. Natural Environment Research Council (*NERC*). This included many *in situ* field experiments, in the form of tracer experiments based on the impulsive injection of the fluorescent dye Rhodamine WT. The resulting impulse response data were then used for *SRIV* identification and estimation of the *ADZ* model parameters.

This research programme led to some interesting scientific conclusions. However, the results of the study have been reported in the above references and so it will suffice here to mention only two of the most significant findings of the study. First, a composite parameter of the model, termed the *ADZ dispersive fraction*, is defined as the ratio of the model time constant T to the sum of T and the advective time delay τ. In many cases, this appears to be an important *flow-invariant* property of the river channel, which is able to define the dispersive nature of the stream channel over most of the normal flow régime. Second, the *SRIV* modelling revealed some evidence of a parallel structure, defined by two first order *ADZ* elements with rather different residence times: one quite short, which appears to describe the dispersive behaviour of the main flow pathways; and the other relatively long, which is believed to be associated with the transport of solute through the cobbled bed of the river.

So far, the *ADZ* model has been estimated in terms of the z^{-1} operator transfer functions. However, since the model has more direct physical meaning when considered as a continuous-time differential equation, there is clearly some advantage in using the δ

operator model. A typical example is given in Fig.13, which shows the results obtained from modelling an incomplete set of data obtained from a tracer study on the River Conder, near Lancaster. Here the model is identified as third order, with a pure time delay of *2* sampling intervals and the following estimated parameters,

$$\hat{a}_1 = 1.0844\ (0.033);\ \hat{a}_2 = 0.3886\ (0.0166);\ \hat{a}_3 = 0.0382\ (0.0013);\ \hat{b}_0 = 0.0384\ (0.00125)$$
$$R_T^2 = 0.999:\ YIC = -14.843$$

This model has a steady state gain of 0.996 which is insignificantly different from unity and indicates that the tracer is 'conservative' (i.e.no loss of tracer occurs during the experiment) and, although the *TF* denominator has a mildly oscillatory mode (the roots are at $-0.4635 \pm 0.1669j;\ -0.1574$), it can be approximated by three first order *ADZ* models *in series*, each with an approximate residence time of 2.8 sampling intervals and pure time delay of 0.667 sampling intervals: i.e., in this case, no parallel activity is identified.

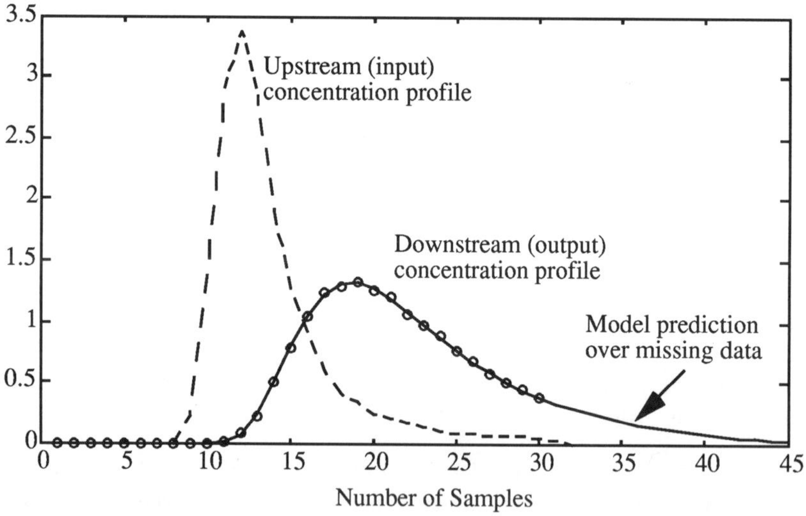

Fig.13 River dispersion example: comparison of *SRIV* estimated δ operator *ADZ* model output (full) with measured output (circles); showing also measured input (dashed) and model prediction of output.

It is interesting to see how, in this example, the objective, data-based approach to modelling proposed in this paper has been able to assist the scientist in making a useful hypothesis about a natural system. In this regard, it is worth noting that when the parallel pathways are revealed by the model, they are not at all obvious, either from observation of the river itself or by visual examination of the experimental concentration-time profiles: the presence of the parallel structure is only revealed by a small elevation in the tail of the measured dye concentration profile, which is rather insignificant to the eye (see Wallis et al, 1989).

The *ADZ* model has been applied in other areas of scientific research and is proving to be similarly useful. One recent example is its use in a related hydrological area; namely the modelling of soil-water processes and, in particular, the transport of solutes through saturated soils (Beven and Young, 1988). Another stimulating example, which we consider next, is the modelling of solute transport in plants, based on data obtained from radioactive tracer experiments.

Modelling of Phloem Translocation and Carbon Partitioning in Plants

Over the past decade, Minchin and his co-workers in New Zealand have developed novel experimental procedures for the *in vivo* measurement and modelling of phloem translocation and carbon partitioning in plants, using the short-lived isotope Carbon-11 as a tracer material (e.g.Minchin and Troughton, 1980; Gould et al., 1988; Minchin and Thorpe, 1988; Minchin and Grusak, 1988). These studies have exploited the recursive methods of time-series analysis discussed in this paper, both for identifying appropriately parameterised z^{-1} operator *TF* models and for estimating time variable parameters in these models. Such *TVP* information has proven useful for the investigation of problems such as the effects of airborne pollutants on the translocation characteristics and the temporal variation of the carbon flows within the plant.

In a particular experiment discussed in Young and Minchin (1991), the input represents the concentration of tracer being exported from the leaf to the rest of the plant; while the output is the concentration of the tracer in the total root system. The experiment involves repeated, impulsive applications of the radioactive tracer and the analysis includes automatic compensation for the natural decay of the tracer. The model is identified from the experimental data as a first order *TF* model in the z^{-1} operator, with no pure time delay and significant variations in both the a_1 and b_0 parameters. The estimates of these parameters were obtained using the simplest version of the fixed interval smoothing algorithm described in Section 3. (equations (14) and (15)). The resulting *TVP* model provides a good fit to the measured output but, more importantly, the variation of the estimated steady state gain is statistically significant and reveals the changes in partitioning occurring over the time of the experiment. In particular, the gain is seen to reduce over the middle period of the experiment, indicating that less labelled material is being transported from the leaf to the root over this period than at the beginning or end of the experiment.

Systems Models of Rainfall-Flow Processes

The modelling of rainfall flow processes is a major task of hydrologists who require such models for applications such as flow and flood forecasting (see e.g. Weyman, 1975; Kraijenhoff and Moll, 1986). The physical processes involved are nonlinear, since "antecedent" rainfall conditions clearly affect the subsequent flow behaviour. In particular, if the prior rainfall has been sufficient to thoroughly wet the soil in the catchment, then river flow will be significantly higher than if the soil had dried out through lack of rainfall. Such "soil moisture" nonlinearity is well known in hydrology and various models have been proposed, from the simple "antecedent precipitation index" (Weyman, 1975) approach, to the construction of large deterministic catchment simulation models such as the "Stanford Watershed Model" (see e.g. Kraijenhoff and

Moll, 1986, p30). The systems contribution to the modelling of such processes is linked with the development of discrete-time *TF* representation of rainfall-flow dynamics. But since such models are essentially linear, it proves necessary to introduce the soil moisture nonlinearity in some manner.

Young (1975), Whitehead et al (1979) and Jakeman et al. (1990), for example, have proposed the following nonlinear model for effective rainfall $u(k)$ at a *kth* discrete instant of time,

$$u(k) = S(k)r^*(k) \qquad \text{(i)}$$

$$S(k) = S(k-1) + [1/T_s]\{r^*(k) - S(k-1)\} \qquad \text{(ii)}$$

$$r^*(k) = \gamma(T_m - T_i)r(k) \qquad \text{(iii)}$$

where $r(k)$ is the rainfall; $S(k)$ is a measure of soil moisture; T_m-T_i is the difference between the overall maximum temperature T_m and the monthly mean temperature T_i for the month in which the *kth* observation is being taken; $r^*(k)$ is a transformed measure of rainfall which allows for temperature induced, evapo-transpiration effects; and γ is a parameter associated with this evapo-transpirative transformation.

The overall effect of this nonlinear transformation is that equation (iii) yields a rainfall measure $r^*(k)$ which is adjusted to allow for the predominantly seasonal evapo-transpiration losses; equation (ii), which is simply the low pass filtering (exponential smoothing) of $r^*(k)$ with time constant T_s, provides a measure of soil moisture with T_s representing a "time constant" associated with the wetting-drying processes; and the multiplicative nonlinear transform in (i) yields an effective rainfall measure which is now compensated for both of the major physical processes thought to be involved in the rainfall-flow dynamic system.

In practice, over shorter periods of time, the evapo-transpirative effect is small, so that $r^*(k)$ in (ii) can simply be replaced by $r(k)$; in which case the nonlinearity is assumed to depend entirely on the soil moisture changes and, in particular, on the time constant T_s. Recognising that river flow $y(k)$ is itself naturally low pass filtered rainfall, the first author (Young, 1990; Young and Beven, 1991) has proposed an alternative and simpler "bilinear" model in which $r^*(k)$ in (i) is replaced by the flow measure $y(k-\tau)$, where τ is normally zero, so that there are no longer any unknown parameters in the nonlinear part of the model. In effect, this is equivalent to assuming that the natural time constant associated with the rainfall-flow process will also reflect the soil moisture dynamics.

Fig. 14 is a plot of the resultant "effective rainfall" or "rainfall excess" for a typical catchment in Wales, with $\tau=0$: in relation to the original measured rainfall (see Young and Minchin, 1991), the main effect of the nonlinear transform is to adjust the magnitudes of the rainfall peaks to reflect the prevailing flow at the same sampling instant, which is plotted as the dashed line in Fig.14; so amplifying peaks where flow is high and attenuating those where flow is low.

Having accounted for most of the nonlinear aspects of the rainfall-flow process, it is now possible to model the remaining *linear dynamic* relationship between the effective

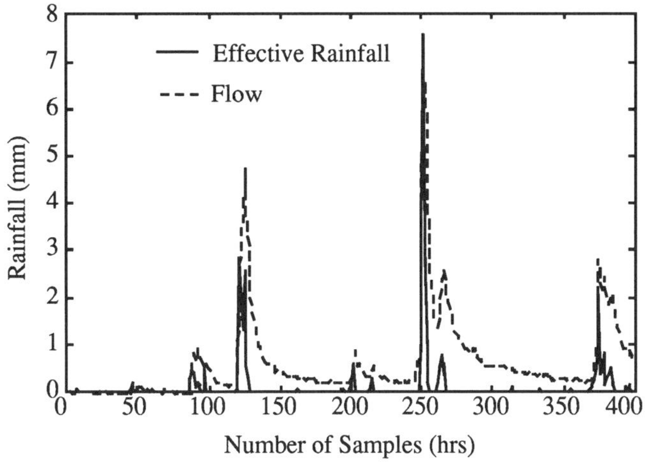

Fig. 14 Effective rainfall measure obtained from nonlinear transformation of the measured rainfall series.

rainfall in Fig.14 and the flow using *SRIV* identification and estimation. The resulting δ operator *TF* model is identified as second order and takes the form,

$$y(k) = \frac{19.5821\,\delta + 0.6927}{\delta^2 + 0.2626\,\delta + 0.00278}\,u(k) \tag{16}$$

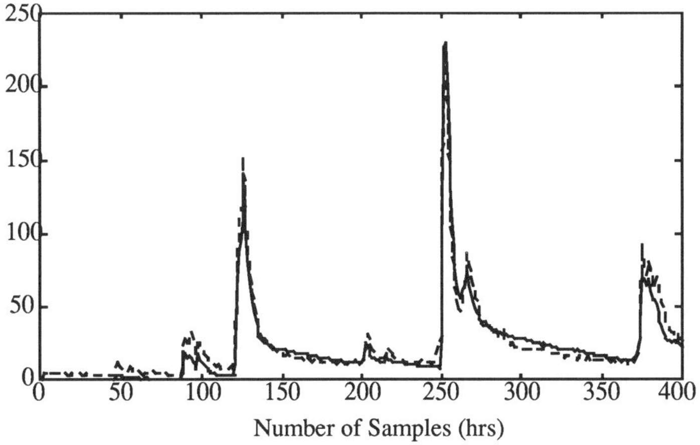

Fig.15 Rainfall-flow example: comparison of *SRIV* estimated delta operator model output (full) with measured output (dashed)

which yields a coefficient of determination $R_T^2=0.943$ and $YIC=-8.2$. A comparison of the model output and the measured flow is given in Fig.15.

This example is particularly interesting since it is easily shown that the transfer

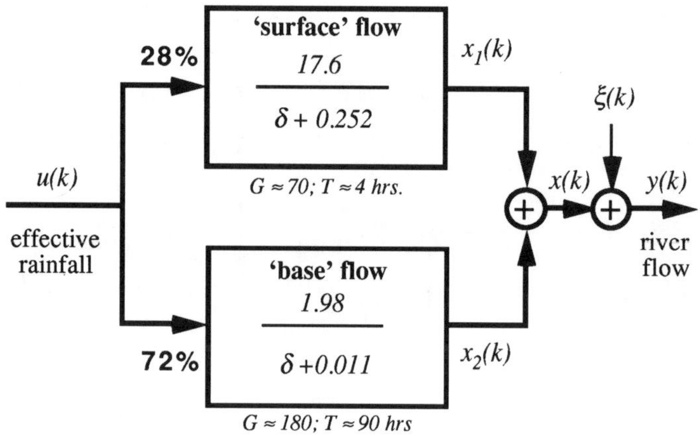

Fig.16 Rainfall-flow example: The delta operator *TF* model considered as a parallel connection of two first order processes (*G* denotes steady state gain; and *T* the time constant or residence time)

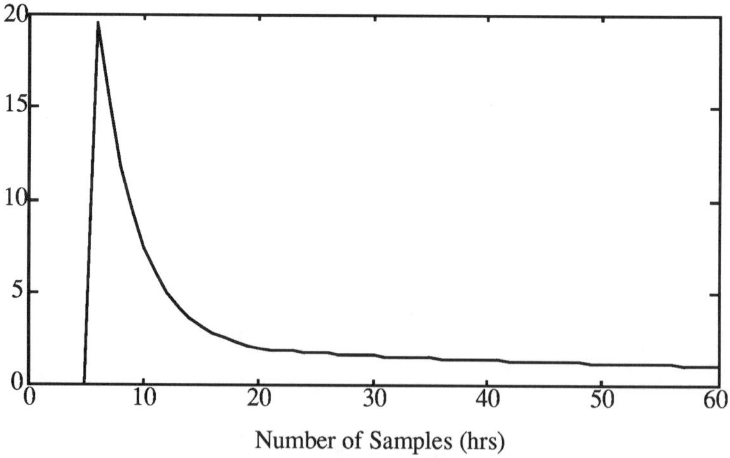

Fig.17 Rainfall-Flow example: impulse response of delta operator model

function can be decomposed by partial fraction expansion into the parallel structure shown in Fig.16, where we see that the model indicates that some 28% of the effective rainfall on the catchment reaches the river rapidly (time constant or residence time *TC=3.97hrs*) probably as "surface" flow; while 72% reaches the river much more slowly (*TC=90.5hrs*) via the groundwater as "baseflow" (however, see the discussion in Young and Beven, 1991). This longer term effect is clearly visible as the elevated 'tail' on the recession part of the model impulse response shown in Fig.17.

This method of modelling rainfall-flow processes has the additional desirable effect of allowing the modeller to objectively estimate the "baseflow" element of the river flow:

this is simply obtained as the output $x_2(k)$ of the lower *TF* in Fig.16. This can supplement other methods of baseflow estimation used in conventional hydrological analysis. A word of caution is necessary, however: the parallel decomposition and partitioning of flows is sensitive to uncertainty on the *TF* parameters, so that the inferences drawn from this δ operator model in this regard are a little different from those obtained using an alternative z^{-1}operator model (see Young and Minchin, 1991).

Example 6: Pilot Scale Nutrient Film Technique (*NFT*) System

As a final example, we will consider the identification and estimation of a δ operator model for a pilot scale *Nutrient Film Technique* (*NFT*) system used in glasshouse horticulture. Previous publications (e.g. Young et al., 1987a; Young et al., 1988; Young, 1989) have discussed the z^{-1} operator modelling and control of this system and full details of the problem can be obtained from these references. Here, we will simply consider the *SRIV* estimation of the δ operator model based on the same experimental input-output data used in these previous studies and shown in Fig.18. This is not an easy system to model and the δ operator estimation yields a marginally worse model than that obtained in the previous z^{-1} operator modelling exercises. However, the reasons for these difficulties can be diagnosed and raise certain interesting questions about the initialisation of δ operator estimation algorithms.

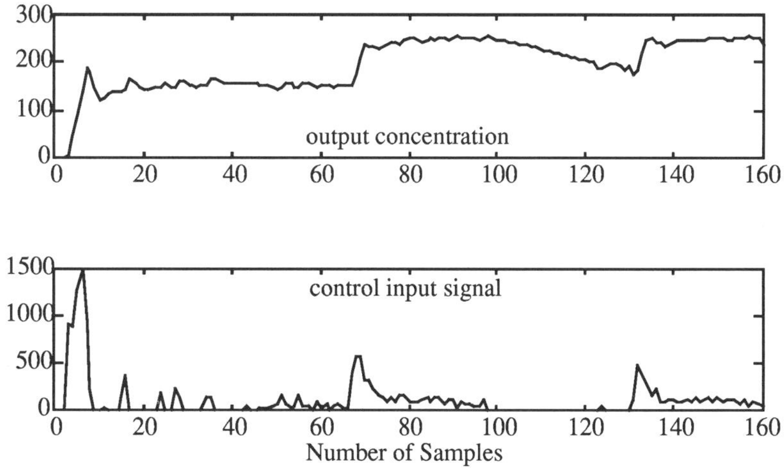

Fig.18 Pilot scale *NFT* system: input and output measurements

As in the z^{-1} operator modelling studies, the best identified δ operator model is third order and is characterised by the following parameter estimates,

$$\hat{a}_1=0.9394\ (0.0912);\ \hat{a}_2=0.5413\ (0.270);\ \hat{a}_3=0.00367\ (0.0018)$$

$$\hat{b}_0=0.0584\ (0.0125);\ \hat{b}_1=0.0540\ (0.0193);\ \hat{b}_2=0.01454\ (0.0073)$$

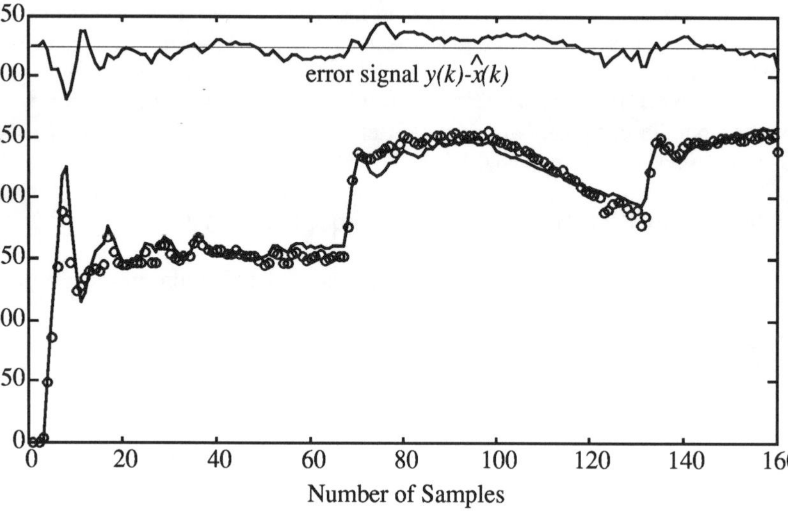

Fig.19 Pilot scale *NFT* system: *SRIV* estimated delta operator model output (full) comp with measured output (circles); error (*+325*) shown above.

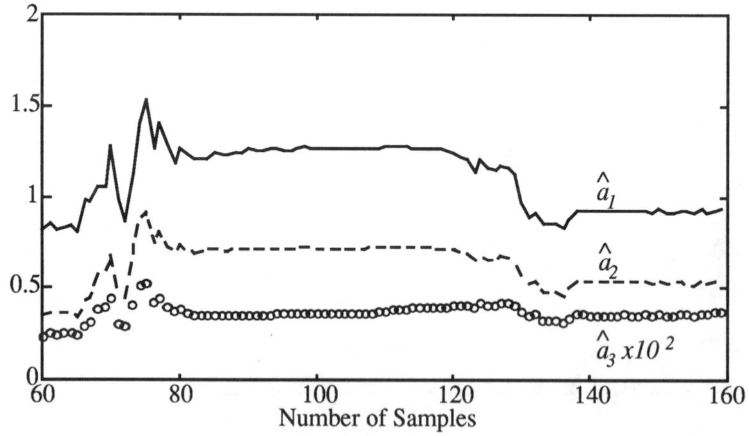

Fig.20 Pilot scale *NFT* system: recursive *SRIV* estimates of *TF* denominator parameters.

with $R_T^2=0.969$; $YIC=-5.476$. The model output is compared with the measured output in Fig.19, which also shows that the modelling error is heavily autocorrelated due to the simulated diurnal uptake of nutrient by the plants, as represented by a controlled leak from the system. This does not cause any estimation problems, however, because of the instrumental variable nature of the *SRIV* algorithm, which avoids the need for concurrent estimation of a model for the residuals, whilst eliminating asymptotic bias.

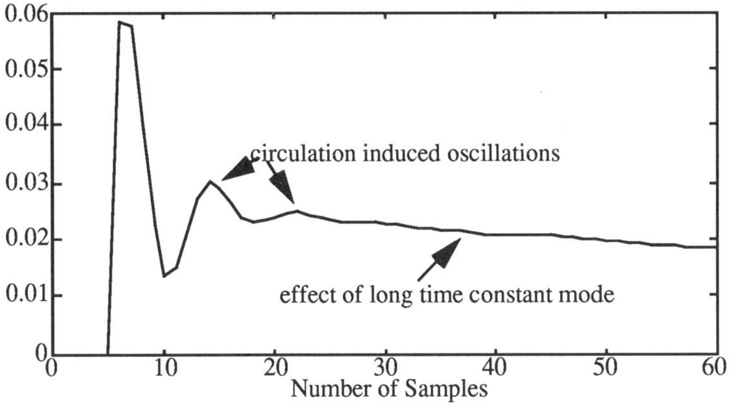

Fig.21 Pilot scale *NFT* system: *SRIV* estimated delta operator model impulse response.

The recursive *SRIV* estimates of the *TF* denominator coefficients are plotted in Fig.20; and the model impulse response in Fig.21; The former figure reveals that the estimates are not particularly well behaved and are certainly more volatile than those obtained in the previous z^{-1} operator modelling analysis applied to the same data. The latter suggests one reason for the problem: following the initial oscillatory response, which is caused by the re-circulation of the nutrient mixture in the *NFT* system, the response is dominated by a mode with a very long time constant (identified in the model above as approximately 146 sampling intervals). This is a situation where one might expect certain initialisation problems with the prefilters, leading to slow removal or even amplification of the initial condition effects. This could be obviated by concurrently estimating the initial conditions. However, the problem is not too severe in this case and the results are quite acceptable for applications such as control system design.

5. True Digital Control and the *PIP* Controller

In addition to their use in the modelling dynamic systems, *RIV* and *SRIV* identification and estimation procedures are also extremely useful in the design of automatic control systems, where they can be used both as a prelude to control system design and, in their on-line form, as a vehicle for implementing self-tuning and self-adaptive control systems. Our own studies in this regard have been concerned with a special, unified approach to *TDC* based on the concept of a Non-Minimum State Space (*NMSS*) model form which is derived directly from the *TF* models obtained from *RIV* or *SRIV* identification and estimation. Previous publications (e.g. Young et al., 1987b; Wang and Young, 1988; Young et al., 1988; Chotai et al., 1991) have described fully the z^{-1} operator version of this *NMSS* approach to control system design and the many attractions of the *Proportional-Integral-Plus* (*PIP*) controller which is synthesised in this manner. As in the previous Sections of the paper, therefore, we will concentrate here on the alternative δ operator analysis, which is more appropriate for rapidly sampled or continuous-time systems.

Naturally, the δ operator *TF* model can be represented by variety of different *NMSS* model forms (see Appendix 1, comment 3.). The one will consider here is chosen both because of its algebraic similarity to the *NMSS* form which has been used previously for z^{-1}operator *PIP* control system design, and also because it is defined directly in terms of the parameters of δ operator *TF* model. The model takes the following form,

$$\delta \boldsymbol{x}(k) = \boldsymbol{F}\boldsymbol{x}(k) + \boldsymbol{g}\, v(k) + \boldsymbol{d}\, y_d(k) \tag{17}$$

where $\boldsymbol{F}$, $\boldsymbol{g}$, and $\boldsymbol{d}$ are defined in the following manner,

$$\boldsymbol{F} = \begin{bmatrix} -a_1 & -a_2 & \cdots & -a_{n-1} & -a_n & b_1 & b_2 & \cdots & b_{m-1} & b_m & 0 \\ 1 & 0 & \cdots & 0 & 0 & 0 & 0 & \cdots & 0 & 0 & 0 \\ 0 & 1 & \cdots & 0 & 0 & 0 & 0 & \cdots & 0 & 0 & 0 \\ \vdots & \vdots & & \vdots & \vdots & \vdots & \vdots & & \vdots & \vdots & \vdots \\ 0 & 0 & \cdots & 1 & 0 & 0 & 0 & \cdots & 0 & 0 & 0 \\ 0 & 0 & \cdots & 0 & 0 & 0 & 0 & \cdots & 0 & 0 & 0 \\ 0 & 0 & \cdots & 0 & 0 & 1 & 0 & \cdots & 0 & 0 & 0 \\ 0 & 0 & \cdots & 0 & 0 & 0 & 1 & \cdots & 0 & 0 & 0 \\ \vdots & \vdots & & \vdots & \vdots & \vdots & \vdots & & \vdots & \vdots & \vdots \\ 0 & 0 & \cdots & 0 & 0 & 0 & 0 & \cdots & 1 & 0 & 0 \\ 0 & 0 & \cdots & 0 & -1 & 0 & 0 & \cdots & 0 & 0 & 0 \end{bmatrix}$$

$$\boldsymbol{g} = [\, b_0 \;\; 0 \; \ldots \; 0 \;\; 1 \;\; 0 \;\; 0 \; \ldots \; 0 \;\; 0 \;\; 0 \,]^T$$

$$\boldsymbol{d} = [\, 0 \;\; 0 \; \ldots \; 0 \;\; 0 \;\; 0 \;\; 0 \; \ldots \; 0 \;\; 0 \;\; 1 \,]^T$$

In this formulation, the control variable is denoted by $v(k)$, which is defined as follows in terms of the control input $u(k)$,

$$v(k) = \delta^m u(k)$$

with the associated state vector $\boldsymbol{x}(k)$ defined as,

$$\boldsymbol{x}(k) = [\delta^{n-1}y(k),\ \delta^{n-2}y(k),\ \ldots,\ \delta y(k),\ y(k),\ \delta^{m-1}u(k),\ \ldots\ldots,\ \delta u(k),\ u(k),\ z(k)]^T$$

In these equations, $z(k)$ is an "integral of error" state, which is introduced to ensure type 1 servomechanism performance and is defined in terms of the the inverse delta operator, or digital integrator δ^{-1}, i.e.,

$$z(k) = \delta^{-1}\{y_d(k) - y(k)\}$$

where $y_d(k)$ is the reference or command input at the kth sampling instant.

The Control Algorithm

As in the z^{-1} operator case (see references quoted previously), the State Variable Feedback (*SVF*) control law is defined in terms of the state variables: in this δ operator situation these are the output and input and their discrete differentials up to the appropriate order, as well as the integral of error state $z(k)$. This *SVF* control law can be written in the form,

$$v(k) = -\mathbf{k}^T\mathbf{x}(k)$$

where,

$$\mathbf{k}^T = [f_{n-1}\ f_{n-2}\ \ldots f_0\ g_{m-1}\ g_{m-2}\ \ldots g_0\ k_I]$$

is the *SVF* control gain vector for the *NMSS* δ operator model form. A block diagram of δ

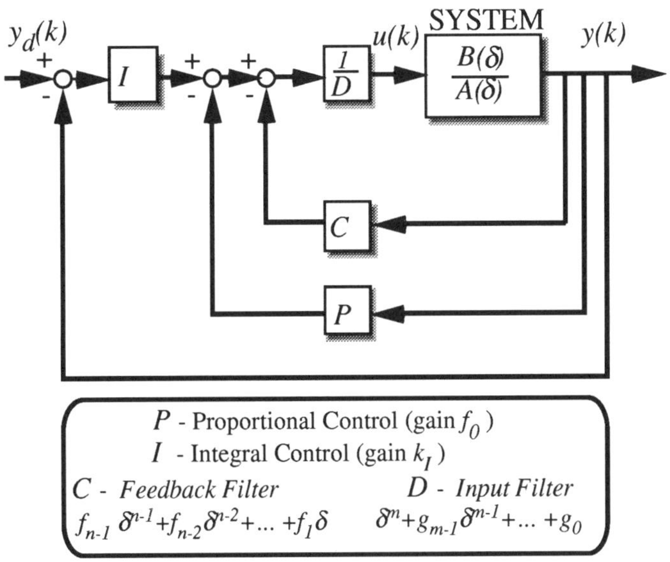

Fig.22 Block Diagram Representation of Delta (δ) Operator *PIP* Control System

operator *PIP* control system is shown in Fig.22. Note that, in order to avoid noise amplification by differencing, the system shown here would normally be converted into either a backward shift or an equivalent realisable delta operator form (e.g. using an observer) and then implemented accordingly (see the later discussion on implementation of δ operator *PIP* controllers).

A Simple Example

Consider a second order process described by the following δ operator *TF* model,

$$y(k) = \frac{\delta+b}{\delta^2}\,u(k) \tag{18}$$

This represents the parallel connection of a digital integrator $1/\delta$ and a double integrator b/δ^2, and is characterised by a single parameter b . For a short sampling interval Δt, this process will behave in a similar manner to the equivalent continuous-time system defined by an algebraically identical *TF*, but with δ replaced by the Laplace operator s. In a later sub-Section, we consider the self-tuning and self-adaptive control of this process for a sampling interval of *0.01* units, using adaptive implementations of a simple *PIP* control system developed below.

The *NMSS* state space model in this case takes the form,

$$\delta\boldsymbol{x}(k) = \begin{bmatrix} 0 & 0 & b & 0 \\ 1 & 0 & 0 & 0 \\ 0 & 0 & 0 & 0 \\ 0 & -1 & 0 & 0 \end{bmatrix} \boldsymbol{x}(k) + \begin{bmatrix} 1 \\ 0 \\ 1 \\ 0 \end{bmatrix} v(k) + \begin{bmatrix} 0 \\ 0 \\ 0 \\ 1 \end{bmatrix} y_d(k)$$

$$y(k) = [\,0 \quad 1 \quad 0 \quad 0\,]\,\boldsymbol{x}(k)$$

where the *NMSS* vector is defined as,

$$\boldsymbol{x}(k) = [\;\delta y(k) \;\; y(k) \;\; u(k) \;\; z(k)]^T$$

and $z(k)$ is the integral-of error state,

$$z(k) = \delta^{-1}\{y_d(k) - y(k)\}$$

The *SVF* control law is then given by,

$$v(k) = -\,\boldsymbol{k}^T\boldsymbol{x}(k)$$

where,

$$\boldsymbol{k}^T = [\,f_1 \; f_0 \; g_0 \; k_I\,]$$

Note that in this δ operator case,

$$v(k) = \delta\, u(k)$$

and the closed loop system will be fourth order.

Let us consider a pole assignment design with all four poles assigned to -10 in the complex δ domain, i.e. the desired closed loop characteristic polynomial $d(\delta)$ will be defined as,

$$d(\delta) = \delta^4 + 40\delta^3 + 600\delta^2 + 4000\delta + 10000$$

This specification yields the following equations for the *SVF* control gains,

$$k_I = 10000/b \;;\; f_0 = (4000 - k_I)/b \;;\; f_1 = (600 - f_0)/b \;;\; g_0 = (40 - f_1)$$

In a rapidly sampled situation, this model can be considered as a close approximation to true continuous-time (i.e. with $\delta \rightarrow s$), so that the design specification is to have the system respond like a model system with unity steady state gain, composed of four first order systems, each with a time constant of 0.1 time units; i.e 'fast' critically damped response. For this specification, a short sampling interval of 0.01 time units, for example, will provide performance which can be compared directly to that of an equivalent, digitised continuous-time system[10]

The Implementation of δ Operator *PIP* control systems

One of the major attractions of the z^{-1} operator *PIP* control system designs is that they are easy to implement in practice because the NMSS vector only involves past input and output signals, all of which are available for direct utilisation. In contrast, the δ operator designs involve the numerical derivatives of the input and output signals and their generation by direct numerical differentiation is clearly not advisable in noisy situations. However, it is easy to convert between the z and δ operator model parameters (details of the matrix transformations are given in Chotai et al., 1990a). As a result, a δ operator *PIP* control system design can be implemented in practice by converting it back to the z^{-1} domain and then implementing the design in these simpler, backward shift terms, thereby avoiding direct numerical differentiation. This approach, which is used in the adaptive simulations described below, needs further investigation, however, since it could be criticised because of possible sensitivity to uncertainty at higher sampling rates. With double precision arithmetic so freely available in microprocessors, this may not be too problematical; nevertheless, it is worth considering other solutions which are implemeted directly in δ operator terms.

The simplest such alternative is to manipulate the controller into the form of *realisable* δ operator transfer functions (i.e. where direct differentiation is avoided); unfortunately, this is not a general solution and can introduce other problems. A superior approach is either to introduce a state reconstruction filter (observer) or its stochastic equivalent, the Kalman filter, and then implement the *PIP* controller by invoking the separation principle and replacing the numerical derivatives in the *PIP* control law by the reconstructed states in the usual manner.

Once the δ operator model parameters have been estimated using the *RIV* or *SRIV* algorithms, it is straightforward to design an observer or a Kalman filter system: such designs in the δ domain are discussed, for example, in Middleton and Goodwin (1990). On the other hand, filtered elements of *NMSS* state vector (i.e. the filtered numerical derivatives of the input and output variables) are generated automatically by the *RIV* and *SRIV* algorithms and we might suspect that these filtered variables are themselves linearly related to the "optimal" state estimates obtained by an observer or Kalman filter. In the case of conventional discrete-time models, the first author (1979; see also comment 2. in Appendix 1) has shown that the Kalman state estimates can indeed be obtained in this manner if the prefilter is chosen correctly. These results carry over immediately to the δ domain and suggest that the filtered derivatives from suitably chosen prefilters can be used directly for implementing the *NMSS* state variable feedback control law of the *PIP* control system. A block diagram of the resulting system is shown in Fig.23, where the

---

10. An optimal LQ design could also be considered in this δ operator example, using a similar approach to that employed in the z^{-1} operator case ; see e.g.Young et al, 1991a.

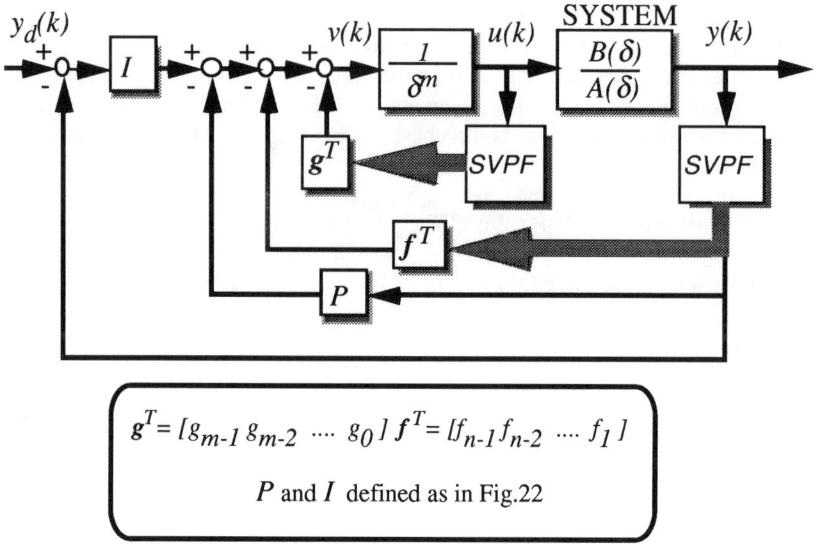

Fig.23 Block Diagram Representation of Delta (δ) Operator *PIP* Control System with stochastic observer based on prefilters (shown as *SVPF)*

"state variable prefilters", which will be of the same *basic* form as that shown in Fig.1 (but not necessarily with the same parameters; see comment 2. in Appendix 1.), are denoted by *SVPF* .

Self-tuning and Self-adaptive *PIP* Control

Adaptive modification of fixed gain *PIP* controllers requires either: (a) 'schedule gain' implementation based on the off-line modelling (this is typical of those adaptive systems used for aircraft autostabilisation, and we have used a similar approach in Chotai et al., 1991); or (b) a complete self adaptive system based on an on-line Time-Variable-Parameter (*TVP*) version of the recursive estimation algorithms discussed in Section 3 (equation (14)). For the type (b) approach, we propose the *TVP* version of the recursive least squares or recursive instrumental variable algorithms used in our previous research on adaptive systems (Young 1969a, 1969b, 1970, 1971, 1981, 1984; Young et al., 1987a, 1988, 1991a; Young, 1989; Chotai et al. 1991). This is a special development of the *TVP* algorithms outlined in Section 3., which allows for the exploitation of prior knowledge on parameter variation. The applications of the resultant algorithm can range from simple adaptive adjustment or on-line "tuning" of a scheduled gain system, to full adaptive control capable of tracking rapid parameter changes.

On-line simplifications and modifications to the *SRIV* algorithm are straightforward. For example, the least squares version of this algorithm might well be appropriate in many low to moderate noise conditions if the prefilters sufficiently reduce the effective noise/signal ratio. Also, in practice, the prefilters would normally be chosen with fixed parameters defined by the prior experimentation (such prior analysis is an essential feature of the *TDC* philosophy). This is possible because, in contrast to off-line estimation where relatively good statistical efficiency is usually required, the accuracy of recursive estimators in adaptive control applications need not be too high, simply because a well designed controller such as PIP will be relatively robust to uncertainty. Consequently, as we shall see in the example below, the performance of the algorithm is not too sensitive to the prefilter parameters, provided that the pass-band encompasses that of the system being modelled.

As a simple example of self-tuning and self-adaptive system design, let us consider the δ operator model system (18) with a sampling interval of 0.01 secs and the associated *PIP* design considered earlier. The adaptive design in this case is quite simple. First, the model is characterised by only the single unknown gain parameter *b*, so that the recursive estimation algorithm is in a scalar form. For additional simplicity, we use the RLS version of the *SRIV* algorithm (i.e. prefiltering but no *IV* modification). Second, the prefilters are made time-invariant, and are related to the desired *closed loop* pole assignment specifications: in particular, the prefilters for the input and output signals are specified as,

$$\frac{100\delta}{\delta^2 + 20\delta + 100}$$

with two poles at -10 in the complex δ plane and a δ term in the numerator. This is a pragmatic choice: it assumes little knowledge of the open loop system (which is unstable in this case and so is not a suitable basis for prefilter design); on the other hand, it has a wide passband and so encompasses the frequency response characteristics of the open loop system, as required. The numerator is chosen to introduce high pass filtering and so remove any drift or bias from the signals: it is equivalent to the assumption that such behaviour can be modelled as a *RW* in the δ domain, i.e. $C(z^{-1})=\delta$; $D(z^{-1})=1.0$ (see Young et al, 1988).

The *NVR* value and the initial $\boldsymbol{P}(0)$ matrix are the only parameters in the recursive algorithm to be specified by designer: $\boldsymbol{P}(0)$ is set to $10^6\boldsymbol{I}$, indicating no *a priori* confidence in the initial estimate of the *b* parameter; and the *NVR* is selected to accommodate the expected parameter variations, as discussed below. Strictly, if we had confidence in the *a priori* value, $P(0)$ should be set to a much lower value. However, this "diffuse prior" is chosen to see how fast the recursive estimate would converge. To avoid initial transients, the controller uses the prior value $\hat{b}(0)$ for the first *10* recursions, before switching to the appropriate recursive estimates after this initial period. For all the simulations below, the recursive estimates are updated every 10 samples (i.e. every 0.1 secs).

Figs.24 to 27 illustrate the self-adaptive results obtained with the *b* parameter reducing in a stepwise exponential manner from an initial value of 70 to a final value of 10 in ten equal time steps. In Fig.24, the command input $y_d(k)$ is in the form of repeated steps at a switching period of about 3 secs and the system is subjected to system noise, introduced by adding a zero mean, white noise disturbance, together with a large, step, load disturbance added to the output over the middle period of the simulation. The figure shows the how the changing nature of the system is reflected in the adaptive adjustment of the control input which maintains satisfactory uniform closed loop response throughout the simulation. This good performance of the adaptive system can be contrasted with that of the equivalent fixed gain *PIP* controller shown in Fig.25. Here, the closed loop behaviour continues to degrade as the system parameter changes and the fixed gain *PIP* controller is unable to adequately compensate for these changes.

The basic system used to generate the results in Fig.26 is similar to that used in Fig.24 except that the input command sequence has a lower frequency and the step load disturbance is replaced by a fixed amplitude, sinusoidal load: once again, the adaptive adjustment is clearly visible in the control input signal.

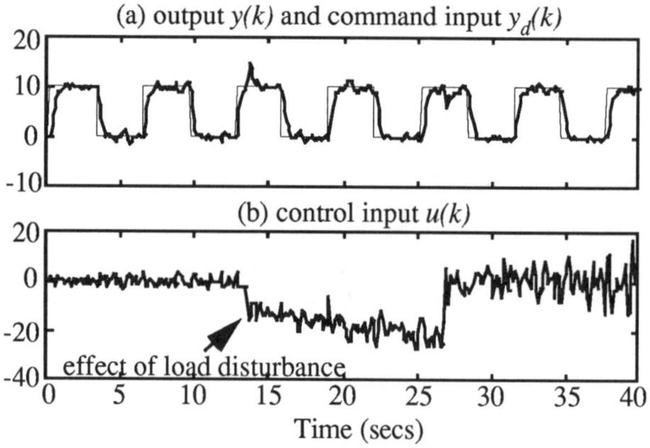

Fig.24 Self-adaptive *PIP* system applied to *TVP* version of double integrator system.

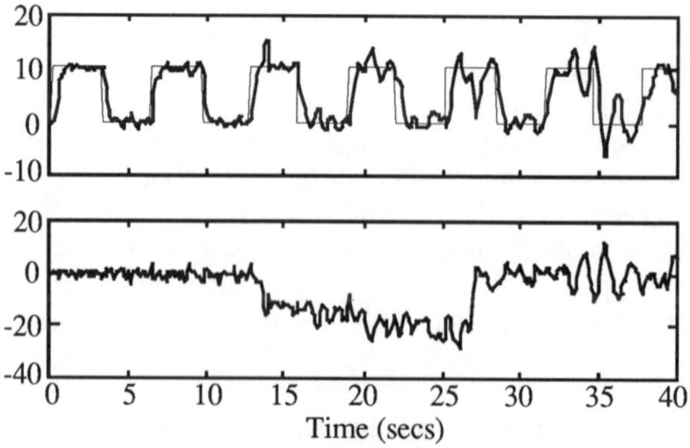

Fig.25 As Fig.24 but using a fixed-gain *PIP* system: showing inability to satisfactorily handle nonstationarity without adaption.

Finally, the recursive estimates of the time variable parameter b used in the adaptive *PIP* controllers for the two adaptive simulations are given in Fig.27: plot (i)

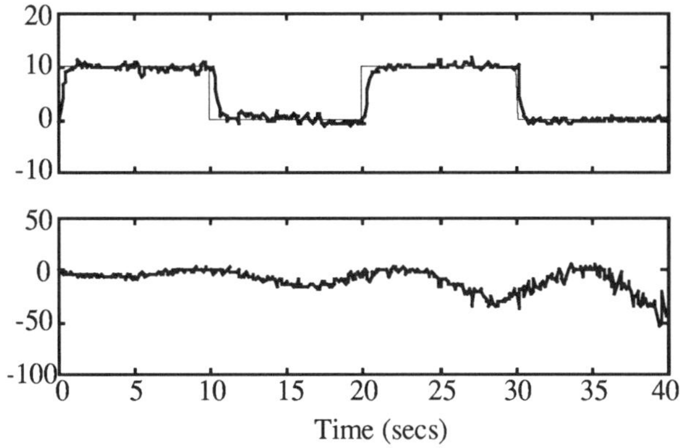

Fig.26 As Fig.24 but with sinusoidal load disturbance.

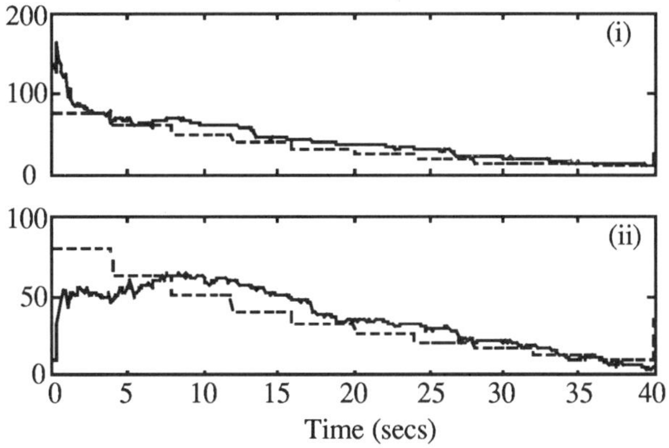

Fig.27 Recursive estimates of time variable parameter : (i) for system in Fig.24; (ii) for system in Fig.26.

shows the recursive estimate associated with the results in Fig.24; while plot (ii) shows that for Fig.26. In these self adaptive runs, the initial estimate of b was set to 50 and the *NVR* was maintained at 0.0000001 throughout each run. Note that, although the parameter trtacking is not particularly rapid, it is quite adequate to maintain tight adaptive control: this demonstrates the effectiveness of the *PIP* control system, which is robust enough to handle this level of uncertainty in the parameter values

The reader may like to compare the results of this simulation example with those given in Goodwin et al. (1988) and Åstrom et al. (1986), who consider the control of a deterministic, second order system with real roots. In the present case, we are controlling a more difficult double integrator system with system noise and load disturbance and yet still achieving excellent closed loop response characteristics.

6. Conclusions

The main aim of this chapter has been to outline a unified, True Digital Control (*TDC*) approach to the modelling and control of rapidly sampled, continuous-time systems whose dynamic behaviour can be characterised by transfer functions in the delta (δ) operator. We believe that this approach, which integrates the Refined Instrumental Variable *(RIV)* methods of model identification and estimation with special Non-Minimum State Space (*NMSS*) methods of control system design, is more unified and robust than previous approaches based on δ operator models (e.g. the identification and control techniques discussed in Middleton and Goodwin, 1990). In particular, the systematic *TDC* design procedure allows the model identification and estimation phases of the analysis to define directly and uniquely the parameters of a suitably dimensioned *NMSS* description of the system, which then immediately yields a Proportional-Integral-Plus (*PIP*) controller that exploits the power of state variable feedback to achieve either pole assignment or optimal *LQ* or *LQG* design objectives.

When combined with the equivalent z^{-1}operator procedures described in our previous publications, this general *TDC* approach provides an excellent basis for Computer Aided Control System Design (*CACSD)* which, we believe, has wide potential for application. A *CACSD* package, based on this approach and appropriately named *TDC* is currently being developed at Lancaster. This package (see Tych et al, 1991) is intended primarily for the SUN™ workstation but, since it is written (effectively as a "Toolbox") within the Pro-Matlab™ program, it is compatible with most of the other computers which are able to operate versions of Matlab, such as MS-DOS and Apple Macintosh microcomputers, as well as mainframes accessed through graphics terminals. Moreover, by shielding the user from the Matlab command line with a simple Graphical User Interface (GUI), while still allowing easy access to Matlab if this is desired, we believe the package provides a good combination of a well designed, systematic procedure for day-to-day design studies and a flexible tool for more novel research exercises.

Although currently limited to single input, single output systems, the *CACSD* package is being extended in various directions. Multivariable model identification and estimation, although complicated, is possible using the multivariable *RIV* or *SRIV* algorithms (Jakeman and Young, 1979; Young and Wang, 1987). And since the *PIP* (pole assignment or optimal *LQ*) control system design method is posed in state-space terms, its stochastic formulation is obvious, involving simply application of the separation theorem, with the *RIV* algorithms providing full stochastic model identification and estimation. Some aspects of multivariable *NMSS* control system design in the z^{-1} operator case, using the left Matrix Fraction Description (*MFD*) models, are discussed in a recent paper (Chotai et al., 1990b). However, the details of control system design are much less obvious in the multivariable case: in particular, while pole assignment or *LQG* control system design is straightforward, they do not address directly how the additional degrees of freedom provided by *SVF* should be used to solve problems such as decoupling. Decoupling control is possible within the *NMSS* context (Wang, 1988), but has not yet been evaluated in practice.

Aknowledgements

The authors are grateful to the U.K. Natural Environment Research Council (NERC), the U.K. Science and Engineering Research Council (SERC) and the AFRC Institute of Engineering, Silsoe, for their help in supporting the research described in this paper.

Appendix 1 δ Operator Models : Some Brief Observations

1. There are two obvious definitions of the discrete derivative operator based on either the forward difference or backward difference operations, i.e.,

Forward Difference: $$\delta = \frac{z - 1}{\Delta t}$$

Backward Difference: $$\delta = \frac{1 - z^{-1}}{\Delta t}$$

The former is the definition suggested by Goodwin (1985), probably because it is the most straightforward for implementation in real time. This is the form used in the present paper in order to allow for direct comparison with other current literature. However, our recent research suggests that the latter may have some advantages in both theory and practice and we are investigating these at the current time.

2. More research is required on the exact form of the estimation model for δ operator systems. In this chapter we have utilised estimation relationships based on prefiltered versions of equation (2), with the highest numerical filtered derivative on the left hand side of the equation. But other choices are clearly possible, although there does not appear to have been any meaningful research on this topic. Amongst these possibilities, the most attractive and promising are model relationships which relate the *measured* (i.e. unfiltered) output $y(k)$ on the left hand side of the equation with the *prefiltered* numerical derivatives of the input and output signals on the right hand side. The author has provided a statistical motivation for this kind of model in the z^{-1} domain by showing how it can be related directly to the Kalman filter (Young, 1979). This analysis carries over immediately to δ operator models, where it is easy to show that asymptotically, as $k \rightarrow \infty$, the innovations form of the Kalman filter can be converted to the *ARMAX* transfer function model form ; i.e., in the present context, equation (4) of Section 3., with $C(\delta)=A(\delta)$. It is then straightforward to develop the following expression for the optimally filtered output, $\hat{y}(k)$,

$$\hat{y}(k) = \frac{D(\delta) - A(\delta)}{D(\delta)}\, y(k) + \frac{B(\delta)}{D(\delta)}\, u(k) \tag{A.1}$$

where, by definition of the innovations process $e(k)$,

$$y(k) = \hat{y}(k) + e(k)$$

As a result, we can immediately obtain the following expression for $y(k)$,

$$y(k) = -\alpha_1 \delta^{n-1} y^*(k) - \ldots - \alpha_n y^*(k) + b_0 \delta^m u^*(k) + \ldots + b_m u^*(k) + e(k) \qquad \text{(A.2)}$$

where $\alpha_i = d_i - a_i$; and the star superscript now denotes prefiltering by $1/D(\delta)$. Other relevant discussion on models of this general type appear in Middleton and Goodwin (1990); and Young (1969a), in a continuous-time context.

Equation (A.2) is clearly an attractive estimation model from both theoretical and practical standpoints. The reason it was not used in the present chapter lies in the form of the prefilter, which demands concurrent estimation of a constrained *ARMA* model for the noise process (i.e. with $C(\delta)=A(\delta)$ in equation (3) of Section 3.), in order to define the prefilter parameters. Nevertheless, this model form deserves further investigation, not only because of its relevance to parameter estimation, but also because it shows how, with appropriate parameterisation, the prefiltered variables provide optimal estimates of the system state variables (see Young, 1979; and the sub-Section on the implementation of δ operator *PIP* control systems in Section 5.).

3. Obviously, the δ operator *TF* model could be converted into a wide variety of different *NMSS* forms and our analysis in Section 5 has simply used one particular representation for the reasons cited there. Probably the most obvious *NMSS* model in this regard is obtained simply by converting the z^{-1}operator *NMSS* representation used in our previous publications (see the references in Section 5) into equivalent δ operator form. For example, if the z^{-1}operator *NMSS* model is written as,

$$\boldsymbol{x}(k+1) = \boldsymbol{A}\,\boldsymbol{x}(k) + \boldsymbol{b}\,u(k) + \boldsymbol{c}\,y_d(k+1) \qquad \text{(A.3)}$$

then introducing the forward shift operator z and substituting $z=1+\Delta t\delta$, we obtain,

$$\delta\boldsymbol{x}(k) = \boldsymbol{F}\boldsymbol{x}(k) + \boldsymbol{g}\,v(k) + \boldsymbol{d}\,y_d(k) \qquad \text{(A.4)}$$

where,

$$\boldsymbol{F} = \frac{\boldsymbol{A} - \boldsymbol{I}}{\Delta t} \;\; ; \;\; \boldsymbol{g} = \frac{\boldsymbol{b}}{\Delta t} \;\; ; \;\; \boldsymbol{d} = \frac{\boldsymbol{c}}{\Delta t} + \delta\boldsymbol{c}$$

This model form has some attractions since, like the z^{-1}operator *NMSS* model, the states are simply the present and past values of the output and input signals, all of which are available directly for use in the *SVF* control law, without resort to state reconstruction. On the other hand, the model parameters have to be obtained from the estimated δ operator *TF* model parameters and the transformation could be sensitive to uncertainty in the estimates.

References

Akaike, H. (1974) "A new look at statistical model identification", I.E.E.E. Trans. on Auto. Control, vol AC-19, pp. 716-722

Astrom, K.J. and Bohlin, T. (1966) "Numerical identification of linear dynamic systems from normal operating records"; appears in P.H.Hammond (ed.). *"Theory of Self Adaptive Control Systems"*, Plenum Press, New York.

Astrom, K.J. and Eykhoff, P. (1971) "System identification: a survey", Automatica, vol. 7, pp. 123.

Astrom, K. J., Newman, L. and Gutman, P.O. (1986) "A comparison between robust and adaptive control of uncertain systems", IFAC Workshop on Adaptive Control, University of Lund, Sweden.

Beven, K.J. and Young, P. C. (1988) "An aggregate mixing zone model of solute transport through porous media", Jnl. Contaminant Hydrol., vol. 3, pp. 129-143.

Box, G.E.P. and Jenkins, G.M. (1970) *"Time Series Analysis, Forecasting and Control"*, Holden Day, San Francisco.

Bryson, A.E. and Ho, Y.C. (1969) *"Applied Optimal Control"*, Blaisdell, Waltham, Mass.

Chotai, A., Young, P.C., and Tych, W. (1990a) "A non-minimum state space approach to true digital control based on the backward shift and delta operator models", appears in M.H. Hamza (ed.), *"Proc. IASTED Conference1990"*, Acta Press, Calgary, pp. 1-4.

Chotai, A., Young, P.C. and Wang, C.L. (1990b) "True digital control of multivariable systems by input/output, state variable feedback", Report No. TR 86/1990, Centre for Research on Environmental Systems, University of Lancaster.

Chotai, A., Young, P.C. and Behzadi, M.A. (1991) "The self-adaptive design of a nonlinear temperature control system", Proc. I.E.E., Pt.D, special Issue on Self Tuning Control, vol. 138, pp, 41-49.

Gawthrop, P.J. (1988) "Implementation of continuous-time cont-rollers"; appears in K. Warwick (ed.). *"Implementation of Self Tuning Controllers"*. Peter Perigrinus, London, 1988, pp. 140-156 .

Gelb, A., Kasper, J.F., Nash, R.A., Price, C.F. and Sutherland, A.A. (1974) *"Applied Optimal Estimation"*, MIT Press for the Analytical Sciences Corp., Cambridge, Mass.

Goodwin, G.C. (1985) "Some observations on robust estimation and control", appears in H.A.Barker and P.C.Young (ed.). *"Identification and System Parameter Estimation 1985, Vols 1 and 2"*, Pergamon, Oxford, pp. 851-859.

Goodwin, G.C. (1988) “Some observations on robust stochastic estimation”; appears in H. F. Chen (ed.). *“Identification and System Parameter Estimation, 1988”*. Pergamon Press, Oxford, pp. 22-32.

Goodwin, G. C., Middleton, R.H. and Salgado, M. (1988) “A unified approach to adaptive control”; appears in K. Warwick (ed.). *“Implementation of Self Tuning Controllers”*, Peter Perigrinus, London, pp. 126-139.

Gould, R., Minchin, P.E.M. and Young, P.C. (1988) “The effects of sulphur dioxide on phloem transport in two cereals”, Jnl. of Exp. Botany, vol. 39, pp. 997-1007.

Jakeman, A.J. and Young, P.C. (1979) “Refined instrumental variable methods of recursive time-series analysis, Part 2: multivariable systems”, Int. Jnl. Control, vol. 29, pp. 621-634.

Jakeman, A.J., Littlewood, I.G. and Whitehead, P.G. (1990) “Computation of the instantaneous unit hydrograph and identifiable component flows with application to two small upland catchments”, Journal of Hydrology, vol. 117, pp. 275-300.

Joseph, P., Lewis, J. and Tou, J. (1961) “Plant identification in the presence of disturbances and application to digital adaptive systems”, A.I.E.E. Trans. on Applications and Industry, vol. 80, pp. 18.

Kalman, R.E. (1958) “Design of a self-optimizing system”, A.S.M.E. Trans., Jnl. of Basic Eng., vol. 80-D, pp. 468-478.

Kraijenhoff, D.A. and Moll, J.R. (1986) *“River Flow Modelling and Forecasting”* (Water Science and Technology Library). D. Reidel, Dordrecht.

Liaw, C.M., Pan, C.T. and Ouyang, M. (1986) “Model reduction of discrete systems using the power decomposition method and the system identification method”, IEE Proc., vol. 133, Pt. D, pp. 30-34,.

Ljung, L. and Soderstrom, T. (1983) *“Theory and Practice of Recursive Identification”*, MIT press, Cambridge, Mass.

Middleton, R.H. and Goodwin, G.C. (1990) *“Digital Control and Estimation: A unified Approach”*, Prentice Hall, Englewood Cliffs, N.J.

Minchin, P.E.H. and Grusak, M.A. (1988) “Continuous in vivo measurement of carbon partitioning within whole plants”, Journal of Experimental Botany, vol 39, pp. 561-571.

Minchin, P.E.H. and Thorpe, M.R. (1988) “Carbon partitioning to whole versus surgically modified ovules of pea: an application of the in vivo measurement of carbon flows over many hours using the short-lived isotope Carbon-11”, Ibid., vol. 40, pp. 781-787.

Minchin, P.E.H. and Troughton, J.H. (1980) “Quantitative interpretation of phloem translocation data”, Annual Review of Plant Physiology, vol. 31, pp. 191-215.

Ng, C.N. and Young, P.C. (1990) “Recursive estimation and forecasting of nonstationary time series”, Jnl. of Forecasting, Special Issue on *"State-Space Forecasting and Seasonal Adjustment"*, ed. by P.C. Young, vol. 9, pp. 173-204.

Norton, J.P. (1975) "Optimal smoothing in the identification of linear time-varying systems", Proc. I.E.E. (U.K.), vol. 122, pp. 663-668.

Norton, J.P. (1986) *"An Introduction to Identification"*, Academic Press, London.

Pierce, D.A. (1972) "Least squares estimation in dynamic disturbance time-series models", Biometrika,vol. 46, pp. 73-78.

Shieh, L.C. (1975) "A mixed method for multivariable system reduction", IEEE Trans. on Auto. Control., vol. AC-20, pp. 429-432.

Soderstrom,T. and Stoica, P.G. (1983) *"Instrumental Variable Methods for System Identification"*. Springer-Verlag, Berlin.

Soderstrom,T. and Stoica, P.G. (1989) *"System Identification"*, Prentice Hall International, Hemel Hempstead.

Tych, W. and Young, P.C. (1990) "A refined instrumental variable approach to model reduction for control systems design", Report No. TR81/1990, Centre for Research on Environmental Systems, University of Lancaster.

Tych, W., Young, P.C. and Chotai, A. (1991) "TDC - A computer aided design package for True Digital Control ", to appear, Proc. IFAC Symposium on Computer Aided Design in Control Systems, Swansea (see also paper by same authors in I.E.E Conference Publication Number 332 "Control 91", Institution of Electrical Engineers, London, pp 288-293).

Wallis, S.G., Young, P.C. and Beven, K.J. (1989) "Experimental investigation of the Aggregated dead zone model for longitudinal solute transport in stream channels", Proc. Inst. Civ. Engrs. (U.K), vol. 87, pp. 1-22.

Wang, C.L. (1988) *"New methods for the Direct Digital Control of Discrete-Time Systems"*, Ph.D. Thesis, Centre for Research on Environmental Systems, University of Lancaster.

Wang, C.L. and Young, P.C. (1988) Direct digital control by input-output, state variable feedback: theoretical background, *Int. Jnl. of Control*, vol. 47, pp. 97-109.

Weyman, D.R. (1975) *"Runoff Processes and Streamflow Modelling"*, Oxford University Press, Oxford.

Whitehead, P.G., Young, P.C. and Hornberger, G.H. (1979) "A systems model of stream flow and water quality in the Bedford-Ouse River; I stream flow modelling", Water Research, vol. 13, pp. 1155-116.

Young, P.C. (1964) "In-flight dynamic checkout", I.E.E.E. Trans. on Aerospace, vol. AS2, pp. 1106-1111.

Young, P.C. (1965) "The determination of the parameters of a dynamic process", Radio and Electronic Engineer, vol. 29, pp. 345-362.

Young, P.C. (1966) "Process parameter estimation and self adaptive control"; appears in P.H.Hammond (ed.). *"Theory of Self Adaptive Control Systems"*, Plenum Press, New York, pp. 118-140.

Young, P.C. (1969a) *"The Differential Equation Error Method of Process Parameter Estimation"*, Ph.D Thesis, Department of Engineering, University of Cambridge.

Young, P.C. (1969b, 1970) An instrumental variable method for real time identification of a noisy process, *Proc. IFAC Congress*, Warsaw (also appears in *Automatica*, vol. 6, pp. 271-287, 1970).

Young, P.C. (1969c) "The use of a priori parameter variation information to enhance the performance of a recursive least squares estimator". Tech. Note 404-90, Naval Weapons Center, California, *(53 pages).*

Young, P.C. (1971) "A second generation adaptive pitch auto-stabilisation system for a missile or aircraft", Tech Note 404-109, Naval Weapons Center, California.

Young, P.C. (1975) "Recursive approaches to time series analysis", Bull. Inst. Math. and Applic., vol. 10, pp. 209-224.

Young, P.C. (1976) "Some observations on instrumental variable methods of time-series analysis", Int. Jnl. of Control, vol. 23, pp. 593-612.

Young, P.C. (1979) "Self-adaptive Kalman filter", Electronics Letters, vol. 15, pp. 358-360.

Young, P.C. (1979, 1981) "Parameter estimation for continuous-time models: a survey", appears in R.Isermann (ed.). *"Identification and System Parameter Estimation"*, Pergamon Press, Oxford, pp. 1073-1086 (see also Automatica, vol. 17, pp. 23-39, 1981)

Young, P.C. (1981) "A second generation adaptive autostabilisation system for airborne vehicles", Automatica, vol. 17, pp. 459-469.

Young, P.C. (1984) *"Recursive Estimation and Time Series Analysis: An Introduction"*, Springer-Verlag, Berlin.

Young, P.C. (1985) "The instrumental variable method: a practical approach to identification and system parameter estimation", appears in H.A.Barker and P.C.Young (ed.). *"Identification and System Parameter Estimation 1985, Vols 1 and 2"*, Pergamon, Oxford, pp. 1-16.

Young, P.C. (1986) "Time-series methods and recursive estimation in hydrological systems analysis", appears in D.A. Kraijenhoff and J.R. Moll (eds.) *"River Flow Modelling and Forecasting"*, D. Reidel, Dordrecht, pp. 129-180.

Young, P.C. (1988) "Recursive extrapolation, interpolation and smoothing of nonstationary time-series", appears in H.F. Chen (ed.) *"Identification and System Parameter Estimation, 1988"*, Pergamon, Oxford, pp. 33-44.

Young, P.C. (1989) "Recursive estimation, forecasting and adaptive control", appears in C.T.Leondes (ed.), *"Control and Dynamic Systems"*, Academic Press, San Diego, pp. 119-166.

Young, P.C. (1990) "Rainfall-flow modelling: some new ideas", Rep. No. TR83 (1990), Centre for Research on Environmental Systems, University of Lancaster (to appear in *"Concise Encyclopedia on Environmental Systems"*, P.C. Young (ed.), Pergamon Press, Oxford, 1991-2)

Young, P.C., and Armitage, P. (1991) *"Recursive Estimation and Time Series Analysis"*, Springer-Verlag, Berlin.

Young, P.C. and Beven, K.J. (1991) "Rainfall-flow modelling using *SRIV* identification and estimation: comments on paper by Jakeman et al.", to appear, Journal of Hydrology.

Young, P.C., Jakeman, A.J. (1979) "Refined instrumental variable methods of recursive time-series analysis, Part 1: single input, single output systems", Int. Jnl. Control, vol. 29, pp. 1-30.

Young, P.C. and Jakeman, A.J. (1980) "Refined instrumental variable methods of recursive time-series analysis, Part 3: extensions", Int. Jnl. Control, vol. 31, pp. 741-764.

Young, P.C. and Minchin, P.E.H. (1991) "Environmetric time-series analysis: Modelling Natural Systems from Experimental Time-Series Data", to appear in the International Jnl. of Biological Macromolecules.

Young, P.C. and Ng, C.N. (1989) "Variance intervention", Jnl. of Forecasting, vol. 8, pp. 399-416.

Young, P.C. and Runkle, D.E. (1989) "Recursive estimation and the modelling of nonstationary and nonlinear time-series", appears in *"Adaptive Systems in Control and Signal Processing"*, Institute of Measurement and Control, London, pp. 49-64.

Young, P.C. and Wallis, S.G. (1985) "Recursive Estimation : A Unified Approach to Identification, Estimation and Forecasting of Hydrological Systems", Applied Mathematics and Computation, vol. 17, pp. 299-334

Young, P.C. and Wallis, S.G. (1986) "The Aggregated Dead Zone (ADZ) model for dispersion in rivers". Proc. BHRA Int. Conf. on *"Water Quality Modelling in the Inland Natural Environment"*, Bournemouth, England.

Young, P.C. and Wang, C.L. (1987) "Identification and estimation of multivariable dynamic systems", appears in J. O'Reilly (ed.) *"Multivariable Control for Industrial Applications"*, Peter Perigrinus, London, pp. 244-279.

Young, P.C., Behzadi, M.A. and Chotai, A. (1988) *"Self tuning and self adaptive PIP control systems"*; appears in K. Warwick (ed.), *"Implementation of Self-Tuning Controllers"*, Peter Perigrinus, London, pp. 220-259.

Young, P.C., Chotai, A. and Tych, W. (1991a) "True Digital Control: A Unified Design Procedure for Linear Sampled Data Systems", to appear as chapter in K.Warwick (ed.), *"Advanced Methods in Adaptive Control for Industrial Applications"*, Springer-Verlag, Berlin.

Young, P.C., Jakeman, A.J. and McMurtrie, R. (1980) "An instrumental variable method for model order identification", Automatica, vol. 16, pp. 281-294.

Young, P.C., Ng, C.N. and Armitage, P. (1989) "A systems approach to economic forecasting and seasonal adjustment", International Journal on Computers and Mathematics with Applications, vol. 18, pp. 481-501.

Young, P.C., Behzadi, M.A., Chotai, A. and Davis, P. (1987a) "The modelling and control of nutrient film systems", appears in J. A. Clark, K. Gregson and R. A. Scafell (eds.), *"Computer Applications in Agricultural Environments"*. Butterworth, London, pp. 21-43.

Young, P.C., Behzadi, M.A., Wang, C.L. and Chotai, A. (1987b) "Direct digital control by input-output, state variable feedback pole assignment", Int. Jnl. Control, vol. 46, pp. 1867-1881.

Young, P.C., Ng, C.N., Lane, K. and Parker, D. (1991b) "Recursive forecasting, smoothing and seasonal adjustment of nonstationary environmental data", Jnl. of Forecasting, vol. 10, pp. 57-89.

Identification of multivariable continuous-time systems

E Boje
Department of Electrical & Control Engineering
University of Durban–Westville
Private Bag X54001, Durban, 4000
South Africa

Abstract

This chapter presents a parameter identification algorithm for multivariable, continuous–time systems. It is shown how to treat systems with (non–zero) initial conditions and measurement offset or bias. The system model structure used is a minimal order input–output representation. Differentiation of measured data is avoided by means of either multiple lowpass filtering or multiple finite time integration. Equations in unknown parameters are set up and then solved by means of the linear least–squares. The singular value decomposition has been used to solve the least–squares problem because of its numerical robustness and because of the extra data it provides for analysis. Examples of up to 14th order are discussed.

1. Introduction

This chapter presents a parameter identification algorithm for multivariable, continuous time systems, based on a paper by the author (Boje, 1988). Systems with non–zero initial conditions and measurement offset (or bias) can be identified from arbitrary, but active input–output data.

The model structure used (Section 2) is a minimal input–output representation in the Popov canonic form (Kailath, 1980), modified to include offset and for computational ease. A benefit of this structure over models with a diagonal output (denominator) matrix (Mathew and Fairman, 1974; Saha and Rao, 1982; Whitfield and Messali, 1987) is that the highest order derivative in any row and the number of free parameters is less (or equal) in the minimal representation. The minimal representation allows immediate construction of a minimal order state–space representation (Appendix 1). Kailath (1980) would provide background on polynomial system representations.

N. K. Sinha and G. P. Rao (eds.), Identification of Continuous-Time Systems, 419–442.

A well known method of preparing measured input–output data for parameter identification is by multiple integration (Mathew and Fairman, 1974; Whitfield and Messali, 1987; Golubev and Horowitz, 1982). In Section 3 it is shown how this idea is extended by replacing integration with low–pass filtering (Saha and Rao, 1982) or with finite time integration (Eitelberg, 1987).

The final stage of the algorithm is a linear least–squares solution for the unknown parameters (Section 4). The singular value decomposition can be used to improve numerical robustness and for analyzing the least–squares problem.

Examples are given in Section 5.

2. A Canonic Linear Multivariable Input–Output Model

Development of linear model

A general linear(ised) model is obtained from the linearisation of an oriented, noise free, time invariant, proper system model with m – inputs, u(t), q – outputs, y(t), and n – states, x(t), described by the non–linear state–space differential equation,

$$p\underline{x}(t) = \underline{f}(\underline{x}(t),\underline{u}(t)) \tag{1}$$

with algebraic output equation,

$$\underline{y}(t) = \underline{g}(\underline{x}(t),\underline{u}(t)) \tag{2}$$

about some steady state operating point (denoted subscript ss),

$$(p\underline{x}^T, \underline{x}^T, \underline{y}^T, \underline{u}^T)^T = (\underline{0}, \underline{x}_{ss}^T, \underline{y}_{ss}^T, \underline{u}_{ss}^T)^T \tag{3}$$

The linearisation is,

$$p\underline{x}(t) = \underline{A}\,\underline{x}(t) + \underline{B}\,\underline{u}(t) + \underline{z}_1 \tag{4a}$$

$$\underline{y}(t) = \underline{C}\,\underline{x}(t) + \underline{D}\,\underline{u}(t) + \underline{z}_2 \tag{4b}$$

with,

$$\begin{bmatrix} \underline{A} & \underline{B} \\ \underline{C} & \underline{D} \end{bmatrix} = \begin{bmatrix} \frac{\partial \underline{f}}{\partial \underline{x}} & \frac{\partial \underline{f}}{\partial \underline{u}} \\ \frac{\partial \underline{g}}{\partial \underline{x}} & \frac{\partial \underline{g}}{\partial \underline{u}} \end{bmatrix} \tag{5}$$

and,

$$\underline{z}_1 = -\underline{A}\,\underline{x}_{ss} - \underline{B}\,\underline{u}_{ss} \tag{6a}$$

$$\underline{z}_2 = \underline{y}_{ss} - \underline{C}\,\underline{x}_{ss} - \underline{D}\,\underline{u}_{ss} \tag{6b}$$

The notation, $p = \frac{d}{dt}$, (the derivative with respect to time) will be used in this chapter. $p^i = \frac{d^i}{dt^i}$, $i \geq 0$ and $p^0 = 1$. The inverse of p or of functions of p are not required. The constant offset terms, $\underline{z}_1$ and $\underline{z}_2$, in eq(4) (which must not be confused with noise terms in stochastic models) are a result of calibration offset or plant non–linearity and can be removed by a change in co–ordinates. Whitfield and Messali (1987) consider identification of offset (as a constant disturbance) but it has been largely ignored in the literature despite its occurrence in many practical problems when the co–ordinate change required to cancel the offset is not known *a priori*.

The discussion in this chapter is restricted to oriented systems where the inputs and outputs can be distinguished from each other. The identification of non–oriented systems can use the same method but has some extra degrees of freedom in the choice of parameters. See Boje (1986) for some suggestions on how to approach such problems.

Development of an input–output description from the (linear) state–space model

Simultaneous identification of the state and system matrices from input–output data using eq(4) would involve setting up and solving non–linear equations (which will be under–determined unless additional constraints are added). A model which describes the input–output behaviour without the state vector (which can later be constructed from the model, inputs and initial state vector) and which is linear in its parameters is more useful for the parameter identification problem.

For most problems, models obtained from input–output data can only model those parts of a system which are both controllable (excited from the measured inputs) and observable (detected in the measured outputs) – comments covering some situations when this is not the case are made in Appendix 2. It will also be assumed that the inputs are sufficiently exciting and that the measurements are suitable for identification in the sense discussed by Saha and Rao (1982).

The simplest method of eliminating the state vector in eq(4) is to write,

$$\det(p\underline{I}-\underline{A})\ \underline{y}(t) = (\underline{C}\ \mathrm{Adj}(p\underline{I}-\underline{A})\ \underline{B} + \det(p\underline{I}-\underline{A})\ \underline{D})\ \underline{u}(t) + \left[\underline{C}\ \mathrm{Adj}(p\underline{I}-\underline{A})\underline{z}_1 + \det(p\underline{I}-\underline{A})\underline{z}_2\right]\Big|_{p=0} \tag{7}$$

(Differentiation of a constant gives zero which allows the substitution, p=0, above.) The derivation of eq(7) requires that signals are sufficiently differentiable but does not require the inverse of p which has not been defined. The proof of this is given in Appendix 3. The more usual approach via the Laplace transform has been avoided to highlight the fact that the problem is set in the time domain. Use of the Laplace transform does not prevent the inclusion of offsets and initial conditions but does require some assumptions for signals which are known only over finite times (Golubev and Horowitz, 1982).

Rewriting eq(7) with more useful notation,

$$\underline{D}(p)\ \underline{y}(t) = \underline{N}(p)\ \underline{u}(t) + \underline{z} \tag{8}$$

where,

$$\underline{D}(p) = \mathrm{diag}\{\det(p\underline{I}-\underline{A})\} \qquad \text{– an } [m\times m] \text{ diagonal matrix} \tag{9a}$$

$$d_{ii}(p) = \sum_{k=0}^{n} d_{ii\,k}\, p^k \qquad \text{– an } n^{th} \text{ degree polynomial in p} \tag{9b}$$

$$\underline{N}(p) = \underline{C}\ \mathrm{Adj}(p\underline{I}-\underline{A})\ \underline{B} + \det(p\underline{I}-\underline{A})\ \underline{D} \qquad \text{– a } [q\times m] \text{ matrix} \tag{10}$$

$$\underline{z} = -(\underline{C}\ \mathrm{Adj}(\underline{A})\ \underline{z}_1 + \det(\underline{A})\ \underline{z}_2) \qquad \text{– a constant } [q] \text{ vector of offsets} \tag{11}$$

The diagonal $\underline{D}(p)$ model of eq(8) which has used by Saha and Rao (1982), and Whitfield and Messali (1987) may be appealing as a model for identification algorithms because of its apparent simplicity. In order to determine the parameters uniquely, one coefficient in each row of eq(8) is fixed (usually the coefficient of the highest order derivative in $d_{ii}(p)$) and common factors in each row removed. The following example shows that this model structure is not necessarily the best for identification problems.

Example 1 : Some difficulties with a diagonal D(p) model structure

Consider the following 2–input, 2–output proper first order system.

$$p\,x = -1\,x + (-1\ 1)\,\underline{u} + 2 \tag{12a}$$

$$\underline{y} = \begin{bmatrix} 1 \\ -3 \end{bmatrix} x + \begin{bmatrix} 2 & 3 \\ -4 & -5 \end{bmatrix} \underline{u} + \begin{bmatrix} 1 \\ 4 \end{bmatrix} \tag{12b}$$

Using eq(7) the input–output representation for this system is,

$$\begin{bmatrix} p+1 & 0 \\ 0 & P+1 \end{bmatrix} \underline{y} = \begin{bmatrix} 2p+1 & 3p+4 \\ -4p-1 & -5p-8 \end{bmatrix} \underline{u} + \begin{bmatrix} 3 \\ -2 \end{bmatrix} \tag{13}$$

Although there are no common factors in any one row, in this simple example (with $x = y - D\,u - z_2$) it is easy to see that,

$$\begin{bmatrix} p+1 & 0 \\ 3 & 1 \end{bmatrix} \underline{y} = \begin{bmatrix} 2p+1 & 3p+4 \\ 2 & 4 \end{bmatrix} \underline{u} + \begin{bmatrix} 3 \\ 7 \end{bmatrix} \tag{14}$$

Between eq(13) and eq(14), a (matrix) common factor of,

$$\underline{L}(p) = \begin{bmatrix} 1 & 0 \\ -3 & p+1 \end{bmatrix} \; (= \mathrm{Adj}(p\underline{I} - \underline{A})) \tag{15}$$

has been removed. (This can be verified by pre–multiplication of eq(14) by $\underline{L}(p)$.)

The points are made:

1) If all but one of the coefficients are free (unknown), eq(13) has 6 free parameters and 1 free initial condition per row. Row 2 of eq(14) has only 4 free parameters and no free initial conditions. (See also eq(28)). Can the larger number of unknowns in row 2 of eq(13) be identified from measured input–output data? (They can be – see Section 5.)

2) After identification, would it be possible (by examining the degree of the determinant of $\underline{D}(p)$) to decide whether eq(13) is a 1st order or a 2nd order system? (In practical problems, terms which should be equal may differ as a result of the effects of noise, non–linearity, limited precision arithmetic and unmodelled dynamics.) An identification result such as eq(14) may reveal underlying system structure more clearly than one such as eq(13).

3) A direct state–space construction for eq(14), having deg(det($\underline{D}$(p))) (=n=1) state is given in Appendix 1. There is no directly applicable state–space construction for eq(13) with fewer than (n×m=2) states.

A minimal, canonic input–output representation

Common left factors and minimal input–output representations

It was seen in Example 1 that by allowing off–diagonal entries in $\underline{D}(p)$, it may be possible to find algebraically common terms on the left and right hand sides of eq(8). Algebraically the common terms may be separated by writing (after Kailath, 1980, p.376),

$$\underline{L}(p)\ \underline{D}'(p)\ \underline{y}(t) = \underline{L}(p)\ \underline{N}'(p)\ \underline{u}(t) + [\underline{L}(p)\ \underline{z}']\big|_{p=0} \tag{16}$$

where,

$$\underline{L}(p)\ \underline{D}'(p) = \underline{D}(p), \qquad \underline{L}(p)\ \underline{N}'(p) = \underline{N}(p), \qquad [\underline{L}(p)\ \underline{z}']\big|_{p=0} = \underline{z}$$

$\underline{L}(p)$ is a *common left factor* (clf) of eq(8). Eq(16) represents the original system as can be seen if it is multiplied out (multiplication of polynomials in p is defined). $\underline{L}(p)$ must be square and must not be strongly singular (i.e. $\det(\underline{L}(p)) \neq 0$ for any p). Note that $\underline{L}(p)$ is not unique.

Given the input–output equation, eq(8) and a clf, $\underline{L}(p)$, satisfying eq(16) then,

$$\underline{D}'(p)\ \underline{y}(t) = \underline{N}'(p)\ \underline{u}(t) + \underline{z}' \tag{17}$$

(eq(16) with the common left factor removed) represents a system with identical external behaviour to the original system, eq(8). (The uniqueness of the solution of linear equations can be used to prove this.)

The highest order derivative in (or degree of) the determinants of output (denominator) matrices from eq(16) is,

$$\deg(\det(\underline{L}(p))) + \deg(\det(\underline{D}'(p))) = \deg(\det(\underline{D}(p))) \tag{18}$$

The reduction in model order shown by eq(18) and corresponding reduction in the number of parameters in both the input–output representation and in the state–space representation motivates finding a *greatest* common left factor (gclf), $\underline{L}_G(p)$, in the sense of having the highest possible degree in its determinant. If $\underline{L}_G(p)$ is a gclf of eq(8), then removing $\underline{L}_G(p)$ (as in eq(17)) gives a *minimal* input–output representation, where $\det(\underline{D}_{minimal}(p))$ has the lowest possible degree.

The modified Popov canonic form

A minimal representation cannot be used as the basis for system identification without further constraints to make it unique (or canonic). (E.g. any row of eq(8) can be multiplied by a non–zero constant without affecting the external behaviour.) A canonic form has been chosen that differs trivially from a canonic form attributed to V M Popov (*ibid* p.481, where it is presented for column–reduced matrices) and which was proposed mostly for discrete time identification by Beghelli and Guidorzi (1976). The Popov form is *row–reduced* (Kailath, 1980, p.385) and this allows the degree (maximum order of derivative) of the entries of $\underline{\mathbf{D}}(p)$ and $\underline{\mathbf{N}}(p)$ to be determined *a priori* from knowledge of the highest order of specific entries in $\underline{\mathbf{D}}(p)$ (see below) and of properness or strict properness. The reason for modifying the definition of the canonic form is simply to make the numerical implementation of the identification algorithm which follows less cumbersome and to include offset and direct transfer terms which have not generally appeared in the literature. The modification does not affect the row–reduced property and as the canonic form can be obtained from any minimal input–output representation by elementary row operations or equivalently by pre–multiplication with a suitably chosen unimodular matrix (a polynomial matrix with a constant, non zero determinant), it does not restrict the generality of the linear model.

The general (and now assumed minimal) representation is,

$$\underline{\mathbf{D}}(p)\,\underline{\mathbf{y}}(t) = \underline{\mathbf{N}}(p)\,\underline{\mathbf{u}}(t) + \underline{\mathbf{z}} \tag{19}$$

In identification problems it is not necessary to be able to construct a minimal representation from a general one. It is sufficient to know that a "minimal" representation (suitably modified to include offset and non–controllable modes) exists and to use its structure to determine which parameters need to be identified. Eq(19) is written with the highest order terms in p written separately as,

$$(\underline{\mathbf{S}}(p)\,\underline{\mathbf{D}}_H + \underline{\mathbf{\Psi}}(p)\,\underline{\mathbf{D}}_L)\,\underline{\mathbf{y}}(t) = (\underline{\mathbf{S}}(p)\,\underline{\mathbf{N}}_H + \underline{\mathbf{\Psi}}(p)\,\underline{\mathbf{N}}_L)\,\underline{\mathbf{u}}(t) + \underline{\mathbf{z}} \tag{20}$$

with,

$$\underline{\mathbf{S}}(p) = \begin{bmatrix} p^{k_1} & & & 0 \\ & p^{k_2} & & \\ & & \ddots & \\ 0 & & & p^{k_q} \end{bmatrix} \quad \text{a diagonal matrix in p}$$

k_i the highest degree (in p) in row i

$\underline{\mathbf{D}}_H$, $\underline{\mathbf{N}}_H$ coefficient matrices of the highest order terms

$$\underline{\Psi}(p) = \left[\begin{array}{cccc|cccc|c|cccc} p^{k_1-1} & \cdots & p^1 & p^0 & 0 & \cdots & 0 & 0 & & & & & \\ 0 & \cdots & 0 & 0 & p^{k_2-1} & \cdots & p^1 & p^0 & & & 0 & & \\ & & & & & & & & \ddots & & & & \\ & & 0 & & & & & & & p^{k_q-1} & \cdots & p^1 & p^0 \end{array}\right]$$

a block diagonal matrix

$\underline{\mathbf{D}}_L$, $\underline{\mathbf{N}}_L$ coefficient matrices of lower order terms.

Assertion : Every minimal input–output representation has a unique modified Popov form with,

$\underline{\mathbf{D}}_H$ in eq(20) lower triangular with unit diagonal elements (21a)

and entries in $\underline{\mathbf{D}}_H$,

$$d_{Hij} = d_{Hji} = 0 \;\text{ if } k_i = k_j \text{ and } i \neq j \tag{21b}$$

The modified Popov form can always be obtained by rearranging the rows of the Popov form so that the *pivot* entries (*ibid*, p.481) lie on the diagonal. Requirement eq(21a) means that $\underline{\mathbf{D}}_H$ is row reduced as $\underline{\mathbf{D}}_H$ is always non–singular. Requirement eq(21b) is always possible to achieve by subtracting a multiple of row i from row j or *vice versa*. Elementary row operations are used to achieve the above requirements on $\underline{\mathbf{D}}(p)$ in eq(19) and by specifying $\underline{\mathbf{D}}(p)$ uniquely, the representation is made unique. In addition to the requirements on $\underline{\mathbf{D}}_H$, some of the elements in $\underline{\mathbf{D}}_L$ are structurally equal to zero – see eq(26) below. Eq(14) is in the modified Popov form.

Row i of eq(19) can be written element by element as,

$$\left[\sum_{j=1}^{q}\sum_{h=0}^{k_i} d_{ij(h)} y_j^{[h]}(t)\right] = \left[\sum_{r=1}^{m}\sum_{h=0}^{k_i} n_{ir(h)} u_j^{[h]}(t)\right] + z_i \tag{22}$$

where (for example) $d_{ij(h)}$ is the coefficient in p^h of element (i,j) of $\underline{\mathbf{D}}(p)$ and, $y_i^{[h]}$ is the hth derivative of output i.

From the canonic requirement, eq(21), the following may be said about the elements of eq(22):

For $\underline{D}_H$ lower triangular, $d_{ij(k_i)} = 0$ if $j > i$ (23)

For monic diagonal elements in $\underline{D}_H$, $d_{ii(k_i)} = 1$ (24)

If row orders are equal, $d_{ij(k_i)} = 0$ if $j < i$ and $k_i = k_j$ (25)

For $\underline{D}(p)$ row reduced in modified Popov form, $d_{ij(h)} = 0$ if $h > k_i$ (26)

If the system is strictly proper, $n_{ir(k_i)} = 0$ (Eq(22) is proper.) (27)

An observer form state space representation corresponding to the modified Popov form is given in Appendix 1.

The number of free parameters in row i of the modified Popov form is,

$$n_{free,i} = \left[\sum_{j=1}^{q} \min\{k_i,k_j\} \right] + (k_i+1) \times m + 1 \qquad (28)$$

for a proper system with offset. In an identification problem, there may also be free initial conditions – see Section 3. (Under the same conditions, row i of the diagonal form has $(n_i'+1)\times(m+1)$ free parameters, where n_i' ($\leq$ n) is the row order after removal of common factors.)

3. Preparing input–output data for parameter identification by means of integral functionals

A direct approach to solving the parameter identification problem is to differentiate the input–output measurements (numerically or with analog "differentiators") as many times as the highest order derivatives in eq(22). At each of a number of discrete time points t_α, $\alpha = 1..N$, a linear equation in the unknown plant model parameters is constructed from eq(22). The resulting set of equations could then be solved, one row of eq(22) at a time. (The canonic form of Section 2, sufficiently exciting inputs and correct choice of t_α ensure that the set of equations is properly determined.) Attempting to obtain high order derivatives makes this approach unsatisfactory because of the amplification of noise and round–off errors.

In this section two alternatives are presented which replace differential equations, eq(22), with exact, continuous–time integral representations. Numerical algorithms have been developed to set up equations in the unknown parameters using these integral representations. The first representation uses low passing functionals (the "Poisson moment functionals" of Saha and Rao (1982)) of the data. This is a generalization of the method of Mathew and Fairman (1974); Golubev and Horowitz (1982); and Whitfield and Messali (1987). This method is a special case of the state variable filter method (Young, 1981) which will not discussed here. The second representation uses so called "macro–differences", based on finite time integrals (Eitelberg, 1987). The use and derivation of these methods is described in more detail in Boje (1986).

As the integral representations are in the continuous time domain, equations need not be set up equidistantly ($t_\alpha - t_{\alpha+1}$ need not be constant). The low–pass or finite time filtering can be achieved using analog filters or high order numerical methods (trapezoidal and Simpson's rules have been used by the author). For accurate numerical filtering, more than one measurement can be made between t_α and $t_{\alpha+1}$. Special algorithms can be devised if measurement of different channels is not coherent as is typical in low cost multi–channel analog to digital converter cards for PC's.

Low passing functionals

Consider a first order low pass filter, with input w(t), output, v(t) and feedback gain, a (a > 0 for a stable filter). Let the filter initial condition (which is not at all related to the initial conditions of the measured plant) at $t_0=t=0$, v_0, be zero. Define the filter output (the convolution of the filter input and impulse response) as $v(t)=L\{w(t), a\}$ when $v_0=0$.

$$v(t) = L\{w(t), a\} \doteq e^{-at} \int_0^t e^{a\tau} w(\tau)\, d\tau \tag{29}$$

A second application of eq(29) to w(t) gives,

$$L^2\{w(t), a\} = L\{L\{w(t), a\}, a\}$$
$$= e^{-at} \int_0^t \int_0^{t_1} e^{a\tau_2}\, w(\tau_2)\, d\tau_2\, d\tau_1 \tag{30}$$

so that, for $k \geq 1$,

$$L^k\{w(t),a\} = e^{-at} \underbrace{\int_0^t \cdot \cdot \cdot \cdot \cdot \int_0^{\tau_{k-1}}}_{k \text{ times}} e^{a\tau_k}\, w(\tau_k)\, d\tau_k \cdot \cdot \cdot d\tau_1 \tag{31}$$

Note that, $L^0\{w(t),a\} = w(t)$. If $L\{\cdot, a\}$ is applied to each element of eq(22), k_i times, the resulting system representation is (Boje, 1986),

$$\sum_{j=1}^{q} \sum_{h=0}^{k_i} d_{ij(h)} \Bigg[\sum_{v=1}^{m} ((-a)^v \tbinom{h}{v})\, L^{(k_i-h+v)}\{y_j(t), a\}) -$$

$$-e^{-at} \sum_{v=0}^{h-1} \left[\frac{t^{(k_i-h+v)}}{(k_i-h+v)!} \sum_{\ell=0}^{v} ((-a)^{(v-\ell)} \tbinom{v}{\ell}\, y_j^{[\ell]}(0)) \right] \Bigg] =$$

$$= \sum_{r=1}^{m} \sum_{h=0}^{k_i} n_{ir(h)} \Bigg[\sum_{v=0}^{h} ((-a)^v \tbinom{h}{v})\, L^{(k_i-h+v)}\{u_r(t), a\}) -$$

$$-e^{-at} \sum_{v=0}^{h-1} \left[\frac{t^{(k_i-h+v)}}{(k_i-h+v)!} \sum_{\ell=0}^{v} ((-a)^{(v-\ell)} \tbinom{v}{\ell}\, u_r^{[\ell]}(0)) \right] \Bigg] +$$

$$+ z_i\, e^{-at}\, a^{-k_i} \left[e^{at} - \sum_{h=0}^{k_i-1} ((at)^h/h!) \right] \tag{32}$$

where,

$$\tbinom{v}{h} = \frac{v!}{h!\ (v-h)!} \tag{33}$$

are binomial coefficients.

In eq(32) the (possibly) unknown initial external conditions, $y_j^{[\ell]}(0)$ and $u_r^{[\ell]}(0)$, occur as linear combinations of one another and are multiplied by the (possibly) unknown system parameters. A linear estimation problem which avoids differentiation (the one–sided numerical differentiation required to obtain derivatives

...ial time would be especially unreliable) is obtained by replacing the ... of the external variables' initial conditions and system parameters (the ... and fourth rows of eq(32) above) with the initial states, $x_{(i;j)}(0)$ $(j=1,\ldots,k_i)$ ... notation, (i;j), is defined in eq(A1.4)) of the observer form state space ...presentation presented in Appendix 1. This substitution yields,

$$\sum_{j=1}^{q}\sum_{h=0}^{k_i} d_{ij(h)}\left[\sum_{v=0}^{h}((-a)^{v}\tbinom{h}{v}\, L^{(k_i-h+v)}\{y_j(t),\, a\})\right] -$$

$$-e^{-at}\sum_{h=0}^{k_i}\left[x_{(i;h+1)}(0)\,\frac{t^h}{h!}\sum_{w=0}^{k_i-h}((-at)^{w}/w!)\right] =$$

$$= \sum_{r=0}^{m}\sum_{h=0}^{m} n_{ir(h)}\left[\sum_{v=0}^{h}((-a)^{v}\tbinom{h}{v}\, L^{(k_i-h+v)}\{u_r(t),a\})\right] +$$

$$+ z_i\, e^{-at}\, a^{-k_i}\left[e^{at} - \sum_{h=0}^{k_i-1}(at)^{h}/h!\right] \qquad (34)$$

Both eq(32) and eq(34) are exact representations of the system, eq(32) but as eq(34) does not contain intrinsically (or algebraically) dependent unknowns, it useful for system parameter identification. If the system under consideration is initially at rest internally ($\underline{x}(t_0) = \underline{0}$) then eq(32) and eq(34) are identical. Some external initial conditions may be non–zero but their effects on the state must cancel.

Example 2 Representation by means of low–pass functionals

Consider the second order SISO system,

$$d_2\, y^{[2]}(t) + d_1\, y^{[1]}(t) + d_0\, y(t) = n_0\, u(t) + z \qquad (35)$$

From the restrictions required for the canonic form, eq(22), $d_2=1$ but for clarity, d_2 will be carried through the example. Only one term in $u^{[j]}$, $(j=1,\ldots,k_i)$ has been used for simplicity. In the notation of eq(32) this is,

$$d_2(y(t)-2aL\{y(t),a\}+a^2L^2\{y(t),a\}) + d_1(L\{y(t),a\}-aL^2\{y(t),a\}) + d_0L^2\{y(t),a\} - e^{-at}(d_2(y(0)+t(y^{[1]}(0)-ay(0)))+d_1ty(0)) = n_0L^2\{u(t),a\} + ze^{-at}(e^{at}-1-at)/(a^2) \tag{36}$$

In the notation of eq(34) this is,

$$d_2(y(t)-2aL\{y(t),a\}+a^2L^2\{y(t),a\}) + d_1(L\{y(t),a\}-aL^2\{y(t),a\}) + d_0L^2\{y(t),a\} - e^{-at}(x_1(0)(1-at)+x_2(0)) = n_0L^2\{u(t),a\} + ze^{-at}(e^{at}-1-at)/(a^2) \tag{37}$$

If the lowpass parameter, a, is chosen equal to zero (a=0), the equations for the representation simplify considerably, reducing the amount of computing required to form equations in the unknown parameters. The resulting representation is the generalization (Mathew and Fairman, 1974; Whitfield and Messali, 1987) of the method of Golubev and Horowitz (1982), but using the canonic form described in Section 2. (The choice of lowpass parameter is discussed briefly in Section 4.)

The method of "macro differences"

Eitelberg (1987) gives another useful system representation which is based on finite time integration of the system inputs and outputs. The low–pass integral, eq(29) can be replaced by a finite time average,

$$v(t) = A_T\{w(t)\} \doteq \frac{1}{T}\int_{t-T}^{t} w(\tau)d\tau \tag{38}$$

$$A_T^2\{w(t)\} = \frac{1}{T^2}\int_{t-T}^{t}\int_{\tau_1-T}^{\tau_1} w(\tau_2)\, d\tau_2\, d\tau_1 \qquad \text{etc.}$$

If $A_T\{\cdot\}$ is applied k_i times to any row of equation eq(22), the result can be written as a macro difference equation,

$$\sum_{j=1}^{q}\sum_{h=0}^{k_i} d_{ij(h)}\left[\sum_{v=0}^{h}((-1)^v\binom{h}{v}\frac{1}{T^h}A_T^{(k_i-h)}\{y_j(t-vT)\})\right] =$$

$$= \sum_{r=1}^{m}\sum_{h=0}^{k_i} n_{ir(h)}\left[\sum_{v=0}^{h}((-1)^v\binom{h}{v}\frac{1}{T^h}A_T^{(k_i-h+v)}\{u_r(t-vT)\})\right] + z_i \tag{39}$$

where, $\binom{h}{v}$ are binomial coefficients as in eq(33).

Eq(39) is an integral representation but requires no initial condition terms to be a complete representation of the original system, eq(22). This feature makes the representation attractive as the basis of an identification algorithm because it results in a smaller number of unknown parameters (unless initial conditions are known to be zero). This representation allows an elegant, exact approach to the problem of using non–contiguous data (Gawthrop, 1984) as data from different experiments can be used together in order to obtain more representative or better conditioned results. The macro differences representation cannot be used to identify initial conditions (which may be required for simulation studies).

4. Solving the parameter estimation problem by linear least squares

Either of the system representations presented in Section 3 can be used to solve for unknown system parameters and initial conditions and offset using the following algorithm:

1) Construct $L^k\{w(t_\alpha),a\}$ or $A_T^k\{w(t_\alpha)\}$ for each measured variable ($w(t_\alpha) = u_r(t_\alpha)$ (r=1..m) or $y_j(t_\alpha)$ (j=1..q)), for k=0..k_{max} (the maximum expected row order) and for each time point, t_α (α=1..N). (Fewer equations than the number of measurement points result from the application of the macro–differences method.) This step only needs to be performed once if the intermediate results are stored.
The lowpass parameter (a) or the macro–interval (T) must be specified. They are single parameters which can be chosen so as to improve the identification result, but a simple mechanism to choose them is open to further research. Generally, larger values of (1/T) or of (a), result in better modelling of high frequencies (at the expense of lower frequencies).

2) For each row of eq(19) (i=1,..q), make use of known parameters and construct a set of linear equations in the unknown parameters from eq(34) or eq(39), using the results from Step 1 above. Eq(23) to eq(27) define those parameter values that are

constrained by canonic requirements; first principles modelling such as a knowledge of steady state gains, initial conditions, offset or high frequency behaviour may allow other model parameters to be fixed.

Eq(34) and eq(39) require that the row orders, k_i, as defined in eq(19) be known *a priori*. If these cannot be determined from first principles understanding of the plant, suitable row orders can be determined from the experimental data by examination of the least squares problem (discussed below) and by simulation studies. Guidorzi (1975) (for discrete time systems) and Eldem and Yildizbayrak (1988) (for adaptive identification) contain suggestions on structure (k_i) selection.

3) Solve the linear equations for the unknown parameters using any convenient method. The author has used unweighted linear least squares. The singular value decomposition (Lawson and Hanson, 1974) has been used to solve the least–squares problem because of its numerical robustness and because ready analysis of the least–squares problem (especially in poorly conditioned cases) is possible. This analysis facilitates row order determination and allows poorly identified parameters to be highlighted, possibly serving as a guide towards further experimentation. The parameter identification could be performed on–line, using a recursive least–squares algorithm (but it cannot surpass the quality of the off–line estimates). The row order selection in its simplest form consists of examination of the conditioning of the least squares problem. Ill conditioning can result from the input signals not being sufficiently exciting, or from incorrect row order selections. For row order selection the possible structural dependence in any one row on row orders of other rows is an unfortunate result from eqs 23, 25 and 26. This means that the row order selection must be achieved in an iterative fashion if it is not known *a priori*.

4) Return to Step 2 until unknown parameters for all rows of eq(19) have been identified.

5. Examples

Example 1 revisited

The model in eq(12) was simulated (using 12 bits to model analog–to–digital conversion) with N=200 steps, a step size, h=0.05s and 5s long unit pulses starting at t=0s (input 1) and t=2.5s (input 2). The initial state was zero but due to the non–canonic offset vector, initial state of the canonic observer form state space representation (Appendix 1) is non–zero and must be identified. Using the method of lowpassing functionals, a=0 and the modified Popov form, the identified model was,

$$\begin{bmatrix} p+0.9992 & 0 \\ 3.000 & 1 \end{bmatrix} \underline{y} = \begin{bmatrix} 2.031p+0.9999 & 2.967p+3.996 \\ 3.000 & 4.000 \end{bmatrix} \underline{u} + \begin{bmatrix} 3.048 \\ 7.000 \end{bmatrix} \qquad (40)$$

(cf. eq(13)). With a diagonal model structure, the identified model was,

$$\begin{bmatrix} p+0.9992 & 0 \\ 0 & p+0.9988 \end{bmatrix} \underline{y} = \begin{bmatrix} 2.031p+.9999 & 2.967p+3.996 \\ -4.096p-1.000 & -4.902p-7.988 \end{bmatrix} \underline{u} + \begin{bmatrix} 3.048 \\ -2.150 \end{bmatrix} \qquad (41)$$

(cf. eq(14)). The condition numbers, κ (=ratio of maximum to minimum singular values), and residual errors (Lawson and Hanson, 1974) of the least squares problems arising from each row of eq(40) and eq(41) are shown in Table 1. Note that it is unusual for a row order to be equal to the system order.

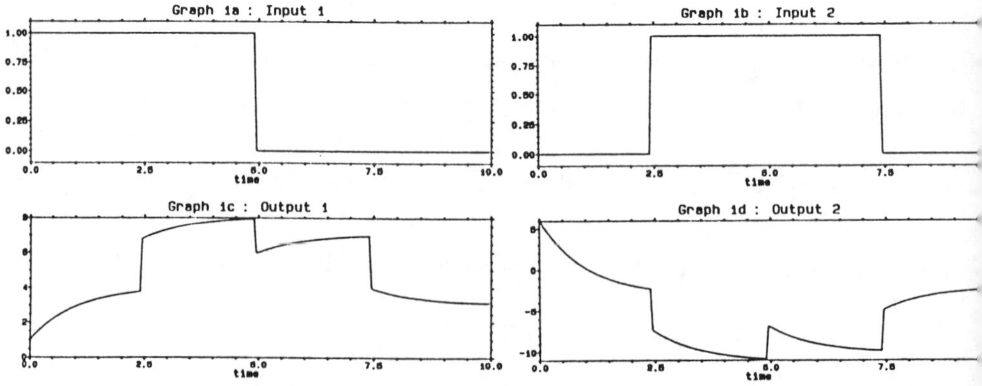

Graph 1 First order example – Inputs and Outputs

Table 1 – Condition numbers and residual errors Example 1

	Popov row 1	Popov row 2	Diagonal row 1	Diagonal row 2
Condition number κ	604.22	49.689	604.22	386.65
L.S. residual error	3.003E–2	5.406E–2	3.002E–2	7.666E–2

Reduced order linear model for two coupled turbogenerators

Data for this example was obtained by simulation of a 30th order, first principles, non–linear model of sub–synchronous resonance in turbogenerators connected to a long, series compensated, transmission line (Balda, 1986, Chapter 6). A low order, linear state–space model was required for shunt reactor controller design. The inputs were unit step changes in the shunt reactance at time t=0.05 (shunt reactor 1) and at t=2.5s (shunt reactor 2). Inter–sample time was h=.005 and N=1000 samples were used, resulting in the outputs shown in Graph 1. The system is strictly proper with zero offset and has zero initial conditions. The non–linear turbogenerator models were identical, but constraining the model parameters to force symmetry increased the least–squares residual errors and prediction (simulation) errors (due to numerical effects in the simulation of the high order plant).

Simulation studies from identification under the conditions summarized in Table 2 are shown. The root mean square simulation error is defined as,

$$\text{sim. err.} = \left[\sum_{\alpha=1}^{N} (y_{meas}(t_\alpha) - y_{sim}(t_\alpha))^2 \right]^{1/2} / N \tag{42}$$

Notes : 1) Graph 2 illustrates the potential for low order model identification and model order reduction.

2) Visually, there is little difference between Graphs 3 and 4 (the simulation errors in Graph 4 are higher), but the difference in condition numbers (Table 2) suggest that the result for a=10 is more trustworthy (more detailed examination of the results is required to confirm this – see Lawson and Hanson, 1974). With 10th order

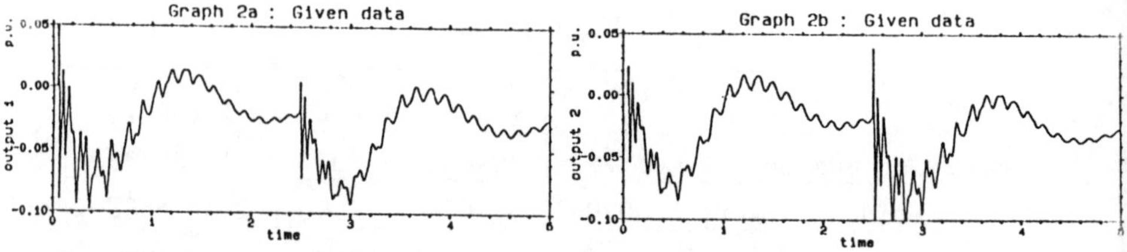

Graph 2 Simulation data from high order non–linear model

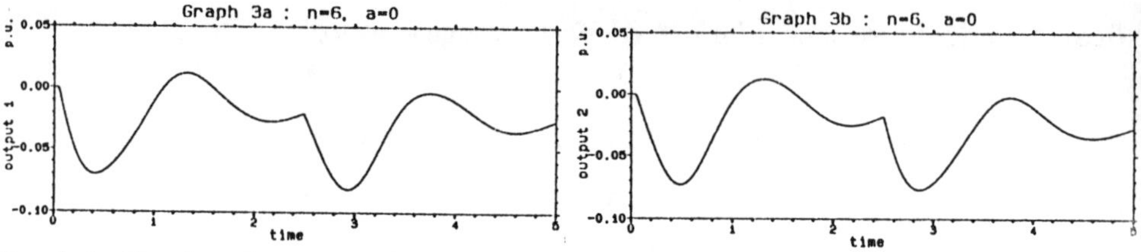

Graph 3 Simulated response for sixth order linear model, a=0

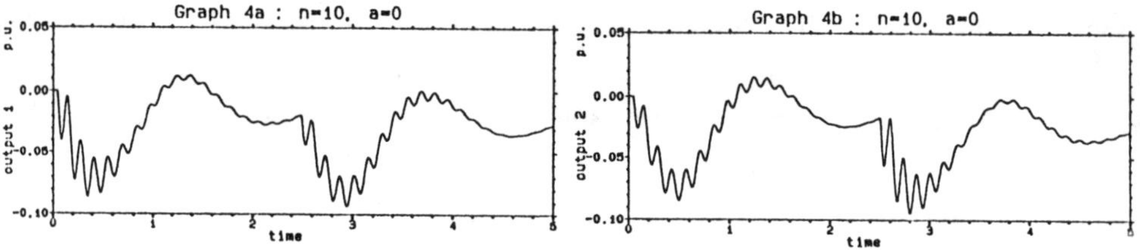

Graph 4 Simulated response for tenth order linear model, a=0

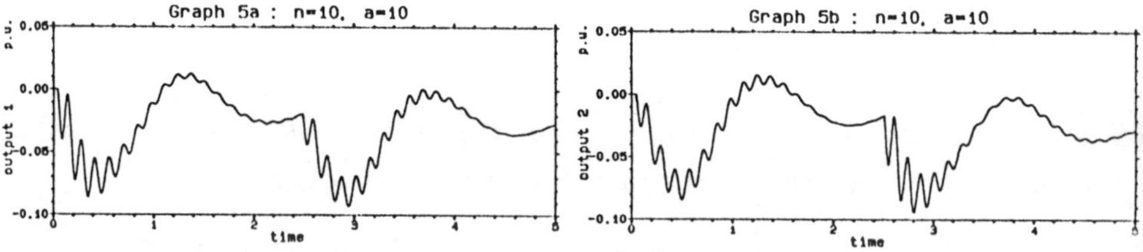

Graph 5 Simulated response for tenth order linear model, a=10

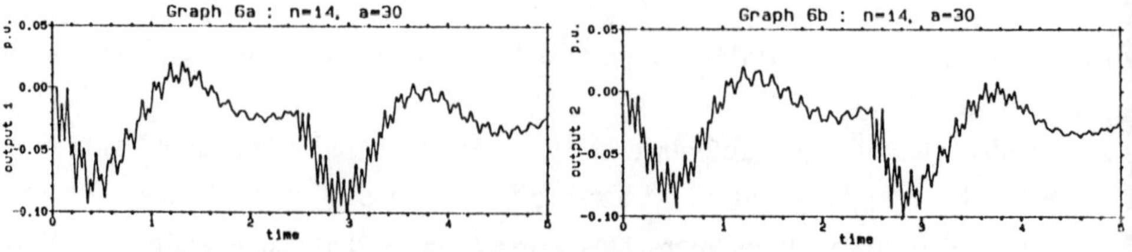

Graph 6 Simulated response for fourteenth order linear model, a=30

Graphs 2–6 Coupled Turbogenerators

Table 2 Results shown for turbogenerator example

Graph number	3	4	5	6
Model order	6	10	10	14
Lowpass param. a	0	0	10	30
LS resid. err. row 1	0.2913	0.1975	0.1889	0.08725
LS resid. err. row 2	0.2884	0.1958	0.1887	0.08727
Condition no. row 1	1.591E+6	2.347E+9	1.032E+6	1.019E+10
Condition no. row 2	1.591E+6	2.347E+9	1.032E+6	1.019E+10
RMS sim. err. o/p 1	9.280E-3	8.755E-3	1.077E-2	7.3160E-3
RMS sim. err. o/p 2	9.185E-3	9.100E-3	1.079E-2	7.2165E-3

models, simulated behaviour was reasonable for a in [0,25].

3) Graph 5 represents the highest order model that could be identified from the given data with acceptable simulated responses. The large condition number, the small improvement in simulation error (from 10th order models) and the fact that reasonable results were only obtained for some values of a (rather than for all a in some interval as in the 10th order case), suggest that this is the limit of the algorithm for this problem and that this result should be viewed with caution.

6. Conclusions

This chapter has presented an identification method that is applicable to a wide class of linearised systems.

7. Acknowledgement

The research on which this chapter is based was carried out under the supervision of Professor E Eitelberg. His advice during the preparation of this chapter is appreciated.

8. References

Balda J C *Subsynchronous resonance of turbogenerators with particular reference to a few selected countermeasures*, PhD Thesis, University of Natal, Durban, South Africa, 1986.

Beghelli S and Guidorzi R " A new input–output canonical form for multivariable systems" *IEEE Trans. Automatic Control*, pp 692–696, October 1976

Boje E S *Linear Model Identification for Multivariable Continuous Time Systems* MSc Thesis, University of Natal, Durban, South Africa, 1986.

Boje E S "Linear Model Identification for Multivariable Continuous Time Systems *8th IFAC/IFORS Symposium on Identification and System Parameter Estimation*, pp 633–638, Beijing, PR China, 27–31 August 1988.

Eitelberg E "Continuous time system representation with exact macro–difference expressions", *Int. J. Control*, Vol 47, No 5, pp 1207–1212, 1988

Eldem V and Yildizbayrak N "Parameter and structure identification of linear multivariable systems", *Automatica*, Vol 24, No 3, pp 365–373, 1988.

Gawthrop P J "Parameter estimation from non–contiguous data", *IEE Proc. D*, Vol 131, No 6, pp 261–266, 1984.

Golubev B and Horowitz I "Plant rational transfer function approximation from input–output data", *Int. J. Control*, Vol 36, No 4, pp 711–723, 1982.

Guidorzi R "Canonical structures in the identification of multivariable systems" *Automatica*, Vol 11, pp 361–374, 1975.

Kailath T *Linear systems*, Prentice Hall, Englewood Cliffs, 1980.

Lawson C L and Hanson R J *Solving Least Squares Problems*, Prentice–Hall, Englewood Cliffs, 1974.

Mathew A V and Fairman F W "Transfer function matrix identification", *IEEE Tran.*, Vol CAS–21, No 5, pp 584–588, 1974

Saha D C and Rao G P "Transfer function matrix identification in MIMO systems via Poisson moment functionals", *Int. J. Control*, Vol 35, No 4, pp 727–738, 1982

Whitfield A H and Messali N "Integral–equation approach to system identification", *Int. J. Control*, Vol 45, No 4, pp 1431–1445, 1987.

Young P C "Parameter estimation for continuous time models: A survey", *Automatica*, Vol 17, pp 23–39, 1981.

Appendix 1 State Space Construction

An observer form state–space representation corresponding to the canonic input–output representation of Section 2 is given below. The construction (described in more detail in Boje, 1986) is based on Kailath (1980, p.414). One reason for requiring a state–space representation is that the initial state is uniquely determined unlike the initial conditions of the external variables which are linear combinations of one another (an nth order differential equation has only n independent initial conditions). As a result, the initial state is identified (see Section 3). State–space methods may be useful in the further analysis of some problems (eg. state–space simulation is convenient) and are more familiar to some engineers. (Note that the canonic requirement eq(21) ensures that $\underline{D}_H^{-1}$ (used below) always exists ($\det(\underline{D}_H)=1$) and is easily constructed using the back substitution step of Gaussian elimination.)

The complete observer form state–space description is,

$$p\underline{\mathbf{x}}(t) = \underline{\mathbf{A}}_O\, \underline{\mathbf{x}}(t) + \underline{\mathbf{B}}_O\, \underline{\mathbf{u}}(t) + \underline{\mathbf{z}}_{O1}$$

$$\underline{\mathbf{y}}(t) = \underline{\mathbf{C}}_O\, \underline{\mathbf{x}}(t) + \underline{\mathbf{D}}_O\, \underline{\mathbf{u}}(t) + \underline{\mathbf{z}}_{O2} \tag{A1.1}$$

with state–space matrices,

$$\underline{\mathbf{A}}_O = \underline{\mathbf{A}}_O^0 - \underline{\mathbf{D}}_L\, \underline{\mathbf{D}}_H^{-1}\, \underline{\mathbf{C}}_O^0, \qquad \underline{\mathbf{B}}_O = \underline{\mathbf{N}}_L - \underline{\mathbf{D}}_L\, \underline{\mathbf{D}}_H^{-1}\, \underline{\mathbf{N}}_H,$$

$$\underline{\mathbf{C}}_O = \underline{\mathbf{D}}_H^{-1}\, \underline{\mathbf{C}}_O^0, \qquad \underline{\mathbf{D}}_O = \underline{\mathbf{D}}_H^{-1}\, \underline{\mathbf{N}}_H$$

and offset vectors,

$$z_{O1(i;j)} = 0 \quad j \neq k_i \text{ or } k_i = 0$$

$$z_{O1(i;j)} = z_i \quad j = k_i \text{ and } k_i \neq 0 \tag{A1.2}$$

and

$$z_{O2\,i} = 0 \quad k_i \neq 0$$

$$z_{O2\,i} = z_i \quad k_i = 0 \tag{A1.3}$$

The subscript, (i;j), in eq(A1.2) above, is used to count the states and decodes,

$$(i;j) = \left[\sum_{h=1}^{i-1} k_h \right] + j \tag{A1.4}$$

The core observer form matrices $\underline{\mathbf{A}}_O^0$ and $\underline{\mathbf{C}}_O^0$ above are (*ibid*, p.414) block diagonal matrices,

$$\underline{\mathbf{A}}_O^0 \doteq \text{block diag}\begin{Bmatrix} 0 & 1 & 0 & \cdot & \cdot & \cdot & 0 \\ \vdots & & \cdot & & 0 & & \vdots \\ \vdots & & & \cdot & & \cdot & \vdots \\ \vdots & & 0 & & & 0 & 1 \\ 0 & \cdot & \cdot & \cdot & \cdot & \cdot & 0 \end{Bmatrix} \quad [k_i \times k_i] \quad i=1..q \tag{A1.5}$$

$$\underline{\mathbf{C}}_O^0 \doteq \text{block diag}\{1\ 0\ \cdot\ \cdot\ \cdot\ \cdot\ 0\} \quad [1 \times k_i] \quad i=1..q \tag{A1.6}$$

The observer form states are given by,

$$x_{(i;k)} = \sum_{j=1}^{q} \sum_{h=0}^{k-1} \left[d_{ij(k_i-h)}\ y_j^{[k-h-1]}(t) \right] - \sum_{r=1}^{m} \sum_{h=0}^{k-1} \left[n_{ij(k_i-h)}\ u_r^{[k-h-1]}(t) \right] \tag{A1.7}$$

Appendix 2 Controllability and Observability

Greatest common left ($\underline{L}(p)$) and right ($\underline{R}(p)$) factors can be found in eq(4), which is then written,

$$\underline{0} = \underline{L}(p)\,(\underline{A}'-p\underline{I})\,\underline{R}(p)\,\underline{x}(t) + \underline{L}(p)\,\underline{B}'\,\underline{u}(t) + [\underline{L}(p)\underline{z}_1']|_{p=0}$$

$$\underline{y}(t) = \underline{C}'\,\underline{R}(p)\,\underline{x}(t) + \underline{D}\,\underline{u}(t) + \underline{z}_2 \qquad (A2.1)$$

where,

$$\underline{L}(p)(\underline{A}'-p\underline{I})\,\underline{R}(p) = (\underline{A}-p\underline{I}), \qquad \underline{L}(p)\,\underline{B}' = \underline{B},$$

$$[\underline{L}(p)\,\underline{z}_1']|_{p=0} = \underline{z}_1 \quad \text{and} \qquad \underline{C}'\,\underline{R}(p) = \underline{C}.$$

Eq(7) would be,

$$\det(p\underline{I}-\underline{A}')\,\underline{y}(t) = (\underline{C}'\,\mathrm{Adj}(p\underline{I}-\underline{A}')\,\underline{B}' + \det(p\underline{I}-\underline{A}')\,\underline{D})\,\underline{u}(t) -$$

$$[\underline{C}\,\mathrm{Adj}(\underline{A}')\,\underline{z}_1' + \det(\underline{A}')\underline{z}_2] \qquad (A2.2)$$

By Theorem 6.2–6 in Kailath (1980, p.366), $\underline{L}(p)$ is unimodular iff $\{\underline{A}, \underline{B}\}$ is controllable and $\underline{R}(p)$ is unimodular iff $\{\underline{A}, \underline{C}\}$ is observable. Unless uncontrollable (but observable) states are sufficiently excited by means other than the measured inputs (e.g by noise, unmeasured inputs or independently set up initial conditions), only the controllable, observable parts of a system's behaviour can be excited and detected in the input–output behaviour.

As offset has not received much attention in the literature, it is noted here that if $\underline{L}(p)$ is a common left factor of ($\underline{A}$–$\underline{I}$p) and $\underline{B}$, it is also a common left factor of z_1 iff, $\underline{L}(0)\,\underline{z}_1' = \underline{z}_1$ is consistent (always the case if $\underline{L}(p)$ is unimodular (when $\{\underline{A}, \underline{B}\}$ is controllable) as $\underline{L}(p)$ is then nonsingular for all p; never the case if the system has an uncontrolled integrator with a constant, non–zero offset at the integrator input).

Appendix 3 – Proof of the identity : $\det(\underline{M}(p))\,\underline{I} = \mathrm{Adj}(\underline{M}(p))\,\underline{M}(p)$

Consider an [n×n] polynomial matrix, $\underline{M}(p)$, with entries $m_{ij}(p)$. Entries of the adjoint matrix are given by,

$$\mathrm{Adj}(\underline{M}(p))_{ij} \doteq (-1)^{(i+j)}\,\mathrm{Cof}(\underline{M}(p))_{ij} \qquad (A3.1)$$

The cofactor is the determinant of the sub–matrix obtained by deleting the ith row and the jth column of $\underline{\mathbf{M}}(p)$.

Examine the entry (i,j) of the product, $\underline{\mathbf{T}}(p) = \mathrm{Adj}(\underline{\mathbf{M}}(p))\ \underline{\mathbf{M}}(p)$,

$$t_{ij}(p) = \sum_{k=1}^{n} (-1)^{(i+k)}\ \mathrm{Cof}(\underline{\mathbf{M}}(p))_{ik}\ m_{kj}(p) \tag{A3.2}$$
$$= 0 \qquad \text{if } i \neq j$$
$$= \det(\underline{\mathbf{M}}(p)) \quad \text{if } i = j$$

Hence,

$$\det(\underline{\mathbf{M}}(p))\ \underline{\mathbf{I}} = \mathrm{Adj}(\underline{\mathbf{M}}(p))\ \underline{\mathbf{M}}(p) \tag{A3.3}$$

Example: Take a [2×2] matrix,

$$\underline{\mathbf{M}}(p) = \begin{bmatrix} m_{11} & m_{12} \\ m_{21} & m_{22} \end{bmatrix}$$

which has,

$$\det(\underline{\mathbf{M}}) = m_{11}\ m_{22} - m_{21}\ m_{12}$$

$$\mathrm{Adj}(\underline{\mathbf{M}}) = \begin{bmatrix} m_{11} & -m_{12} \\ -m_{21} & m_{22} \end{bmatrix}$$

and

$$\mathrm{Adj}(\underline{\mathbf{M}})\ \underline{\mathbf{M}} = \det(\underline{\mathbf{M}})\ \underline{\mathbf{I}}$$

without requiring the inverse of operator, p.

Use of Pseudo-Observability Indices In Identification of Continuous-Time Multivariable Models

S. Bingulac and D.L. Cooper

Bradley Department of Electrical Engineering
Virginia Polytechnic Institute and State University
Blacksburg, VA 24061

Abstract

A procedure for identifying linear continuous-time multivariable systems is presented. The scheme is based on a discrete identification method which requires no structural identification. This is accomplished by the use of pseudo-observable forms whose structure is defined by their corresponding pseudo-observability indices. A First-Order Hold transformation of this discrete-time model to an equivalent continuous-time model is then performed. First, the state transition matrix is transformed to an equivalent continuous-time system matrix using a procedure which is insensitive to the size of the spectral radii of the system. The remaining system matrices are then transformed using a generalized First-Order Hold transformation technique which places no restrictions on the singularity of the continuous-time system.

1. Introduction

The presented identification procedure belongs to the class of indirect identification procedures (Sinha and Kuszta 1983) consisting of two phases. In the first phase, using available sequences of sampled input/output data, a discrete-time MIMO state space representation is determined. In the second phase a corresponding continuous-time representation is derived. Special, unique features of this identification procedure are as follows:

- The discrete-time model identification procedure is based on the use of an admissible set of pseudo-observability indices. Therefore, it is not necessary to perform structural identification, i.e. to determine in advance the unique set of observability indices. The discrete-time model is represented by one of many possible pseudo-observable forms.

- Moreover, the discrete-time model identification algorithm does not requre that the system to be identified is initially at rest, since, in addition to matrices in state space representation, it also determines the corresponding initial conditions.

N. K. Sinha and G. P. Rao (eds.), Identification of Continuous-Time Systems, 443–471.

- The corresponding continuous-time MIMO model is obtained using the "Ramp-invariant", i.e. First-Order Hold (FOH) approximation, which is clearly superior over the commonly used "Step-invariant" or Zero-Order Hold (ZOH) approximation. The applied procedure of deriving the FOH continuous-time model can equally treat models with both zero and multiple eigenvalues, which is not the case in some of the existing FOH approximation techniques.

- Calculation of the natrual logarithm of the discrete-time system matrix, which is inevitable in all indirect identification methods, is done by a modification of existing procedures which is not limited by the values of spectral radii of some matrices involved in the calculation. In this way, it is possible to obtain a desired accuracy in important cases of unstable systems as well as when the sampling interval is relatively large.

This paper is organized as follows:

After the introduction, Section 2 contains a brief review of pseudo-observable forms, giving emphasis on the fact that there are more pseudo-observable forms describing a given system, thus avoiding the necessity of structural identification. In Section 3, the algorithm for identification of discrete-time models in pseudo-observable forms is presented. It is shown how both the system matrices and the initial conditions are determined. Section 4 describes the First-Order Hold approximation approach. Particular attention is given to the cases of multiple eigenvalues and singularity of the continuous-time system matrix. The used procedure of calculating the natrual logarithm of the discrete-time system matrix is given in Section 5. The cases of matrices with spectral radii greater than one is particularly stressed. Finally, a simple computational example is given in Section 6 illustrating all features of the proposed procedure of identifying continuous-time state space MIMO models from samples of input/output data.

2. Pseudo-Observable Forms

Consider a linear time-invariant continuous system

$$\begin{aligned} \dot{x}(t) &= A\,x(t) + B\,u(t) \\ y(t) &= C\,x(t) + D\,u(t) \\ x(0) &= x_0 \end{aligned} \tag{2.1}$$

where $x \in \mathbb{R}^n$, $u \in \mathbb{R}^m$, and $y \in \mathbb{R}^p$, are the state, input, and output vectors respectively, while A, B, C, and D are matrices of compatible dimensions. This system may be represented by an equivalent discrete-time system (Brogan 1985) given by

$$\begin{aligned} x(k+1) &= F\,x(k) + G\,u(k) \\ y(k) &= C\,x(k) + H\,u(k) \\ x(0) &= x_0 \end{aligned} \tag{2.2}$$

It is known (Gevers and Wertz 1984, Bingulac and Krtolica 1988, Bingulac and Krtolica 1987, Correa and Glover 1986) that any n^{th} order observable MIMO discrete system can be represented by the following pseudo-observable canonical form

$$\begin{aligned} x(k+1) &= F_o\, x(k) + G_o\, u(k) \\ y(k) &= C_o\, x(k) + H_o\, u(k) \\ x(0) &= x_0 \end{aligned} \tag{2.3}$$

which is based on a selected set of pseudo-observability indices $\eta = \{\eta_1, \eta_2, ..., \eta_p\}$, where p is the number of outputs of the system. It is also know (Bingulac and Krtolica 1987) that this representation is possible only for an admissible set of pseudo-observability indices and that the total number of sets of admissible pseudo-observability indices is less than or equal to

$$I = \frac{(n-1)!}{(p-1)!\,(n-p)!} \tag{2.4}$$

According to the admissibility condition specified by Bingulac and Krtolica (1987), the pseudo-observability index η_i that corresponds to a particular component y_i of the output vector should not be greater than its upper bound specified by the individual observability index for this output component, i.e. by the number of modes observable from y_i individually.

The pseudo-observable forms of the matrices F_o and C_o are characterized by the following structures.

$$F_o = \begin{bmatrix} 0 & \cdots & 0 & 1 & & & & \\ f_{11} & \cdots & & f_{1j} & & \cdots & & f_{1n} \\ 0 & \cdots & & 0 & 1 & & & \\ f_{21} & \cdots & & f_{2j} & & \cdots & & f_{2n} \\ 0 & \cdots & & & 0 & 1 & & \\ & & & & & & \ddots & \\ 0 & \cdots & & & & & 0 & 1 \\ f_{p1} & \cdots & & f_{pj} & & \cdots & & f_{pn} \end{bmatrix} \tag{2.5}$$

$$C_o = \begin{bmatrix} 1 & 0 & \cdots & & & & \cdots & 0 \\ 0 & 1 & 0 & \cdots & & & \cdots & 0 \\ \vdots & & \ddots & \ddots & & & & \vdots \\ 0 & \cdots & 0 & 1 & 0 & \cdots & \cdots & 0 \end{bmatrix} \tag{2.6}$$

It can be seen from (2.5) that F_o has only p rows with non-zero and non-unity elements. The locations of these rows, $\underline{s} = \{s_1, s_2, ..., s_p\}$, are uniquely determined by the assumed set of admissible pseudo-observability indices η. The remaining (n - p) rows of F_o correspond to the last (n - p) rows of the (n x n) Identity matrix I_n. From (2.6) it can be noted that the rows in C_o correspond to the first p rows of I_n. Matrices G_o and H_o have no particular structure, as shown below.

$$G_o = \begin{bmatrix} g_{11} & \cdots & g_{1m} \\ \vdots & & \vdots \\ g_{n1} & \cdots & g_{nm} \end{bmatrix} \qquad H_o = \begin{bmatrix} h_{11} & \cdots & h_{1m} \\ \vdots & & \vdots \\ h_{p1} & \cdots & h_{pm} \end{bmatrix} \tag{2.7}$$

It may be easily verified that a pseudo-observable form, given by (2.5) - (2.6) and based on a set of admissible pseudo-observability indices $\underline{\eta} = \{ \eta_1, ..., \eta_p \}$, may be obtained from (2.2) by applying a state space transformation T containing n rows from the observability matrix

$$Q_o = \begin{bmatrix} C \\ C\ F \\ \vdots \\ C\ F^{n-p} \end{bmatrix} \tag{2.8}$$

corresponding to n locations $\underline{h} = \{ h_1, ..., h_n \}$ which are uniquely determined by the selected set of admissible pseudo-observability indices $\underline{\eta}$. As it was defined in (Bingulac and Krtolica 1987), the admissibility condition guarantees that the n rows selected from Q_o into T are linearly independent, i.e. that the transformation matrix T is nonsingular. More discussion on the relationship between indices $\underline{\eta}$ and locations $\underline{h}$ is given in Section 3.

3. Identification of Discrete-Time Models

As it was mentioned before, the first step in the identification approach is to identify an nth order equivalent discrete-time representation from the available input-output sample sequences {u(k), y(k)}; k=0, 1, ..., N-1; $u(k) = u_k$; $y(k) = y_k$. The discrete representation should satisfy (2.3), where F_o, G_o, C_o, and H_o are given by (2.5) - (2.7).

The following equation is obtained from (2.3).

$$Y(k) = Q_o\ x(k) + R\ U(k) \tag{3.1}$$

where Y(k) and U(k) are (q+1)p and (q+1)m column vectors and Q_o and R are matrices of the dimensions ((q+1)p x n) and ((q+1)p x (q+1)m), where q = n-p+1. The matrix Q_o is the observability matrix calculated from the matrices F_o and C_o given in (2.5) and (2.6), respectively. The vectors Y(k) and U(k) have the following structure.

$$Y(k) \triangleq Y_k = \begin{bmatrix} y_k \\ \vdots \\ y_{k+i} \\ \vdots \\ y_{k+q} \end{bmatrix} \qquad U(k) \triangleq U_k = \begin{bmatrix} u_k \\ \vdots \\ u_{k+i} \\ \vdots \\ u_{k+q} \end{bmatrix} \tag{3.2}$$

The matrices Q_o and R have the structure depicted below.

$$Q_o = \begin{bmatrix} C_o \\ C_oF_o \\ \vdots \\ C_oF_o^{\,i} \\ \vdots \\ C_oF_o^{\,q} \end{bmatrix} \tag{3.3}$$

$$R = \begin{bmatrix} H_o & 0 & \cdots & & & 0 \\ C_oG_o & H_o & 0 & & & \vdots \\ C_oF_oG_o & C_oG_o & \ddots & \ddots & & \\ \vdots & \ddots & & & & \\ & & & & H_o & 0 \\ C_oF_o^{\,q-1}G_o & & \cdots & & C_oG_o & H_o \end{bmatrix} \tag{3.4}$$

The elements of the vectors Y(k), U(k), and x(k) are related to the elements in the matrices F_o, G_o, C_o, and H_o by the set of (q+1)p scalar equations described by (3.1). It may be easily verified that the observability matrix Q_o (3.3) contains n rows equal to the rows in the (n x n) Identity matrix I_n and p rows which contain the non-zero non-unity rows from F_o. As mentioned previously, the locations of the rows corresponding to I_n and the rows corresponding to F_o are uniquely determined by the assumed set of admissible pseudo-observability indices. These locations are denoted by sets $\underline{h} = \{ h_1, ..., h_n \}$ and $\underline{r} = \{ r_1, ..., r_n \}$, respectively.

Subvectors Y_{1k} and Y_{2k} are now selected from Y_k corresponding to the n rows in Q_o containing rows equal to the rows of I_n and the p rows in Q_o which contain non-zero non-unity rows from F_o, respectively. Submatrices R_1 and R_2 are selected from the corresponding rows of matrix R. Using these submatrices, equation (3.1) can be rewritten as

$$Y_{1k} = I\, x_k + R_1\, U_k \tag{3.5}$$

$$Y_{2k} = \tilde{F}\, x_k + R_2\, U_k \tag{3.6}$$

where

$$\tilde{F} = \begin{bmatrix} f_{11} & f_{1n} \\ \vdots & \vdots \\ f_{p1} & f_{pn} \end{bmatrix} \tag{3.7}$$

Solving for x_k in (3.5) and substituting this equation into (3.6) yields

$$Y_{2k} = \tilde{F}\, Y_{1k} - \tilde{F}\, R_1\, U_k + R_2\, U_k \tag{3.8}$$

which may be written in matrix form as

$$Y_{2k} = \begin{bmatrix} \tilde{G} & \vdots & \tilde{F} \end{bmatrix} \begin{bmatrix} U_{1k} \\ \cdots \\ Y_{1k} \end{bmatrix} \tag{3.9}$$

where U_{1k} contains the first (η_M+1)m rows from U_k; $\eta_M = \max\{\eta_i\}$ for i = 1, ..., p. The matrix $\tilde{G}$ in (3.9) is given by $\tilde{G} = R_2 - \tilde{F}\, R_1$. We now define:

$$Y = \begin{bmatrix} Y_{2k} & \vdots & Y_{2(k+1)} & \vdots & \cdots & \vdots & Y_{2(k+q-1)} \end{bmatrix} \tag{3.10}$$

$$Z = \begin{bmatrix} U_{1k} & U_{1(k+1)} & \cdots & U_{1(k+q-1)} \\ \cdots & \cdots & \cdots & \cdots \\ Y_{1k} & Y_{1(k+1)} & & Y_{1(k+q-1)} \end{bmatrix} = \begin{bmatrix} U_q \\ \cdots \\ Y_q \end{bmatrix} \tag{3.11}$$

where q, q ≥ n+(η_M+1)m, represents the number of sets of available measurements. The dimensions of Y and Z are (p x q) and $(((\eta_M+1)m+n) \times q)$ respectively. It is known that the matrix U_q is of full row rank if the input signal u(k) is sufficiently rich (Bingulac and Farias 1977, Bingulac 1978) and that the assumed set of pseudo-observability indices is admissible if matrix Z is of full rank (Bingulac and Krtolica 1988). Among all possible sets of admissible pseudo-observability indices it is advisable to use the set which leads to the smallest condition number of the matrix Z.

Using the Least Squares Method (Sinha and Kuszta 1983), the parameter matrix $[\,\tilde{G} \vdots \tilde{F}\,]$ which depends only on the elements of F_o, G_o, C_o, and H_o may be expressed as

$$\begin{bmatrix} \tilde{G} & \vdots & \tilde{F} \end{bmatrix} = Y\, Z^T\, (\, Z\, Z^T)^{-1} \tag{3.12}$$

The matrix F_o is determined directly from (3.12) and (2.5) while C_o has been previously assumed (2.6).

Since F_o and C_o have been determined, G_o and H_o may now be calculated. It can be shown that

$$G_o = Q_c{}^e\, G^* \tag{3.13}$$

where

$$Q_c{}^e = \begin{bmatrix} G_e & \vdots & F_o G_e & \vdots & \cdots & \vdots & F_o{}^{\eta_M}\, G_e \end{bmatrix} \tag{3.14}$$

is the $(n \times (\eta_M+1)p)$ controllability matrix formed with F_o and the (n x p) "equivalent" input matrix G_e which contains p rows from the (p x p) Identity matrix I_p at locations $\underline{s}$ where F_o has non-unity non-zero rows (Bingulac and Farias 1977). All other rows in G_e contain only zero elements. The $((\eta_M+1)p \times m)$ matrix G^* is defined by partitioning the $((\eta_M+1)m \times p)$ matrix $\tilde{G}$ into (p x m) submatrices in the following form.

$$\tilde{G} = [\ \tilde{G}_0 \ \vdots\ \tilde{G}_1 \ \vdots\ \cdots \ \vdots\ \tilde{G}_{\eta_M}\ ] \tag{3.15}$$

and

$$G^* = \begin{bmatrix} \tilde{G}_0 \\ \tilde{G}_1 \\ \vdots \\ \tilde{G}_{\eta_M} \end{bmatrix} \tag{3.16}$$

To determine H_o, consider the definition of the transfer function matrix of the system (2.3) described by

$$W(z) = C_o\ (Iz - F_o)^{-1}\ G_o + H_o \tag{3.17}$$

The polynomial matrix W(z) could also be written as

$$W(z) = P^{-1}(z)\ N(z) \tag{3.18}$$

where N(z) and P(z) are the (p x m) and (p x p) co-prime polynomial matrices (Chen 1984), respectively, defined by

$$N(z) = \sum_{i=0}^{n_M} N_i z^i \qquad P(z) = \sum_{i=0}^{n_M} P_i z^i \tag{3.19}$$

Rewriting (3.17) and (3.18) yields

$$H_o = P^{-1}(z)\ N(z) - C_o\ (Iz - F_o)^{-1}\ G_o \tag{3.20}$$

Since H_o is independent of the Z-transform variable z, it could be calculated for any arbitrary value of z. If the value of z is chosen to be zero, then P_o and F_o are always non-singular, where P_o is given by the first p columns of $\tilde{F}$. Also, it can be shown (Chen 1984) that $N_o = \tilde{G}_o$(see (3.9) and (3.15)). Thus replacing z=0 into (3.20) yields

$$H_o = P_o^{-1}\ N_o + C_o\ F_o^{-1}\ G_o \tag{3.21}$$

The initial condition vector x_o, corresponding to (2.3) could be calculated directly from (3.5) by setting k = 0, i.e.

$$x_o = Y_{10} - R_1\ U_0 \tag{3.22}$$

where Y_{10} and U_0 are the first columns from matrices Y_{1k} and U_k.

An algorithm which, using the assumed set of admissible psuedo-observability indices, $\underline{\eta}$, uniquely determines the location vectors $\underline{s}$, $\underline{s}_c$, $\underline{h}$, and $\underline{r}$, specifying

a) the locations $\underline{s}$ of the p non-zero non-unity rows in F_o,
b) the locations $\underline{s}_c$ of the last n-p rows of I_n in F_o,
c) the locations $\underline{h}$ of the n rows of I_n in the observability matrix Q_o, and
d) the locations $\underline{r}$ of the p rows of non-zero, non-unity rows of F_o in Q_o

respectively, has been given in (Bingulac and Farias 1977, Gorti, Bingulac, and

VanLandingham 1990). This algorithm is as follows.

1. Define a set $\underline{\eta} = \{ \eta_1, ..., \eta_p \}$, of admissible pseudo-observability indices, where p is the number of outputs of the system.
2. Set $n = \sum_{i=1}^{p} \eta_i$; n = system order, $\eta_M = \max \{\eta_i\}$, $\eta_P = (\eta_M+1)p$.
3. Set i = 1.
4. Set k = i.
5. For j = 1 through $\eta_M + 1$, Set $V(k) = \eta_i + 1 - j$, k = k + p.
6. Set i = i + 1.
7. If i > p, go to 8; else go to 4.
8. Set $i_1 = 0$, $i_2 = 0$, k = 1.
9. If $V(k) \leq 0$, go to 11; else go to 10.
10. Set $i_1 = i_1 + 1$, $\underline{h}(i_1) = k$.
11. If $V(k) \neq 0$, go to 13; else go to 12.
12. Set $i_2 = i_2 + 1$, $\underline{r}(i_2) = k$.
13. Set k = k + 1.
14. If $k \leq \eta_P$, go to 9; else go to 15.
15. Set k = p + 1, $i_a = 0$, $i_p = 0$, $i_i = 0$.
16. Set $i_p = i_p + 1$.
17. If V(k) < 0, Set $i_p = i_p - 1$ and go to 21; else go to 18.
18. If $V(k) \neq 0$, go to 19; else go to 20.
19. Set $i_i = i_i + 1$, $\underline{s}_c(i_i) = i_p$, go to 21.
20. Set $i_a = i_a + 1$, $\underline{s}(i_a) = i_p$.
21. Set k = k + 1.
22. if $k \leq \eta_P$, go to 16; else STOP.

A more intuitive explaination of the above algorithm is given by Cooper (1990).

Consider, for example, the case where p = 3 and $\underline{\eta} = \{ 1, 4, 2 \}$. The algorithm given above gives the following location vectors.

$\underline{s} = \{ 1, 5, 7 \}$ $\qquad$ $\underline{s}_c = \{ 2, 3, 4, 6 \}$

$\underline{h} = \{ 1, 2, 3, 5, 6, 8, 11 \}$ $\qquad$ $\underline{r} = \{ 4, 9, 14 \}$

In the case of another set of admissible pseudo-observability indices given by $\underline{\eta} = \{ 2, 3, 2 \}$, the location vectors $\underline{s}$, $\underline{s}_c$, $\underline{h}$ and $\underline{r}$ become:

$\underline{s} = \{ 4, 6, 7 \}$ $\qquad$ $\underline{s}_c = \{ 1, 2, 3, 5 \}$

$\underline{h} = \{ 1, 2, 3, 4, 5, 6, 8 \}$ $\qquad$ $\underline{r} = \{ 7, 9, 11 \}$

Note that the location set $\underline{s}_c$ is the "complement" of the set $\underline{s}$ in the sense that among n integers in the set I = 1, 2, ..., n, some p integers are in $\underline{s}$ while the remaining n - p integers are in $\underline{s}_c$.

4. First-Order-Hold Transformation

As mentioned earlier, the second phase in the proposed procedure for identification of continuous-time models is to transform the identified discrete-time model into an equivalent continuous-time model (Sinha and Kuszta 1983). The major advantage to this approach is that considerable literature is already available on the first part of the problem (Bingulac and Krtolica 1988, Gorti, Bingulac, and VanLandingham 1990, Guidorzi 1975). The second sub-problem has been studied for the SISO case in several papers (Hsia 1972, Sinha 1972) and the MIMO case has been researched in later papers (Bingulac and Sinha 1989, Bingulac and Cooper 1990, Cooper and Bingulac 1990, Lastman, Puthenpura, and Sinha 1984, Puthenpura and Sinha 1984, Sinha and Lastman 1981, Strmčnik and Bremšak 1979).

In some of these latter multivariable cases, the continuous-time model is determined from the corresponding discrete-time model based on the assumption that the input signal to the continuous-time system is held constant between sampling intervals. This is a valid assumption if the system is actually sampled using a zero-order hold or if the sampling interval is sufficiently small so as to validate this assumption. However, for practical cases where the input actually varies between sampling instants, satisfactory results may not be produced from a model obtained under this assumption. For these cases it may be more reasonable to assume that the input varies linearly between sampling instants. This is referred to as a "ramp-invariant" or first-order hold (FOH) transformation and has been considered for SISO cases in earlier papers (Haykin 1972, Sinha 1972, Keviczky 1977). Keviczky uses transfer function techniques to derive new transformation algorithms for SISO models in the state-space form, whereas most methods had previously considered only transfer function representations of the transformed system.

The multivariable case of the ramp-invariant transformation has been previously considered by Strmčnik and Bremšak (1979), Bingulac and Sinha (1989), and Bingulac and Cooper (1990). Strmčnik and Bremšak developed equations for the ramp-invariant transformation of MIMO systems but the expressions, as presented, are not applicable in the case when the rank of the continuous system matrix A_c is less than the system order, (i.e. when the continuous system matrix has at least one zero eigenvalue). The method proposed by Bingulac and Sinha (1989) is applicable for the singular continuous system matrix, but was designed for a specific system structure. Bingulac and Cooper (1990) have developed a method of performing the ramp-invariant transformation which is much more general and is valid for any linear system. The procedure uses polynomial matrices to reduce the calculation of the ramp-invariant system to the solution of an overdetermined system of linear algebraic equations. This section will be concerned with the ramp-invariant transformation method which has been recently proposed by Bingulac and Cooper (1990).

As it was explained in Section 3, given the set of sampled input and output vectors $\{u_j, y_j\}$ of a continuous MIMO system, the system can be identified by a discrete-time representation in pseudo-observable form. The state-space representation of the system is given by

$$\begin{aligned} x_{k+1} &= F\, x_k + G\, u_k \\ y_k &= C\, x_k + H\, u_k \end{aligned} \tag{4.1}$$

where, $x \in \mathbb{R}^n$, $u \in \mathbb{R}^m$, and $y \in \mathbb{R}^p$ are the state, input and output vectors respectively while F, G, C, and H are matrices of compatible dimensions. The sampled input vector is given by : $u_k \triangleq u(kT)$. The system can also be represented by the Z-domain transfer function matrix equation

$$y(z) = [\, C\, (\, Iz - F)^{-1}\, G + H\,]\, u(z) \tag{4.2}$$

Consider now a continuous-time linear multivariable system, described by the equations

$$\begin{aligned} \dot{x}(t) &= A\, x(t) + B\, u(t) \\ y(t) &= C\, x(t) + D\, u(t) \end{aligned} \tag{4.3}$$

where $x(t) \in \mathbb{R}^n$, $u(t) \in \mathbb{R}^m$, and $y(t) \in \mathbb{R}^p$ are the state, input and output vectors, respectively, while A, B, C, and D are matrices of compatible dimensions.

Assuming that the state of the system is known at time $t = t_i$, its value at $t_2 > t_1$ is easily obtained as

$$x(t_2) = e^{A(t_2 - t_1)}\, x(t_1) \quad + \int_{t_1}^{t_2} e^{A(t_2 - \tau)}\, B\, u(\tau)\, d\tau \tag{4.4}$$

Now, given the sampling interval as T, the values t_1 and t_2 are selected as

$$t_1 = kT \qquad\qquad t_2 = (k + 1)T \tag{4.5}$$

where k is any positive integer. Equation (4.4) can now be written as

$$x_{k+1} = e^{AT}\, x_k \quad + \int_{kT}^{(k+1)T} e^{A(kT+T-\tau)}\, B\, u(\tau)\, d\tau \tag{4.6}$$

where

$$x_k \triangleq x(kT) \qquad y_k \triangleq y(kT) = C\, x_k + H\, u_k \tag{4.7}$$

Equation (4.6) can not be integrated in the present form to determine the state of the system and then obtain the output, since the input $u(t)$ is known only at the sampling instants.

In the step-invariant transformation the input to the system is assumed to be held constant during the sampling interval. This is a valid assumption if the system is actually sampled using a Zero-order hold, but it is not reasonable for a system in which the input to the system is not held constant over the sampling interval. It may be more reasonable to assume that the input is varying linearly between sampling times. This leads to the following equation for the system input

$$u(t) = u_k \left(\frac{(k+1)T - t}{T} \right) + u_{k+1} \left(\frac{t - kT}{T} \right) \tag{4.8}$$

which represents a first-order hold and leads to a ramp-invariant transformation.

It has been shown by Bingulac and Sinha (1989) and Strmčnik and Bremšak (1979) that by replacing $u(\tau)$ in (4.6) by (4.8) the state space representation of the discrete system could be written as

$$\begin{aligned} x_{k+1} &= F\, x_k + G_0\, u_k + G_1\, u_{k+1} \\ y_k &= C\, x_k + \tilde{D}\, u_k \end{aligned} \tag{4.9}$$

or in the Z-domain as

$$\begin{aligned} z \cdot x(z) &= F\, x(z) + G_0\, u(z) + G_1\, z\, u(z) \\ y(z) &= C\, x(z) + \tilde{D}\, u(z) \end{aligned} \tag{4.10}$$

where the (n x n) matrix F and (n x m) matrices G_0 and G_1 are given by

$$F = e^{AT}, \quad G_0 = M_0 B\, T\,, \; G_1 = M_1 B\, T \tag{4.11}$$

Matrices M_0 and M_1 are given by

$$M_0 = (A\, T\, F - F + I_n)\,(A\, T)^{-2} \qquad M_1 = (F - A\, T - I_n)\,(A\, T)^{-2} \tag{4.12}$$

where I_n represents the (n x n) Identity matrix. From (4.11) we obtain

$$M_1^{-1} G_1 = M_0^{-1} G_o \tag{4.13}$$

and solving for G_1 yields

$$G_1 = M_1\, M_0^{-1}\, G_0 \quad \text{or} \quad G_1 = P\, G_0 \tag{4.14}$$

where

$$P = M_1\, M_0^{-1} \tag{4.15}$$

Equations (4.12) cannot be used to calculate M_0 and M_1 if the matrix A is singular, i.e. if the continuous model contains at least one zero eigenvalue, (Bingulac and Sinha 1989). In this case, the equations for M_0 and M_1 are given by

$$M_0 = \sum_{i=0}^{\infty} (i + 1)\, \frac{(A\, T)^i}{(i + 2)!} \qquad M_1 = \sum_{i=0}^{\infty} \frac{(A\, T)^i}{(i + 2)!} \tag{4.16}$$

Expressions similar to (4.16) were derived by VanLandingham (1985) and Brogan (1985), among others, in the context of digital filter design.

In order to obtain a continuous time model {A, B, C, D}, given by (4.3), corresponding to the ramp-invariant transformation, the problem now becomes determining the matrices G_0 and $\tilde{D}$ in (4.10). Note that A and B are given by equation (4.11), C is the same for the discrete-time system given in (4.10), and D is equal to $\tilde{D}$ in (4.9). To solve this problem, equation (4.10), using (4.14), becomes

$$(\, I_n z - F\,)\; x(z) = G_0\, u(z) + P\, G_0\, z\, u(z) \tag{4.17}$$

which leads to

$$(\, I_n z - F\,)\; x(z) = (\, I_n + P\, z\,)\; G_0\, u(z) \tag{4.18}$$

The Z-domain transfer function matrix for this representation can be expressed as

$$y(z) = [\ C\ (\ I_n\ z\ -\ F)^{-1}\ (\ I_n + P\ z)\ G_0 + \tilde{D}\ ]\ u(z) \tag{4.19}$$

Equation (4.19) can also be written as

$$y(z) = [\ C\ (\ I_n\ z\ -\ F)_{adj}\ (\ I_n + P\ z\ )\ \vdots\ \Delta(z)\ I_p\ ] \begin{bmatrix} G_0 \\ \cdots \\ \tilde{D} \end{bmatrix} \frac{1}{\Delta(z)}\ u(z) \tag{4.20}$$

where $(X)_{adj}$ represents the adjoint of the matrix X, $\Delta(z) \triangleq$ the characteristic polynomial of the matrix A, and I_p represents the (p x p) Identity matrix. Similarly, equation (4.2) can also be written as

$$y(z) = [\ C\ (\ I_n\ z\ -\ F)_{adj}\ G + \ \Delta(z)\ H\ ]\ \ \frac{1}{\Delta(z)}\ u(z) \tag{4.21}$$

Equations (4.20) and (4.21) are discrete-time representations of the same system. Therefore, the polynomial matrix equality

$$[\ C\ (\ I_n z\ -\ F)_{adj}\ (\ I_n + P\ z\ )\ \vdots\ \Delta(z)\ I_p\ ] \begin{bmatrix} G_0 \\ \cdots \\ \tilde{D} \end{bmatrix} = [\ C(\ I_n z\ -\ F)_{adj}\ G + \Delta(z)\ H] \tag{4.22}$$

should hold.

To simplify equation (4.22), the (p x m) polynomial matrix

$$W(z) = C\ (\ I_n\ z\ -\ F\ )_{adj}\ G \tag{4.23}$$

will be written as

$$W(z) = \sum_{j=0}^{n-1} W_j\ z^j \tag{4.24}$$

where

$$W_j = \begin{bmatrix} w_{11j} & \cdots & w_{1mj} \\ \vdots & \ddots & \vdots \\ w_{p1j} & \cdots & w_{pmj} \end{bmatrix} \tag{4.25}$$

is a (p x m) real matrix containing coefficients w_{ikj} of the polynomials

$$w_{ik}(z) = \sum_{j=0}^{n-1} w_{ikj}\ z^j \tag{4.26}$$

in the polynomial matrix $W(z) = \{w_{ik}(z)\}$. Similarly, the (p x n) polynomial matrix

$$V(z) = C\ (\ I_n\ z\ -\ F\ )_{adj}\ I_n \tag{4.27}$$

will be expressed as

$$V(z) = \sum_{j=0}^{n-1} V_j\ z^j \tag{4.28}$$

where

$$V_j = \begin{bmatrix} v_{11j} & \cdots & v_{1mj} \\ \vdots & \ddots & \vdots \\ v_{p1j} & \cdots & v_{pmj} \end{bmatrix} \tag{4.29}$$

is a (p x n) real matrix containing coefficients v_{ikj} of the polynomials

$$v_{ik}(z) = \sum_{j=0}^{n-1} v_{ikj}\, z^j \tag{4.30}$$

in the polynomial matrix $V(z) = \{v_{ik}(z)\}$.

Substituting equations (4.24) and (4.28) into equation (4.22) yields

$$\left[\, V(z)\,(\, I_n + P\, z\,) \;\vdots\; \Delta(z)\, I_p \right] \begin{bmatrix} G_0 \\ \cdots \\ \tilde{D} \end{bmatrix} = [\, W(z) + \Delta(z)\, H] \tag{4.31}$$

The characteristic polynomial, $\Delta(z)$, is a monic polynomial of the form

$$\Delta(z) = p_0 + p_1\, z + \ldots + p_{n-1}\, z^{n-1} + 1\, z^n = \sum_{i=0}^{n-1} p_i\, z^i + z^n \tag{4.32}$$

Now, substituting equations (4.24), (4.28), and (4.32), equation (4.31) becomes

$$\left[\left(\sum_{j=0}^{n-1} V_j\, z^j \right) (\, I_n + Pz\,) \;\vdots\; \sum_{j=0}^{n-1} p_j\, z^j\, I_p + z^n\, I_p \right] \begin{bmatrix} G_0 \\ \cdots \\ \tilde{D} \end{bmatrix} = \left[\sum_{j=0}^{n-1} W_j\, z^j + \sum_{j=0}^{n-1} p_j\, H\, z^j + z^n\, H \right] \tag{4.33}$$

leading to

$$\left[\left(\sum_{j=0}^{n-1} V_j\, z^j \right) (\, I_n + Pz\,) \;\vdots\; \sum_{j=0}^{n-1} p_j\, z^j\, I_p + z^n\, I_p \right] \begin{bmatrix} G_0 \\ \cdots \\ \tilde{D} \end{bmatrix} = \left[\sum_{j=0}^{n-1} \left(W_j + p_j\, H \right) z^j + z^n\, H \right] \tag{4.34}$$

Equation (4.34) can now be expressed as

$$\left[\sum_{j=0}^{n} \left(V_j + V_{j-1}\, P \right) z^j \;\vdots\; \sum_{j=0}^{n} p_j\, I_p\, z^j \right] \begin{bmatrix} G_0 \\ \cdots \\ \tilde{D} \end{bmatrix} = \left[\sum_{j=0}^{n} \left(W_j + p_j\, H \right) z^j \right] \tag{4.35}$$

where $V_{-1} = 0$, $V_n = 0$, $W_n = 0$, and $p_n = 1$.

Combining terms associated with z^j ($j = 0, \ldots, n$), equation (4.35) can formally be written as

$$\left[I_p z^0 \vdots I_p z^1 \vdots \cdots \vdots I_p z^n \right] \left[\tilde{A}\ X \right] = \left[I_p z^0 \vdots I_p z^1 \vdots \cdots \vdots I_p z^n \right] \left[\tilde{B} \right] \tag{4.36}$$

where the [(n+1)p x (n+p)], [(n+p) x m], and [(n+1)p x m] matrices $\tilde{A}$, X, and $\tilde{B}$ respectively, are given by

$$\tilde{A} = \left[\begin{bmatrix} V_0 \\ V_1 \\ \vdots \\ V_{n-1} \\ 0 \end{bmatrix} + \begin{bmatrix} 0 \\ V_0 \\ \vdots \\ V_{n-2} \\ V_{n-1} \end{bmatrix} P \begin{matrix} \vdots & p_0 I_p \\ \vdots & p_1 I_p \\ \vdots & \vdots \\ \vdots & p_{n-1} I_p \\ \vdots & I_p \end{matrix} \right] \tag{4.37}$$

$$\tilde{B} = \left[\begin{bmatrix} W_0 \\ W_1 \\ \vdots \\ W_{n-1} \\ 0 \end{bmatrix} + \begin{bmatrix} p_0\ H \\ p_1\ H \\ \vdots \\ p_{n-1}\ H \\ H \end{bmatrix} \right] \quad \text{and} \quad X = \begin{bmatrix} G_0 \\ \cdots \\ \tilde{D} \end{bmatrix} \tag{4.38}$$

Equation (4.36) will be satisfied for all z if

$$\tilde{A}X = \tilde{B} \tag{4.39}$$

Since $(n+1)p \geq n + p$, (the equality holds only if $p = 1$) equation (4.39) represents an overdetermined system of linear algebraic equations. The matrices $\tilde{A}$ and $\tilde{B}$ are dependent on the known variables V_i, P, p_i, W_i, and H while the unknown matrix X contains the matrices G_0 and $\tilde{D}$ to be determined. It can be easily verified that $\tilde{B} \subset \mathbb{R}(\tilde{A})$ thus a unique solution exists for this overdetermined system of linear algebraic equations.

Finally, having determined the matrices G_0 and $\tilde{D}$, the ramp-invariant or first-order hold (FOH) equivalent continuous-time representation is given by

$$\dot{x} = A\ x(t) + B\ u(t) \tag{4.40}$$

$$y = C\ x(t) + D\ u(t) \tag{4.41}$$

where

$$A = \frac{1}{T} \ln(F) \tag{4.42}$$

$$B = \frac{1}{T} M_0^{-1} G_0 \tag{4.43}$$

$$C = C \quad D = \tilde{D} \tag{4.44}$$

5. The Log of a Square Matrix

Equation (4.42) is commonly called the natural log of the matrix F since, for the case where F is a scalar, the scalar variable A can be calculated directly as the logarithm of F divided by T. But this is the trivial case. The calculation of A in the case where F is a square matrix is not quite as direct.

For the case where the matrix F has distinct, non-zero eigenvalues, it can be shown (Sinha and Lastman 1982) that matrix A may be calculated as

$$A = M * \text{diag}\left\{ \frac{1}{T}\ln f_1,\ \frac{1}{T}\ln f_2,\ \ldots,\ \frac{1}{T}\ln f_n \right\} * M^{-1} \tag{5.1}$$

with $M = \{ v_1, v_2, \ldots, v_n\}$ where $v_1, v_2, \ldots, v_n$ represent the eigenvectors of F corresponding to the eigenvalues $f_1, f_2, \ldots, f_n$. The requirement on the eigenvalues of F to be distinct makes this method much too restrictive.

In an earlier work (Sinha and Lastman 1981), a more general iterative method for performing this calculation was given. This method utilized the infinite series definition of the natural logarithm in the calculation. As an improvement to this algorithm, the Chebyshev minimax theory was later used to improve the accuracy of the initial approximation (Puthenpura and Sinha 1984).

One of the latest and most accurate algorithms has been proposed by Lastman, Puthenpura, and Sinha (1984) and is based on the summing series

$$AT = \sum_{k=1}^{\infty} \frac{(-1)^{k+1} L^k}{k} \tag{5.2}$$

This algorithm is convergent in the case where $L = F - I_n$ has a spectral radius less than one, where F is the (n x n) state transition matrix and I_n represents the (n x n) Identity matrix. If the spectral radius is between one half and one, i.e. $0.5 \leq \rho(L) \leq 1$, convergence can take numerous iterations causing the algorithm to be painfully slow. Cooper and Bingulac (1990) have proposed an improvement to this algorithm to obtain convergence for spectral radii greater than one and to speed up the iterative process by reducing the spectral radius to or below one half.

To obtain convergence in the case when $\rho(L) > 1$, consider the matrix F given as

$$F = e^{AT} \tag{5.3}$$

which could also be expressed as

$$A = \frac{1}{T}\ln(F) \tag{5.4}$$

as in equation (4.42). Now let

$$T = n \cdot t, \tag{5.5}$$

leading to

$$t = \frac{T}{n} \tag{5.6}$$

where $n = 2^m$, $m = 1, \ldots, q$. Substituting equation (5.5) into equation (5.3), we obtain

$$F = e^{An \cdot t} \tag{5.7}$$

Now let

$$\check{F} = (F)^{1/2^m} \tag{5.8}$$

Consequently,

$$\check{F} = e^{At} = e^{\check{A}T} \tag{5.9}$$

Equation (5.4) can now be represented by

$$A = \frac{n}{T} \ln(\check{F}) \tag{5.10}$$

leading to

$$\check{A} = \frac{1}{T} \ln(\check{F}) \tag{5.11}$$

Rewriting equation (5.2) in variables $\check{A}$ and $\check{F}$ leads to

$$\ln(\check{F}) = \check{A}T = \sum_{k=1}^{\infty} \frac{(\check{L})^k}{k}(-1)^{k+1} \tag{5.12}$$

where $\check{L} = \check{F}$ - I.

This infinite series is convergent, as given in equation (5.2), if the spectral radius of $\check{L}$ is less than one. It is also noted that less iterations are needed for smaller spectral radii. This algorithm reduces the spectral radius of the matrix $\check{L}$ sufficiently so as to reduce the number of iterations required by the algorithm, and in the case where the spectral radius is greater than one, reduces the spectral radius so this infinite sum based algorithm will converge. Also, the reduction is limited so as to avoid introducing serious round-off error into the calculations.

The condition to be met is to find a value of m such that the spectral radius of $\check{L}$ = ($\check{F}$ - I_n), where $\check{F}$ is calculated as in equation (5.8), is less than 0.5. The algorithm is described as follows.

1. Select a positive integer N and a small positive number ϵ.
2. Set $\check{Q}_1 = \check{L}$ and $\check{S}_1 = \check{L}$ where $\check{L}$ = (F - I_n)
3. Set k = 0, and j = 0
4. a: j = j + 1
5. Calculate the spectral radius of the matrix $\check{L}$, $\rho(\check{L})$
6. If $\rho(\check{L}) \leq 0.5$, go to b:
7. If $\rho(\check{L}) > 0.5$, then let $\check{L} = (F^{1/2^j} - I_n)$ and go to a:
8. b: k = k + 1
9. Set $\check{Q}_{k+1} = \dfrac{-k\ \check{L}\check{Q}_{k+1}}{(k+1)}$
10. Set $\check{S}_{k+1} = \check{S}_k + \check{Q}_{k+1}$
11. If k > N or $(d(\check{S}_{k+1}, \check{S}_k)) \leq \epsilon$ then go to c:
12. Else, go to b:
13. c: Calculate matrix $A = \check{S}_{k+1} \cdot \frac{1}{T} \cdot 2^{j-1}$

The function d(x,y) can be any suitable measure of the realitive difference or closeness of $\check{S}_{k+1}$ and $\check{S}_k$. One such measure is given by

$$d(\check{S}_{k+1}, \check{S}_k) = \frac{\max_{i,j} | \check{S}_{(k+1)_{i,j}} - \check{S}_{(k)_{i,j}} |}{\max_{i,j} | \check{S}_{(k+1)_{i,j}} |} \tag{5.13}$$

The variable N defines the maximum number of iterations carried out in calculating $\check{S}_{k+1}$ and the value of ϵ can be used to control the accuracy of the algorithm.

A possible algorithm to calculate the square root of a general matrix A is given below.

1. Select a small positive number ϵ
2. Set $X_1 = A$, and $i = 0$
3. a: $i = i + 1$
4. Set $X_{i+1} = 0.5(X_i + AX_i^{-1})$
5. If $d(X_{i+1}, X_1) < \epsilon$ then go to b:
6. Else, go to a:
7. b: Set $\sqrt{A} = X_{i+1}$

This algorithm is based on the standard procedure $x_{i+1} = 0.5(x_i + \frac{a}{x_i})$ for calculating the square root, x, of the scalar a where $x = \sqrt{a}$.

This algorithm yields excellent approximations to the matrix A when the spectral radius of $\check{L}$ is less than one. Just as importantly, however, is the fact that in the case where the spectral radius of $\check{L}$ is greater than one, this algorithm will converge to a sufficiently accurate approximation to the continuous-time system matrix.

6. Illustrative Examples

To illustrate the indirect, continuous-time model identification method based on the use of pseudo-observability indices presented in this paper, two examples will be considered. The simple steps in the simulation proceedure presented here were as follows:

1. Stimulate the continuous-time system with an input signal and obtain samples of the system response.
2. Use the samples of the continuous-time system input and system response to identify a discrete-time model using two admissible sets of pseudo-observability indices.
3. Use the step-invariant and ramp-invariant transformation techniques to transform the discrete-time models to equivalent continuous-time representations.
4. Compare the accuracy of both procedures.

In the following examples, the ramp-invariant transformation is compared to the classic step-invariant transformation. The responses of both systems are compared to the response of the sampled system to be identified. It can be seen that the ramp-invariant transformation produces a more accurate model than the step-invariant transformation, especially at larger sampling periods.

All calculations and simulations were done either on an IBM PC©, a PC clone or a Sun© workstation in double precision using the computer package LAS© (Bingulac 1988, Bingulac and Cooper 1990a).

The simulation of a true continuous-time system was accomplished by solving the state-space differential equation at points 0.01 seconds apart. These "samples" were then considered to be the continuous-time input and output data which was sampled at different sample times to identify. The continuous-time models which were calculated were then given the original system input and the response of these identified models then calculated, again every 0.01 seconds.

The following fourth-order model, with one input and two outputs (n = 4, m = 1, p = 2) was chosen as a continuous-time system to be identified

$$A = \begin{bmatrix} -5 & 10 & 0 & 0 \\ 0 & -5 & 10 & 0 \\ 0 & 0 & -1.5 & 6 \\ 0 & 0 & 0 & 0 \end{bmatrix}; \qquad B = \begin{bmatrix} 1 \\ 1 \\ 1 \\ 1 \end{bmatrix}$$

$$C = \begin{bmatrix} 1 & 0 & 0 & 0 \\ 0 & 0 & 4 & 0 \end{bmatrix}; \qquad D = \begin{bmatrix} 0 \\ 0 \end{bmatrix} \tag{6.1}$$

where the matrices A, B, C, and D are defined by

$$\begin{aligned} \dot{x}(t) &= A\ x(t) + B\ u(t) \\ y(t) &= C\ x(t) + D\ u(t) \end{aligned} \tag{6.2}$$

The input to this system is shown in Figure 1. This sinusoidal input was chosen since it is more general than the input signal previously used by Bingulac and Sinha (1989). Sampling times of 0.5, 0.25, and 0.1 seconds were used to sample the available input-output data over a 5 second period.

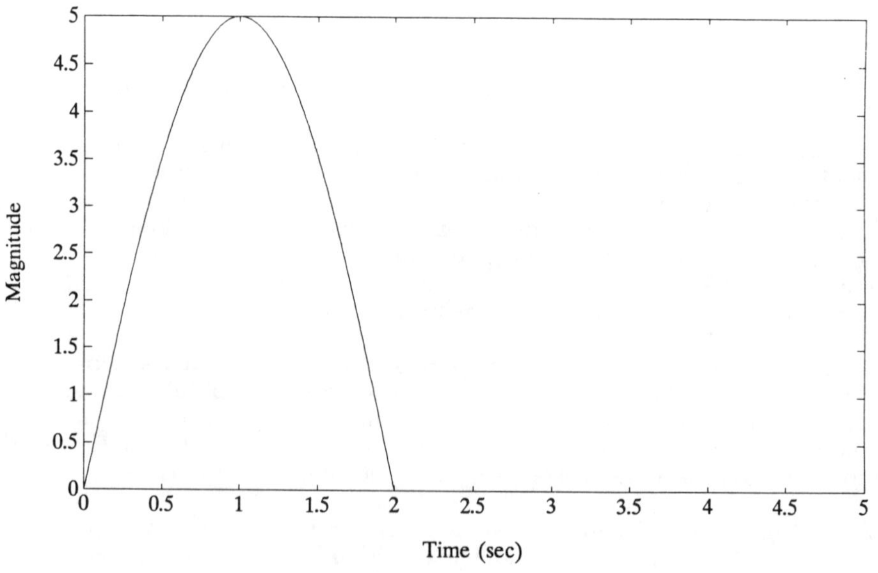

Figure 1. Input Signal u(t)

Consider now the observability matrix

$$Q_o = \begin{bmatrix} C \\ CA \\ \vdots \\ CA^{n-p} \end{bmatrix} = \begin{bmatrix} 1 & 0 & 0 & 0 \\ 0 & 0 & 4 & 0 \\ -5 & 10 & 0 & 0 \\ 0 & 0 & -6 & 24 \\ 25 & -100 & 100 & 0 \\ 0 & 0 & 9 & -36 \end{bmatrix} \tag{6.3}$$

The possible sets of pseudo-observability indices are $\underline{n}_1 = \{3, 1\}$, $\underline{n}_2 = \{2, 2\}$, and $\underline{n}_3 = \{1, 3\}$. The set $\underline{n}_3$ was determined to be non-admissible. Therefore, for the two examples in this section, the following sets of pseudo-observability indices given by $\underline{n}_1 = \{3, 1\}$ and $\underline{n}_2 = \{2, 2\}$ were used.

The first step is to identify a discrete-time model given by

$$\begin{aligned} x(k+1) &= F_o\, x(k) + G_o\, u(k) \\ y(k) &= C_o\, x(k) + H_o\, u(k) \\ x(0) &= x_0 \end{aligned} \tag{6.4}$$

where the matrices F_o, G_o, C_o, and H_o are in pseudo-observable form. In the identification procedure, first the matrices Y and Z were built to use in solving equation (3.12) for the parameter matrix

$$[\ \tilde{G} \ \vdots\ \tilde{F}\ ] = Y\, Z^T\, (\, Z\, Z^T)^{-1} \tag{6.5}$$

In the case of using $\underline{\eta}_1 = \{3,1\}$ and a sampling time of 0.5 seconds, the following parameter matrix was obtained

$$[\ \tilde{G} \ \vdots\ \tilde{F}\ ] = \begin{bmatrix} .011 & -1.213 & -.615 & 0 & .007 & .165 & -.165 & .993 \\ -.018 & 1.166 & 2.155 & .619 & .009 & -.259 & -.223 & 1.473 \end{bmatrix} \tag{6.6}$$

leading to

$$F_o = \begin{bmatrix} 0 & 0 & 1 & 0 \\ .007 & .165 & -.165 & .993 \\ 0 & 0 & 0 & 1 \\ .009 & -.259 & -.223 & 1.473 \end{bmatrix} \tag{6.7}$$

Using (3.13) G_o was calculated as

$$G_o = \begin{bmatrix} .586 \\ 1.361 \\ 1.157 \\ 1.771 \end{bmatrix} \tag{6.8}$$

Matrix C_o is assumed to be know and given by

$$C_o = \begin{bmatrix} 1 & 0 & 0 & 0 \\ 0 & 1 & 0 & 0 \end{bmatrix} \tag{6.9}$$

Using (3.21) H_o was calculated as

$$H_o = \begin{bmatrix} .619 \\ 1.730 \end{bmatrix} \tag{6.10}$$

The next step in the identification procedure is to transform this discrete-time representation, given by (6.7) - (6.10). The equivalent continuous-time system matrix A is calculated as

$$A = \frac{1}{T} \ln (F_o) = \begin{bmatrix} -7.036 & -9.123 & 25.680 & -9.521 \\ .021 & -2.369 & -.477 & 2.825 \\ -.148 & .963 & -3.412 & 2.597 \\ .030 & -.514 & -.886 & 1.370 \end{bmatrix} \tag{6.11}$$

which has eigenvalues at $\lambda_i = \{ -4.973 \pm 0.2i, -1.5, 0.0 \}$.

Since A is singular, the matices M_0 and M_1, used in calculating G_0 and G_1 as in equation (4.11) must be calculated by the power series defined in equation (4.16). These matrices are given below.

$$M_0 = \begin{bmatrix} .055 & -.112 & .739 & -.181 \\ .003 & .197 & -.060 & .361 \\ -.002 & .028 & .114 & .360 \\ .003 & -.089 & -.087 & .673 \end{bmatrix} \tag{6.12}$$

$$M_1 = \begin{bmatrix} .196 & -.174 & .712 & -.235 \\ .001 & .334 & -.033 & .197 \\ -.003 & .032 & .277 & .194 \\ .002 & -.045 & -.052 & .595 \end{bmatrix} \tag{6.13}$$

To calculate the FOH system, equation (4.39) is implemented, solving for G_o and D. Using equations (4.43) to (4.44), matrices B, C and D of the First-order hold system are calculated and given below.

$$B = \begin{bmatrix} 1.104 \\ 4.201 \\ 6.172 \\ 11.285 \end{bmatrix}; C = \begin{bmatrix} 1 & 0 & 0 & 0 \\ 0 & 1 & 0 & 0 \end{bmatrix}; D = \begin{bmatrix} .003 \\ .016 \end{bmatrix} \tag{6.14}$$

The response of this ramp-invariant system to the input given by Figure 1 is shown in Figure 2.

For comparison, a step-invariant (ZOH) system is also calculated. For this system, matrix A in the same as (6.11). The remaining matrices are given below.

$$B_z = \begin{bmatrix} 3.336 \\ 7.959 \\ 8.915 \\ 12.888 \end{bmatrix}; C_z = \begin{bmatrix} 1 & 0 & 0 & 0 \\ 0 & 1 & 0 & 0 \end{bmatrix}; D_z = \begin{bmatrix} .619 \\ 1.730 \end{bmatrix} \tag{6.15}$$

Figure 3 shows the response of this ZOH system plotted with the sampled system response.

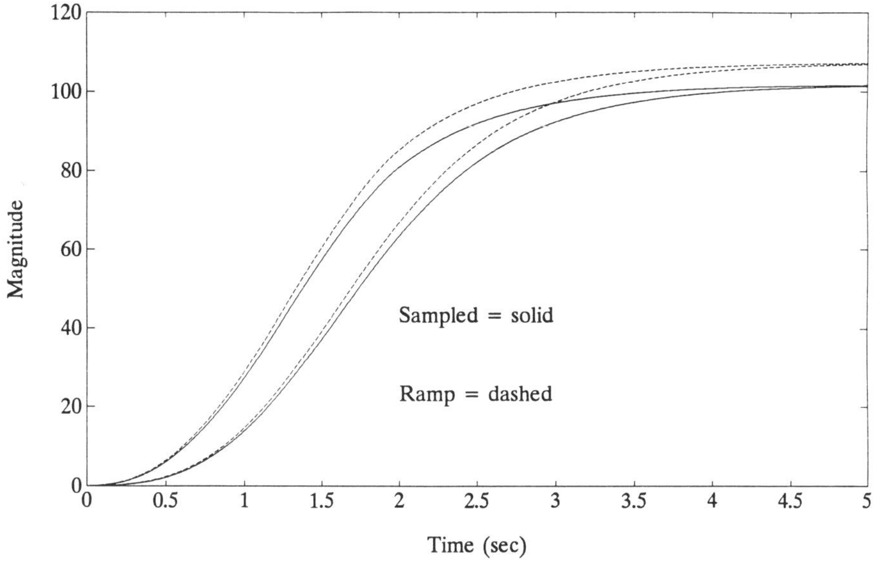

Figure 2. Outputs y(t) of Sampled and Ramp-Invariant Continuous-Time Models; $\Delta t = 0.5$ sec

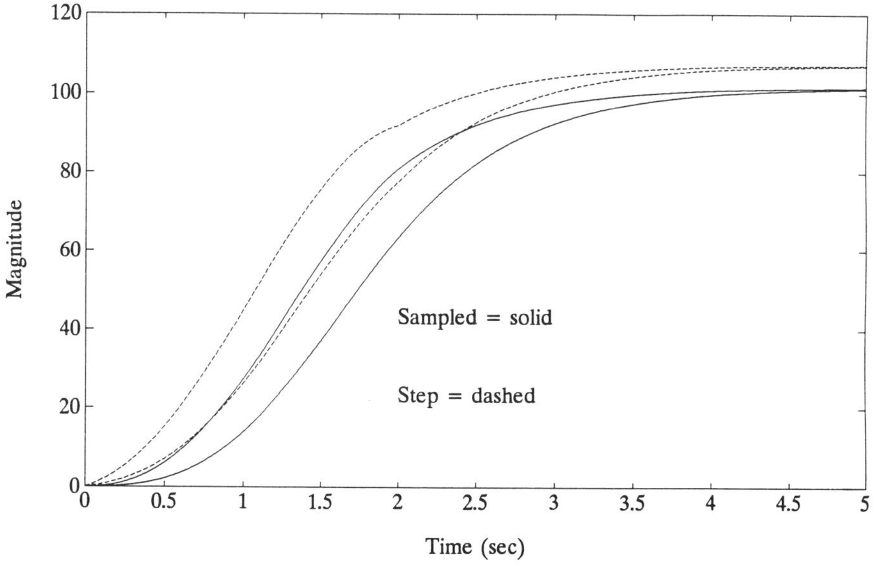

Figure 3. Outputs y(t) of Sampled and Step-Invariant Continuous-Time Models; $\Delta t = 0.5$ sec

Now using a sampling time of 0.25 seconds, the output of the system given by (6.2) was sampled and a discrete-time representation was obtained. From this model, the matrix A was obtained and is given below.

$$A = \frac{1}{T}\ln(F_o) = \begin{bmatrix} -9.312 & -3.044 & 16.131 & -3.755 \\ .669 & -4.235 & -4.714 & 8.280 \\ -.916 & .718 & -2.607 & 2.805 \\ .463 & -.720 & -4.374 & 4.631 \end{bmatrix} \tag{6.16}$$

For the ramp-invariant system, the remaining matrices were calculated as

$$B = \begin{bmatrix} 1.011 \\ 4.049 \\ 3.053 \\ 5.932 \end{bmatrix}; \; C = \begin{bmatrix} 1 & 0 & 0 & 0 \\ 0 & 1 & 0 & 0 \end{bmatrix}; \; D = \begin{bmatrix} 0 \\ .001 \end{bmatrix} \tag{6.17}$$

The step-invariant (ZOH) system is given below.

$$B_z = \begin{bmatrix} 1.913 \\ 6.070 \\ 4.472 \\ 7.333 \end{bmatrix}; \; C_z = \begin{bmatrix} 1 & 0 & 0 & 0 \\ 0 & 1 & 0 & 0 \end{bmatrix}; \; D_z = \begin{bmatrix} .197 \\ .681 \end{bmatrix} \tag{6.18}$$

Figures 4 and 5 show the responses of these ramp-invariant and step-invariant systems, respectively, plotted with the response of the sampled system.

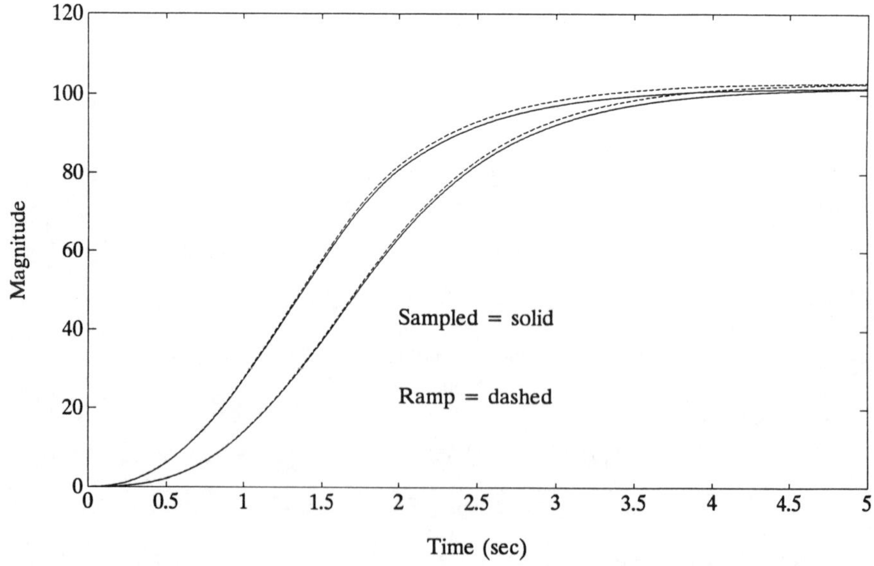

Figure 4. Outputs y(t) of Sampled and Ramp-Invariant Continuous-Time Models; $\Delta t = 0.25$ sec

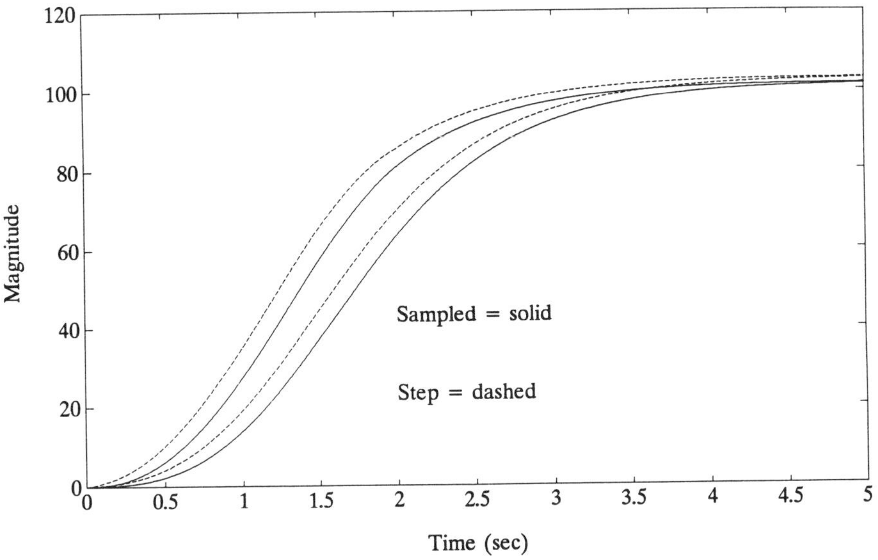

Figure 5. Outputs y(t) of Sampled and Step-Invariant Continuous-Time Models; $\Delta t = 0.25$ sec

Finally, a sample time of 0.1 seconds was used. The continuous-time matrix A is given below.

$$A = \frac{1}{T} \ln (F_o) = \begin{bmatrix} -9.312 & -3.044 & 16.131 & -3.755 \\ .669 & -4.235 & -4.714 & 8.280 \\ -.916 & .718 & -2.607 & 2.805 \\ .463 & -.720 & -4.374 & 4.631 \end{bmatrix} \tag{6.19}$$

The ramp-invariant system was calculated as

$$B = \begin{bmatrix} 1.001 \\ 4.006 \\ 1.634 \\ 2.510 \end{bmatrix}; C = \begin{bmatrix} 1 & 0 & 0 & 0 \\ 0 & 1 & 0 & 0 \end{bmatrix}; D = \begin{bmatrix} 0 \\ 0 \end{bmatrix} \tag{6.20}$$

and the step-invariant system for this sampling time was found to be

$$B_z = \begin{bmatrix} 1.296 \\ 4.865 \\ 2.054 \\ 3.024 \end{bmatrix}; C_z = \begin{bmatrix} 1 & 0 & 0 & 0 \\ 0 & 1 & 0 & 0 \end{bmatrix}; D_z = \begin{bmatrix} .060 \\ .230 \end{bmatrix} \tag{6.21}$$

Figures 6 and 7 show the responses of these ramp-invariant and step-invariant systems, respectively, plotted with the sampled system response.

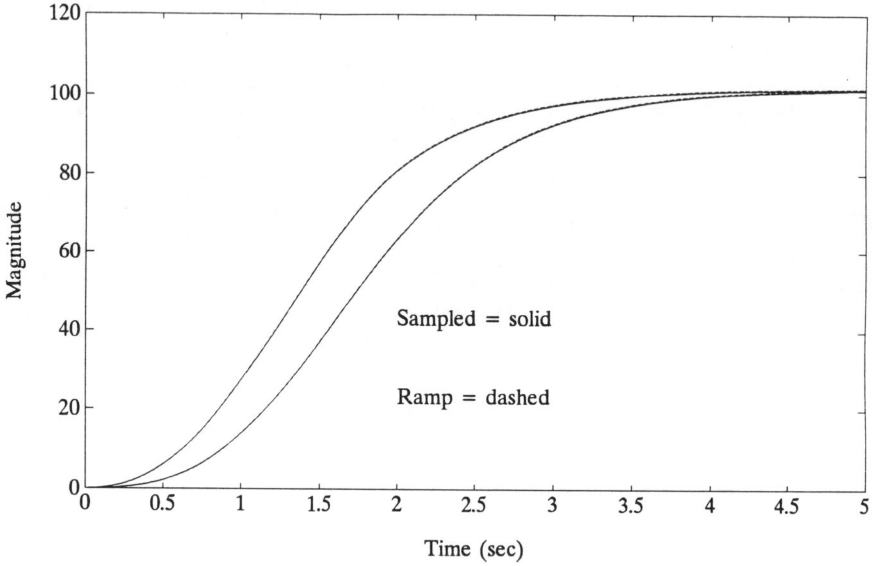

Figure 6. Outputs y(t) of Sampled and Ramp-Invariant Continuous-Time Models; $\Delta t = 0.1$ sec

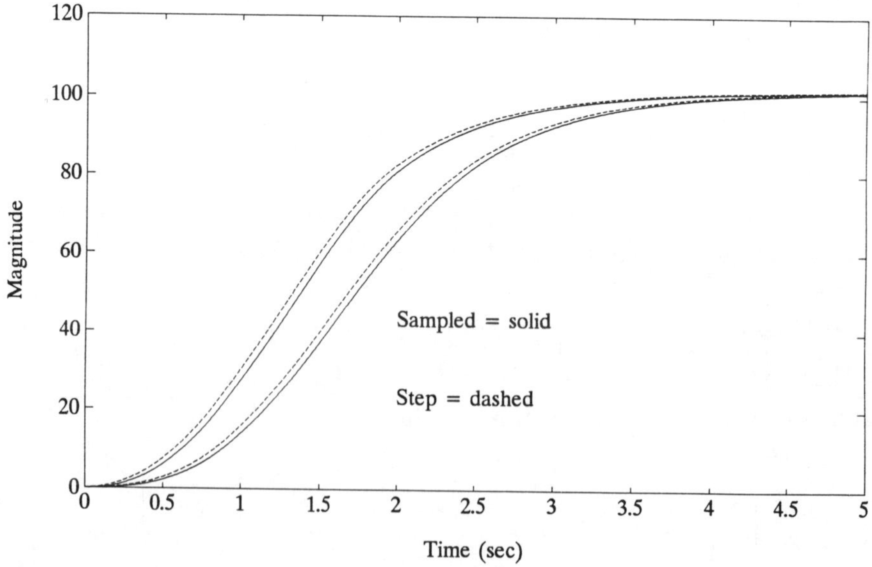

Figure 7. Outputs y(t) of Sampled and Step-Invariant Continuous-Time Models; $\Delta t = 0.1$ sec

To lend more insight into the accuracy of the ramp-invariant transformation as compared with the classic step-invariant transformation, Figure 8 shows the error of the ramp-invariant system for each sample time, which was calculated as the difference between the response of the sampled system and the response of the step-invariant transformed system response. Since the errors of both output vectors of the multi-output system given in (6.1) are approximately equal, only the error of the first output, $y_1(t)$, is shown. Figure 9 shows the error of the step-invariant system for each sample time. It is obvious from these graphs, Figures 8 and 9, that the ramp-invariant system more accurately approximates the original system for every sampling time for this type of input signal. Note that the Magnitude scale (y-axis) in Figure 8 and Figure 9 are different.

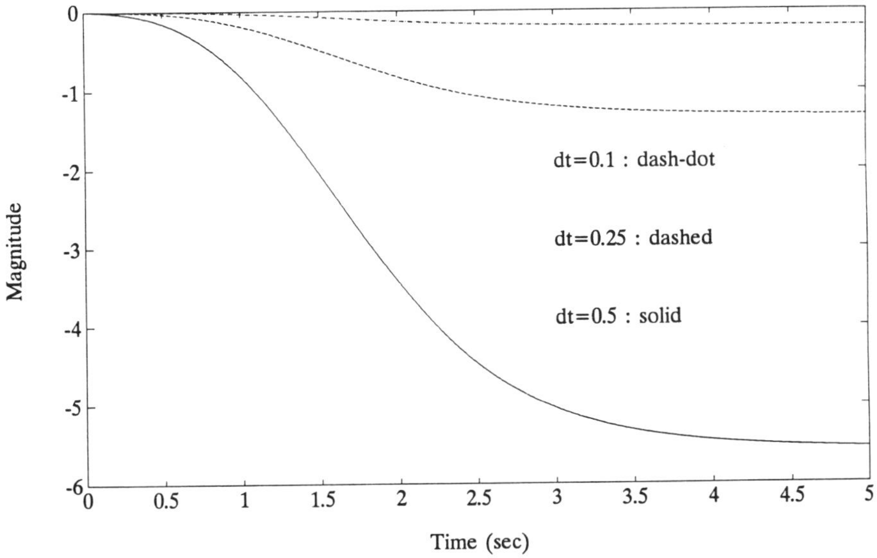

Figure 8. Differences between Sampled and Ramp-Invariant System Responses; $\Delta t = 0.5, 0.25, 0.1$ sec

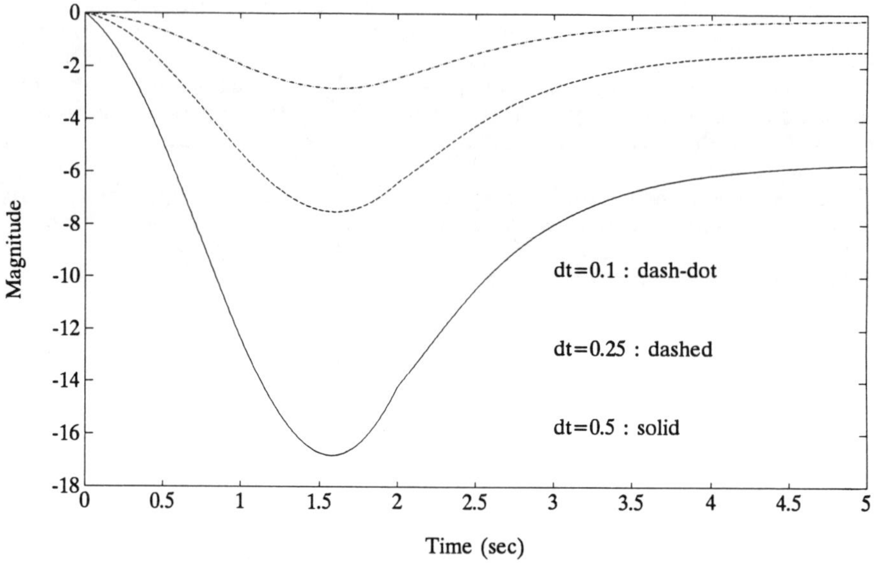

Figure 9. Differences between Sampled and Step-Invariant System Responses; $\Delta t = 0.5$, 0.25. 0.1 sec

Using the second set of admissible pseudo-observability indices, i.e. $\eta = \{2, 2\}$, and a sampling time of 0.5 sec, the following matrices were obtained:

$$F_o = \begin{bmatrix} 0 & 0 & 1 & 0 \\ 0 & 0 & 0 & 1 \\ -.008 & -.173 & .179 & 1.002 \\ -.003 & -.485 & .023 & 1.464 \end{bmatrix} \qquad G_o = \begin{bmatrix} 3.06 \\ 5.30 \\ 5.54 \\ 6.95 \end{bmatrix} \qquad H_o = \begin{bmatrix} .619 \\ 1.730 \end{bmatrix}$$

$$A = \frac{1}{T} \ln (F_o) = \begin{bmatrix} -6.868 & -7.021 & 25.22 & -9.330 \\ .022 & -2.735 & -.185 & 2.899 \\ -.174 & .4971 & -2.920 & 2.596 \\ -.007 & -1.375 & .0576 & 1.324 \end{bmatrix} \tag{6.22}$$

$$B = \begin{bmatrix} 1.194 \\ 4.209 \\ 6.170 \\ 10.888 \end{bmatrix}; \qquad D = \begin{bmatrix} -.002 \\ .016 \end{bmatrix}$$

which should be compared with (6.7), (6.8), (6.10), (6.11), and (6.14) obtained in the case of $\eta = \{3,1\}$. Obviously, these matrices are different, but according to the definition of pseudo-observability indices both sets of matrices $\{F_o, G_o, C_o, H_o\}$ represent the same system. Consequently, the response of system (6.2) with matrices {A, B, C, D} from (6.22) are equal to the response using {A, B, C, D} in the case of $\eta = \{3,1\}$. This verifies that the use of pseudo-observable forms does not require the structural identification, since any admissible set of pseudo-observability indices may be used in representing a given multivariable system.

7. Conclusions

Here we presented a procedure for identification of continuous-time state space models of linear multivariable systems from samples of input/output observations. Some of the unique features of the presented procedure are as follows. The necessity of structural identification is avoided by using one of many possible pseudo-observable state space forms. The system to be identified does not need to be at rest. A continuous-time model is determined using the First-Order Hold, i.e. Ramp invariant, assumption. The calculation of the natural log of the discrete-time system matrix is done by a modification of existing procedures which allows achieving a desired accuracy in cases of unstable systems and relatively large sampling intervals.

References

Bingulac, S. (1978), "Identification of multivariable dynamic systems", "*Identification and System Parameter Estimation*", Rajbman, N. S., (editor), North-Holland Publishing Co., pp. 1979-1987.

Bingulac, S. (1988), "CAD package L-A-S: A research and educational tool in systems and control", Proc. 20th Southeastern Symposium on System Theory, Charlotte, NC, pp. 44-49.

Bingulac, S. and Farias, M. A. C. (1977), "Identification and minimal realization of multivariable systems", Proc. of the IFAC Symposium on Multivariable Technological Systems, (Fredericton, Canada) pp. 95.1-95.5.

Bingulac, S. and Krtolica, R. (1987), "On admissibility of pseudoobservability indexes", *IEEE Transactions on Automatic Control*, vol. AC-32, pp. 920-922.

Bingulac, S. and Krtolica, R. (1988), "An algorithm for simultaneous order and parameter identification in multivariable systems", Proc. 8th IFAC Symposium on Identification and System Parameter Estimation, (Beijing, China), pp. 542-547.

Bingulac, S. and Sinha, N. K. (1989), "On the identification of continuous-time multivariable systems from samples of input-output data", Proceedings of the 7th International Conference on Mathematical and Computer Modelling, (Chicago, Ill, August), pp. 231-239.

Bingulac, S. and Cooper, D. L. (1990), "Derivation of discrete- and continuous-time ramp invariant representations", *Electronics Letters*, vol. 26, no. 10, pp. 664-666.

Bingulac, S. and Cooper, D. L. (1990a), "*L-A-S User's Manual*", Virginia Polytechnic Institute and State University.

Brogan, W. L. (1985), "*Modern control theory*", Prentice-Hall, New Jersey.

Chen, C. T. (1984), "*Linear System Theory and Design*', Holt, Rinehart, and Winston, Inc., New York.

Cooper, D. L. (1990), "Continuous-Time Multivariable System Identification", M.S. Thesis, Virginia Polytechnic Institute and State University, Blacksburg.

Cooper, D. L., and Bingulac, S. (1990), "Computational improvement in the calculation of the natural log of a square matrix", *Electronics Letters*, vol. 26, no. 13, pp. 861-862.

Correa, G. and Glover, K. (1986), "On the choice of parametrization for identification", *IEEE Trans. Auto. Control*, vol. AC-31, no. 1, pp. 8-15.

Gevers, M. and Wertz, V. (1984), "Uniquely identifiable state-space and ARMA parametrizations for multivariable linear systems", *Automatica*, vol. 20, pp. 333-347.

Gorti, B. M., Bingulac, S., and Vanlandingham, H. F. (1990), "Deterministic identification of linear multivariable systems", Proc. of the IEEE 22nd Southeast Symp. Syst. Theory, (Cookville, TN, March), pp. 126-131.

Guidorzi, R. (1975), "Canonical structures in the identification of multivariable systems", *Automatica*, vol. 11, pp. 361-374.

Haykin, S. S. (1972), "A unified treatment of recursive digital filtering", *IEEE Transactions on Automatic Control*, vol. AC-17, pp. 113-116.

Hsia, T. C. (1972), "On sampled data approach to parameter identification of continuous linear systems", *IEEE Trans. Auto. Control*, vol. AC-17, no. 2, pp. 247-249.

Kailath, T. (1980), "*Linear Systems*", Prentice-Hall, New Jersey.

Keviczky, L. (1977), "On the equivalence of discrete and continuous transfer functions", *Problems of Control and Information Theory*, vol. 6, no. 2, pp. 111-128.

Lastman, G. J., Puthenpura, S. C., and Sinha, N. K. (1984), "Algorithm for the identification of continuous-time multivarible systems from their discrete-time models", *Electronic Letters*, vol. 20, no. 22, pp. 918-919.

Puthenpura, S. C. and Sinha, N. K. (1984), "Transformation of continuous-time model of a linear multivariable system from its discrete-time model", *Electronic Letters*, vol. 20, no. 18, pp. 737-738.

Sinha, N. K. (1972), "Estimation of transfer function of continuous system from sampled data", Proceedings IEE, vol. 19, pp. 612-614.

Sinha, N. K. and Lastman, G. J. (1981), "Transformation algorithm for identification of continuous-time multivariable systems from discrete data", *Electronics Letters*, vol. 17, pp. 779-780.

Sinha, N. K. and Lastman, G. J. (1982), "Identification of continuous-time multivariable systems from sampled data", *International Journal of Control*, vol. 35, pp. 117-126.

Sinha, N. K. and Kusta, B. (1983), "*Modeling and Identification of Dynamic Systems*", Van Nostrand-Reinhold, New York.

Strmčnik, S. and Bremšak, F. (1979), "Some new transformation algorithms in the identification of continuous-time multivariable systems using discrete identification methods', Preprints, 5th IFAC Symposium on Identification and System Parameter Estimation, (Darmstadt, Germany),pp. 397-405.

VanLandingham, H. F. (1985), "*Introduction to Digital Control Systems*", Macmillan Publishing Co., New York.

SVD-based subspace methods for multivariable continuous-time systems identification

Marc Moonen, Bart De Moor, Joos Vandewalle
ESAT Katholieke Universiteit Leuven
K.Mercierlaan 94, 3001 Heverlee, Belgium

Abstract

Recently several subspace methods have appeared in the literature for multivariable discrete-time state space identification, *where state space models are computed directly from input/output data. These state space identification methods are viewed as the better alternatives to polynomial model identification, owing to the better numerical conditioning associated with state space models, especially for high-order multivariable systems.*

In this contribution, a similar method is described for continuous-time state space identification. *Here also, the key tool is the singular value decomposition (SVD), a numerical technique known to be very robust and accurate when dealing with noisy data. The noise coloring is compensated for by using a generalization of the SVD, namely the quotient SVD. The resulting identification scheme is then shown to give consistent results under certain conditions.*

N. K. Sinha and G. P. Rao (eds.), Identification of Continuous-Time Systems, 473–488.

1 Introduction

The greater part of the systems identification literature is concerned with computing polynomial system models, which are however known to typically give rise to numerically ill-conditioned mathematical problems. While state space models on the other hand show much more desirable numerical properties, direct state space identification is clearly a much less established area. Most commonly known are the realization algorithms of [Ho and Kalman 1966, Zeiger and Mc Ewen 1974, Kung 1978], where a discrete-time state space model is computed from a block Hankel matrix with Markov parameters. The Markov parameters might be hard to obtain though[1]*, and furthermore –as far as we know– the latter algorithms do not allow for treating the effect of inexact data. An alternative direct identification scheme is described in [Moonen et al. 1989a] and [Moonen & Vandewalle 1990], where a discrete-time state space model is computed from a block Hankel matrix with I/O data. The key tool is the singular value decomposition (SVD), a numerical technique, known to be very robust and accurate when dealing with noisy data. In the colored noise case, the noise coloring is compensated for by using a generalization of the SVD, namely the quotient SVD (QSVD) [De Moor & Golub 1989]. These algorithms are readily convertible into adaptive identification algorithms for model updating, where use is made of (Q)SVD updating techniques.*

In this contribution, a similar method is described for continuous-time state space identification. Due to the necessary use of a certain pre-filtering, the procedure itself is seen to always introduce a noise coloring, so that a QSVD is needed in any case. First, an algorithm for state space identification in the noise-free case is developed in sections 2 and 3. In section 4, we then consider the case where the data are corrupted by noise and describe the general identification set-up. The derivation of these algorithms is analogous to the derivation of the discrete-time algorithms in [Moonen et al. 1989a] and [Moonen & Vandewalle 1990]. We therefore refer to those reports for further details.

[1] for infinite impulse response systems, especially with unstable poles, computing a deconvolution directly –i.e. by solving a set of linear equations– is impossible

2 Preliminaries

For the time being, we consider linear time invariant multivariable systems with state space description as follows

$$\begin{aligned} \dot{x}(t) &= A \cdot x(t) + B \cdot u(t) \\ y(t) &= C \cdot x(t) + D \cdot u(t) \end{aligned}$$

where $u(t), y(t)$ and $x(t)$ denote the input (m-vector), output (l-vector) and state vector at time t, the dimension of $x(t)$ being the minimal system order n. A, B, C and D are the unknown system matrices to be identified.

In the sequel, we often make use of matrices which are constructed with samples of the in- and outputs and their derivatives. We let $u^{(k)}$ and $y^{(k)}$ denote the k-th derivative of u and y, and define

$$\underbrace{U}_{2mi \times j} = \begin{bmatrix} u(t_1) & u(t_2) & \dots & u(t_j) \\ u^{(1)}(t_1) & u^{(1)}(t_2) & \dots & u^{(1)}(t_j) \\ u^{(2)}(t_1) & u^{(2)}(t_2) & \dots & u^{(2)}(t_j) \\ \vdots & \vdots & & \vdots \\ u^{(2i-1)}(t_1) & u^{(2i-1)}(t_2) & \dots & u^{(2i-1)}(t_j) \end{bmatrix}$$

$$\underbrace{Y}_{2li \times j} = \begin{bmatrix} y(t_1) & y(t_2) & \dots & y(t_j) \\ y^{(1)}(t_1) & y^{(1)}(t_2) & \dots & y^{(1)}(t_j) \\ y^{(2)}(t_1) & y^{(2)}(t_2) & \dots & y^{(2)}(t_j) \\ \vdots & \vdots & & \vdots \\ y^{(2i-1)}(t_1) & y^{(2i-1)}(t_2) & \dots & y^{(2i-1)}(t_j) \end{bmatrix}.$$

Here, the sampling intervals $t_2 - t_1$, $t_3 - t_2$, etc. need not even be equidistant. The problems associated with computing derivatives are tackled in section 4. For the time being we thus assume that these matrices can indeed be constructed. We now briefly state three important theorems which are used in the sequel for developing our identification algorithms.

Theorem 1
Time sequences $u(t), y(t), x(t)$ that satisfy the above state space equations, also satisfy the following general structured I/O-equation :

$$Y = \Gamma_{2i} \cdot X + T_{2i} \cdot U.$$

Here U and Y are the above defined data matrices, and X contains a corresponding sampling of the state vector :

$$X = \left[\begin{array}{cccc} x(t_1) & x(t_2) & \dots & x(t_j) \end{array} \right].$$

Finally

$$\Gamma_{2i} = \begin{bmatrix} C \\ CA \\ CA^2 \\ \dots \\ CA^{2i-1} \end{bmatrix}, \qquad T_{2i} = \begin{bmatrix} D & 0 & \dots & 0 \\ CB & D & \dots & 0 \\ CAB & CB & \dots & 0 \\ \vdots & \vdots & & \vdots \\ CA^{2i-2}B & CA^{2i-3}B & \dots & D \end{bmatrix}.$$

The proof of this is straightforward by repeated substitution of the state space equations.

□

Instead of going into details, we loosely state that i and j should be chosen 'sufficiently large', so that Y and U contain enough information on the system, and in particular $j \gg \max(2mi, 2li)$ ('short fat' matrices), as this reduces the noise sensitivity (see below).

Theorem 2

Let Y, U and X be defined as in the previous theorem, and let H denote the concatenation of Y and U

$$H = \begin{bmatrix} U \\ Y \end{bmatrix}$$

then, under the conditions that

(i) $\text{rank}\{X\} = n$ *where n is the minimal system order, and*

(ii) $\text{span}_{\text{row}}\{X\} \cap \text{span}_{\text{row}}\{U\} = \emptyset$ *,*

the following rank property holds :

$$\text{rank}\{H\} = \text{rank}\{U\} + n.$$

In other words the row space of Y adds n dimensions to the row space of U. Also, when

(iii) $\text{rank}\{U\} = 2mi =$ *number of rows in U ,*

this rank property reduces to

$$\text{rank}\{H\} = 2mi + n.$$

This theorem allows to estimate the system order, prior to further identification of the system matrices. For a proof of this, we refer to [Moonen et al. 1989], where a similar proof is given for the discrete-time case.

□

Note on condition (i) : rank$\{X\}$ $= n$ *means that all modes are sufficiently excited (persistent excitation). When certain modes are not sufficiently excited, i.e. unobservable in the I/O-data, they cannot be identified either and application of the above rank property will reveal too low a system order. This problem is inherent in system identification.*

Note on condition (ii) : *When this condition is not satisfied, application of the rank property again reveals an underestimation of the system order. However it can be verified that rank cancellation is not generic*[2]*, and the probability that rank cancellation occurs, decreases for fixed* $2i$ *(number of block rows in* U*) with increasing* j *(number of columns in* U *and* X *).*

Note on condition (iii) : *Similar to the previous ones, this third condition is generically satisfied when the input is sufficiently rich (inherent in the identification problem).*

In the following, it is always assumed that these three conditions are satisfied.

Suppose we would now cut the above matrices U *and* Y *in half as follows*

$$U = \begin{bmatrix} U_1 \\ U_2 \end{bmatrix} \quad Y = \begin{bmatrix} Y_1 \\ Y_2 \end{bmatrix}$$

with

$$\underbrace{U_1}_{mi\times j} = \begin{bmatrix} u(t_1) & u(t_2) & \dots & u(t_j) \\ u^{(1)}(t_1) & u^{(1)}(t_2) & \dots & u^{(1)}(t_j) \\ \vdots & \vdots & & \vdots \\ u^{(i-1)}(t_1) & u^{(i-1)}(t_2) & \dots & u^{(i-1)}(t_j) \end{bmatrix}$$

$$\underbrace{U_2}_{mi\times j} = \begin{bmatrix} u^{(i)}(t_1) & u^{(i)}(t_2) & \dots & u^{(i)}(t_j) \\ u^{(i+1)}(t_1) & u^{(i+1)}(t_2) & \dots & u^{(i+1)}(t_j) \\ \vdots & \vdots & & \vdots \\ u^{(2i-1)}(t_1) & u^{(2i-1)}(t_2) & \dots & u^{(2i-1)}(t_j) \end{bmatrix}$$

[2]unless the system is controlled by state feedback $u_k = -Fx_k$

and Y_1, Y_2 similarly constructed. Both matrix pairs satisfy an I/O-equation as follows:

$$\begin{aligned} Y_1 &= \Gamma_i \cdot X + T_i \cdot U_1 \\ Y_2 &= \Gamma_i \cdot X^{(i)} + T_i \cdot U_2 \end{aligned}$$

with

$$X^{(i)} = \left[\begin{array}{cccc} x^{(i)}(t_1) & x^{(i)}(t_2) & \dots & x^{(i)}(t_j) \end{array} \right].$$

The corresponding derivatives satisfy

$$\begin{aligned} Y_1^{(1)} &= \Gamma_i \cdot X^{(1)} + T_i \cdot U_1^{(1)} \\ Y_2^{(1)} &= \Gamma_i \cdot X^{(i+1)} + T_i \cdot U_2^{(1)} \end{aligned}$$

with

$$X^{(i+1)} = \left[\begin{array}{cccc} x^{(i+1)}(t_1) & x^{(i+1)}(t_2) & \dots & x^{(i+1)}(t_j) \end{array} \right].$$

Theorem 3

With $H_1, H_2, X^{(i)}$ defined as above

$$\text{span}_{\text{row}}\{X^{(i)}\} = \text{span}_{\text{row}}\{H_1\} \cap \text{span}_{\text{row}}\{H_2\}$$

so that any basis for this intersection constitutes a valid state vector sequence $X^{(i)}$, with the basis vectors as the consecutive row vectors. Similarly, we have

$$\text{span}_{\text{row}}\{X^{(i+1)}\} = \text{span}_{\text{row}}\{H_1^{(1)}\} \cap \text{span}_{\text{row}}\{H_2^{(1)}\}.$$

Note that different choices for a basis differ in a transformation matrix P that transforms a model A,B,C,D into an equivalent model $P^{-1}AP$, $P^{-1}B$, CP, D. For a proof of the above theorem, we again refer to [Moonen et al. 1989], where a similar proof is given for the discrete-time case.

□

The above theorem allows to calculate derivatives of the state vector sequence, by making use of the measured I/O-data and their derivatives only. Once these sequences are known, the system matrices are readily identified from a set of linear equations as follows

$$\left[\begin{array}{ccc} x^{(i+1)}(t_1) & \dots & x^{(i+1)}(t_j) \\ y^{(i)}(t_1) & \dots & y^{(i)}(t_j) \end{array} \right] = \left[\begin{array}{cc} A & B \\ C & D \end{array} \right] \cdot \left[\begin{array}{ccc} x^{(i)}(t_1) & \dots & x^{(i)}(t_j) \\ u^{(i)}(t_1) & \dots & u^{(i)}(t_j) \end{array} \right].$$

The above results constitute the heart of a two-step identification scheme. First the state vector sequences are computed as the intersection of the row spaces of certain I/O data matrices. Then the system matrices are obtained all at once as the solution of a set of linear equations. In the next sections, this procedure is built into a practical computational scheme.

3 An SVD based algorithm for the noise free case

In this section, we derive a computational scheme for the case where there is no noise on the data. We also assume this in that case computing derivatives poses no problems. In the next section, the more general case with additive noise is considered.

For compact notation, it is useful to first redefine matrices H_1 *and* H_2 *in the following way :*

$$H_1 = \begin{bmatrix} u(t_1) & u(t_2) & \dots & u(t_j) \\ y(t_1) & y(t_2) & \dots & y(t_j) \\ u^{(1)}(t_1) & u^{(1)}(t_2) & \dots & u^{(1)}(t_j) \\ y^{(1)}(t_1) & y^{(1)}(t_2) & \dots & y^{(1)}(t_j) \\ \vdots & \vdots & & \vdots \\ u^{(i-1)}(t_1) & u^{(i-1)}(t_2) & \dots & u^{(i-1)}(t_j) \\ y^{(i-1)}(t_1) & y^{(i-1)}(t_2) & \dots & y^{(i-1)}(t_j) \end{bmatrix}$$

$$H_2 = \begin{bmatrix} u^{(i)}(t_1) & u^{(i)}(t_2) & \dots & u^{(i)}(t_j) \\ y^{(i)}(t_1) & y^{(i)}(t_2) & \dots & y^{(i)}(t_j) \\ u^{(i+1)}(t_1) & u^{(i+1)}(t_2) & \dots & u^{(i+1)}(t_j) \\ y^{(i+1)}(t_1) & y^{(i+1)}(t_2) & \dots & y^{(i+1)}(t_j) \\ \vdots & \vdots & & \vdots \\ u^{(2i-1)}(t_1) & u^{(2i-1)}(t_2) & \dots & u^{(2i-1)}(t_j) \\ y^{(2i-1)}(t_1) & y^{(2i-1)}(t_2) & \dots & y^{(2i-1)}(t_j) \end{bmatrix}.$$

Notice that theorem 3 remains valid. We also introduce the following notation :

$M\{p:q,r:s\}$ *is the submatrix of* M *at the intersection of rows* $p,p+1,\dots,q$ *and columns* $r,r+1,\dots,s$

$M\{:,r:s\}$ *is the submatrix of* M *containing columns* $r,r+1,\dots,s$

$M\{p:q,:\}$ *is the submatrix of* M *containing rows* $p,p+1,\dots,q$.

As an example, referring to theorem 3,

$$H_1^{(1)} = H\{m+l+1:(i+1)(m+l),:\}$$

where H *is the concatenation* $\begin{bmatrix} H_1 \\ H_2 \end{bmatrix}$.

The intersection of the row spaces of H_1 and H_2, can be computed from the SVD of H as follows

$$\begin{aligned} H &= \begin{bmatrix} H_1 \\ H_2 \end{bmatrix} \\ &= Q \cdot \Sigma \cdot V^T \\ &= \begin{bmatrix} Q_{11} & Q_{12} \\ Q_{21} & Q_{22} \end{bmatrix} \cdot \begin{bmatrix} \Sigma_{11} & 0 \\ 0 & 0 \end{bmatrix} \cdot V^T \end{aligned}$$

$$\begin{aligned} \dim\{Q_{11}\} = \dim\{Q_{21}\} &= (mi + li) \times (2mi + n) \\ \dim\{Q_{12}\} = \dim\{Q_{22}\} &= (mi + li) \times (2li - n) \\ \dim\{\Sigma_{11}\} &= (2mi + n) \times (2mi + n). \end{aligned}$$

Here Q and V are orthogonal matrices, Σ is a diagonal matrix. From

$$Q_{12}^T \cdot H_1 = -Q_{22}^T \cdot H_2$$

it follows that the row space of $Q_{12}^T \cdot H_1$ equals the required intersection. However, $Q_{12}^T \cdot H_1$ contains $2li - n$ row vectors , only n of which are linearly independent (dimension of the intersection). Thus, it remains to select n suitable combinations of these row vectors. Making use of a CS-decomposition [Golub & Van Loan 1983], one easily shows that

$$Q_{12} = \begin{bmatrix} \underbrace{Q'_{12}}_{li-n} & \underbrace{Q''_{12}}_{n} & \underbrace{Q'''_{12}}_{li-n} \end{bmatrix} \cdot \begin{bmatrix} I & & \\ & C & \\ & & 0 \end{bmatrix} \cdot W^T$$

$$Q_{22} = \begin{bmatrix} \underbrace{Q'_{22}}_{li-n} & \underbrace{Q''_{22}}_{n} & \underbrace{Q'''_{22}}_{li-n} \end{bmatrix} \cdot \begin{bmatrix} O & & \\ & S & \\ & & I \end{bmatrix} \cdot W^T$$

$$\begin{aligned} C &= diag\{c_1, \ldots, c_n\} \\ S &= diag\{s_1, \ldots, s_n\} \\ C^2 + S^2 &= I_{n \times n}. \end{aligned}$$

Clearly, only Q''_{12} delivers useful combinations for the computation of the intersection, and we can take

$$\begin{aligned} X^{(i)} &= (Q''_{12})^T \cdot H_1 \\ &= (Q''_{12})^T \cdot H\{1 : i(m+l), :\}. \end{aligned}$$

The above expressions for Q_{12} and Q_{22} are in itself SVD's of these matrices, and can be computed as such. It thus suffices to compute e.g. the SVD of Q_{12}, and select left singular vectors Q''_{12} corresponding to singular values $\sigma \neq 0, 1$. The computation of the required intersection then reduces to the computation of two successive SVD's (for H and Q_{12} respectively). Finally, from the above formula, it also follows that

$$\begin{aligned} X^{(i+1)} &= (Q''_{12})^T \cdot H_1^{(1)} \\ &= (Q''_{12})^T \cdot H\{m+l+1:(i+1)(m+l),:\}. \end{aligned}$$

In the second step, the system matrices are identified from a set of linear equations. Much as was done in [Moonen et al. 1989], it can be shown that the system matrices can be computed from a reduced set as follows (obtained after discarding the common orthogonal factor V)

$$\begin{bmatrix} (Q''_{12})^T \cdot Q\{m+l+1:(i+1)(m+l), 1:2mi+n\} \cdot \Sigma_{11} \\ Q\{mi+li+m+1:(m+l)(i+1), 1:2mi+n\} \cdot \Sigma_{11} \end{bmatrix}$$
$$= \begin{bmatrix} A & B \\ C & D \end{bmatrix} \begin{bmatrix} (Q''_{12})^T \cdot Q\{1:mi+li, 1:2mi+n\} \cdot \Sigma_{11} \\ Q\{mi+li+1:mi+li+m, 1:2mi+n\} \cdot \Sigma_{11} \end{bmatrix}.$$

Here, the fact that the system matrices can be computed from a reduced set of equations introduces a significant computational saving. The largest dimension in the set of equations is now $2mi+n$ instead of j, where $j \gg 2mi+n$ ('short fat' matrices). As only Q and Σ are needed from the first (largest) SVD of H, the computational effort in the first stage is cut down as well.

4 A QSVD based algorithm for the colored noise case

In the previous section we assumed noise free data, and derived a fairly simple SVD based identification scheme. In practice however, we cannot afford this luxury, so we will have to take the noise into account explicitly, when deriving a computational scheme. When dealing with noisy I/O data, one major problem is of course the computation of the derivatives. In this section, it is shown how the problem is solved by incorporating a pre-filter, and compensating for the corresponding noise coloring with a QSVD instead of an SVD.

First, it is readily seen that the dynamic relation between inputs and outputs is not changed if all in- and outputs are filtered with one and the same linear filter $\frac{F(s)}{E(s)}$. In other words, we also have

$$\begin{aligned}\dot{x}_f(t) &= A \cdot x_f(t) + B \cdot u_f(t)\\ y_f(t) &= C \cdot x_f(t) + D \cdot u_f(t)\end{aligned}$$

with unchanged A, B, C, D-matrices, and

$$\begin{aligned}U_f(s) &= \text{diag}\{\frac{F(s)}{E(s)}\} \cdot U(s)\\ Y_f(s) &= \text{diag}\{\frac{F(s)}{E(s)}\} \cdot Y(s)\\ X_f(s) &= \text{diag}\{\frac{F(s)}{E(s)}\} \cdot X(s).\end{aligned}$$

Here $\frac{F(s)}{E(s)}$ is a scalar valued rational function, which we specify as follows

$$\frac{F(s)}{E(s)} = \frac{f_p s^p + f_{p-1} s^{p-1} + \ldots f_1 s^1 + f_0}{s^{p+2i} + e_{p+2i-1} s^{p+2i-1} + \ldots e_1 s^1 + e_0}.$$

Typically $\frac{F(s)}{E(s)}$ serves to get rid of the differentiation. The advantage of having $2i$ zeroes at infinity is apparent from **Figure 1**, *where a realization of the pre-filter is exhibited. For an arbitrary input $z(t)$ (scalar valued), the output $z_f(t)$ is produced, as well as its derivatives up to the $(2i-1)th$. The additional flexibility in choosing the other filter coefficients can e.g. be used to construct a bandpass filter if a model in a limited frequency range is searched for. Note that the pre-filter can be implemented as a digital filter as well.*

In principle, one could now use the computational scheme of the previous sections, where however the filtered data are used throughout. For instance, one would use

$$H_{1_f} = \begin{bmatrix} u_f(t_1) & u_f(t_2) & \dots & u_f(t_j) \\ y_f(t_1) & y_f(t_2) & \dots & y_f(t_j) \\ u_f^{(1)}(t_1) & u_f^{(1)}(t_2) & \dots & u_f^{(1)}(t_j) \\ y_f^{(1)}(t_1) & y_f^{(1)}(t_2) & \dots & y_f^{(1)}(t_j) \\ \vdots & \vdots & & \vdots \\ u_f^{(i-1)}(t_1) & u_f^{(i-1)}(t_2) & \dots & u_f^{(i-1)}(t_j) \\ y_f^{(i-1)}(t_1) & y_f^{(i-1)}(t_2) & \dots & y_f^{(i-1)}(t_j) \end{bmatrix}$$

and a similar H_{2_f}, *etcetera. Although the problem of computing derivatives is apparently solved, the SVD scheme still fails to give consistent results, the reason for this being the noise coloring introduced by the pre-filter. Suppose the data matrix* H_f *is corrupted by additive noise*

$$H_f = H_{f_{\text{exact}}} + H_{f_{\text{noise}}}$$

then it is well known that the SVD scheme delivers consistent results in the 'white noise case', where

$$E\{H_{f_{\text{noise}}} \cdot H^T_{f_{\text{noise}}}\} = \sigma \cdot I.$$

Here $E\{\cdot\}$ *is the expectation operator and* I *is the identity matrix. In the general case, we have*

$$E\{H_{f_{\text{noise}}} \cdot H^T_{f_{\text{noise}}}\} = \sigma \cdot L_{\text{noise}} \cdot L^T_{\text{noise}}$$

where L_{noise} *is an (assumed known) lower triangular factor. The noise coloring can then be compensated for, by using a QSVD for the matrix pair* $\{H, L_{\text{noise}}\}$ *instead of an SVD of* H *[Moonen & Vandewalle 1990, De Moor & Golub 1989].*

So the first thing is to compute the L_{noise} *for our case. Therefore, we convert the filter model to state space form, which reads as follows*

$$\dot{x}(t) = \underbrace{\begin{bmatrix} 0 & I & & & & & & \\ & \ddots & \ddots & & & & & \\ & & \ddots & \ddots & & & & \\ & & & 0 & I & & & \\ & & & & & & & \\ & & & & -E_{2i+p-1} & I & & \\ & & & & \vdots & & \ddots & \\ & & & & -E_{2i} & & & I \\ -E_0 & -E_1 & \dots & -E_{2i-2} & -E_{2i-1} & & & 0 \end{bmatrix}}_{A_{\text{filter}}} \cdot x(t) + \underbrace{\begin{bmatrix} 0 \\ 0 \\ \vdots \\ 0 \\ \\ F_p \\ \vdots \\ F_1 \\ F_0 \end{bmatrix}}_{B_{\text{filter}}} \cdot \begin{bmatrix} u(t) \\ y(t) \end{bmatrix}$$

$$\begin{bmatrix} u_f(t) \\ y_f(t) \end{bmatrix} = \begin{bmatrix} I & 0 & \dots & 0 \end{bmatrix} \cdot x(t)$$

where

$$\begin{aligned} E_i &\stackrel{\text{def}}{=} e_i \cdot I_{(m+l)\times(m+l)} \\ F_i &\stackrel{\text{def}}{=} f_i \cdot I_{(m+l)\times(m+l)}. \end{aligned}$$

Here, for compact notation, we consider the pre-filtering of all the in- and outputs at once. The first $2i(m+l)$ components of the state vector constitute the filter outputs and their derivatives. Suppose now that the filter inputs are subject to additive wide-sense stationary noise with covariance V, i.e.

$$E\{\begin{bmatrix} u^T(t)_{\text{noise}} & y^T(t)_{\text{noise}} \end{bmatrix} \cdot \begin{bmatrix} u(t)_{\text{noise}} \\ y(t)_{\text{noise}} \end{bmatrix}\} = V.$$

The covariance matrix for the noise on the state vector (for $t \to \infty$) is then known to be proportional to $L \cdot L^T$, the solution of the Lyapunov equation

$$A_{\text{filter}} \cdot (L \cdot L^T) + (L \cdot L^T) \cdot A_{\text{filter}}^T + B_{\text{filter}} \cdot V \cdot B_{\text{filter}}^T = 0.$$

This equation can be solved for the lower triangular factor L directly, as described in e.g. [Hammarling 1982]. The upper left block in L, then equals L_{noise}

$$L_{\text{noise}} = L\{1 : 2i(m+l), 1 : 2i(m+l)\}.$$

The required matrix L_{noise} can thus be computed from the knowledge of V and the filter coefficients. Once L_{noise} is known, this matrix can be used in a QSVD based scheme, which is similar to the original SVD scheme. First the QSVD of the matrix pair $\{H_f, L_{\text{noise}}\}$ is computed

$$\begin{aligned} H_f &= Q^{-T} \cdot \Sigma \cdot V^T \\ L_{\text{noise}} &= Q^{-T} \cdot \Sigma_L \cdot V_L^T \end{aligned}$$

where now Q is a square non-singular matrix (not necessarily orthogonal), V and V_L are orthogonal, and Σ is a diagonal matrix with non-increasing entries along the diagonal. With the same partitioning for Q as in the SVD scheme, we now use a second QSVD for $\{Q_{12}, Q_{22}\}$ (instead of a CS-decomposition), to give

$$Q_{12} = \left[\begin{array}{ccc} Q'_{12} & Q''_{12} & Q'''_{12} \end{array} \right] \cdot \left[\begin{array}{ccc} I & & \\ & C & \\ & & 0 \end{array} \right] \cdot W^T$$

$$Q_{22} = \left[\begin{array}{ccc} Q'_{22} & Q''_{22} & Q'''_{22} \end{array} \right] \cdot \left[\begin{array}{ccc} O & & \\ & S & \\ & & I \end{array} \right] \cdot W^T$$

$$\begin{aligned} C &= diag\{c_1, \ldots, c_n\} \\ S &= diag\{s_1, \ldots, s_n\} \\ C^2 + S^2 &= I_{n \times n}. \end{aligned}$$

Again it can be shown that Q''_{12} delivers the useful combinations for computing the required intersections (see [Moonen & Vandewalle 1990] for details). The system matrices are finally computed from

$$\left[\begin{array}{c} (Q''_{12})^T \cdot Q^{-T}\{m+l+1 : (i+1)(m+l), 1 : 2mi+n\} \cdot \Sigma_{11} \\ Q^{-T}\{mi+li+m+1 : (m+l)(i+1), 1 : 2mi+n\} \cdot \Sigma_{11} \end{array} \right]$$
$$= \left[\begin{array}{cc} A & B \\ C & D \end{array} \right] \left[\begin{array}{c} (Q''_{12})^T \cdot Q^{-T}\{1 : mi+li, 1 : 2mi+n\} \cdot \Sigma_{11} \\ Q^{-T}\{mi+li+1 : mi+li+m, 1 : 2mi+n\} \cdot \Sigma_{11} \end{array} \right].$$

Similar to what we had in the SVD case, only the Q and Σ matrices are needed from the first (largest) QSVD. This not only cuts down the computational effort in the first stage, but furthermore it allows for constructing *efficient adaptive algorithms, where use is made of QSVD updating techniques. We refer to [Moonen & Vandewalle 1990] for details.*

Finally, it can be shown that the above double QSVD scheme delivers consistent results for infinite data sequences, if we let $j \to \infty$. For details we refer to [Moonen & Vandewalle 1990], were similar results for the discrete-time case are derived, and [De Moor 1990].

5 Conclusions

It has been demonstrated how state space identification techniques for multivariable discrete-time systems are translated into identification techniques for continuous-time systems. The key tool is the singular value decomposition (SVD), a numerical technique known to be very robust and accurate when dealing with noisy data. The noise coloring is compensated for by using a generalization of the SVD, namely the quotient SVD. The resulting identification scheme is shown to give consistent results under certain conditions.

References

[1] B. De Moor and G.H. Golub 1989. Generalized Singular Value Decompositions: A proposal for a standardized nomenclature. *Numerical Analysis Project.* Manuscript **NA-89-04**. Dept. of Comp. Sc., Stanford Univ.

[2] B. De Moor 1990. The SVD and long and short spaces of noisy matrices. *ESAT-SISTA report 1990-38.* Dept. of E.E., Katholieke Universiteit Leuven, Belgium.

[3] G.H. Golub and C.F. Van Loan 1988. *Matrix computations.* Second Edition. North Oxford academic Publishing Co., Johns Hopkins Press.

[4] S.J. Hammarling 1982. Numerical solution of the stable non-negative definite Lyapunov equation. *IMA Journal of Numerical Analysis* **2**, pp 303-323.

[5] B.L. Ho and R.E. Kalman 1966. Effective construction of linear state variable models from input output functions. *Regelungstechnik* **14**, pp 545-548.

[6] S.Y. Kung 1978. A new identification and model reduction algorithm via singular value decomposition. *Proc. 12th Asilomar Conf. on Circuits, Systems and Computers, Pacific Grove*, pp 705-714.

[7] M. Moonen, B. De Moor, L. Vandenberghe, J. Vandewalle 1989a. On- and off-line identification of linear state space models. *Int. Journal of Control* **49**, No. 1, pp 219-232.

[8] M. Moonen, J. Vandewalle 1990b. A QSVD approach to on- and off-line state space identification. *Int. Journal of Control* **51**, No. 5, pp 1133-1146.

[9] C.C. Paige and M. Saunders 1981. Towards a generalized singular value decomposition. *SIAM J. Numer. Anal.*, **18**, No. 3, pp 398-405.

[10] C.F. Van Loan 1976. Generalizing the singular value decomposition. *SIAM J. Numer. Anal.* **13**, No. 1, pp 76-83.

[11] H.P. Zeiger and A.J. Mc Ewen 1974. Approximate linear realizations of given dimensions via Ho's algorithm. *IEEE Trans.* **AC-19**, p 153.

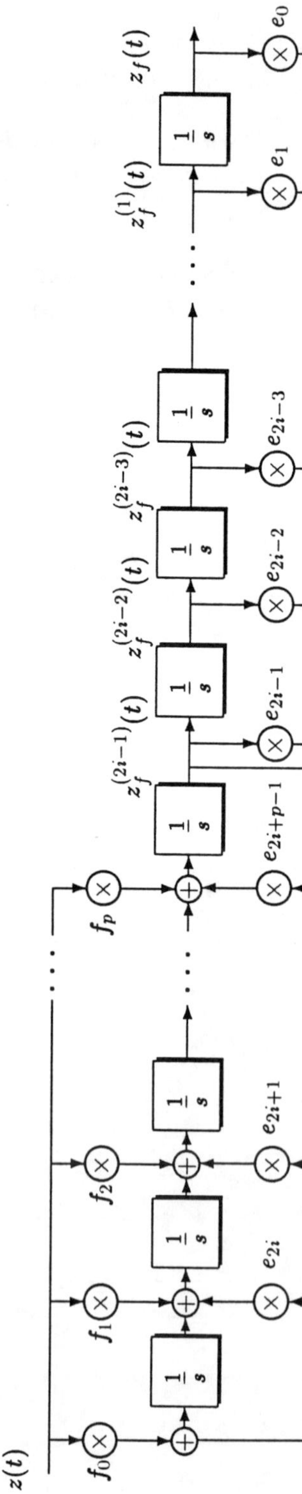
$z(t)$
f_0
f_1
f_2
f_p
$\frac{1}{s}$
e_{2i}
e_{2i+1}
e_{2i+p-1}
$z_f^{(2i-1)}(t)$
e_{2i-1}
$z_f^{(2i-2)}(t)$
e_{2i-2}
$z_f^{(2i-3)}(t)$
e_{2i-3}
$z_f^{(1)}(t)$
e_1
$z_f(t)$
e_0

Identification of continuous-time systems using multiharmonic test signals

A. van den Bos
Department of Applied Physics
Delft University of Technology
2600 GA Delft, The Netherlands

Abstract

The estimation of the parameters of continuous-time linear dynamic systems from error disturbed responses to periodic, multiharmonic test signals is discussed. Closed-form, linear instrumental variable estimators and weighted least squares estimators are proposed and the relation between these estimators is established. Expressions for the asymptotic covariance matrices of the estimators are presented. From these expressions optimal instrumental sequences and minimum variance least squares estimators are derived.

1. Introduction

The purpose of this chapter is to show how the parameters of linear continuous-time systems can be estimated from error disturbed steady state responses to periodic, multiharmonic test signals in a simple, flexible and, at every stage of the estimation process, controlled way.

The most important assumptions in this chapter are that a test signal may be introduced into the system and that this signal is periodic and multiharmonic. The differences of the methods proposed with classical frequency response are that the system is simultaneously investigated at a number of harmonic frequencies of the test signal and that the parameters of the system transfer function are estimated.

As to the proposed test signals, an important difference with and advantage over standard test signals (Unbehauen and Rao 1987) is the relative freedom of choice of the spectrum both with respect to the amplitudes of the harmonics and their location. An advantage of the use of any periodic, multiharmonic test signal is that, prior to the estimation of the system parameters, the often lengthy input-output records can be transformed into a relatively small number of estimated Fourier coefficients. This small number of newly created observations does not increase with the number of periods observed and makes extensive numerical and statistical manipulation feasible. In addition, as an intermediate step, from the Fourier coefficient estimates, the value of the system transfer function may be computed for the frequencies of the harmonics used. Thus the usefulness of these estimates for the

N. K. Sinha and G. P. Rao (eds.), Identification of Continuous-Time Systems, 489–508.

ultimate purpose, the estimation of the system parameters, can be checked. Finally, the use of the estimated Fourier coefficients as observations instead of the original time domain records will be seen to make the construction of linear, closed form estimators of the system parameters feasible.

The outline of this chapter is as follows. Section 2 is devoted to a brief review of periodic test signals, terminology and definitions. A relevant property of digitally generated periodic test signals and the relation between the Fourier coefficients of a periodic input and the corresponding steady state response of a linear system are also described. In section 3, properties of estimated Fourier coefficients are discussed, bias and variance in particular. Section 4 describes the properties of the residuals of the system differential equation in the frequency domain. The estimation of the system parameters, the coefficients of the system differential equation, is discussed in section 5. For that purpose, closed-form and linear, instrumental variable and weighted least squares estimators are proposed. Their interrelation is also established. The properties of the residuals, described in section 4, are used in section 6 to derive expressions for the asymptotic covariance matrices of the proposed estimators. From these expressions optimal instrumental sequences and optimal weights for the proposed least squares estimator are derived. These sequences and weights show how the experimenter may use a priori knowledge of system and errors to enhance the precision. They also show how a second step may be added to the proposed least squares estimator to the same effect. Finally, in section 7 a discussion of important characteristics and useful extensions of the estimators is presented.

2. Multiharmonic test signals

The essential assumptions made in this chapter are that a test signal may be introduced into the system and that this test signal is periodic and multiharmonic. Multiharmonicity implies that the signal consists of more than one non-zero harmonic. Since any real harmonic represents two degrees of freedom, the number of parameters to be estimated should in any case not exceed twice the number of non-zero real harmonics (Van den Bos 1974).

The most important examples of multiharmonic signals are maximum length binary sequences (Godfrey 1980) and multifrequency signals (Van den Bos 1974, Van den Bos and Krol 1979, Van den Bos 1991). Maximum length binary sequences are outside the scope of this chapter since for a particular length their harmonic content is fixed. Multifrequency signals are periodic signals that have their power, or at least the major part of it, distributed in a specified way over a, usually relatively small, number of harmonics. Important examples are sums of harmonically related sinusoids and binary multifrequency signals. The latter are two-level signals that in the sense of a chosen criterion optimally approximate a given harmonic spectrum (Van den Bos and Krol 1979).

Throughout, the following notation will be used. Let x(t) be a function of the time t and be periodic with T. Then the complex Fourier coefficient of the kth harmonic of x(t) is defined as:

$$\gamma_{kx} = \frac{1}{T}\int_0^T x(t)\, \exp(-j2\pi kt/T)\, dt \tag{2.1}$$

where j = $\sqrt{-1}$. Furthermore, if x(t) is real

$$\gamma_{-kx} = \bar{\gamma}_{kx} \tag{2.2}$$

where the bar denotes complex conjugation. Next suppose that the system to be identified is described by the following differential equation:

$$\begin{aligned} a_R y^{(R)}(t) + a_{R-1} y^{(R-1)}(t) + \ldots + a_0 y(t) &= \\ b_S u^{(S)}(t) + \ldots + b_1 u^{(1)}(t) + u(t) \end{aligned} \tag{2.3}$$

where u(t) is the input, y(t) is the response, $y^{(\ell)}(t)$ is the ℓth order derivative of y(t) with respect to t, and the derivatives of u(t) are denoted analogously. Suppose that u(t) is periodic and that y(t) is the steady state response to u(t). Then it is easily shown that the Fourier coefficients of u(t) and the corresponding ones of y(t) satisfy:

$$A(j\omega_k)\, \gamma_{ky} - B(j\omega_k)\, \gamma_{ku} = 0 \tag{2.4}$$

where

$$A(j\omega) = a_R(j\omega)^R + a_{R-1}(j\omega)^{R-1} + \ldots + a_0 , \tag{2.5}$$

$$B(j\omega) = b_S(j\omega)^S + b_{S-1}(j\omega)^{S-1} + \ldots + 1 \tag{2.6}$$

and $\omega_k = 2\pi k/T$. Notice that (2.4) constitutes two real linear equations in the system parameters:

$$\mathbf{c} = (\, a_0 \ldots a_R\; b_1 \ldots b_S\,)' \tag{2.7}$$

where the prime denotes transposition. It will be assumed throughout that **c** is the parameter vector to be estimated.

Practical periodic test signals are digitally generated. As a consequence, they are usually discrete-interval signals. This means that they can only change amplitude in a finite number of equidistant points in a period. Suppose that the time origin is chosen such that these points are described by $t_n = (n+\frac{1}{2})\Delta$ with Δ a fixed time interval and $n = 0, \ldots, N-1$. Then it can be shown that the Fourier coefficients of a periodic input u(t) thus defined satisfy:

$$\gamma_{ku} = S_{ku}\, sinc(\pi k/N) \tag{2.8}$$

where S_{ku} is the discrete Fourier transform of the sequence $u(0), u(\Delta), \ldots, u((N-1)\Delta)$ with k taken modulo N, and $sinc(x) = \sin(x)/x$. Furthermore, since u(t) is supposed real:

$$S_{ku} = \bar{S}_{(N-k)u} \tag{2.9}$$

where the bar denotes complex conjugation. So, it is concluded from (2.8) and (2.9) that the higher harmonic content of u(t) is completely defined by the first [N/2]+1 terms of the sequence S_{ku} where [N/2] denotes the integral part of N/2. The expressions (2.8) and (2.9) will be used in the next section to compute the bias due to aliasing of practical Fourier coefficient estimators.

3. Estimation of Fourier coefficients

Let again x(t) be a function of time and be periodic with T. Suppose that observations are available described by:

$$z(t) = x(t) + d(t) \qquad 0 \le t < MT \tag{3.1}$$

where d(t) is a stationary stochastic process and M is an integral number. So, MT is the observation interval. Then it is easily shown that under general conditions

$$\frac{1}{MT}\int_0^{MT} z(t)\exp(-j2\pi kt/T)\, dt \tag{3.2}$$

is an unbiased least squares estimator and converges in the mean square to γ_{kx} (Levin 1959).

In practice, γ_{kx} will be estimated from samples of z(t). Suppose that the sampling interval is Δ and that $J\Delta = MT$, where J is an integral number denoting the total number of samples. Then the discrete version of (3.2) is:

$$\gamma_{kz} = \frac{1}{J} \sum_{n=0}^{J-1} z(n) \exp(-j2\pi kn/N) \tag{3.3}$$

where $z(n) = z(n\Delta)$ and $N = J/M$ is the number of samples in a period. However, this is only an unbiased estimator of γ_{kx} if no aliasing occurs. This condition is met if all harmonics with harmonic numbers $k \pm \ell N$, $\ell = 1, 2, \ldots$ are equal to zero. Bias will be returned to later in this section.

Now consider the vector of estimated Fourier coefficients:

$$\gamma_z = \left(\gamma_{1z}\ \gamma_{2z} \cdots \gamma_{Kz}\ \gamma_{-1z}\ \gamma_{-2z} \cdots \gamma_{-Kz} \right)' \tag{3.4}$$

and suppose that these have been computed from the same observations. Also assume that the conditions for unbiasedness are met. Then the 2K×2K autocovariance matrix of γ_z is equal to:

$$E\left[(\gamma_z - \gamma_x)\ (\gamma_z - \gamma_x)^*\right] \tag{3.5}$$

where the asterisk denotes complex conjugate transposition. Next, to find the asymptotic expression for the autocovariance matrix (3.5) use is made of the following result derived by Van den Bos (1974):

$$\lim_{M \to \infty} cov(\sqrt{MT}\gamma_{kz}, \sqrt{MT}\gamma_{\ell z}) = \begin{cases} S_{dd}(j\omega_k) & \ell = k \\ 0 & \ell \neq k \end{cases} \tag{3.6}$$

where $S_{dd}(j\omega)$ is the, possibly aliased, power density spectrum of d(t). From this expression it follows that asymptotically the autocovariance matrix (3.5) is diagonal with k-th and (k+K)-th diagonal elements:

$$\frac{1}{MT} S_{dd}(j\omega_k) \tag{3.7}$$

where $k = 1, \ldots, K$. The asymptotic formulation (3.6) suggests that (3.7) is valid only if M is very large. However, the detailed non-asymptotic expression for the variance of the γ_{kz} shows that (3.6) may be safely used whenever the width of the autocovariance function of d(t) is small compared with MT. The expression (3.6) is exact for all M if the errors d(n) are non-covariant. Furthermore, non-asymptotic expressions for the covariances of the γ_{kz} and $\gamma_{\ell z}$ with $k \neq \ell$ show that these covariances are already small for a small number of periods if $S_{dd}(j\omega)$ is not too steep around ω_k and ω_ℓ. Also, they are inversely proportional to $|k - \ell|$.

It has been stated above that the estimator (3.3) is unbiased if no aliasing occurs. However, in section 2 it has been shown that practical, discrete-interval, periodic test signals are not bandwidth limited. So, there will always be aliasing and, consequently, bias to some extent. Now let x(t) be discrete-interval. Then the computation of the bias in $|\gamma_{kz}|^2$ from (2.8) and (2.9) is relatively easy if the Poisson sum formula (Papoulis 1962) is used to sum the aliases. For example, suppose that all $|\gamma_{kx}|^2$ are equal to one. Then Table 1 shows the expectations of $|\gamma_{kz}|^2$ for ℓN samples in a period with ℓ = 1, 2 and 4.

Table 1

k/N	$\lvert\gamma_{kx}\rvert^2$	$E[\lvert\gamma_{kz}\rvert^2]$		
		$\ell = 1$	$\ell = 2$	$\ell = 4$
1/16	1.000	1.013	1.003	1.001
1/8	1.000	1.053	1.013	1.003
1/4	1.000	1.233	1.053	1.013

Table 1 shows how the bias due to aliasing is reduced by selecting an adequate number of discrete intervals in a period and a sufficiently high sampling rate. In many practical situations, the input is the test signal itself. So, the input Fourier coefficients are known and only the Fourier coefficients of the response have to be estimated. Then the limited bandwidth of practical systems will substantially reduce the undesirable higher harmonics. So, the bias will then become negligible if the harmonic spectrum of the test signal, the number of intervals in a period, and the sampling rate are carefully chosen. Table 1 gives an indication how this can be done. This procedure may be preferable to the use of an anti-aliasing filter since this influences the system coefficient estimates. The construction of discrete-interval multifrequency signals with a desired harmonic spectrum is described in (Van den Bos and Krol 1979). This reference also addresses synthesis of binary (on-off) discrete-interval multifrequency signals. Methods for synthesis of low peak factor, discrete-interval multifrequency signals are described in (Van den Bos 1987) and (Van der Ouderaa, Schoukens and Renneboog 1988).

4. Properties of the residuals

Suppose that the observations of the periodic input u(t) and those of the corresponding steady state, periodic response y(t) are described by, respectively:

$$v(t) = u(t) + g(t) \tag{4.1}$$

and

$$w(t) = y(t) + h(t) \tag{4.2}$$

with $t \in [0, MT)$. In these expressions

$$g(t) = g_1(t) + g_2(t) \tag{4.3}$$

and

$$h(t) = h_1(t) + h_2(t) + h_3(t) \tag{4.4}$$

where $g_1(t)$, $g_2(t)$, $h_1(t)$, $h_2(t)$ and $h_3(t)$ are stationary stochastic processes modelling the errors in the observations. In (4.1), u(t) is not necessarily the test signal itself. It may be any signal having a linear, static or steady state dynamic relationship with the test signal. For example, if the system is in closed loop, the test signal may be superimposed on the set point. Then the steady state, periodic input is different from the test signal. Furthermore, in (4.3), the process $g_1(t)$ is an additional input to the system. It may be the normal operating process. The process $g_2(t)$ represents input measurement errors. In (4.4) the process $h_1(t)$ is the stationary response to $g_1(t)$. It is, therefore, covariant with this process. In the same expression, $h_2(t)$ represents response measurement errors. It will be assumed throughout that the measurement errors $g_2(t)$ and $h_2(t)$ are neither mutually covariant nor covariant with any of the other error processes. Finally, the process $h_3(t)$ represents errors covariant with $g_1(t)$ and, therefore, with $h_1(t)$ but not causally related by the system with $g_1(t)$. The process $h_3(t)$ has been included in h(t) to cover the case that the system is in closed loop.

To keep the notation simple, from now on all results will concern the case that the 2K harmonics with harmonic numbers k = 1 , ... , K, -1 , ... , -K are all used for the estimation of the system parameters. In practice, certain harmonics may not be present or may not be used. This will change the notation, not the results. Anyhow, it will be assumed that a number of L different (k,-k)-pairs of complex harmonics will actually be used and that K is the highest harmonic present. The 2L complex harmonics actually used will be called *relevant* harmonics. It will also be assumed that

$2L \geq R+S+1$, that is , the number of parameters to be estimated is smaller than or equal to twice the number of (k,-k)-pairs of relevant harmonics.

Now, for the moment, suppose that the Fourier coefficients γ_{ku} and γ_{ky} , $k = \pm 1 , \ldots , \pm K$ are exactly known. Then by (2.4) they satisfy:

$$A(s_k)\ \gamma_{ky} - B(s_k)\ \gamma_{ku} = 0 \tag{4.5}$$

where $s_k = j\omega_k$. These equations are equivalent to the following system of linear equations in $\mathbf{c}$:

$$\Phi c = \delta \ . \tag{4.6}$$

In this expression the elements of the $2K\times(R+S+1)$ matrix Φ are defined as:

$$\varphi_{km} = s_k^{m-1}\ \gamma_{ky} \qquad m = 1 , \ldots , R+1 , \tag{4.7}$$

$$\varphi_{km} = -s_k^{m-R-1}\ \gamma_{ku} \qquad m = R+2 , \ldots , R+S+1 \tag{4.8}$$

and $\varphi_{K+k,m} = \varphi_{-k,m}$ for $k = 1 , \ldots , K$. Notice that by (2.2) the matrix consisting of the lower K rows of Φ is equal to the complex conjugate of the matrix consisting of the upper K rows. It has been shown by Van den Bos (1974) that Φ is full rank if $2L \geq R+S+1$. Furthermore, in (4.6), the $2K\times 1$ vector $\boldsymbol{\delta}$ is defined as:

$$\delta = (\gamma_{1u} \cdots \gamma_{Ku}\ \gamma_{-1u} \cdots \gamma_{-Ku})' \tag{4.9}$$

where the prime denotes transposition.

Next suppose that the errors g(t) and h(t) are non-zero and that γ_{kv} and γ_{kw} are taken as estimators of γ_{ku} and γ_{ky} . Then substitution of γ_{kv} and γ_{kw} in (4.6)-(4.8) yields the $2K\times 1$ vector of residuals:

$$e = Pc - d \ . \tag{4.10}$$

Thus the elements of the $2K\times(R+S+1)$ matrix $\mathbf{P}$ are obtained by substituting γ_{kw} for γ_{ky} and γ_{kv} for γ_{ku} in (4.7) and (4.8), respectively. Hence, $E[\mathbf{P}] = \Phi$, since γ_{kw} and γ_{kv} are, by assumption, unbiased estimators of γ_{ky} and γ_{ku}. Similarly, the $2K\times 1$ vector $\mathbf{d}$ in (4.10) is obtained by substituting γ_{kv} for γ_{ku} in (4.9) and $E[\mathbf{d}] = \boldsymbol{\delta}$. Hence,

$$E[e] = E[P]c - E[d] = \Phi c - \delta = \mathbf{0} \tag{4.11}$$

where use has been made of (4.6) and from the fact that **c** in (4.10) is exact. So, the expectation of the residuals is equal to zero for all M. This property will be used in section 6 where the asymptotic covariance matrix of estimators of the system coefficients **c** will be derived. Furthermore, in what follows it will be assumed that **P** retains the full rank property of Φ.

From (4.1)-(4.4) and (3.3), it follows that γ_{kv} and γ_{kw} may be described as:

$$\gamma_{kv} = \gamma_{ku} + \gamma_{kg} = \gamma_{ku} + (\gamma_{kg})_1 + (\gamma_{kg})_2 \tag{4.12}$$

and

$$\gamma_{kw} = \gamma_{ky} + \gamma_{kh} = \gamma_{ky} + (\gamma_{kh})_1 + (\gamma_{kh})_2 + (\gamma_{kh})_3 \tag{4.13}$$

where the subscripts of the parenthesized expressions refer to the various components of g(t) and h(t). By (4.5), γ_{ku} and γ_{ky} do not contribute to the element e_k of **e**. The contributions $(\gamma_{kg})_1$ and $(\gamma_{kh})_1$ may also be left out since the processes $g_1(t)$ and $h_1(t)$ are causally related by the system (Van den Bos 1974). Hence, substituting (4.12) and (4.13) in (4.10) yields:

$$e_k = A(s_k)\{(\gamma_{kh})_2 + (\gamma_{kh})_3\} - B(s_k)(\gamma_{kg})_2 \,. \tag{4.14}$$

By definition, $h_2(t)$, $h_3(t)$ and $g_2(t)$ are mutually non-covariant. Then it is easily shown that the same is true with respect to $(\gamma_{kg})_2$, $(\gamma_{kh})_2$ and $(\gamma_{kh})_3$. Using, in addition, the asymptotic covariance properties of the Fourier coefficient estimates described in section 3, the computation of the 2K×2K asymptotic covariance matrix of **e** from (4.14) is straightforward. The result for the k-th diagonal element is:

$$\frac{1}{MT}\lambda_k \tag{4.15}$$

with

$$\lambda_k = |A(s_k)|^2 \left[\{S_{hh}(s_k)\}_2 + \{S_{hh}(s_k)\}_3\right] + |B(s_k)|^2\{S_{gg}(s_k)\}_2 \tag{4.16}$$

and $\lambda_{K+k} = \lambda_k$ for k = 1 , ... , K. Furthermore, by (3.6), the off-diagonal elements are all equal to zero. Thus the asymptotic covariance matrix of **e** is described by:

$$\frac{1}{MT}\Lambda \tag{4.17}$$

with

$$\Lambda = diag\ (\lambda_1 \ldots \lambda_K\ \lambda_1 \ldots \lambda_K)\ . \tag{4.18}$$

In conclusion, asymptotically the residuals are non-covariant and have different variances. This implies that to make the residuals *white* in the usual sense, it is sufficient to normalize them with respect to their standard deviation.

5. Estimating the system parameters

In this section, instrumental variable estimators and least squares estimators for the system parameters are derived and their interrelationship is established. The most important property of both classes of estimators is that they are solutions of systems of linear equations and are, consequently, closed form.

Instrumental variable estimators

Since in (4.1) and (4.2), u(t) and y(t) are periodic with T, the selection of instrumental sequences will be seen to be particularly simple. Let $f_m(t)$, m = 1 , ... , R+S+1 be linearly independent sequences, periodic with T and let the harmonic numbers of the non-zero harmonics of the sequences be harmonic numbers of relevant harmonics of u(t). Then these sequences qualify as instrumental sequences. The reason is that sequences thus defined are correlated with u(t) and y(t) but are generally not covariant with the stochastic error processes g(t) and h(t). Furthermore, because of their linear independence, they produce consistent instrumental variable estimators as will be seen below.

Suppose that $f_m(t)$, m = 1 , ... , R+S+1 are instrumental sequences meeting the above conditions and let

$$(\ \gamma_{1f} \cdots \gamma_{Kf}\ \gamma_{-1f} \cdots \gamma_{-Kf}\ )'_m \tag{5.1}$$

be the vector of Fourier coefficients of the m-th sequence. Next define the complex, frequency domain, instrumental variable matrix Ψ as the 2K×(R+S+1) matrix having (5.1) as its m-th column. So, the elements of Ψ are described by:

$$\psi_{km} = (\gamma_{kf})_m \tag{5.2}$$

and $\psi_{K+k,m} = (\gamma_{-kf})_m$ for k = 1 , ... , K . Notice that Ψ is full rank since its columns (5.1) have been taken linearly independent. Then it follows from (4.6) that:

$$\Psi^*(\Phi c - \delta) = \mathbf{0}\ . \tag{5.3}$$

Simple calculus shows that (5.3) may be considered to be a set of R+S+1 real, linear equations in the R+S+1 elements of **c** . This set is nonsingular since both Ψ and Φ are full rank under the assumptions made.

Subsequent substitution of the estimators γ_{kv} and γ_{kw} for the corresponding γ_{ku} and γ_{ky} in (5.3) produces the frequency domain instrumental variable estimator $\mathbf{c}_{IV}$ for **c** :

$$\Psi^* (P c_{IV} - d) = \mathbf{0} \,. \tag{5.4}$$

So, different from the usual *time domain* instrumental variable estimator, (5.4) exploits the absence of covariance of the *frequency domain* residuals **Pc** - **d** with the *frequency domain* instrumental sequences (5.1). Like (5.3), (5.4) is a set of R+S+1 real equations in the R+S+1 unknown elements of $\mathbf{c}_{IV}$. This set is nonsingular since under the assumptions made both Ψ and **P** are full rank. Also notice that under these conditions $\mathbf{c}_{IV}$ being a function of the estimators γ_{kv} and γ_{kw} is consistent whenever these estimators are.

Least squares estimators

Minimizing the residuals **e** in (4.10) in the weighted least squares sense with respect to **c** yields:

$$P^* \Omega (P c_{LS} - d) = \mathbf{0} \tag{5.5}$$

where $\mathbf{c}_{LS}$ is the weighted least squares estimator of **c** and Ω is a 2K×2K positive definite Hermitian weighting matrix.

It is observed that (5.5) is a set of R+S+1 real linear equations in the R+S+1 unknown element of $\mathbf{c}_{LS}$. This set is nonsingular since $\mathbf{P}^*\Omega\mathbf{P}$ is nonsingular because, by definition, Ω is positive definite Hermitian and, by assumption, **P** is full rank. Under these conditions, $\mathbf{c}_{LS}$ being a function of the estimators γ_{kv} and γ_{kw} is consistent if these estimators are.

Interrelationship of the instrumental variable and the least squares estimators

A comparison of the equations (5.4) defining the instrumental variable estimator with the equations (5.5) defining the weighted least squares estimator shows that the latter estimator is obtained from the former by replacing Ψ^* by $\mathbf{P}^*\Omega$. Hence, $\mathbf{P}^*\Omega$ might be interpreted as an instrumental matrix. Let, for the moment, Ω be the identity matrix. Then (5.5) defines the ordinary least squares estimator. Now equating $\boldsymbol{\Psi}$ to $\mathbf{P}$ yields by (4.7), (4.8) and (5.2) :

$$(\gamma_{kf})_m = s_k^{m-1} \gamma_{kw} \qquad m = 1 , \dots , R+1 \tag{5.6}$$

and

$$(\gamma_{kf})_m = -s_k^{m-R-1} \gamma_{kv} \qquad m = R+2 , \dots , R+S+1 . \tag{5.7}$$

From these equations the following conclusions may be drawn. Suppose that the first R+1 time domain instrumental sequences are taken as the derivatives of order 0 , ... , R+1 of the periodic signal having the γ_{kw} , $k = \pm 1 , \dots , \pm K$ as its Fourier coefficients. Furthermore, suppose that the last S time domain instrumental sequences are taken as the derivatives of order 1 , ... , S of the periodic signal having the $-\gamma_{kv}$, $k = \pm 1 , \dots , \pm K$ as its Fourier coefficients. Then the ordinary least squares estimator and the instrumental variable estimator are identical. So, the *least squares equivalent* instrumental sequences are equal to the sum of the estimated relevant harmonics of the input, the sum of the same harmonics of the response and their time derivatives.

For what follows in section 6, it is useful to next consider the case that $\boldsymbol{\Omega}$ is diagonal. It is easily shown that then the weighted least squares estimator and the instrumental variable estimator are identical if

$$(\gamma_{kf})_m = \omega_{kk} s_k^{m-1} \gamma_{kw} \qquad m = 1 , \dots , R+1 \tag{5.8}$$

and

$$(\gamma_{kf})_m = -\omega_{kk} s_k^{m-R-1} \gamma_{kv} \qquad m = R+2 , \dots , R+S+1 . \tag{5.9}$$

Hence, the instrumental sequences are now derivatives of the periodic signal having $\omega_{kk} \gamma_{kw}$ as its Fourier coefficients and of that having $-\omega_{kk} \gamma_{kv}$ as it Fourier coefficients.

6. The covariance matrices of the instrumental variable and the least squares estimators

In this section, expressions will be derived for the asymptotic covariance matrix of the instrumental variable estimator $\mathbf{c}_{IV}$ defined by (5.4) and that of the least squares estimator $\mathbf{c}_{LS}$ defined by (5.5). In both cases, use will be made of a result by Goldberger (1964), pp. 124-125. Next, the expressions for the asymptotic covariance matrices will be used to derive optimal weights for the least squares estimator and optimal instrumental sequences for the instrumental variable estimator.

Slightly modified, Goldberger's result may be described as follows. Let $\{ \mathbf{x}^{(M)} \}$

be a sequence of stochastic vectors with $\{ E[\mathbf{x}^{(M)}] \} = \{ \xi \}$ and let:

$$\lim_{M\to\infty} E[\sqrt{M}(x^{(M)} - \xi)\ \sqrt{M}(x^{(M)} - \xi)^*] = V \tag{6.1}$$

where the elements of ξ and $\mathbf{V}$ are constants. Also suppose that the joint moments of the elements of $M^{1/2}(\mathbf{x}^{(M)} - \xi)$ of order higher than two are asymptotically equal to zero. Furthermore, let $\mathbf{y}(\mathbf{x})$ be a differentiable, vector valued function of $\mathbf{x}$ and define the sequence $\{ \mathbf{y}^{(M)} \}$ as $\{ \mathbf{y}(\mathbf{x}^{(M)} \}$ and $\boldsymbol{\eta}$ as $\mathbf{y}(\xi)$. Then

$$\lim_{M\to\infty} E[\sqrt{M}\ (y^{(M)} - \eta)\ \sqrt{M}\ (y^{(M)} - \eta)^*] = \frac{\partial y}{\partial x}\ V\ (\frac{\partial y}{\partial x})^* \tag{6.2}$$

where the (i,j)-th element of $\partial\mathbf{y}/\partial\mathbf{x}$ is $\partial y_i(\mathbf{x})/\partial x_j$ and the derivatives are evaluated at $\mathbf{x} = \xi$.

First, this result is used to compute the asymptotic covariance matrix of the instrumental variable estimator $\mathbf{c}_{IV}$. By (4.10) and (5.4) :

$$\Psi^* e = -\Psi^* P(c_{IV} - c)\ . \tag{6.3}$$

Now the matrix **P** is rewritten as:

$$P = \Phi + \Delta\Phi \tag{6.4}$$

where the elements of the matrix $\Delta\Phi$ are described by:

$$\Delta\varphi_{km} = s_k^{m-1}\ \gamma_{kh} \qquad m = 1\ ,\ \dots\ ,\ R{+}1 \tag{6.5}$$

and

$$\Delta\varphi_{km} = -s_k^{m-R-1}\ \gamma_{kg} \qquad m = R{+}2\ ,\ \dots\ ,\ R{+}S{+}1 \tag{6.6}$$

where the expectations of γ_{kh} and γ_{kg} are equal to zero. Substituting (6.4) in (6.3) and rearranging yields:

$$c_{IV} - c = -\{\Psi^*(\Phi + \Delta\Phi)\}^{-1}\Psi^* e\ . \tag{6.7}$$

Thus, the stochastic vector $\mathbf{c}_{IV} - \mathbf{c}$ is expressed in the residuals $\mathbf{e}$ and in the stochastic errors γ_{kh} and γ_{kg} in the estimated Fourier coefficients present in $\Delta\Phi$ as described by (6.5) and (6.6). Next, in Goldberger's result, the $(R+S+1)\times 1$ vector $\mathbf{c}_{IV} - \mathbf{c}$ is identified with $\mathbf{y}^{(M)}$ and the $6K\times 1$ vector of residuals $\mathbf{e}$ and errors γ_{kh} and γ_{kg} in the estimated Fourier coefficients is identified with $\mathbf{x}^{(M)}$. Since the expectations of all elements of the latter vector are equal to zero, it follows from (6.7) that the matrix $\partial \mathbf{y}/\partial \mathbf{x}$ in (6.2) is equal to:

$$\left(-(\Psi^* \Phi)^{-1} \Psi^* \quad O \right) \tag{6.8}$$

where the first partition is (R+S+1)×2K and **O** is the (R+S+1)×4K null matrix. The matrix **V** in Goldberger's result is the asymptotic covariance matrix of the 6K×1 vector consisting of the 2K residuals and the 4K errors in the Fourier coefficients, all multiplied by $M^{1/2}$. The submatrix consisting of the first 2K elements of the first 2K rows of **V** is, by definition, the asymptotic covariance matrix of $M^{1/2}\mathbf{e}$ and is, therefore, by (4.17), equal to Λ/T. All further elements of **V** do, as a result of the special form of $\partial \mathbf{y}/\partial \mathbf{x}$ described by (6.8), not contribute to the product (6.2) which is described by:

$$\lim_{M\to\infty} E\left[\sqrt{M}(c_{IV} - c)\, \sqrt{M}(c_{IV} - c)'\right] = \frac{1}{T} (\Psi^* \Phi)^{-1} \Psi^* \Lambda \Psi (\Phi^* \Psi)^{-1} . \tag{6.9}$$

So, the asymptotic covariance matrix of $\mathbf{c}_{IV}$ - **c** is:

$$\frac{1}{MT} (\Psi^* \Phi)^{-1} \Psi^* \Lambda \Psi (\Phi^* \Psi)^{-1} . \tag{6.10}$$

The derivation of the asymptotic covariance matrix of the least squares estimator is similar. Simple algebra now shows that the least squares analogue of the matrix (6.8) is described by:

$$\left(-(\Phi^* \Omega \Phi)^{-1} \Phi^* \Omega \quad O \right) . \tag{6.11}$$

This produces the asymptotic covariance matrix of $\mathbf{c}_{LS}$ - **c** :

$$\frac{1}{MT} (\Phi^* \Omega \Phi)^{-1} \Phi^* \Omega \Lambda \Omega \Phi (\Phi \Omega \Phi^*)^{-1} . \tag{6.12}$$

A comparison of expression (6.12) with the expression for the covariance matrix of the weighted least squares estimator in the linear regression model (Eykhoff 1974) shows that these expressions have the same structure if the covariance matrix of the errors in this model is identified with the asymptotic covariance matrix of the residuals and the matrix of regressors with Φ , respectively. From regression theory, it is known that (6.12) is smallest if $\Omega = \Lambda^{-1}$. Then (6.12) is described by:

$$\frac{1}{MT} (\Phi^* \Lambda^{-1} \Phi)^{-1} . \tag{6.13}$$

Here the expression *smallest* means that the difference of (6.12) with arbitrary positive definite Hermitian Ω and (6.13) is positive semi-definite. Hence, the least squares estimator (5.5) with weighting matrix $\Omega = \Lambda^{-1}$ is minimum variance within its class. So, in this sense, this weighting matrix is optimal and will be referred to as such in what follows. Moreover, it has been shown by Van den Bos (1974) that (6.13) is equal to the Cramér Rao lower bound on the variance of unbiased estimators of **c** if g(t) and h(t) are normally distributed.

The diagonal elements of the diagonal weighting matrix Λ^{-1} are the reciprocals of the λ_k defined by (4.16). This expression shows that these weights may be far from uniform. For example, if $g(t) \equiv 0$ and $S_{hh}(j\omega)$ is flat up to ω_K , the weights will be equal to:

$$|A(s_k)|^{-2} . \tag{6.14}$$

Generally, the expression (4.16) shows that the selecting of approximately optimal weights requires knowledge of the transfer function of the system and of the power spectra of the errors. Knowledge of the system transfer function may be acquired in the form of estimated system parameters in a first, uniformly weighted, least squares step. If the spectra of $g_2(t)$, $h_2(t)$ and $h_3(t)$ are known or have separately been measured, these and the system parameters estimated in the first step may be used to compute approximations of the optimal weights for the second least squares step. An alternative two-step procedure is proposed by Van den Bos (1974). In this procedure, it is supposed that first the discrete Fourier transforms (3.3) of w(n) and v(n) are computed for $k = 0 , \dots , J\text{-}1$. This produces the desired Fourier coefficient estimates for the values of k corresponding to the harmonic frequencies and, in addition, the discrete Fourier transforms of g(t) and h(t) for frequencies in between these harmonic frequencies. Again, uniformly weighted least squares estimates of the system parameters are computed in the first step. Then these estimated parameters and the discrete Fourier transforms are substituted in (4.10) to compute a cluster of residuals around each of the relevant harmonic frequencies. Next the variance of the residuals in each of the clusters is estimated. Finally, the system parameters are estimated again but now the reciprocals of the estimated residual variances are used as approximations of the optimal weights. As the number of observations increases, the number of residuals in between the relevant harmonic frequencies increases. This offers the opportunity to reduce bias and variance of the variance of the residuals in the same way as in classical spectral analysis (Jenkins and Watts 1968).

From the above considerations concerning the optimal weighting matrix for the least squares estimator, the choice of the optimal instrumental sequences to achieve the same result is now also clear. Substituting the optimal weights in (5.8) and (5.9) yields:

$$(\gamma_{kf})_m = \frac{1}{\lambda_k} s_k^{m-1} \gamma_{ky} \qquad m = 1 , \dots , R+1 \tag{6.15}$$

and

$$(\gamma_{kf})_m = -\frac{1}{\lambda_k} s_k^{m-R-1} \gamma_{ku} \qquad m = R+2 , \dots , R+S+1 . \tag{6.16}$$

These expressions provide the experimenter with a guideline as to how to choose the sequences and use the available a priori knowledge. Notice that the time domain equivalents of the R+1 sequences defined by (6.15) and the S sequences defined by (6.16) are all differentiated versions of $f_1(t)$ and $f_{R+2}(t)$, respectively.

7. Discussion and extensions

The preceding sections are a sketch of a theoretical framework for the proposed methods for estimation of continuous time system parameters using multiharmonic test signals. Details have been left out. The purpose of this section is to comment on the main characteristics of the methods and to show how these characteristics may be used to extend their possibilities.

The main characteristic of the proposed methods is that they are linear and closed form. As a consequence, the solutions are unique. Also, the solutions are obtained in a single computational step. No iterations are needed. Thus the, often problematic, specification of initial conditions required by iterative procedures is avoided. Also, the computation time is fixed which may be important in on line applications. Furthermore, for the solution of the linear sets of equations required by the proposed estimators, robust numerical methods have been developed (Goodwin and Payne 1977). The required software is accessible to any user (Dongarra *e.a.* 1980, Press *e.a.* 1988). Also, compared with iterative schemes, the computation time is modest. This is not only attractive from an efficiency point of view, it also makes important extensions of the proposed methods feasible. For example, if the system to be identified contains a pure time delay, the system parameters may easily be solved for for all desired values of the time delay and the best solution in the sense of the criterion selected. Notice that the introduction of a time delay is equivalent to an appropriate phase shift of the measured input or response Fourier coefficients. A further extension may be the selection of the order of the left hand and right hand member of the system differential equation. Again the computation of the system coefficients for a number of orders only requires the solution of the same number of sets of linear equations. The residuals for the various orders could then be exploited for statistical order selection making use of the statistical properties of the residuals described in this chapter. Also, although in simulations the finite sample bias of the proposed least squares estimators has been found to be small compared with their standard deviation, refinements such as Quenouille's bias removal schemes (Kendall and Stuart 1979) could be contemplated in applications where unbiasedness is desirable. In the most important of these schemes, the system parameters are first estimated from all available M periods. These estimates are subsequently combined with the estimates from all subsets of M-1 periods to remove first order bias. Again, application of this scheme has become feasible through the efficiency of the system parameter estimators proposed.

A further characteristic of the proposed least squares estimator is its *formal similarity* to the classical statistical linear regression problem. It is emphasized that from a statistical point of view the statistical models used in this chapter are certainly *not* regression models. However, the formal similarity implies the applicability of results originally derived for regression problems. The benefits of this insight have already become apparent in section 6 of this chapter where the optimal weights for the least squares estimator were derived. A further example is the possibility of construction of optimal test signal power density spectra using experimental design methods (Goodwin and Payne 1977).

A third important characteristic of the proposed continuous-time estimators described in this chapter is that they have analogous discrete-time counterparts (Van den Bos 1991). The discrete analogues are obtained by replacing the continuous Fourier transforms by discrete Fourier transforms. Then the shift operator $\exp(-j\omega\Delta)$ replaces the differentiation operator $j\omega$. Also the least squares equivalent instrumental sequences described in section 5 and 6 become shifted versions of one the same sequence instead of derivatives, and so on. This analogy is of more than theoretical importance since it also extends to the software for continuous time and for discrete time estimation. This may be exploited in the software development.

In section 3 of this chapter, it has been assumed that the Fourier coefficients of the periodic input and the steady state response are estimated using discrete Fourier transforms. Usually, this choice of estimator is taken for granted. However, a discrete Fourier transform estimator is a uniformly weighted linear least squares estimator. This estimator is optimal in the sense that it is the maximum likelihood estimator if the errors in the observations are non-covariant and normally distributed. Under these conditions, it also attains the Cramér Rao lower bound for any number of periods. Strictly, to retain these favourable properties, this least squares estimator must be weighted with the inverse of the covariance function of the errors if the errors are covariant and normally distributed. Clearly, this estimator is much more complicated than a simple discrete Fourier transform. However, it has been found by Van den Bos (1974) that *asymptotically* it makes no difference if uniform weights are taken instead. This is probably a consequence of the fact that the covariance matrix of the errors becomes increasingly sparse as the number of observed periods increases.

The conditions are completely different if the errors are not normally distributed. The maximum likelihood estimator of the Fourier coefficients is then usually essentially different from least squares estimators of any kind. For example, if the errors are Laplacean, that is, doubly exponentially distributed and independent, the maximum likelihood estimator of the Fourier coefficients is the least moduli estimator. The variance of least squares estimators is known to be quite sensitive to errors with long tailed distributions such as the Laplacean distribution and to outliers. Using the discrete Fourier transform, that is, least squares then produces much larger errors in the Fourier coefficient estimates than would have been obtained with the maximum likelihood estimator.

In this context, the question also arises how to estimate the Fourier coefficients if the knowledge of the probability density function of the errors is only partial. This means that all that is known is that the probability density function belongs to a certain class. For example, it may only be known that the errors are symmetrically distributed and never exceed a given amplitude level. During the last decade, so-called *robust* estimators have become available (Poljak and Tsypkin 1980, Zypkin 1987) that allow the experimenter to exploit such partial knowledge. The basic idea is to first determine the *least favorable* error probability density function. This is, within the allowable class, the probability density function corresponding with the maximum likelihood estimator having the maximum variance. Then the criterion of goodness of fit is taken as that producing maximum likelihood estimates in the presence of least favorably distributed errors. So, this criterion produces smallest variance under the worst conditions. It can also be shown that using this criterion the variance of the estimator for all other error distributions is smaller. Much of the existing theory of robust estimation applies to observations linear in the parameters. and is, therefore, particularly suitable for Fourier coefficient estimation. For a short description of properties of Fourier coefficient estimators for non-normal errors and of robust estimators, see (Van den Bos 1989).

Up to now, the probability density function of the Fourier coefficient estimates has not been addressed. However, under general conditions, these estimates are asymptotically normally distributed and non-covariant. This fact is used by Schoukens, Pintelon and Renneboog (1988) to derive a maximum likelihood estimator using least squares Fourier coefficient estimates as observations. The resulting estimator is nonlinear and requires an iterative numerical approach.

A final comment concerns multi-input multi-output systems. Here, the use of multiharmonic signals offers a unique opportunity: the simultaneous application of orthogonal signals to the various inputs. This form of labelling provides the experimenter with the possibility to discern the transfers for each input output pair. The construction of orthogonal test signals is relatively simple since signals that do not share non-zero harmonics are by definition orthogonal. Examples of the construction of orthogonal binary multiharmonic signals are described by Van den Bos (1974).

References

Dongarra, J.J., Moler, C.B., Bunck, J.R. and Stewart, G.W. (1980), "*LINPACK Users' Guide*", SIAM, Philadelphia.

Eykhoff, P. (1974), "*System Identification - Parameter and State Estimation*", Wiley, London.

Godfrey, K.R. (1980), "Correlation methods", *Automatica*, vol. 16, number 5, pp. 527-534.

Goldberger, A.S. (1964), "*Econometric Theory*", Wiley, London.

Goodwin, G.C. and Payne, R.L. (1977), "*Dynamic System Identification: Experiment Design and Data Analysis*", Academic Press, New York.

Jenkins, G.M. and Watts, D.G. (1968), "*Spectral Analysis and its Applications*", Holden-Day, San Francisco.

Kendall, M.G. and Stuart, A. (1979), "*The Advanced Theory of Statistics*", Vol. 2: Inference and Relationship, Griffin, London.

Levin, M.J. (1959), "Estimation of the characteristics of linear systems in the presence of noise", Doctor of Engineering Science Thesis, Columbia University, New York.

Papoulis, A. (1962), "*The Fourier Integral and its Applications*", Wiley, New York.

Poljak, B.T. and Tsypkin, Ya. Z. (1980), "Robust identification", *Automatica*, vol. 16, pp. 53-63.

Press, W.H., Flannery, B.P., Teukolsky, S.A. and Vetterling, W.T. (1988), "*Numerical Recipes: The Art of Scientific Computing*", Cambridge University Press, Cambridge.

Schoukens, J., Pintelon, R. and Renneboog, J. (1988), "A maximum likelihood estimator for linear and nonlinear systems - A practical application of estimation techniques in measurement problems", *IEEE Transactions on Instrumentation and Measurement*, vol. IM-37, pp. 10-17.

Unbehauen, H. and Rao, G.P. (1987), "*Identification of Continuous Systems*", North-Holland, Amsterdam.

Van den Bos, A. (1974), "*Estimation of Parameters of Linear Systems Using Periodic Test Signals*", Doctor of Technical Sciences Thesis, Delft University of Technology, Delft University Press, Delft.

Van den Bos, A. and Krol, R.G. (1979), "Synthesis of discrete-interval binary signals with specified Fourier amplitude spectra", *International Journal of Control*, vol. 30, pp. 871-884.

Van den Bos, A. (1987), "A method for synthesis of low-peak-factor signals", *IEEE Transactions on Acoustics, Speech, and Signal Processing*, vol. ASSP-35, pp. 120-122.

Van den Bos, A. (1989), "Estimation of Fourier Coefficients", *IEEE Transactions on Instrumentation and Measurement*, vol. IM-38, pp. 1005-1007.

Van den Bos (1991), "Periodic test signals - Properties and use", in: Godfrey, K.R., editor, "*Perturbation Signals for System Identification*", to appear.

Van der Ouderaa, E., Schoukens, J. and Renneboog, J. (1988), "Peak factor minimization using a time-frequency domain swapping algorithm", *IEEE Transactions on Instrumentation and Measurement*, vol. IM-37, pp. 146-147.

Zypkin, Ja. S. (1987), "*Grundlagen der informationellen Theorie der Identifikation*", VEB Verlag Technik, Berlin.

ADAPTIVE MODEL APPROACHES

H. Unbehauen
Department of Electrical Engineering
Ruhr-University
4630 Bochum, Germany

In this Chaper algorithms for identification of continuous-time systems based on model reference principles will be presented. The main idea behind the model reference principle is to set up a procedure for adjustment, based on a comparison between the performance of two units, one of which serves as a reference for the other to follow. The adjustment is made on the unit which follows the reference unit in such a way that the discrepancy between the performance of the two units is minimized or reduced to zero. In the case of the identification problem the plant to be identified assumes the role of the reference model and the estimated model assumes the role of the adjustable unit. It is this dual character of the model reference principle that enables us to apply several techniques of the well known Model Reference Adaptive Control (MRAC) Systems (Landau 1979) to the identification problem. In order to assign distinct identity to this class of adaptive model techniques for identification, we will call them *System Reference Adaptive Model* (SRAM) techniques. These techniques may be further classified on the basis of the following features:

a) *Structure:* In accordance with the disposition of the adjustable model with respect to the reference system the SRAM schemes can be classified as i) parallel, ii) series and iii) series-parallel schemes.

b) *Design method:* There are mainly three design methods viz.
   1) Local optimization (e.g. gradient methods),
   2) Liapunov's second method, and
   3) Hyperstable design method.

N. K. Sinha and G. P. Rao (eds.), Identification of Continuous-Time Systems, 509–548.

The mathematical treatment of the SRAM schemes can be either in the state space or in the input-output transfer function form. The state space treatment is applicable when all the states of the process are available in the identification set up. Adaptive model methods of system identification are related to the deterministic output error (OE) approaches. Extensive surveys on the deterministic OE approaches are available (Landau 1972 and 1979, Eykhoff 1974, Unbehauen and Rao 1987). Stability and convergence in these aproaches are important in the context of these schemes. Several investigations have been made on these aspects in recent years (Dugard and Goodwin 1985, Anderson et al. 1985, Unbehauen 1981, Parks 1966, Landau 1979).

1. Model Adaptation via Gradient Methods

1.1. The parallel model approach

This method of the SRAM techniques is based on the system structure described in Fig. 1. The identification problem consists in determining the parameters of a linear model in parallel with the system to be identified. As the model and the system are driven by the same input

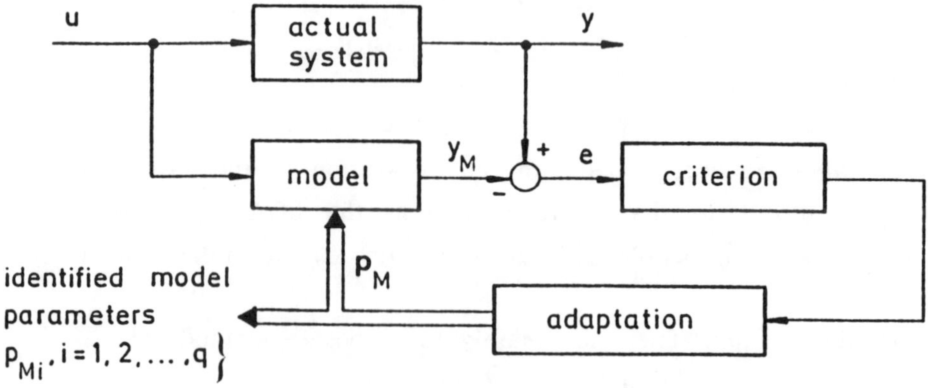

Fig. 1. The parallel model scheme in SRAM approach

u(t), the parameters p_{Mi} of the model transfer function $G_M(s)$ which define the model parameter vector $\mathbf{p}_M = \{p_{Mi};\ i = 1, 2, ..., q\}$ should be so chosen as to minimize a chosen measure of the output error. The output error is defined as

$$e(t,\mathbf{p}_M) = y(t) - y_M(t,\mathbf{p}_M). \tag{1.1}$$

This output error will be used in the formation of a scalar functional

$$I(\mathbf{p}_M) = f[e(t,\mathbf{p}_M)]. \tag{1.2}$$

This functional depends only on the model parameter vector $\mathbf{p}_M$ and thus on a function of the parameters to be identified. This cost function is minimized with respect to the parameter vector, and under the minimal condition we have

$$\mathbf{p}_M \equiv \mathbf{p}_{Mopt}. \tag{1.3}$$

The minimum of $I(\mathbf{p}_M)$ is determined by using the gradient method. In this method the parameter vector $\mathbf{p}_M(t)$ must be changed until $I(\mathbf{p}_M)$ attains its minimum value. The resulting *adaptation law* (Unbehauen and Rao 1987) for the model parameter vector follows as

$$\mathbf{p}_M(t) = \mathbf{p}_M(0) - h \int_0^t \nabla I(\mathbf{p}_M)d\tau. \tag{1.4}$$

To proceed further the mean square value of the error, namely

$$I(\mathbf{p}_M) = \overline{e^2(t,\mathbf{p}_M)} \tag{1.5}$$

is used. Thus the gradient method corresponding to Eq.(1.4) yields the following adaptation law for the required parameter vector (Rake 1966)

$$\mathbf{p}_M(t) = \mathbf{p}_M(0) + 2\ h \overline{\int_0^t e(\tau)\ \nabla y_M\ (\tau,\mathbf{p}_M)d\tau}. \tag{1.6}$$

Eq.(1.6) governs the computational process by which the parallel model matches itself with the system to be identified. For this time-domain optimization we require the signals contained in the vector $\nabla y_M(t,\mathbf{p}_M)$. These necessary signals are generated by a treatment in the frequency-domain. When the model parameter variations are slow in comparison with its own dynamics, we can describe the model by the transfer function $G_M(s,\mathbf{p}_M)$. With this the model output

$$Y_M(s,\mathbf{p}_M) = U(s)\ G_M(s,\mathbf{p}_M). \tag{1.7}$$

The partial derivative w.r.t. $\mathbf{p}_M$

$$\nabla Y_M(s,\mathbf{p}_M) = \nabla G_M(s,\mathbf{p}_M)\ U(s). \tag{1.8}$$

Consider the general case of a model in a rational n-th order form

$$G_M(s,\mathbf{p}_M) = \frac{\sum_{i=0}^{m} b_{Mi}s^i}{\sum_{i=0}^{n-1} a_{Mi}s^i + s^n} = \frac{B_M(s,\mathbf{b}_M)}{A_M(s,\mathbf{a}_M)}\ . \tag{1.9}$$

Defining the parameter vector as

$$\mathbf{p}_M = \left[\mathbf{a}_M^T \mid \mathbf{b}_M^T\right]^T \tag{1.10}$$

then the elements of the vector $\nabla G_M(s,\mathbf{p}_M)$ take the form

$$\frac{\partial G_M(s,\mathbf{p}_M)}{\partial p_{Mi}} = \frac{A_M(s,\mathbf{a}_M)\ \frac{\partial B_M}{\partial p_{Mi}}(s,\mathbf{b}_M) - B_M(s,\mathbf{b}_M)\ \frac{\partial A_M(s,\mathbf{a}_M)}{\partial p_{Mi}}}{\left[A_M(s,\mathbf{a}_M)\right]^2}\ . \tag{1.11}$$

The vector $\nabla G_M(s,\mathbf{p}_M)$ is also partitioned corresponding to its two parts as

$$\nabla G_M(s,\mathbf{p}_M) = \left[(\nabla_{\mathbf{a}_M} G_M(s,\mathbf{p}_M))^T \mid (\nabla_{\mathbf{b}_M} G_M(s,\mathbf{p}_M))^T\right]^T. \tag{1.12}$$

Thus from Eq.(1.11) we directly get for the two subvectors

$$\nabla_{\mathbf{a}_M} G_M(s,\mathbf{p}_M) = - \frac{G_M(s,\mathbf{p}_M)}{A_M(s,\mathbf{a}_M)} \mathbf{d}_{n-1}(s) \tag{1.13}$$

and

$$\nabla_{\mathbf{b}_M} G_M(s,\mathbf{p}_M) = \frac{1}{A_M(s,\mathbf{a}_M)} \mathbf{d}_m(s). \tag{1.14}$$

by introducing the general vector

$$\mathbf{d}_\nu(s) = [1 \quad s \quad s^2 \quad \ldots \quad s^\nu]^T. \tag{1.15}$$

The block diagram for the self-adjustment of the parameter vector $\mathbf{p}_M(t)$ according to Eq.(1.6) can now be developed as shown for a SISO-system in Fig. 2 in view of Eqs.(1.13) to (1.15). It is to be observed that the averaging process in Eq.(1.6) is performed only over a finite period and thus can be realized through a first order element. The filter for generating ∇y_M out of u(t) contains as indicated in

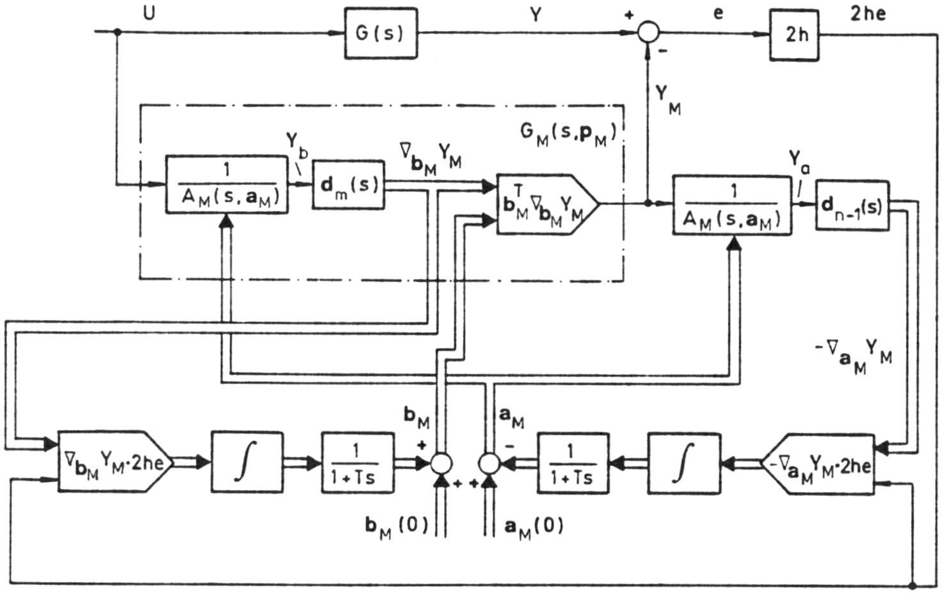

Fig. 2. Block diagram for the identification of a linear SISO-system using the parallel model approach

Eqs.(1.13) and (1.14) the denominator of the model transfer function as an essential part. As the parameter vector $\mathbf{p}_M$ is matched, this must be multiplicatively modified in both the model sections. In this method for the $q = m+n+1$ adjustable values $3n+2m+2$ multiplications are implied. The necessary derivatives of the auxilary signals $y_a(t)$ and $y_b(t)$ can be directly taken from the circuitry realizing $A_M(s,\mathbf{a}_M)$, i.e. the blocks characterized by the vectors $\mathbf{d}_m$ and $\mathbf{d}_{n-1}$ have only symbolic meaning.

Example:

Consider a plant which can be modelled by a transfer function of the form

$$G_M(s,\mathbf{p}_M) = \frac{b_{M0} + b_{M1}s}{a_{M0} + s}.$$

Define the parameter vector according to Eq.(1.6) as

$$\mathbf{p}_M = [a_{M0} \quad b_{M0} \quad b_{M1}]^T.$$

At first the filter transfer functions are obtained from $G_M(s,\mathbf{p}_M)$ as follows

$$\frac{\partial G_M(s,\mathbf{p}_M)}{\partial a_{M0}} = \frac{-(b_{M0} + b_{M1}s)}{(a_{M0} + s)^2} = -G_M(s,\mathbf{p}_M)\,\frac{1}{a_{M0} + s},$$

$$\frac{\partial G_M(s,\mathbf{p}_M)}{\partial b_{M0}} = \frac{a_{M0} + s}{(a_{M0} + s)^2} = \frac{1}{a_{M0} + s},$$

$$\frac{\partial G_M(s,\mathbf{p}_M)}{\partial b_{M1}} = \frac{(a_{M0} + s)s}{(a_{M0} + s)^2} = \frac{s}{a_{M0} + s}.$$

By letting the input signal u(t) into filters which possess these transfer functions, the partial derivatives in the vector $\nabla y_M(t,\mathbf{p}_M)$ are obtained as the outputs of the filters. Thus, now, Eq.(1.6) can be directly realized. Fig. 3 shows the block diagram of the identification system for the chosen example. ■

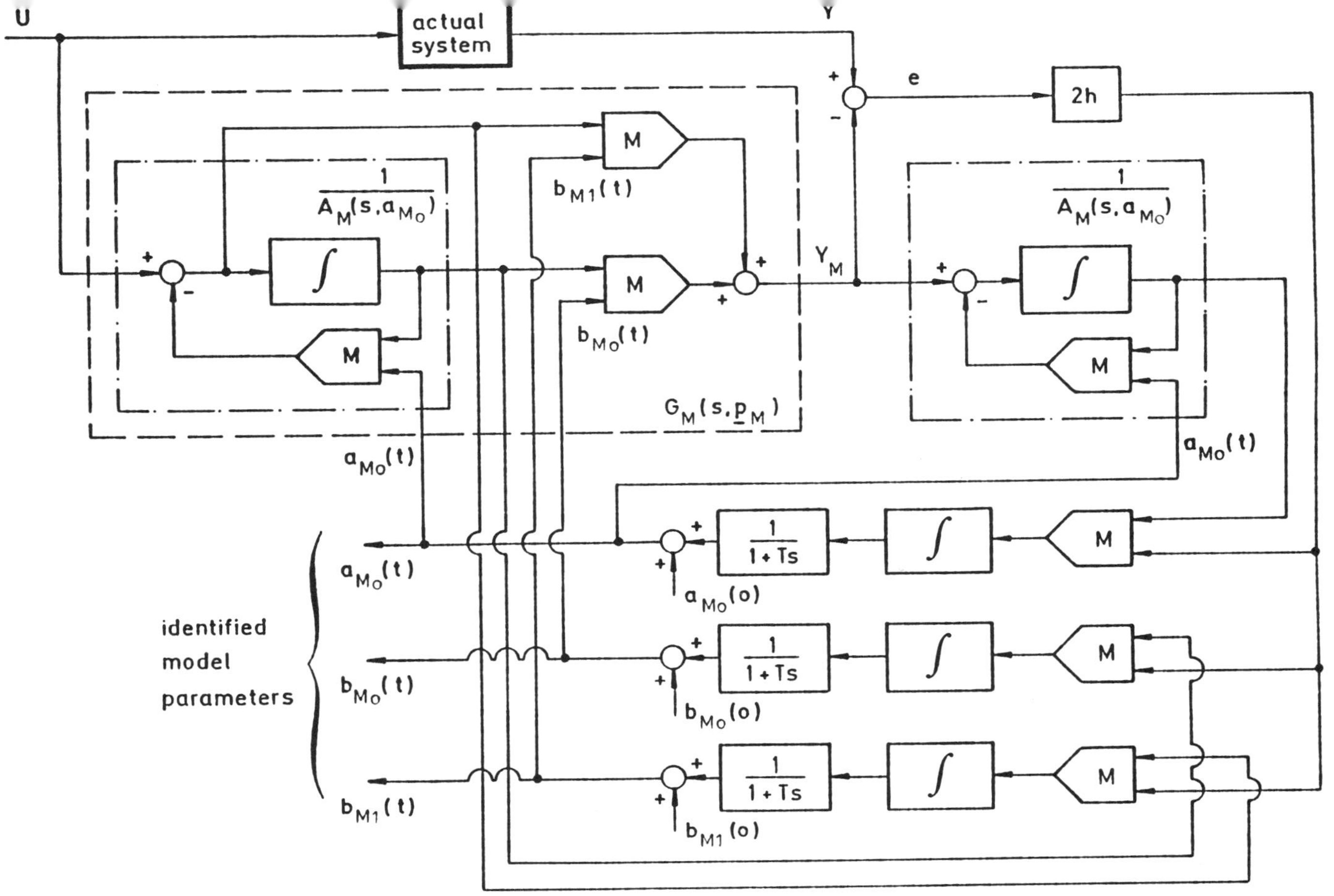

Fig. 3. Block diagram for model adaptation for the chosen example

1.2. The series (reciprocal) model approach

This method was originally developed for identifying transfer functions having only an n-th order denominator polynomial (Marsik 1966). It can also be generalized for systems with m zeros, m<n. Fig. 4 shows the corresponding scheme in which a reciprocal model having transfer function $G_M(s)$ is connected in series with the system. The overall transfer function $G(s)\ G_M(s)$ will be compared with that of the parallel model $G_V(s)$. In the adapted state the q coefficients p_{Mi} of the polynomials $B_M(s)$ and $A_M(s)$ of $G_M(s)$ ought to correspond to those of B(s) and A(s) of G(s). Thus the system coefficients are identified.

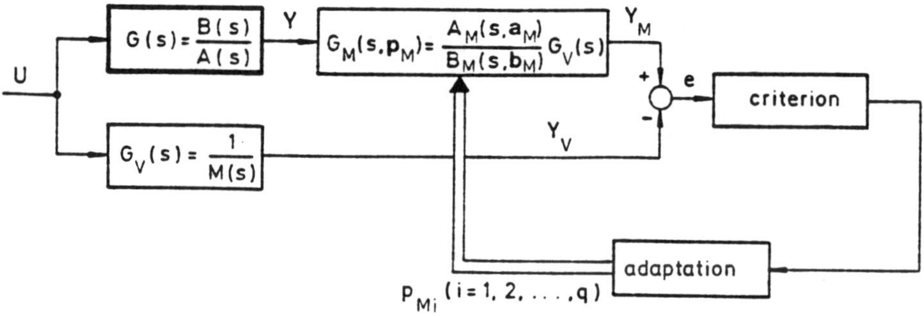

Fig. 4. Identification of a linear SISO-system via series (reciprocal) model approach

As the polynomials $A_M(s)$ and $B_M(s)$ are of orders n and m, on the grounds of realizability of the series model, the denominator polynomial of the transfer function $G_V(s)$, i.e.

$$M(s) = \sum_{i=0}^{r} m_i s^i \tag{1.16}$$

must be of the order $r \geqq n\text{-}m$. With the model output error, defined here as

$$e(t,p_M) = y_M(t,p_M) - y_V(t), \tag{1.17}$$

and the error criterion of Eq.(1.5), the adaptation law for the model parameters can be written nearly identical to Eq.(1.6) where only the integral term becomes a negative sign. To compute the partial derivative of y_M with respect to the parameter vector $\mathbf{p}_M$ we have

$$\nabla Y_M(s,\mathbf{p}_M) = \nabla G_M(s,\mathbf{p}_M)\ Y(s). \qquad (1.18)$$

As in the method with parallel model, the vector $\nabla G_M(s,\mathbf{p}_M)$ can be partitioned into two subvectors

$$\nabla_{\mathbf{a}_M} G_M(s,\mathbf{p}_M) = \frac{G_V(s)}{B_M(s,\mathbf{b}_M)}\ \mathbf{d}_{n-1}(s) \qquad (1.19)$$

and

$$\nabla_{\mathbf{b}_M} G_M(s,\mathbf{p}_M) = \frac{-G_M(s,\mathbf{p}_M)}{B_M(s,\mathbf{b}_M)}\ \mathbf{d}_m(s), \qquad (1.20)$$

wherein the vectors $\mathbf{d}_{n-1}(s)$ and $\mathbf{d}_m(s)$ are according to Eq.(1.15). The realization of Eq.(1.6) under the conditions given in Eqs.(1.19) and (1.20) is shown in the block diagram of Fig. 5. From this it is evident that for $q = m+n+1$ adjustable parameters, $(2n+3m+3)$ multiplications are involved. This method has the advantage that the adaptation can be realized with comparatively small effort. The choice of the coefficients m_i of the stable polynomial $M(s)$ can be arbitrarily made from a large region.

1.3. The series-parallel model approach

The basic idea behind this method is a series-parallel model combination as shown in Fig. 6 In this the numerator polynomial $B_M(s,\mathbf{b}_M)$ of the adjustable model is in parallel with the system to be identified while the denominator polynomial $A_M(s,\mathbf{a}_M)$ is in series with it. On the grounds of realizability of the two submodels, a polynomial

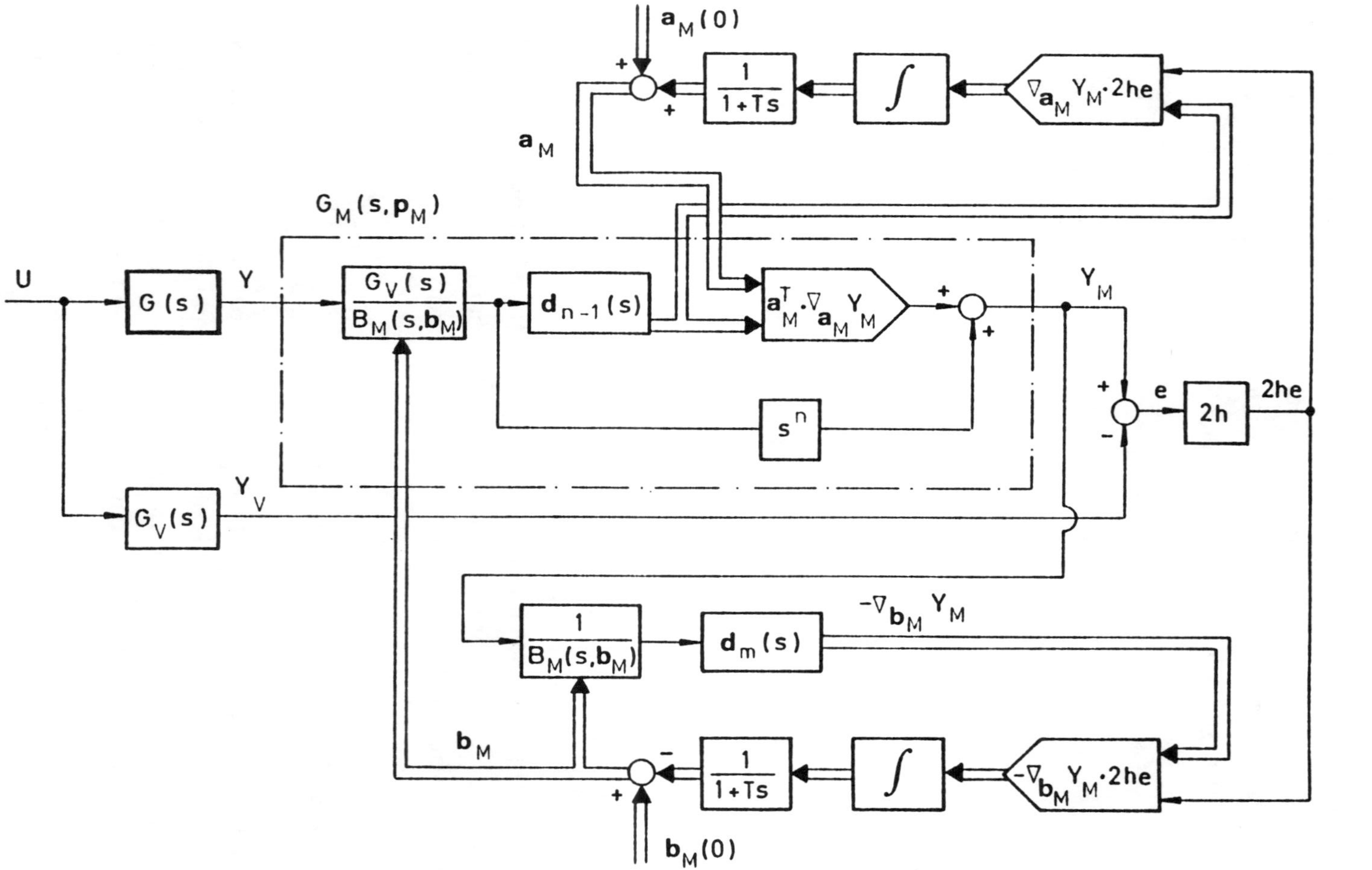

Fig. 5. Block diagram for the identification of a linear SISO-system by series (reciprocal) model approach

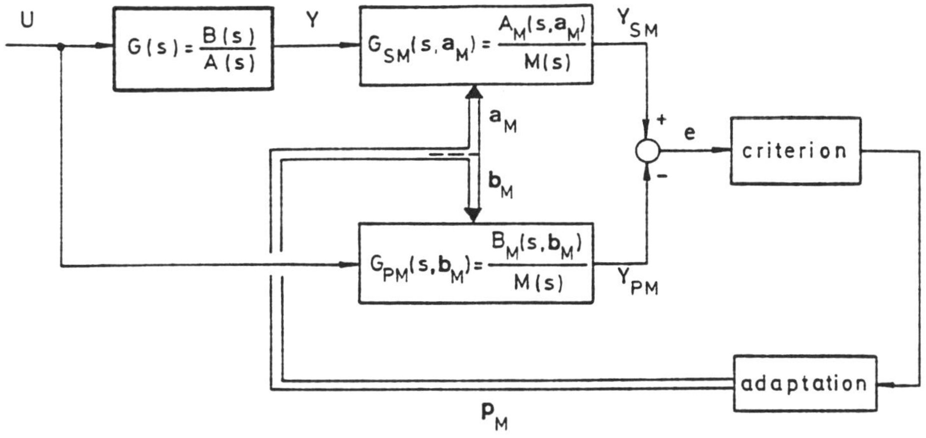

Fig. 6. Identification of a linear SISO-system by the series-parallel model approach ($\mathbf{p}_M$ is split into two portions $\mathbf{a}_M$ and $\mathbf{b}_M$)

$$M(s) = \sum_{i=0}^{n-1} m_i s^i + s^n \tag{1.21}$$

must in addition be included in the denominator of each. This, however, needs no adaptation but should be stable. In the adapted state the numerator coefficients of G_{SM} are equal to those of A(s) and likewise the numerator coefficients of G_{PM} are equal to those of B(s). Thus the system is fully identified.

Choosing an error function

$$e(t,\mathbf{p}_M) = y_{SM}(t,\mathbf{a}_M) - y_{PM}(t,\mathbf{b}_M) \tag{1.22}$$

and a cost functional as in Eq.(1.5), we get the adaptation law as

$$\mathbf{p}_M(t) = \mathbf{p}_M(0) - 2h \int_0^t e(\tau,\mathbf{p}_M)\, \nabla e(\tau,\mathbf{p}_M) d\tau. \tag{1.23}$$

Since the parameters for the numerator and denominator polynomials of the adapting model occur in two separate signal branches, we obtain

$$\nabla_{\mathbf{a}_M} e(t,\mathbf{p}_M) = \nabla_{\mathbf{a}_M} y_{SM}(t,\mathbf{a}_M) \tag{1.24}$$

and

$$\nabla_{\mathbf{b}_M} e(t,\mathbf{p}_M) = \nabla_{\mathbf{b}_M} y_{PM}(t,\mathbf{b}_M). \tag{1.25}$$

As Eqs.(1.24) and (1.25) are linear in the derivated parameters, calculation of the elements of these two vectors is very simple. This particularly simplifies treatment in frequency-domain in which the two vectors are given by the relations

$$\nabla_{\mathbf{a}_M} Y_{SM}(s,\mathbf{p}_M) = \frac{1}{M(s)} \nabla_{\mathbf{a}_M} A_M(s,\mathbf{a}_M)\; Y(s) \tag{1.26}$$

and

$$\nabla_{\mathbf{b}_M} Y_{PM}(s,\mathbf{p}_M) = \frac{1}{M(s)} \nabla_{\mathbf{b}_M} B_M(s,\mathbf{b}_M)\; U(s). \tag{1.27}$$

As one directly recognizes that

$$\nabla_{\mathbf{a}_M} A_M(s,\mathbf{a}_M) = \mathbf{d}_{n-1}(s) \tag{1.28}$$

and

$$\nabla_{\mathbf{b}_M} B_M(s,\mathbf{b}_M) = \mathbf{d}_m(s) \tag{1.29}$$

the filter transfer functions in the present case for obtaining the signals according to Eqs.(1.24) and (1.25) are particularly simple, because they are already available through the subsystems and do not require additional adjustments. The realization effort in this method is comparatively very small. From Fig. 7 it is to be seen that for the $q = m+n+1$ adjustable values, $(2n+2m+2)$ multiplications are required.

1.4. Stability of model adaptation using gradient methods

In the following a simple example for the identification of a plant consisting only of a constant but unknown gain, K, will show the problems related to the convergence and stability of the corresponding adaptation law. In the case of applying the parallel model approach with

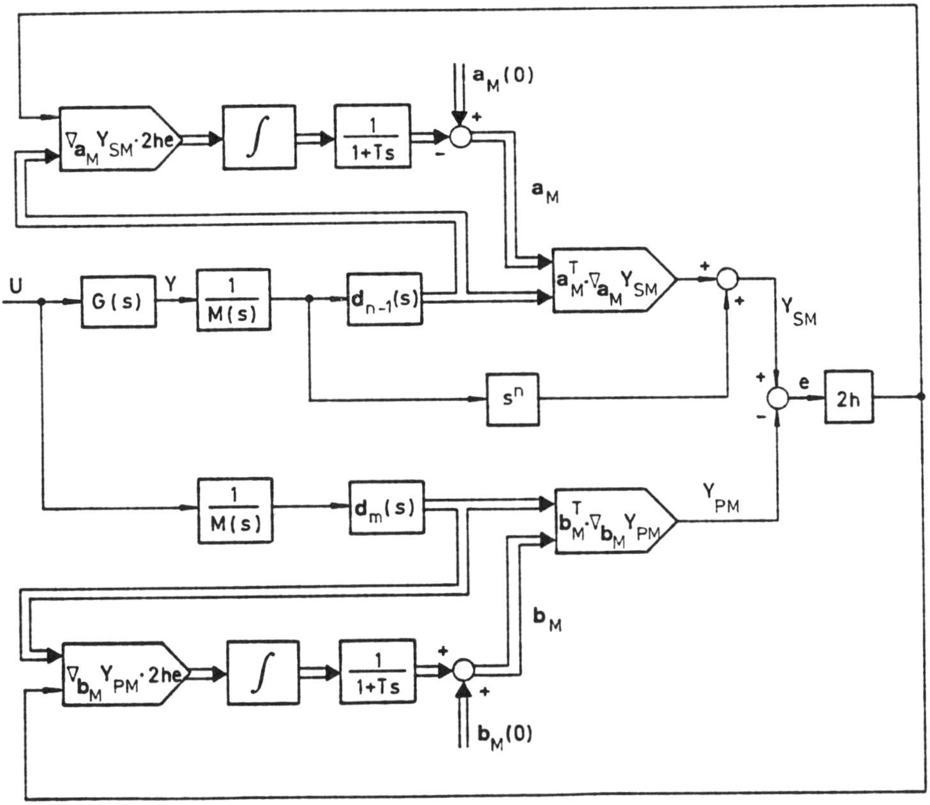

Fig. 7. Block diagram for the identification of a linear SISO-system by the series-parallel model approach

$$y_M(t) = K_M(t)\ u(t) \tag{1.30}$$

we obtain from Eq.(1.6) the adaptation law

$$K_M(t) = K_M(0) + 2h \int_0^t e(\tau)\ u(\tau)\ d\tau, \tag{1.31}$$

where for a deterministic input u(t) the averaging of the integral can be omitted. From

$$e(t) = -[K_M(t) - K]\ u(t) \tag{1.32}$$

and $dK/dt = 0$ follows the derivative of (1.31)

$$\frac{d}{dt}\,[K_M(t)-K] = -2h\,[K_M(t)-K]\,u^2(t) \tag{1.33}$$

or by introducing the parameter error

$$\tilde{K}(t) = K_M(t) - K \tag{1.34}$$

finally the nonlinear differential equation

$$\frac{d\tilde{K}}{dt} = -2h\,\tilde{K}(t)\,u^2(t), \tag{1.35}$$

which has the solution

$$\tilde{K}(t) = K(0)\,e^{-2h\int_0^t u^2(\tau)d\tau} \tag{1.36a}$$

or

$$K_M(t) = K + [K_M(0) - K]\,e^{-2h\int_0^t u^2(\tau)d\tau}\,. \tag{1.36b}$$

For the investigation of convergence and stability Eq.(1.36) is more appropriate than Eq.(1.31).

From Eq.(1.36) follows that $K_M(t)$ converges to K (or $\tilde{K}(t)\to 0$), iff

$$\lim_{t\to\infty} \int_0^t u^2(\tau)\,d\tau \to \infty. \tag{1.37}$$

This *exitation condition* is only met if u(t) is permanently excited, e.g. u(t) = 1. For decreasing $\tilde{K}(t)$ the persistency of excitation is fulfilled if u(t) represents a signal of 'infinite' energy and if in any time-interval, Δt, the condition

$$\int_t^{t+\Delta t} u^2(\tau)\,d\tau > 0$$

holds.

For the proof of stability of Eq.(1.35) an appropriate Liapunov function is selected as

$$V[\tilde{K}(t)] = V[\tilde{K}(0)] - 2h \int_0^t e^2(\tau, \tilde{K})\, d\tau, \tag{1.38}$$

where

$$\frac{dV(K)}{dt} = -2h\, e^2(t, \tilde{K}). \tag{1.39}$$

Asymptotic stability of the adaptation law for $K_M(t)$ is obained under persistent excitation if

a) $V[\tilde{K}(0)] > 2h \int_0^t e^2(\tau, \tilde{K})\, d\tau \Rightarrow \int_0^\infty e^2 d\tau < \infty$ (1.40a)

i.e. e(t) is bounded,

b) $V[\tilde{K}(t) = 0] = 0,$ (1.40b)

c) $\dot{V}[\tilde{K}(t)] < 0$ for $\tilde{K} \neq 0,$ (1.40c)

d) $\dot{V}[\tilde{K}(t)] = 0$ for $\tilde{K} = 0.$ (1.40d)

From Eqs.(1.36) and (1.37 follows that fast convergence would be obtained for a large adaptation gain h. Also, if the plant gain is time-varying, i.e. $K = K(t) \neq$ const, better tracking would be obtained for large h. However, it can easily be shown that in the case of a stochastically disturbed plant output signal y(t), the effect of the additional noise is smaller for lower values of h. Therefore a reasonable value of h has to be selected as a compromise between fast convergence and low noise immunity.

It is easy to understand that the convergence and stability analysis, and especially the selection of an appropriate Liapunov function, will become more difficult if the structure of the system to be identified is more complicated than in the case discussed here. Therefore it seems reasonable to introduce model-based identification approaches which directly rely on stability theory, as discussed in the following sections.

2. Model Adaptation Using Liapunov's Stability Theory

2.1. The basics of the technique

The starting point for the derivation of the adaptation law in this method is not a definite cost function, but an error differential equation of the overall system for identification. The adaptation law is to be so designed that the overall system attains a globally asymptotically stable steady state. The application of the direct method of Liapunov (Shackloth and Butchart 1965, Winsor and Roy 1968) will at first be demonstrated with the aid of a first order system. The differential equation of the system to be identified is

$$\dot{y}(t) + a\ y(t) = K\ u(t), \tag{2.1}$$

with the unknown parameters a and K, and the parallel model is given by

$$\dot{y}_M(t) + a_M(t)\ y_M(t) = K_M(t)\ u(t). \tag{2.2}$$

If we consider the model error

$$e(t) = y(t) - y_M(t) \tag{2.3}$$

then, from Eqs.(2.1) and (2.2) the error differential equation may be directly written as

$$e(t) + a\ e(t) = [K-K_M(t)]\ u(t) + [a_M(t)-a]\ y_M(t) \tag{2.4}$$

which for $\tilde{K}(t) = K-K_M(t) = 0$, $\tilde{a}(t) = a_M(t)-a = 0$ and $a > 0$ has the trajectory $e(t) = 0$ as a stable equilibrium position. $\tilde{K}$ and $\tilde{a}$ are the parameter errors, which should vanish fully in the adapted state. The adjustment must now be so designed that the above trajectory $(e(t) = 0,\ \tilde{K} = 0,\ \tilde{a} = 0)$ is the globally asymptotically stable steady state. A possible Liapunov function (e.g. Unbehauen 1989) is the quadratic form

$$V(e, \tilde{K}, \tilde{a}, t) = \frac{1}{2} e^2(t) + \frac{1}{2\alpha} \tilde{K}^2(t) + \frac{1}{2\beta} \tilde{a}^2(t), \qquad (2.5)$$

whose time derivative is given by

$$\dot{V}(e, \tilde{K}, \tilde{a}, t) = e(t) \dot{e}(t) + \frac{1}{\alpha} \tilde{K}(t) \dot{\tilde{K}}(t) + \frac{1}{\beta} \tilde{a}(t) \dot{\tilde{a}}(t). \qquad (2.6)$$

Inserting $\dot{e}$ from Eq.(2.4) in Eq.(2.6) we get

$$\dot{V}(e, \tilde{K}, \tilde{a}, t) = -a\, e^2(t) + \tilde{K}(t)\, u(t)\, e(t) + \frac{1}{\alpha} \tilde{K}(t) \dot{\tilde{K}}(t)$$
$$+ \tilde{a}(t)\, y_M(t)\, e(t) + \frac{1}{\beta} \tilde{a}(t) \dot{\tilde{a}}(t). \qquad (2.7)$$

The first term on the RHS is negative definite for a stable dynamical system. $\dot{V}$ is then certainly negative definite, if

$$u(t)\, e(t) + \frac{1}{\alpha} \dot{\tilde{K}}(t) = 0 \qquad (2.8)$$

and

$$y_M(t)\, e(t) + \frac{1}{\beta} \dot{\tilde{a}}(t) = 0. \qquad (2.9)$$

With $\dot{\tilde{K}} = -\dot{K}_M$ and $\dot{\tilde{a}} = \dot{a}_M$, Eqs.(2.8) and (2.9) through integration give rise directly to the adaptation relations

$$K_M(t) = K_M(0) + \alpha \int_0^t u(\tau)\, e(\tau)\, d\tau \qquad (2.10)$$

and

$$a_M(t) = a_M(0) - \beta \int_0^t y_M(\tau)\, e(\tau)\, d\tau. \qquad (2.11)$$

The adaptation law is globally asymptotically stable for every positive α and β.

2.2. A General design method for the series–parallel model approach

Next the application of the identification method described earlier will now be made to systems of higher order. A general and simple method (Pazdera and Pottinger 1969) starts with the state space description of the system to be identified

$$\dot{\mathbf{x}}(t) = \mathbf{A}\ \mathbf{x}(t) + \mathbf{b}\ u(t). \tag{2.12}$$

The parameters to be identified are the elements of the system matrix $\mathbf{A}$ and the vector $\mathbf{b}$, and the state $\mathbf{x}(t)$ for $t \geqslant 0$ is known. A model of the same order and fixed homogeneous behaviour is connected across the system in series-parallel as shown in Fig. 8. An additional connection of the system state vector gives the parallel model

$$\dot{\mathbf{x}}_M(t) = \mathbf{D}\ \mathbf{x}_M(t) + \mathbf{b}_M(t)\ u(t) + [\mathbf{A}_M(t)-\mathbf{D}]\ \mathbf{x}(t), \tag{2.13}$$

which, with the state error vector

$$\mathbf{e}(t) = \mathbf{x}(t) - \mathbf{x}_M(t) \tag{2.14}$$

can also be represented in the form of a feedback type parallel model

$$\dot{\mathbf{x}}_M(t) = \mathbf{A}_M(t)\ \mathbf{x}(t) + \mathbf{b}_M(t)\ u(t) - \mathbf{D}\ \mathbf{e}(t). \tag{2.15}$$

Subtracting Eq.(2.15) from Eq.(2.12) and considering Eq.(2.14), with the parameter errors

$$\tilde{\mathbf{A}}(t) = \mathbf{A} - \mathbf{A}_M(t) \tag{2.16}$$

and

$$\tilde{\mathbf{b}}(t) = \mathbf{b} - \mathbf{b}_M(t) \tag{2.17}$$

the error differential equation follows as

$$\dot{\mathbf{e}}(t) = \mathbf{D}\ \mathbf{e}(t) + \tilde{\mathbf{A}}(t)\ \mathbf{x}(t) + \tilde{\mathbf{b}}(t)\ u(t). \tag{2.18}$$

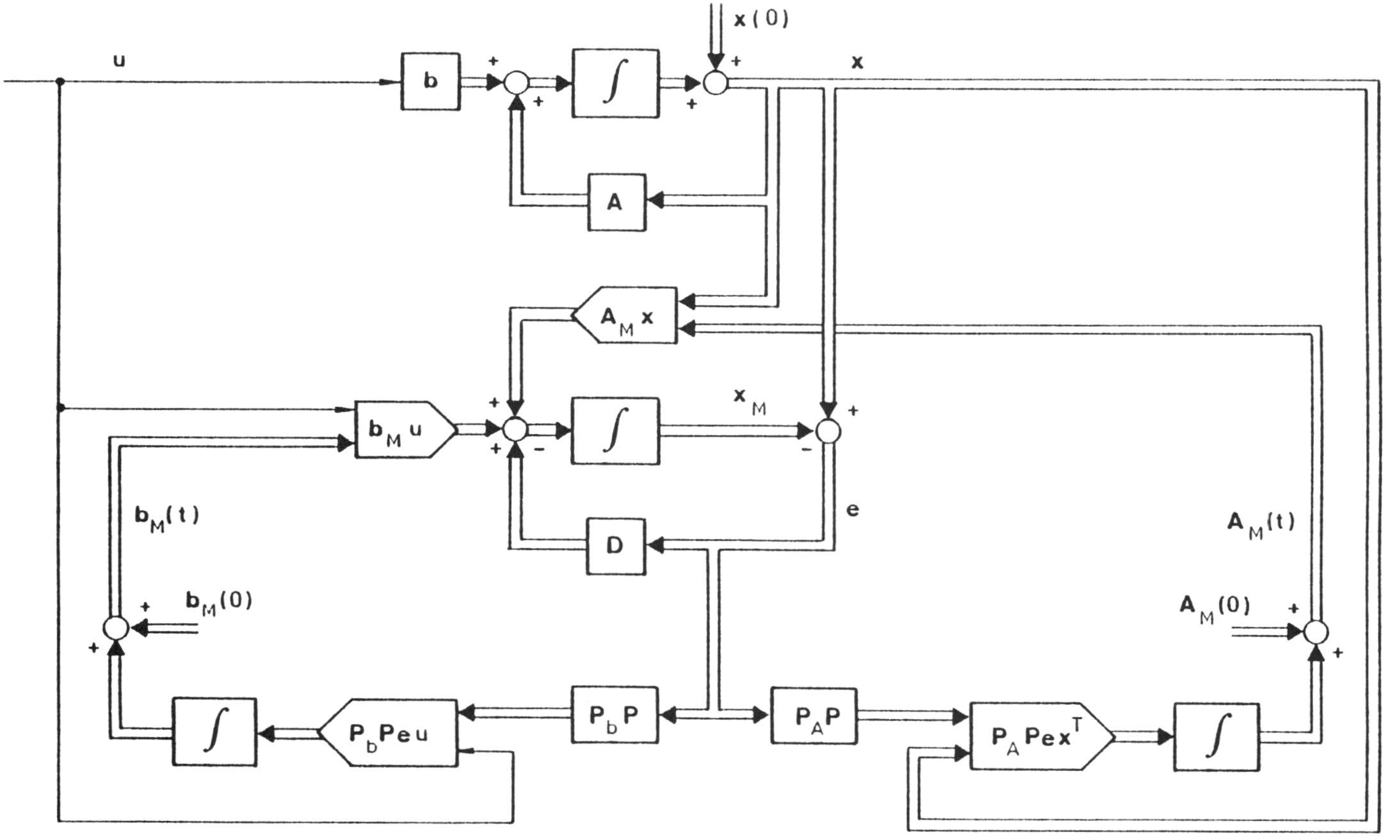

Fig. 8. Block diagram for the identification of a linear SISO-system by series-parallel model adaptation using Liapunov's stability theory

In the adapted condition $\mathbf{A} = \mathbf{A}_M$ and $\mathbf{b} = \mathbf{b}_M$, $\lim_{t\to\infty} \mathbf{e}(t) = \mathbf{0}$ under the assumption that $\mathbf{D}$ is a matrix with stable eigenvalues. If the steady state ($\mathbf{e} = \mathbf{0}$, $\tilde{\mathbf{A}} = \mathbf{0}$, $\tilde{\mathbf{b}} = \mathbf{0}$) of the overall system should be globally asymptotically stable, then a possible Liapunov function is to be found, which guarantees stability in the whole space spanned by the elements of $\mathbf{e}$, $\tilde{\mathbf{A}}$ and $\tilde{\mathbf{b}}$. In view of this we take here a quadratic form

$$V(\mathbf{e}, \tilde{\mathbf{A}}, \tilde{\mathbf{b}}, t) = \frac{1}{2}\, \mathrm{tr}(\tilde{\mathbf{A}}^T\, \mathbf{P}_A^{-1}\, \tilde{\mathbf{A}}) + \frac{1}{2}\, \tilde{\mathbf{b}}^T\, \mathbf{P}_b^{-1}\, \tilde{\mathbf{b}} + \frac{1}{2}\, \mathbf{e}^T\, \mathbf{P}\, \mathbf{e} \quad (2.19)$$

with positive definite symmetric weighting matrices $\mathbf{P}_A$, $\mathbf{P}_b$ and $\mathbf{P}$. The time derivative of this relation

$$\dot{V}(\mathbf{e}, \tilde{\mathbf{A}}\, \tilde{\mathbf{b}}, t) = \mathrm{tr}(\tilde{\mathbf{A}}^T\, \mathbf{P}_A^{-1}\, \dot{\tilde{\mathbf{A}}}) + \tilde{\mathbf{b}}^T\, \mathbf{P}_b^{-1}\, \dot{\tilde{\mathbf{b}}} + \frac{1}{2}\, \dot{\mathbf{e}}^T\, \mathbf{P}\, \mathbf{e} + \frac{1}{2}\, \mathbf{e}^T\, \mathbf{P}\, \dot{\mathbf{e}}. \quad (2.20)$$

With Eq.(2.18) and the derivatives

$$\dot{\tilde{\mathbf{A}}}(t) = -\, \dot{\mathbf{A}}_M(t) \text{ and } \dot{\tilde{\mathbf{b}}}(t) = -\, \dot{\mathbf{b}}_M(t)$$

follows from Eq.(2.20)

$$\begin{aligned}\dot{V}(\mathbf{e}, \tilde{\mathbf{A}}, \tilde{\mathbf{b}}, t) = &- \mathrm{tr}(\tilde{\mathbf{A}}^T\, \mathbf{P}_A^{-1}\, \dot{\mathbf{A}}_M) - \tilde{\mathbf{b}}^T\, \mathbf{P}_b^{-1}\, \dot{\mathbf{b}}_M + \frac{1}{2}\, \mathbf{e}^T(\mathbf{D}^T\, \mathbf{P} + \mathbf{P}\, \mathbf{D})\mathbf{e} \\ &+ \frac{1}{2}\, \mathbf{e}^T\, \mathbf{P}\, \tilde{\mathbf{A}}\, \mathbf{x} + \frac{1}{2}\, \mathbf{e}^T\, \mathbf{P}\, \tilde{\mathbf{b}}\, u + \frac{1}{2}\, \mathbf{x}^T \tilde{\mathbf{A}}^T\, \mathbf{P}\, \mathbf{e} \\ &+ \frac{1}{2}\, \tilde{\mathbf{b}}^T\, \mathbf{P}\, \mathbf{e}\, u. \end{aligned} \quad (2.21)$$

Since $\mathbf{a}^T\, \mathbf{b} = \mathbf{b}^T\, \mathbf{a} = \mathrm{tr}(\mathbf{a}\, \mathbf{b}^T) = \mathrm{tr}(\mathbf{b}\, \mathbf{a}^T)$, Eq.(2.21) simplifies itself to

$$\begin{aligned}\dot{V}(\mathbf{e}, \tilde{\mathbf{A}}, \tilde{\mathbf{b}}, t) = &\ \mathrm{tr}[\tilde{\mathbf{A}}^T(\mathbf{P}\, \mathbf{e}\, \mathbf{x}^T - \mathbf{P}_A^{-1}\, \dot{\mathbf{A}}_M)] + \tilde{\mathbf{b}}^T(\mathbf{P}\, \mathbf{e}\, u - \mathbf{P}_b^{-1}\, \dot{\mathbf{b}}_M) \\ &+ \frac{1}{2}\, \mathbf{e}^T(\mathbf{D}^T\, \mathbf{P} + \mathbf{P}\, \mathbf{D})\, \mathbf{e}. \end{aligned} \quad (2.22)$$

The derivative is negative definite if the third term in Eq.(2.22) is negative definite and the first and the second terms vanish. That is, one should have

$$\mathbf{D}^T \mathbf{P} + \mathbf{P} \mathbf{D} = -\mathbf{Q} \tag{2.23}$$

for a symmetric and positive definite matrix $\mathbf{Q}$ and further

$$\mathbf{P} \mathbf{e} \mathbf{x}^T - \mathbf{P}_A^{-1} \dot{\mathbf{A}}_M = 0 \tag{2.24}$$

$$\mathbf{P} \mathbf{e} \mathbf{u} - \mathbf{P}_b^{-1} \dot{\mathbf{b}}_M = 0. \tag{2.25}$$

If $\mathbf{P}$ is a solution of the matrix Liapunov equation (2.23), then from Eqs.(2.24) and (2.25) through integration we directly get the adaptation laws for the model parameters as

$$\mathbf{A}_M(t) = \mathbf{A}_M(0) + \mathbf{P}_A\mathbf{P} \int_0^t \mathbf{e}(\tau)\ \mathbf{x}^T(\tau)d\tau \tag{2.26a}$$

and

$$\mathbf{b}_M(t) = \mathbf{b}_M(0) + \mathbf{P}_b\mathbf{P} \int_0^t \mathbf{e}(\tau)\ \mathbf{u}(\tau)d\tau. \tag{2.26b}$$

Fig. 8 shows the corresponding block diagram. For this identification method stability is well guaranteed. It is however applicable only to the cases in which all the state variables are known or measurable.

The treatment given above was with an arbitrary state space description. Considering special canonical forms, the adaptation laws are simplified. If $\mathbf{D}$ and $\mathbf{P}_A$ are diagonal with eigenvalues $-d_i$ and p_{ai}(i = 1, 2, ..., n) and in addition if $\mathbf{A}_M$ is in Frobenius form

$$\mathbf{A}_M = \left[\begin{array}{c|c} \mathbf{0} & \mathbf{I}_{n-1} \\ \hline \multicolumn{2}{c}{-\mathbf{a}_M^T} \end{array} \right],$$

then, for instance with $\mathbf{Q} = 2\ \mathbf{I}$ in Eq.(2.23) the adaptation law becomes

$$a_{Mi}(t) = a_{Mi}(0) - \beta \int_0^t e_n(t)\ x_i(\tau)d\tau \qquad (2.26c)$$

with

$$\beta = p_{an}/d_n.$$

2.3. A general design method for the parallel model approach

For the representation of *MIMO-systems* the second term on the right hand side of Eq.(2.12) has to be replaced by $\mathbf{B}\ \mathbf{u}(t)$. This modified equation in connection with the corresponding parallel model in state space representation,

$$\dot{\mathbf{x}}_M(t) = \mathbf{A}_M(t)\ \mathbf{x}_M(t) + \mathbf{B}_M(t)\ \mathbf{u}(t), \qquad (2.27)$$

and the state error vector $\mathbf{e}(t) = \mathbf{x}(t) - \mathbf{x}_M(t)$ leads to the error differential equation

$$\dot{\mathbf{e}}(t) = \mathbf{A}\ \mathbf{e}(t) + \tilde{\mathbf{A}}(t)\ \mathbf{x}_M(t) + \tilde{\mathbf{B}}(t)\ \mathbf{u}(t), \qquad (2.28)$$

where

$$\tilde{\mathbf{A}}(t) = \mathbf{A} - \mathbf{A}_M(t) \text{ and } \tilde{\mathbf{B}}(t) = \mathbf{B} - \mathbf{B}_M(t).$$

We consider a Liapunov function just as in Eq.(2.19) with $\tilde{\mathbf{b}}$ and u replaced by $\tilde{\mathbf{B}}$ and $\mathbf{u}$ respectively, to get to the corresponding matrix Liapunov equation

$$\mathbf{A}^T\ \mathbf{P} + \mathbf{P}\ \mathbf{A} = -\mathbf{Q}. \qquad (2.29)$$

The solution $\mathbf{P}$ of this equation is required to get parameter adaptation laws in the form of Eqs.(2.26a,b) with $\tilde{\mathbf{B}}$ and $\mathbf{u}$ appearing in the places of $\tilde{\mathbf{b}}$ and u respectively. Unlike in the case of Eq.(2.23) wherein the matrix $\mathbf{D}$ is known, in the present case, the solution of Eq.(2.29) is connected with $\mathbf{A}$, the plant matrix, which is yet to be determined. In this situation we insert a reasonable approximation for $\mathbf{A}$ based on a priori knowledge. This can be done, for instance by setting $A = A_M(0)$ in Eq.(2.29). From here follows the advantage of the series-parallel model approach as discussed in section 2.1.2.

3. Model Adaptation Using Hyperstability Theory

3.1. The basic idea

Similar to the concept of absolute stability (e.g. Unbehauen 1989), that of hyperstability (e.g. Hang 1973, Landau 1979) encompasses simultaneous stability of a whole class of systems. The theory of hyperstability can be interpreted in terms of the general theory of passive one-port networks. Let us assume that such a system is described by the input signal u(t), the output signal y(t) and its state vector $\mathbf{x}(t)$ then the following energy principle holds:

$$\int_{t_0}^{t} u(\tau)\, y(\tau)d\tau + \frac{1}{2}\, \mathbf{x}^T(t_0)\, \mathbf{x}(t_0) \geq \frac{1}{2}\, \mathbf{x}^T(t)\, \mathbf{x}(t). \tag{3.1}$$

The first term on the LHS represents the energy supplied during $t_0 \leq \tau \leq t$ to the system, the second term the energy at the initial state and the RHS the instantaneous stored energy. The difference between the left and right hand sides is the dissipated energy. Since the enery contained in a passive system can never be greater than the energy externally supplied to it plus the energy at the initial instant of time, such systems are always stable. Since the output y(t) being a (in general nonlinear and time-varying) function of the input u(t), i.e.,

$$y(t) = f[u(t),t]$$

Eq.(3.1) can be written in a somewhat generalized form as

$$\int_{t_0}^{t} f[u(\tau),t]\, u(\tau)d\tau \geq -\gamma_0^2 + \beta_1\, \|\mathbf{x}(t)\|^2, \qquad \beta_1 > 0, \tag{3.2}$$

where

$$\gamma_o = \sqrt{\frac{1}{2}\, \mathbf{x}^T(t_o)\, \mathbf{x}(t_o)} \tag{3.3}$$

represents a constant dependent upon the initial conditions. On the grounds of energy principle Eq.(3.1) holds not only for passive linear networks, but for all real passive networks, i.e., also for those which are nonlinear and/or time-varying. It can be conversely established that every dynamical system satisfying inequality (3.1) or (3.2) exhibits the same stability properties as the passive network discussed above.

The general hyperstability theory was developed and proved by V. Popov (1963, 1973). Since the adaptive model approach for system identification is characterized by strong nonlinearities, this theory in the recent years has proved to be an exceptionally effective tool helpful in the design of stable adaptive models. As is the case with Liapunov's method the hyperstability theory, however, gives only sufficient but not necessary stability conditions.

Definition of hyperstability. Consider a system described by the state space representation

$$\dot{\mathbf{x}}(t) = \mathbf{f}[\mathbf{x}(t), \mathbf{u}(t), t] \tag{3.4}$$

$$\mathbf{y}(t) = \mathbf{g}[\mathbf{x}(t), \mathbf{u}(t), t] \tag{3.5}$$

with state vector $\mathbf{x}$, input vector $\mathbf{u}$, and the output vector $\mathbf{y}$, wherein $\mathbf{x}$ has dimension (nx1) and $\mathbf{u}$ and likewise $\mathbf{y}$ the dimension (px1). The hyperstability of a system according to Popov now states that the state vector remains bounded if the input vector $\mathbf{u}$ fulfills the inequality (3.2). For the sake of conciseness in further treatment we use in (3.2) the scalar variable

$$\eta(0,t) = \int_0^t \mathbf{u}^T(\tau)\ \mathbf{y}(\tau)d\tau \tag{3.6}$$

for the integral term, wherein without loss of generality the initial time $t_0 = 0$. Although the stability is a property of the steady state of a system, it will be referred to as a system property in accordance with Popov.

In the following, some definitions of hyperstability will be introduced:

Definition of hyperstability: The system described by Eqs.(3.4) and (3.5) is said to be hyperstable if there exist two constants $\beta_1 > 0$ and $\beta_0 \geq 0$, such that the inequality

$$\eta(0,t) + \beta_0\ \|\mathbf{x}(0)\|^2 \geq \beta_1\ \|\mathbf{x}(t)\|^2 \text{ for all } t \geq 0 \tag{3.7}$$

is fulfilled.

For $\beta_0 = \frac{1}{2}$, $\beta_1 = \frac{1}{2}$ and $\|\mathbf{x}\|^2 = \mathbf{x}^T\ \mathbf{x}$ Eq.(3.7) is identical with Eq.(3.1). The inequality (3.7) can be interpreted as a generalized energy relation. Besides this definition there is another fully equivalent definition (Popov 1973):

Definition of asymptotic hyperstability: The system described by Eqs.(3.4) and (3.5) is said to be asymptotically hyperstable, if it is hyperstable, and is globally asymptotically stable for $\mathbf{u} = 0$.

The above definitions contain terms with state variables. If an input/output description in the form

$$\mathbf{y}(t) = \mathbf{f}[\mathbf{u}(\tau),t] \qquad \tau \leq t \text{ for all } t \geq 0 \tag{3.8}$$

is given, then instead of Eq.(3.7) the following mild condition for hyperstability is stated:

Definition of weak hyperstability: The system described by Eqs.(3.4) and (3.5) is said to be weakly hyperstable, if there exists a finite constant $\gamma_0^2 \geqslant 0$ depending only on initial conditions, such that the inequality

$$\eta(0,t) \geqslant -\gamma_0^2 \quad \text{for all } t \geqslant 0 \tag{3.9}$$

is fulfilled.

The definition of weak hyperstability and the inequality (3.9) are especially important for the design of adaptive models, as many designs are based on this definition.

Properties of hyperstable systems. The meaning of hyperstability theory is related essentially to the particular properties, which result by dynamic systems from the above given definitions. The most important of these are summarized in the following. For the proof see (Popov 1973).

Property 1: The hyperstability includes the global stability of Liapunov for $\mathbf{u} = \mathbf{0}$.

Property 2: A hyperstable system has the property that for bounded input the output also remains bounded (termed as BIBO stability).

Property 3: A new system arising out of a parallel connection of two (weakly) hyperstable systems is also (weakly) hyperstable.

Property 4: If a (weakly) hyperstable system is connected in feedback with a second (weakly) hyperstable system, the overall system is again (weakly) hyperstable.

Property 5: If a hyperstable system is connected in feedback with a weakly hyperstable system, the overall system is hyperstable.

The case of time-variation in the feedback as shown in Fig. 9,

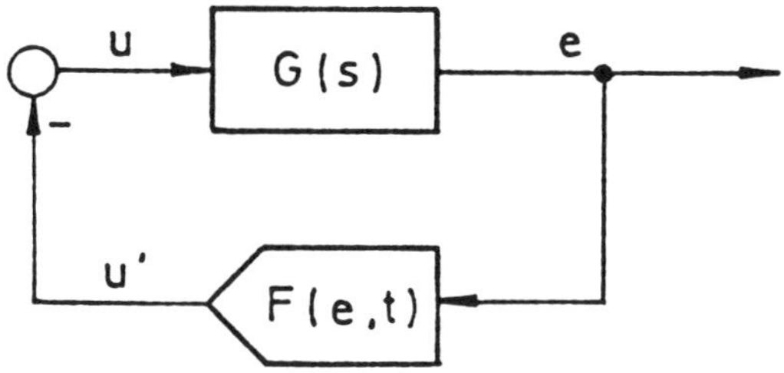

Fig. 9. Standard structure of a nonlinear time-variant system

should still be explicitly pointed out. Also model adaptive systems lend themselves to transformation into such a standard nonlinear structure. According to *Property 5*, stated above, such a system is hyperstable, if on the one hand the transfer function G(s) is strictly positive real (Popov 1973), i.e.

$$\mathrm{Re}\ \{G(j\omega)\} > 0 \text{ for all } \omega, \tag{3.10}$$

and on the other hand

$$\int_0^t e(\tau)\ u'(\tau) \geqslant -\gamma_0^2 \tag{3.11}$$

or in other words the nonlinear time-variant element is weakly hyperstable. Several adaptive models which will be designed with the help of hyperstability theory are based directly on this property.

Hyperstability theorem for linear systems: A linear time-invariant system is hyperstable if and only if it is described by a strictly positive real transfer function. It is weakly hyperstable, if it is described by a positive real but not a strictly positive real transfer function, i.e. $\mathrm{Re}\{G(j\omega)\} \geqslant \omega$ for all ω.

From this theorem and the other definitions stated above we can develop directly the design method for adaptive models for identification based on hyperstability concept in the following steps:

1. Derivation of system equations;
2. Transformation into the standard form of Fig. 9;
3. Division into possible subsystems for proving hyperstability; the free design parameters should be concentrated in the fewest possible subsystems;
4. Design of still free hyperstable subsystems;
5. Inverse transformation into the original system representation.

3.2. Stable identification with adaptive models on the basis of hyperstability theory

If we apply the results of hyperstability theory introduced in the previous section to the case of parallel reference model, the hyperstability concept permits a systematic procedure in several steps.

In the *first step* of this design method the system equations are derived. With this we will proceed to the adaptive model of Fig. 10. The difference e(t) between the output signal of the system to be identified (the plant) and that of the parallel model will be fed at first into a linear transfer system (compensator) yet unknown. The output v(t) of this compensator is the input to a subsystem which contains the actually nonlinear adaptation algorithm for matching the parallel model

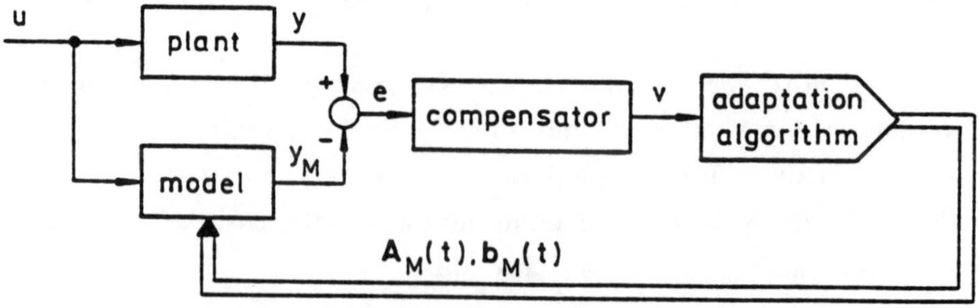

Fig. 10. Structure of the identification method based on hyperstability theory

with the plant. The linear compensator and the nonlinear subsystem should now contain all the free design parameters of the method. They are to be so designed that the steady state of the overall system is globally asymptotically stable.

For a system to be identified with

$$G(s) = \frac{\sum_{i=0}^{m} b_i s^i}{\sum_{i=0}^{n-1} a_i s^i + s^n} = \frac{B(s)}{A(s)} \tag{3.12}$$

and a parallel model with a transfer function as in Eq.(1.9) a state space representation in Frobenius form can be found if the input signal u(t) is m-times differentiable, that is, for example if it is the output signal of a controller whose order is greater than or equal to m. Then the state space representation is

$$\dot{\mathbf{x}}(t) = \mathbf{A}\ \mathbf{x}(t) + \mathbf{B}\ \mathbf{u}(t) \tag{3.13a}$$

$$y(t) = \mathbf{c}^T\ \mathbf{x}(t) \tag{3.13b}$$

with

$$\mathbf{A} = \left[\begin{array}{c|c} 0 & \mathbf{I}_{n-1} \\ \hline \multicolumn{2}{c}{-\mathbf{a}^T} \end{array}\right] \text{ and } \mathbf{B} = \left[\begin{array}{c} 0 \\ \hline \mathbf{b}^T \end{array}\right], \tag{3.14a,b}$$

where the vectors **a** and **b** contain the numerator and denominator coefficients of G(s) as in Eq.(3.12). Besides, we have the vectors

$$\mathbf{c} = \left[\begin{array}{c} 1 \\ \hline 0 \end{array}\right]; \quad \mathbf{u}(t) = \left[u(t)\ \frac{du(t)}{dt}\ \cdots\ \frac{d^m u(t)}{dt^m} \right]^T . \tag{3.15a,b}$$

Correspondingly for the parallel model we have

$$\dot{\mathbf{x}}_M(t) = \mathbf{A}_M(t)\ \mathbf{x}_M(t) + \mathbf{B}_M(t)\ \mathbf{u}(t) \tag{3.16a}$$

$$y_M(t) = \mathbf{c}^T\ \mathbf{x}_M(t) \tag{3.16b}$$

$$\mathbf{A}_M(t) = \left[\begin{array}{c|c} 0 & \mathbf{I}_{n-1} \\ \hline \multicolumn{2}{c}{-\mathbf{a}_M^T(t)} \end{array}\right]; \quad \mathbf{B}_M(t) = \left[\begin{array}{c} 0 \\ \hline \mathbf{b}_M^T(t) \end{array}\right]. \tag{3.17a,b}$$

The transfer function

$$D(s) = \frac{\mathfrak{L}\{v(t)\}}{\mathfrak{L}\{e(t)\}} \tag{3.18}$$

of the unknown linear compensator will be determined with the help of the model error

$$e(t) = \mathbf{c}^T\ \mathbf{e}(t) \tag{3.19}$$

and the state error

$$\mathbf{e}(t) = \mathbf{x}(t) - \mathbf{x}_M(t). \tag{3.20}$$

With these, for the description of the nonlinear subsystem of the adaptation algorithm we get, according to Fig. 10 the relations

$$\mathbf{A}_M = \mathbf{A}_M[v(\tau),t] \tag{3.21}$$

and

$$\mathbf{B}_M = \mathbf{B}_M[v(\tau),t] \tag{3.22}$$

for $\tau \leqslant t$.

In the *second step* of the design method the overall system should be brought into the standard form of Fig. 9 through a linear transformation of the state error as in Eq.(3.20). In the *third step* these subsystems are further separated. This permits the application of the stability theorem of Section 3.1.

The result of this transformation should be a nonhomogeneous vector error differential equation of the form

$$\dot{\mathbf{e}}(t) = \mathbf{f}[\mathbf{e}(t), \tilde{\mathbf{A}}(t), \tilde{\mathbf{B}}(t)]$$

whose nonhomogeneous part vanishes with negligible parameter errors $\tilde{\mathbf{A}}(t) = \mathbf{A}-\mathbf{A}_M(t)$ and $\tilde{\mathbf{B}}(t) = \mathbf{B}-\mathbf{B}_M(t)$. We get this relation by subtracting Eq.(3.16a) from Eq.(3.13a) and by considering Eq.(3.20)

$$\dot{\mathbf{e}}(t) = \mathbf{A}\ \mathbf{e}(t) + \tilde{\mathbf{B}}(t)\ \mathbf{u}(t) + \tilde{\mathbf{A}}(t)\ \mathbf{x}_M(t). \tag{3.23}$$

In view of the canonical form mentioned earlier Eq.(3.23) simplifies to

$$\dot{\mathbf{e}}(t) = \tilde{\mathbf{A}}\ \mathbf{e}(t) + \boldsymbol{\phi}\ [\tilde{\mathbf{b}}^T(t)\ \mathbf{u}(t) - \tilde{\mathbf{a}}^T(t)\ \mathbf{x}_M(t)] \tag{3.24}$$

with the parameter error vectors $\tilde{\mathbf{a}} = \mathbf{a}-\mathbf{a}_M$, and $\tilde{\mathbf{b}} = \mathbf{b}-\mathbf{b}_M$ and similarly

$$\boldsymbol{\phi} = \begin{bmatrix} 0 \\ --- \\ 1 \end{bmatrix}.$$

As $\mathbf{A}_M$ and $\mathbf{B}_M$ are functions of $v(\tau)$ corresponding to Eqs.(3.21) and (3.22), $\tilde{\mathbf{a}}$ and $\tilde{\mathbf{b}}$ are also functions of v. Denoting the scalar in the rectangular brackets of Eq.(3.24) as u_1, the error differential equation

$$\dot{\mathbf{e}}(t) = \mathbf{A}\ \mathbf{e}(t) + \boldsymbol{\phi}\ u_1(t) \tag{3.25}$$

describes a linear system with input signal

$$u_1(t) = \tilde{\mathbf{b}}^T[v(\tau),t]\ \mathbf{u}(t) - \tilde{\mathbf{a}}^T[v(\tau),t]\ \mathbf{x}_M(t) \quad \text{for } \tau \leqslant t, \tag{3.26}$$

which is dependent on v on the grounds mentioned above. Laplace transformation of the linear error system, Eq.(3.25), after a slight rearrangement leads to

$$\frac{\mathfrak{L}\{\mathbf{e}(t)\}}{\mathfrak{L}\{u_1(t)\}} = (s\mathbf{I}-\mathbf{A})^{-1}\ \boldsymbol{\phi}. \tag{3.27}$$

With this relation and with the help of Eqs.(3.18) and (3.19) the required transfer function can be directly written as

$$G_L(s) = \frac{\mathfrak{L}\{v(t)\}}{\mathfrak{L}\{u_1(t)\}} = D(s)\ \mathbf{c}^T(s\mathbf{I}-\mathbf{A})^{-1}\ \boldsymbol{\phi}. \tag{3.28}$$

Because of the special canonical forms chosen here, as one can easily verify, for instance with a second order system, the general result follows as

$$G_L(s) = \frac{D(s)}{A(s)}, \tag{3.29}$$

wherein D(s) may be a simple polynomial in s. The transfer function $G_L(s)$ in Eq.(3.29) describes the dynamic behaviour of the linear subsystem of Fig. 9. The nonlinear subsystem will be determined through Eq.(3.26). The output signal of this subsystem is separated into the following quantities

$$u_1(t) = -u'(t) = -u_a[\mathbf{x}_M, v(\tau), \tilde{\mathbf{a}}, t] - u_b[\mathbf{u}, v(\tau), \tilde{\mathbf{b}}, t] \text{ for } \tau \leq t, \tag{3.30}$$

wherein the signals u_a and u_b represent the outputs of a pair of parallel nonlinear subsystems excited by v. Thus the transformed structure results in the form shown in Fig. 11.

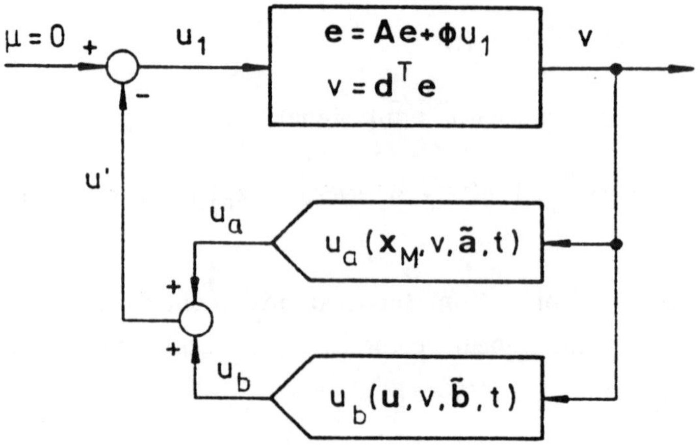

Fig. 11. Structure of the transformed system

In the *fourth step* the stability criteria of Section 3.1 are applied to the system represented in Fig. 11. This system is asymptotically stable, if for $\mu = 0$ the linear subsystem according to Eq.(3.29) is asymptotically hyperstable and the nonlinear subsystem of Eq.(3.30) presents weak hyperstability. The latter is also weakly hyperstable if each of the two parallel nonlinear subsystems taken individually is weakly hyperstable. Thus the three subsystems can be designed independently of the others. The linear subsystem is asymptotically hyperstable if the transfer function $G_L(s)$ of Eq.(3.28) or Eq.(3.29) is strictly positive real. Letting the compensator part of the transfer function be

$$D(s) = \sum_{i=0}^{n-1} d_i s^i \tag{3.31}$$

then we obtain

$$G_L(s) = \mathbf{d}^T(s\mathbf{I}-\mathbf{A})^{-1} \boldsymbol{\phi} \tag{3.32}$$

with $\mathbf{d} = [d_0 d_1 \ldots d_{n-1}]^T$. For the computation of the unknown coefficients d_i the Meyer-Kalman-Yacubovich lemma (Meyer 1965) can be recalled. According to this lemma $G_L(s)$ is strictly positive real if the coefficients

$$d_{i-1} = p_{n_i} \quad \text{for } i = 1, 2, \ldots, n \tag{3.33}$$

are the elements of the last row of the solution matrix $\mathbf{P}$ of the matrix Liapunov equation (Unbehauen and Rao 1987)

$$\mathbf{A}^T \mathbf{P} + \mathbf{P} \mathbf{A} = -\mathbf{Q} - \mathbf{q}\mathbf{q}^T \tag{3.34}$$

with a positive definite symmetric matrix $\mathbf{Q}$. In the design one states $\mathbf{Q}$ and $\mathbf{q}$ and computes $\mathbf{P}$ from Eq.(3.34). Thus corresponding to Eq.(3.33) the coefficients of the compensator have been determined.

The two nonlinear subsystems with outputs u_b and u_a are, in view of Eq.(3.26), of similar structure. Consequently the design of these subsystems can be carried out on a general nonlinear subsystem

$$u_p = u_p[z,\ v(\tau),\ \tilde{p},\ t] \quad \text{for } \tau \leq t. \tag{3.35}$$

For the first subsystem

$$z = u\ ,\quad \tilde{p} = \tilde{b} \tag{3.36}$$

and for the second

$$z = -x_M\ ,\quad \tilde{p} = \tilde{a}\ . \tag{3.37}$$

The nonlinear subsystem of Eq.(3.35) is weakly hyperstable if u_p satisfies the inequality (3.9), i.e.

$$\eta_p(0,t) = \int_0^t u_p[z,\ v(\theta),\ \tilde{p},\ \tau]\ v(\tau)\ d\tau \geq -\gamma_{op}^2 \quad \text{for } 0 \leq \tau.$$

With Eq.(3.26) in view of the change of sign of $u_1 = -u'$, for the above equation

$$\eta_p(0,t) = -\int_0^t \tilde{p}^T[v(\theta),\ \tau]\ z(\tau)\ v(\tau)\ d\tau \geq -\gamma_{op}^2 \quad \text{for } \theta \leq \tau. \tag{3.38}$$

Choosing the parameter error vector

$$\tilde{p}(t) = p - p_M(t)$$

and the proportional set up

$$\tilde{p}^P(t) = -\alpha_p\ z(t)\ v(t), \tag{3.39}$$

the inequality (3.38) is always fulfilled for proportional factors $\alpha_p > 0$, since the integral

$$\eta_p^P(0,t) = \alpha_p \int_0^t z^T(\tau)\ z(\tau)\ v^2(\tau)\ d\tau$$

is always positive. One finds the proportional adaptation law for the model parameter vector p_M through transformation of Eq.(3.39) in the original representation under the consideration of

$$\dot{p}_M^P = -\dot{\tilde{p}}^P = \alpha_p \frac{d}{dt} [z(t)\ v(t)], \tag{3.40}$$

which leads after integration finally to

$$p_M^P(t) = p_M^P(0) + \alpha_p\ z(t)\ v(t). \tag{3.41}$$

For asymptotic stability

$$\lim_{t\to\infty} \|e(t)\| = \lim_{t\to\infty} u_1(t) = \lim_{t\to\infty} v(t) = 0. \tag{3.42}$$

This is the case for a proportional set up only for $\lim_{t\to\infty} \|\tilde{p}^P(t)\| = 0$. In view of Eq.(3.41), for the limit of $\tilde{p} = p - p_M$ in the proportional set up the relation

$$\lim_{t\to\infty} \|\tilde{p}^P(t)\| = \|p - p_M^P(0)\| \neq 0$$

holds. Therefore only simple hyperstability is present. Asymptotic hyperstability is achieved on the other hand by an integral set up

$$\dot{\tilde{p}}^I(t) = -\beta_p\ \dot{\psi}(t) \tag{3.43}$$

with a freely chosen integral function $\psi(t)$. Integrating Eq.(3.43)

$$\tilde{p}^I(t) = \tilde{p}^I(0) - \beta_p\ \psi(t) \tag{3.44}$$

and inserting the result in inequality (3.38), one obtains

$$\eta_p^I(0,t) = -\int_0^t [\tilde{p}^I(0)]^T\ z(\tau)\ v(\tau)\ d\tau + \beta_p \int_0^t \psi^T(\tau)\ z(\tau)\ v(\tau)\ d\tau \geq \gamma_{op}^2. \tag{3.45}$$

The unknown function $\psi(t)$ is chosen such that the second integral in Eq.(3.45) can be directly integrated. For the simple set up

$$\psi(t) = \int_0^t z(\tau)\ v(\tau)\ d\tau, \tag{3.46}$$

we get

$$\eta_p^I(0,t) = -[\tilde{p}^I(0)]^T \psi(t) + \beta_p \int_0^t \psi^T(\tau) \dot{\psi}(\tau) d\tau$$

$$= -[\tilde{p}^T(0)]^T \psi(t) + \frac{\beta_p}{2} \psi^T(t) \psi(t)$$

or by quadratic expansion

$$\eta_p^I(0,t) = \frac{\beta_p}{2} \left\{ \psi^T(t) \psi(t) - \frac{2}{\beta_p} [\tilde{p}^I(0)]^T \psi(t) + \frac{1}{\beta_p^2} [\tilde{p}^I(0)]^T \tilde{p}^I(0) \right\}$$

$$- \frac{\beta_p}{2} [\tilde{p}^I(0)]^T \tilde{p}^I(0)$$

$$= \frac{\beta_p}{2} [\psi(t) - \frac{1}{\beta_p} \tilde{p}^I(0)]^T [\psi(t) - \frac{1}{\beta_p} \tilde{p}^I(0)] - \frac{1}{2\beta_p} [\tilde{p}^I(0)]^T \tilde{p}^I(0). \tag{3.47}$$

Equations (3.47) and the inequality (3.45) give the relation

$$\eta_p^I(0,t) = \frac{\beta_p}{2} \| \psi(t) - \frac{1}{\beta_p} \tilde{p}^I(0) \|^2 - \frac{1}{2\beta_p} \| \tilde{p}^I(0) \|^2 \geqslant - \gamma_{op}^2$$

or after rearrangement

$$\frac{\beta_p}{2} \| \psi(t) - \frac{1}{\beta_p} \tilde{p}^I(0) \|^2 \geqq \frac{1}{2\beta_p} \| \tilde{p}^I(0) \|^2 - \gamma_{op}^2. \tag{3.48}$$

The inequality (3.48) is satisfied for every positive factor β_p. The set up of Eq.(3.43) therefore guarantees weak hyperstability. If now a subsystem with proportional set up is added in parallel to the subsystem with integral set up then the PI-law

$$\tilde{p}(t) = \tilde{p}^P(t) + \tilde{p}^I(t)$$

or using Eqs.(3.39), (3.44) and (3.46)

$$\tilde{p}(t) = \tilde{p}(0) - \alpha_p \dot{\psi}(t) - \beta_p \psi(t) \tag{3.49}$$

is again hyperstable.

The reverse transformation into the original form in the *fifth step* gives the adaptation equations or the nonlinear adaptation law for the adjustment of model parameters. The linear relation $\dot{\tilde{p}}(t) = -\dot{p}_M(t)$ directly resulting from the parameter error vector $\tilde{p}(t) = p - p_M(t)$ for time-invariant systems, here from Eq.(3.49), directly gives the differential equation for the adaptation law

$$\dot{p}_M(t) = -\dot{\tilde{p}}(t) = \alpha_p \ddot{\psi}(t) + \beta_p \dot{\psi}(t). \tag{3.50}$$

Integrating Eq.(3.50) we get the general adaptation law, using Eq.(3.46), as

$$p_M(t) = p_M(0) + \alpha_p z(t) v(t) + \beta_p \int_0^t z(\tau) v(\tau) d\tau. \tag{3.51}$$

For the actual model parameters a_M and b_M, with Eqs.(3.36) and (3.37), the adaptation law is

$$b_M(t) = b_M(0) + \alpha_b u(t) v(t) + \beta_b \int_0^t u(\tau) v(\tau) d\tau \tag{3.52}$$

$$a_M(t) = a_M(0) - \alpha_a x_M(t) v(t) - \beta_a \int_0^t x_M(\tau) v(\tau) d\tau. \tag{3.53}$$

Fig. 12 shows finally the block diagram of the overall system on the basis of the above relations.

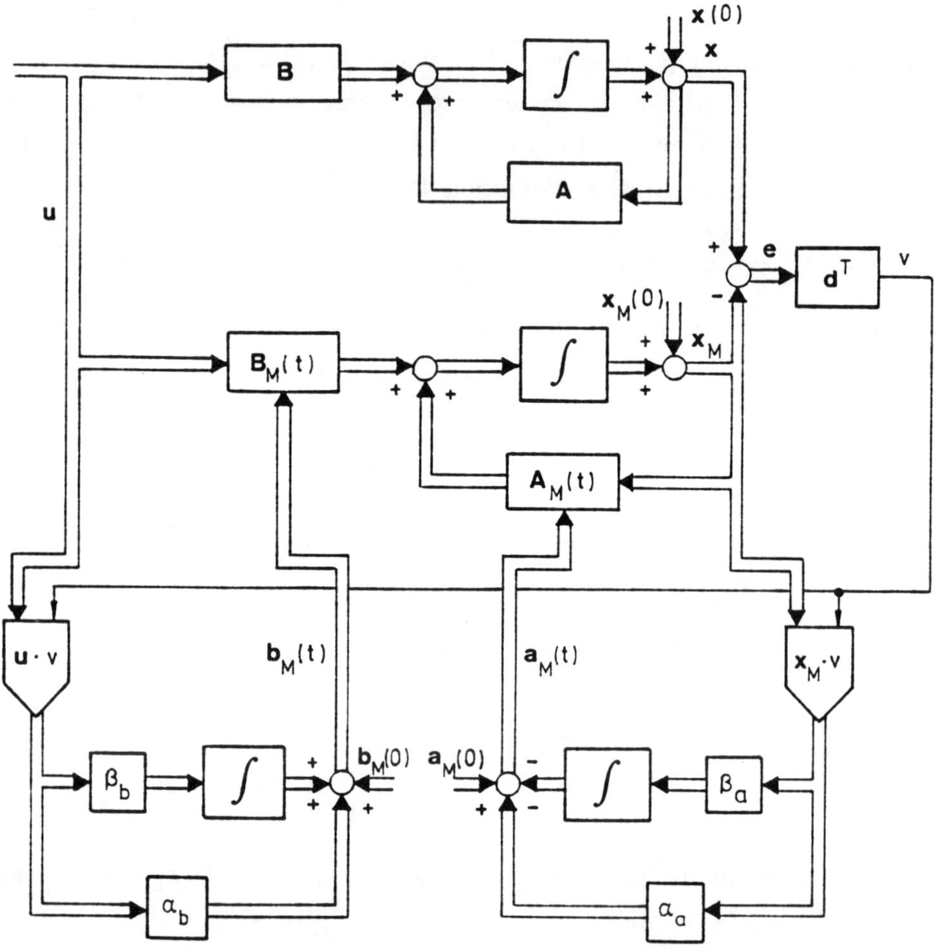

Fig. 12. Block diagram realizing hyperstable SRAM identification method using a parallel model

References

Anderson, B. et al.: Stability of adaptive systems: Positivity and averaging analysis. MIT Press, Cambridge (Mass., USA) 1986.

Dugard, L. and G. Goodwin: Global convergence of Landau's "Output error with adjustable compensator" adaptive algorithm. IEEE Trans. AC-30 (1985), pp. 593-595.

Eykhoff, P.: System Identification. Wiley, New York 1974.

Hang, C.C.: The design of model reference parameter estimation systems using hyperstability theories. Prepr. IFAC-Symposium, Den Haag 1973, pp. 741-744.

Landau, I.D.: Adaptive control (The model reference approach). Marcel Dekker Inc., New York 1979.

Landau, I.D.: Model reference adaptive systems - A survey. ASME J. Dyn. Syst. Measurement and Control, 94, pp. 119-132, 1972.

Lion, P.M.: Rapid identification of linear and nonlinear systems. AIAA Journal 5 (1967), pp. 1835-1842.

Marsik, I.: Versuche mit einem selbsteinstellenden Modell zur automatischen Kennwertermittlung von Regelstrecken. Zeitschrift messen, steuern, regeln 9 (1966), pp. 210-213.

Meyer, K.: On the existence of Liapunov functions for the problem of Lur'e. J. SIAM Control Ser. A3 (1965), pp. 373-383.

Parks, P.C.: Liapunov redesign of model reference adaptive control systems. IEEE Trans. on Aut. Control AC-11 (1966), pp. 362-368.

Pazdera, J.S. und H.J. Pottinger: Linear system identification via Liapunov design techniques. Proceed. JAC, Boulder 1969, pp. 795-801.

Popov, V.: Hyperstability of control systems. Springer-Verlag, Berlin 1973.

Popov, V.: The solution of a new stability problem for controlled systems. Automation and Remote Control. Translated from: Automatika i Telemekhanika 24 (1963), pp. 7-26.

Rake, H.: Automatische Prozeßidentifizierung durch ein selbsteinstellendes System. Zeitschrift messen, steuern, regeln 9 (1966), pp. 213-216.

Shackcloth, B. and R.L. Butchart: Synthesis of model reference adaptive systems by Liapunov's second method. In: P.H. Hammond (Ed.): Theory of self adaptive control systems. Plenum Press, New York 1965.

Unbehauen, H.: Regelungstechnik II. Vieweg-Verlag, Braunschweig/ Wiesbaden 1989 (5. Auflage).

Unbehauen, H.: Regelungstechnik III. Vieweg-Verlag, Braunschweig/ Wiesbaden 1985.

Unbehauen, H.: Systematic design of discrete model reference adaptive systems. In: Harris, C. and S. Billings (Ed.): Self-tuning and adaptive control: Theory and application. P. Peregrinus Ltd., London 1981, pp. 166-203.

Winsor, C. and R. Roy: Design of model reference adaptive control systems by Liapunov's second method. IEEE Trans. on Aut. Control AC-13 (1968), p. 204.

Nonparametric approaches to identification

D. Matko

Faculty of Electrical and Computer Engineering
University of Ljubljana
61000 Ljubljana, Yugoslavia

Abstract

Nonparametric identification methods, i.e. the methods for determining the frequency and the impulse responses from the measured data are presented. Based on a brief review of system and signal processing theory the Fourier analysis, the Spectral analysis, the Frequency response analysis and the Correlation analysis are described. Also bias and convergence analysis is given for all methods. Finally some problems with practical application of the methods on a digital computer are discussed.

1. Introduction

Observing the real-life actual system's input-output behaviour and inferring models from measured data is, what *identification* is about.

Mathematical models, which of course are an idealisation of the "true" systems, are used for pragmatic reasons. Nonlinear time varying partial differential equations are probably the best approximation of the real life but for practical reasons (our knowledge of mathematics, available tools etc.) the models expressed with ordinary linear differential equations with constant parameters are used in practical applications of identification procedures. A detailed validation of the corresponding simplification steps is given in the modelling part of this book or in any good reference on modelling or identification. The coefficients of the differential equations are the parameters of the system and so the differential equation represents one of **parametric** models. There are however also other ways to describe the model's dynamic behaviour as its impulse, step or frequency response. These descriptions do not (explicitly) involve a finite dimensional parameter vector as they are curves or functions and for this reason the corresponding models are called **nonparametric** models.

N. K. Sinha and G. P. Rao (eds.), Identification of Continuous-Time Systems, 549–574.

This contribution is dealing with the identification of nonparametric models and is organised as follows: In Section 2. the concepts of signal and system theory are reviewed shortly. The identification methods are given in Section 3. where special emphasis is given to the analysis of bias and convergence. Section 4. deals with practical applications of the methods with digital computers.

2. Concepts of system and signal processing theory

The basic parametric model is the transfer function which is obtained from ordinary differential equation by means of the Laplace transformation

$$X(s) = \int_{o}^{\infty} x(t)e^{-st}dt \tag{1}$$

if all initial conditions are zero, i.e. if the system doesn't possess any initial energy. Transfer function is the quotient of the system output's Laplace transform over its input's Laplace transform

$$G(s) = \frac{Y(s)}{U(s)} \tag{2}$$

and can be expressed as a fraction of two polynomials of the complex variable

$$G(s) = \frac{b_0 s^n + b_1 s^{n-1} + \ldots + b_{n-1}s + b_n}{s^n + a_1 s^{n-1} + \ldots + a_{n-1}s + a_n}. \tag{3}$$

2.1. Nonparametric models

According to Eq. (2) the Laplace transform of the output ($Y(s)$) is the product of the transfer function ($G(s)$) and the Laplace transform of the input($U(s)$). As the product in the complex variable (s) domain corresponds to the convolution in the time (t) domain, the time domain counterpart of the Eq. (2) can be written in the following form

$$y(t) = \int_{0}^{\infty} g(\tau)u(t-\tau)d\tau \tag{4}$$

where $g(\tau)$ is the inverse Laplace transform of the transfer function $G(s)$. If a unit impulse function (Dirac impulse) $u(t) = \delta(t)$ is applied to the input of the system, its output becomes $g(t)$. The inverse Laplace transform of the transfer function thus equals to the **impulse response** (called also weighting function) and represents a nonparametric model of the system. The impulse response $g(t)$ can be according to Eq. (4) interpreted also as deconvolution of the system's output $y(t)$ over system's input $u(t)$. As the system is linear, its response to the input's function integral corresponds to the output's function integral and so the **step response** (step function being the integral of the unit impulse function) is the integral of the impulse response $h(t) = \int_0^t g(\tau)d\tau$. As signals are supposed to be zero for negative time, the lower limits in Eqns. (1) and (4) can be set to $-\infty$.

Another nonparametric model is the **frequency response.** Frequency response actually describes how the system converts particular frequencies. If a sinusoidal/cosinusoidal signal of the frequency ω_0 is applied to the input

$$u(t) = U_0 \sin(\omega_0 t + \varphi_u) \tag{5}$$

only the same frequency ω_0 will, due to the linearity superposition, appear on the output after the transient response has vanished

$$y(t) = Y_0 \sin(\omega_0 t + \varphi_y). \tag{6}$$

The output signal however will be in general of different amplitude and phase as the input signal. The ratio of the output and input signal's amplitudes $(\frac{Y_0}{U_0})$ is called the *amplitude response* while the difference of their phases $(\varphi_y - \varphi_u)$ is called the *phase response.*

The mathematical evaluation of the frequency response is obtained by introducing the expressions for the sinusoidal input signal Eq. (5) into Eq. (4) yielding

$$y(t) = \frac{U_0}{2\imath} \int_0^\infty g(\tau)\{e^{\imath[\omega_0(t-\tau)+\varphi_u]} - e^{-\imath[\omega_0(t-\tau)+\varphi_u]}\}d\tau \tag{7}$$

where $\sin(\omega_0 t) = \frac{e^{\imath\omega_0 t} - e^{-\imath\omega_0 t}}{2\imath}$ was used. Applying Eq. (1) on (7) the following equation is obtained

$$\begin{aligned} y(t) &= U_0 \frac{e^{\imath(\omega_0 t+\varphi_u)}G(\imath\omega_0) - e^{-\imath(\omega_0 t+\varphi_u)}G(-\imath\omega_0)}{2\imath} = \\ &= U_0|G(\imath\omega_0)| \frac{e^{\imath(\omega_0 t+\varphi_u)}e^{\imath argG(\imath\omega_0)} - e^{-\imath(\omega_0 t+\varphi_u)}e^{-\imath argG(\imath\omega_0)}}{2\imath} = \\ &= U_0|G(\imath\omega_0)| \sin[\omega_0 t + \varphi_u + argG(\imath\omega_0)] \end{aligned} \tag{8}$$

and by comparing this equation with Eq. (6) the following relations between input and output amplitudes and phases respectively are obtained

$$\begin{aligned} Y_0 &= |G(\imath\omega_0)|U_0 \\ \varphi_y &= \varphi_u + arg\ G(\imath\omega_0)\ . \end{aligned} \tag{9}$$

If several frequencies are applied, the amplitude and the phase response can be represented as a function of the frequency ω and the frequency response as the complex function which corresponds to $G(\imath\omega)$

$$Frequency\ response = G(\imath\omega) = \lim_{s \to \imath\omega} G(s). \tag{10}$$

As Fourier transform is the limit of the Laplace transform for the real part of the complex variable s converging towards zero,

$$X(\omega) = \mathcal{F}[x(t)] = \int_{-\infty}^{\infty} x(t)e^{-\imath\omega t}dt = \lim_{s \to \imath\omega} X(s) \tag{11}$$

the impulse response and the frequency response are related by the Fourier transformation

$$G(\imath\omega) = \mathcal{F}\{g(t)\}. \tag{12}$$

Note that the lower integral's limit in Eq. (11) was again replaced by $-\infty$ due to the fact that the signals are zero at negative time. On the other hand the frequency response can be according to Eq. (2) interpreted as the quotient of the process output and input Fourier transforms

$$G(\imath\omega) = \frac{Y(\omega)}{U(\omega)}. \tag{13}$$

2.2. Correlation functions and spectra

The measured data are often corrupted by noise and in this case the theory of stochastic signals must be applied. The autocorrelation and the crosscorrelation functions

$$\begin{aligned} \phi_{uu}(\tau) &= E\{u(t)u(t+\tau)\} = \lim_{T\to\infty} \frac{1}{T}\int_{-\frac{T}{2}}^{\frac{T}{2}} u(t)u(t+\tau)dt \\ \phi_{uy}(\tau) &= E\{u(t)y(t+\tau)\} = \lim_{T\to\infty} \frac{1}{T}\int_{-\frac{T}{2}}^{\frac{T}{2}} u(t)y(t+\tau)dt \end{aligned} \tag{14}$$

play an important role in mathematical treatment of stochastic signals. In Eq. (14) the mathematical expectation operator $E\{\}$ was replaced by time averaging due to the ergodic hypothesis.

If the correlation operator $\lim_{T\to\infty} \frac{1}{T} \int_{-\frac{T}{2}}^{\frac{T}{2}} u(t)y(t+\tau)dt$ is applied to Eq. (4) after interchanging the outer and the inner integral the following basic relations are obtained

$$\begin{aligned}\phi_{uy}(\tau) &= \int_0^\infty g(t)\phi_{uu}(\tau - t)dt \\ \phi_{yy}(\tau) &= \int_0^\infty g(t)\phi_{yu}(\tau - t)dt.\end{aligned} \tag{15}$$

According to the first of the upper equations the crosscorrelation function is the convolution of the impulse response and the autocorrelation of the input signal, so the impulse response $g(t)$ can be interpreted also as deconvolution of the crosscorrelation function $\phi_{uy}(\tau)$ over the autocorrelation function $\phi_{uu}(\tau)$.

The Fourier transforms of the correlation functions are the *power spectral density functions*

$$\begin{aligned}\Phi_{uy}(\imath\omega) &= \mathcal{F}[\phi_{uy}(\tau)] \\ \Phi_{uu}(\imath\omega) &= \mathcal{F}[\phi_{uu}(\tau)].\end{aligned} \tag{16}$$

By introducing the Fourier transform definition, Eq. (11), the crosscorrelation function definition Eq. (14) and the convolution equation Eq. (4) the following relations joining the frequency response and power spectral density functions are obtained

$$\begin{aligned}\Phi_{uy}(\omega) &= G(\imath\omega)\Phi_{uu}(\omega) \\ \Phi_{yy}(\omega) &= G(\imath\omega)\Phi_{yu}(\omega) \\ \Phi_{yy}(\omega) &= |G(\imath\omega)|^2\Phi_{uu}(\omega).\end{aligned} \tag{17}$$

According to the first one of these three equations, the frequency response is the quotient of the of the input-output signal's crosspower spectral density function over the input signal's power spectral density function.

There is only one relation left to be shown, i.e. the relation between the Fourier transforms of the signals and their power spectral density functions:

$$\begin{aligned}\Phi_{uy}(\omega) &= \int_{-\infty}^{\infty} \phi_{uy}(\tau)e^{-\imath\omega\tau}d\tau = \int_{-\infty}^{\infty} \lim_{T\to\infty}\frac{1}{T}\int_{-\frac{T}{2}}^{\frac{T}{2}} u(t)y(t+\tau)dte^{-\imath\omega\tau}d\tau = \\ &= \lim_{T\to\infty}\frac{1}{T}\int_{-\frac{T}{2}}^{\frac{T}{2}} u(t)e^{\imath\omega t}\int_{-\infty}^{\infty} y(t+\tau)e^{-\imath\omega(t+\tau)}d\tau dt = \\ &= \lim_{T\to\infty}\frac{1}{T}\int_{-\frac{T}{2}}^{\frac{T}{2}} u(t)e^{\imath\omega t}Y(\omega)dt = \\ &= \lim_{T\to\infty}\frac{1}{T}U(-\omega)Y(\omega) \\ \Phi_{uu}(\omega) &= lim_{T\to\infty}\frac{1}{T}|U(\omega)|^2 .\end{aligned} \tag{18}$$

All relations among input-output data of the process and their nonparametric models are summarized in Fig. 1.

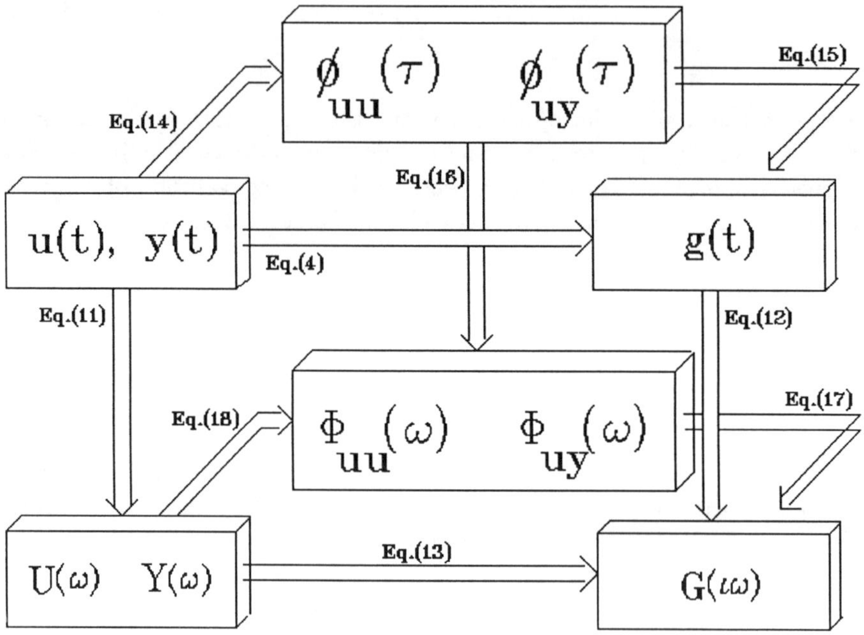

Figure 1: Relations among i/o data and nonparametric models

2.3. The Concept of disturbances

Another important item in identification is the concept of disturbances. Under this term all unmodelled system dynamics, all unmeasurable influences and measurements errors are considered. As linear systems are supposed, all disturbances except measurement errors of the input signal can be merged together into a single disturbance term n which is added to the undisturbed output of the system y_0

$$y(t) = y_0(t) + n(t) \ . \tag{19}$$

Although the disturbances are per definition unpredictable, four different kinds of disturbances are observed in practice, Isermann (1988):

1. high frequency quasistationary stochastic signals,
2. low frequency unstationary disturbance signals such as drifts, steps (a step disturbance is caused also by a biased estimate of the working point),
3. sinusoidal disturbances and
4. unknown disturbances such as spikes etc.

Only the first three items can be treated by the signal processing theory. In the quasistationary stochastic signals treatment the *white noise* plays an important role. The white noise $v(t)$ is a signal with statistically independent values. Its covariance function is a unit impulse function. Usually the mean value of the noise is supposed to be zero, so its autocorrelation function can be written in the following form

$$\phi_{vv}(\tau) = \Phi_0 \delta(\tau) \tag{20}$$

where Φ_0 is the white noise power spectral density, which is per definition constant

$$\Phi_{vv}(\omega) = \Phi_0. \tag{21}$$

Due to this feature white noise is also a preferable signal for system excitation, but due to

$$\lim_{T\to\infty} \frac{1}{T} \int_{-\frac{T}{2}}^{\frac{T}{2}} v^2(t)dt = \phi_{vv}(0) = \infty \tag{22}$$

its mean power is infinite and so it is unrealisable. Instead the *broadband noise* ($n(t)$) is used in practice and for simplified mathematical treatment the broadband noise is supposed to be produced by filtering of the white noise on a broadband low pass filter G_n referred usually as the noise filter and shown in Fig. 2. Broadband

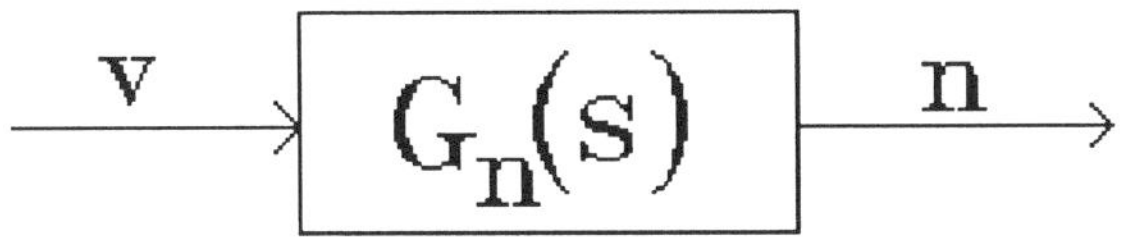

Figure 2: Noise filter

noise is a special form of the *coloured noise* which is supposed to be produced by filtering the white noise with a general filter. It should be noted that the coloured noise generation in Fig. 2 is only a fiction for simplified mathematical treatment and not a way to generate it. The power spectral density of the broadband noise can be expressed using Eq. (16) as

$$\Phi_{nn}(\omega) = |G_n(\imath\omega)|^2 \Phi_0 \tag{23}$$

and with first order noise filter $G_n(\imath\omega) = \frac{1}{1+\imath\omega T_g}$ its autocorrelation function becomes

$$\phi_{nn}(\tau) = \frac{\Phi_0}{2T_g} e^{\frac{-|\tau|}{T_g}} . \tag{24}$$

2.4. Bias and convergence concepts

The limits of integrals in several formulae are infinity. In practice this limits are replaced by a finite value. The resulting quantity in this case is actually an estimate of the true quantity and it is denoted by a hat (^). Two questions arise in the presence of stochastic disturbances:

- Is the expected value equal to the true value? If this is the case, the estimation procedure is said to be unbiased. A consistent estimate is an estimate, which improves as observation interval increases, and tends to the true value, as observation interval tends to infinity.
- Does the variance of the estimated quantity tend toward zero as the observation interval tends to infinity? If this is the case and the estimate is consistent, the estimate is said to be consistent in the mean square.

Let us shortly review the consistency of the correlation function's estimate

$$\hat{\phi}_{uy}(\tau) = \frac{1}{T}\int_{-\frac{T}{2}}^{\frac{T}{2}} u(t)y(t+\tau)dt \tag{25}$$

which is obtained from correlation function definition (14) by neglecting the limitation procedure. We shall suppose that the output signal is disturbed according to Eq. (19) and that the input signal is corrupted by some measurement noise

$$u(t) = u_0(t) + n_u(t) . \tag{26}$$

It is supposed that both noises n and n_u are zero mean and uncorrelated with the signals u_0 and y_0. The expected value of Eq. (25) is obtained using the definition of the correlation function i.e. Eq. (14) and by interchanging the integrals and expectation operators as follows

$$\begin{aligned} E\{\hat{\phi}_{uy}(\tau)\} &= \frac{1}{T}\int_{-\frac{T}{2}}^{\frac{T}{2}} E\{[u_0(t)+n_u(t)][y_0(t+\tau)+n(t+\tau)]\}dt = \\ &= E\{u_0(t)y_0(t+\tau)\} + E\{n_u(t)y_0(t+\tau)\} + \\ &\quad +E\{u_0(t)n(t+\tau)\} + E\{n_u(t)n(t+\tau)\} = \\ &= \phi_{u_0y_0}(\tau) + \phi_{n_un}(\tau) . \end{aligned} \tag{27}$$

The crosscorrelation function is unbiased only if both noises n_u and n are uncorrelated. The bias of the autocorrelation function $\phi_{uu}(\tau)$ is $\phi_{n_u n_u}(\tau)$

The variance of the correlation estimate however can be expressed for large observation intervals T and unbiased estimates as follows, Isermann (1988)

$$\begin{aligned}\sigma^2_{\phi_{uy}}(\tau) &= \frac{1}{T}\int_{-T}^{T}[\phi_{u_0u_0}(\xi)\phi_{y_0y_0}(\xi) + \phi_{u_0y_0}(\tau+\xi)\phi_{u_0y_0}(\tau-\xi) + \\ &+\phi_{u_0u_0}(\xi)\phi_{nn}(\xi) + \phi_{y_0y_0}(\xi)\phi_{n_un_u}(\xi) + \phi_{n_un_u}(\xi)\phi_{nn}(\xi)]d\xi \; . \qquad (28)\end{aligned}$$

3. Nonparametric identification methods

Theoretically every path between measured data represented in the left upper front box and one of the nonparametric models on the front right side in Fig. 1 constitutes an identification method. However different methods are more or less sensitive to the disturbing noise, so only the most popular methods, i.e. the **Fourier analysis**, the **Spectral analysis**, the **Frequency response analysis** (direct and by means of correlation functions) and the **Correlation analysis** will be reviewed here.

A very important issue of the identification procedure is the shape (spectrum) of the *input signal* which excites the system. Mainly three kinds of input signals are used: *aperiodic*, *sinusoidal* and *noise* signals.

3.1. Fourier analysis

The most straightforward identification method is the use of Eqns. (11) and (13) i.e. to compute the Fourier transforms of output and input signals and to divide them. Aperiodic signals, i.e. pulses of various length (step=long pulse) are most commonly used as input signals with this method.

Convergence analysis

If the output signal $y(t)$ is corrupted by noise $n(t)$, according to Eq. (19), the resulting frequency response becomes

$$G(\imath\omega) = \frac{\mathcal{F}[y_0(t) + n(t)]}{F[u(t)]} = G_0(\imath\omega) + \Delta G_n(\imath\omega) \tag{29}$$

where

$$\Delta G_n(\imath\omega) = \frac{\mathcal{F}[n(t)]}{\mathcal{F}[u(t)]} = \frac{N(\omega)}{U(\omega)} \,. \tag{30}$$

If the noise $n(t)$ is not correlated with the input and output signals $y_0(t)$ and $u_0(t)$ respectively and if it is zero mean $E\{n(t)\} = 0$ the expected value of the frequency response error $\Delta G_n(\imath\omega)$ is

$$E\{\Delta G_n(\imath\omega)\} = E\{\frac{\mathcal{F}[n(t)]}{\mathcal{F}[u(t)]}\} = \frac{E\{\mathcal{F}[n(t)]\}}{E\{\mathcal{F}[u(t)]\}} = \frac{\mathcal{F}[E\{n(t)\}]}{\mathcal{F}[E\{u(t)\}]} = \frac{0}{U(\omega)} \,. \tag{31}$$

The resulting frequency response is unbiased for all excited frequencies $(U(\omega) \neq 0)$.

The variance of the frequency response error $\Delta G_n(\imath\omega)$ is according to Eqns. (17) and (18)

$$\begin{aligned} E\{|\Delta G_n(\imath\omega)|^2\} &= \frac{E\{\mathcal{F}[n(t)]\overline{\mathcal{F}[n(t)]}\}}{E\{\mathcal{F}[u(t)]\overline{\mathcal{F}[u(t)]}\}} = \frac{E\{\int_{-\infty}^{\infty} n(t)e^{-\imath\omega t}dt \int_{-\infty}^{\infty} n(\tau)e^{+\imath\omega\tau}d\tau\}}{E\{\int_{-\infty}^{\infty} u(t)e^{-\imath\omega t}dt \int_{-\infty}^{\infty} u(\tau)e^{+\imath\omega\tau}d\tau\}} = \\ &= \frac{\int_{-\infty}^{\infty}\int_{-\infty}^{\infty} E\{n(t)n(\tau)\}e^{-\imath\omega(t-\tau)}dtd\tau}{\int_{-\infty}^{\infty}\int_{-\infty}^{\infty} E\{u(t)u(\tau)\}e^{-\imath\omega(t-\tau)}dtd\tau} = \\ &= \frac{\int_{-\infty}^{\infty} \phi_{nn}(t-\tau)e^{-\imath\omega(t-\tau)}dtd\tau}{\int_{-\infty}^{\infty} \phi_{uu}(t-\tau)e^{-\imath\omega(t-\tau)}dtd\tau} = \\ &= \frac{\Phi_{nn}(\omega)}{\Phi_{uu}(\omega)} \approx \frac{\Phi_{nn}(\omega)T_A}{|U(\omega)|^2} \end{aligned} \tag{32}$$

where the overbar denotes the conjugate complex value and T_A is the observation time of the transient response, which is large enough to embody all dynamics of the signals. It should be observed that the variance of the frequency response error increases with the increasing power spectral density of the noise and with the increasing observation interval and decreases with the increasing Fourier transform of the input signal. The error variance increase with the increasing observation interval is due to the fact that the prolongation of the observation interval doesn't contribute any new information about the transient response (and consequently useful output spectrum) while the noise component corrupts the spectrum of the output signal. It should be also noted that the variance of the frequency response error for nonexcited frequencies ($|U(\imath\omega)| \to 0$) tends to infinity.

For white noise the variance of the frequency response becomes using Eqns. (32) and (21)

$$E\{|\Delta G_v(\imath\omega)|^2\} = \frac{\Phi_0 T_A}{|U(\omega)|^2}. \tag{33}$$

There is however also an other source of errors while using aperiodic signals, i.e. the error in estimating steady state value of signals. The signals $u(t)$ and $y(t)$ namely represent only the deviations of the absolute signals $U(t)$ and $Y(t)$ from their steady state values (called also operating or working point)

$$\begin{aligned} u(t) &= U(t) - U_{00} \\ y(t) &= Y(t) - Y_{00}. \end{aligned} \tag{34}$$

The steady state value (Y_{00}) of the output signal is estimated before and after the transient response (both values may differ if step inputs or integrating type systems are used). As an estimate of the steady state value the mean value of the absolute output signal ($\hat{Y}_{00} = E\{Y(t)\}$) is used and is obtained by averaging the output signal over a time interval T_B

$$\hat{Y}_{00} = \frac{1}{T_B}\int_{-T_B}^{0} Y(t)dt. \tag{35}$$

The expected value of the steady state estimation error

$$\Delta Y_{00} = \hat{Y}_{00} - Y_{00} \tag{36}$$

is

$$E\{\Delta Y_{00}\} = E\{\frac{1}{T_B}\int_{-T_B}^{0}[Y_{00} + y_0(t) + n(t)]dt - Y_{00}\} = \frac{1}{T_B}\int_{-T_B}^{0} E\{n(t)\} = 0 \tag{37}$$

and the estimate is not biased. The variance of the estimation error becomes using Eqns. (35), (34) and (36)

$$\begin{aligned} E\{\Delta Y_{00}^2\} &= E\{[\frac{1}{T_B}\int_{-T_B}^{0} n(t)dt]^2\} = \frac{1}{T_B^2}\int_{-T_B}^{0}\int_{-T_B}^{0} E\{n(t)n(\tau)\}dt d\tau = \\ &= \frac{1}{T_B^2}\int_{-T_B}^{0}\int_{-T_B}^{0} \phi_{nn}(t-\tau)dt d\tau \; . \end{aligned} \tag{38}$$

For white noise ($n(t) = v(t)$) this expression becomes

$$E\{\Delta Y_{00}^2\} = \frac{\Phi_0}{T_B}. \tag{39}$$

The frequency response error caused by the steady state estimation error can be evaluated according to Eq (30) as a step noise signal with the magnitude ΔY_{00} in the following form

$$\Delta G_{\Delta Y}(\imath\omega) = \frac{\Delta Y_{00}}{\imath\omega U(\omega)} \; . \tag{40}$$

The variance of the frequency response error for white noise is obtained using Eqns. (40) and (39) as follows

$$E\{|\Delta G_{\Delta Y}(\imath\omega)|^2\} = \frac{\Phi_0}{\omega^2 |U(\omega)|^2 T_B} \; . \tag{41}$$

If several (m) measurements (applications of the aperiodic input signal, recordings of input and output signals and evaluations of the frequency response) are performed, the power spectral density of the input signal becomes $m\frac{|U(\imath\omega)|^2}{T_A}$ and the variance of the frequency response error is reduced m times. The total variance of the frequency response error for white noise disturbances is thus

$$E\{|\Delta G(\imath\omega)|^2\} = \frac{\Phi_0}{m|U(\omega)|^2}(T_A + \frac{1}{\omega^2 T_B}) \; . \tag{42}$$

Example 1. An aperiodic pulse was applied to the second order process and the output was corrupted with (approximately) white noise. As our example was simulated digitally, the digital white noise with the standard deviation 10% of input's signal magnitude was used. As discussed in Section 4. such a noise corresponds to the broadband noise with approximately constant spectral power density. Fig. 3 delineates the time history of the input and output signal and the corresponding Fourier transforms are shown in Fig. 4. Figs. 5 and 6 depict the real and the

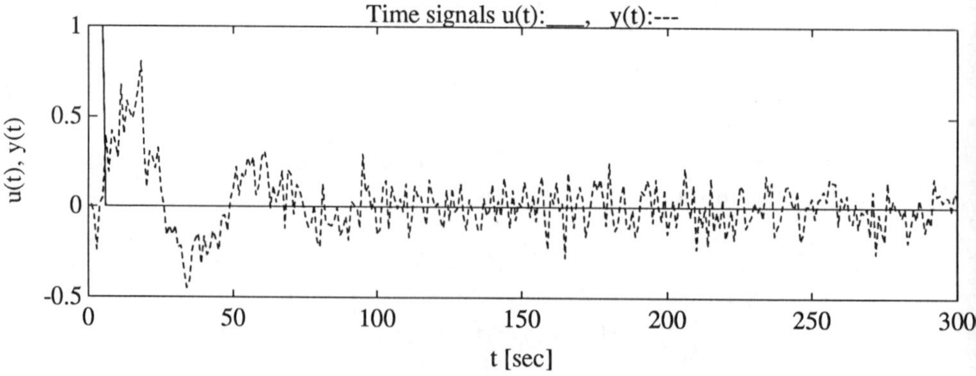

Figure 3: Time history of input and output signals

imaginary part of the estimated frequency response (the quotient of Fourier transforms in Fig. 4), while Fig. 7 shows the corresponding amplitude response. In

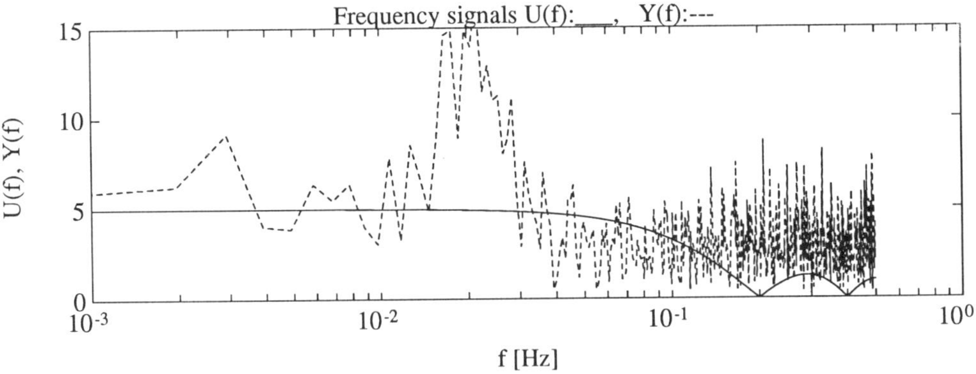

Figure 4: Fourier transforms of input and output signals

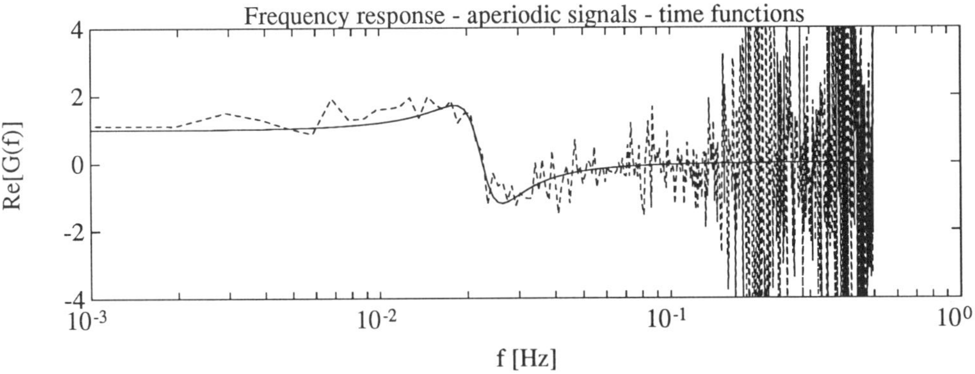

Figure 5: Frequency response - real component

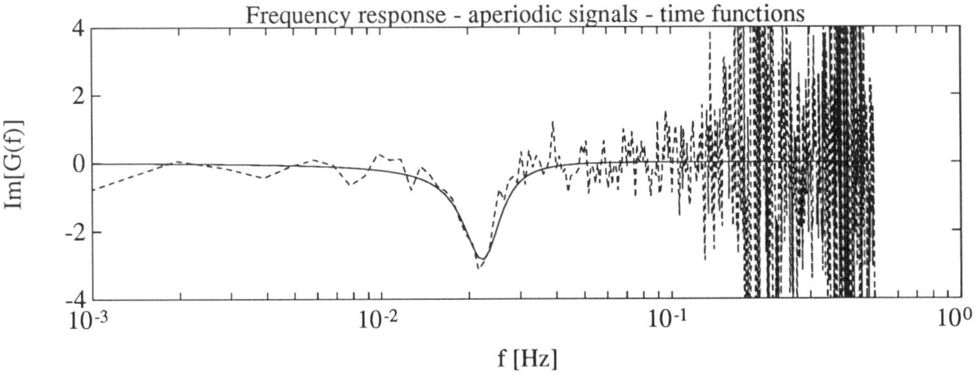

Figure 6: Frequency response - imaginary component

the last three Figures besides estimated responses (dotted line) also the actual responses (solid line) are shown. As seen from Figs. 5 and 6 the estimated frequency response is unbiased. However the amplitude response in Fig. 7 seems to be biased. The reason for this appearance is the fact that the noise component is predomi-

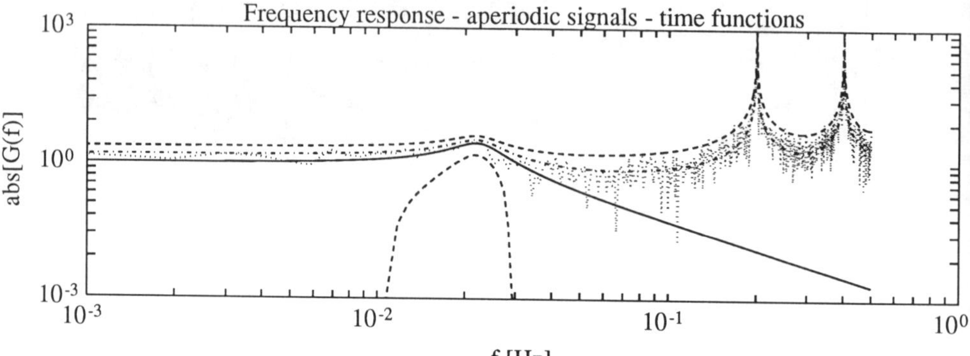

Figure 7: Amplitude response

nant to the actual amplitude response (which is especially at high frequencies very small), the mean absolute value of the zero mean noise however is not zero, but $\int_{-\infty}^{\infty} |v| p(v) dv$, where $p(v)$ is the noise distribution. For gaussian noise this value is 0.7979 of its standard deviation and in Fig. 7 the dash - dotted line denotes the amplitude response shifted for 0.7979 of the frequency response error standard deviation obtained by Eq. (33). In Fig. 7 also the band of possible amplitude responses (true frequency response $\pm$ twice the standard deviation according to Eq. (33) - surrounded by dashed lines) is shown.

□

It is observed that the frequency spectrum of the input (excitation) signal is extremely important for the variance of the estimated frequency response and that the error variance is even with as small noise as in Example 1 very high. So this method can be used only in noise free cases and with very rich excitation.

3.2. Spectral analysis

Another possibility to get the frequency response is to divide the input - output crosspower spectral density function with the input power spectral density function according to Eq. (17). Original ideas of this method, i.e. to measure the power spectral density functions directly by bandpass filtering and integrating the time signals, have lost their applicability as Fourier transform tools became available, Isermann (1988). Power spectral density functions could be obtained from Fourier transforms according to Eq. (18), but a better way to get them is to perform the Fourier transformation on the correlation functions (Eq. 16). Correlation (Eq. 14) is namely the method to eliminate noise. The path from the time signals to the frequency response on Fig. 1 goes in this case through Eqns. (14), (16) and (17). The variance of the frequency response error can be however improved by smoothing the power spectral density functions. Instead of filtering the time signals, as proposed originally, the filtering in the frequency domain is performed. A bandpass frequency window $W(\omega)$ is put over the original power density functions and the evaluated total power is assigned to the power spectral density of the frequency window's centre ω_0. Mathematically this is expressed in the following form

$$\begin{aligned} \Phi_{uy}^{W}(\omega_0) &= \int_{-\infty}^{\infty} W(\omega-\omega_0)\Phi_{uy}(\omega)d\omega \\ \Phi_{uu}^{W}(\omega_0) &= \int_{-\infty}^{\infty} W(\omega-\omega_0)\Phi_{uu}(\omega)d\omega \end{aligned} \tag{43}$$

what represents the convolution due to symmetric windows ($W(\omega) = W(-\omega)$).

If the window is moved over entire frequency range, smoothed power spectral densities are obtained. Windows of different shapes can be used and the reader may find the corresponding review in Ljung (1987) or Söderström and Stoica (1989).

Convergence Analysis

The frequency response obtained from original nonsmoothed power density functions is unbiased, while the smoothing introduces a bias because the power spectral densities and of course the frequency response are not constant.

The variance of the unsmoothed frequency response can be obtained using Eqns. (28) and (32). The smoothing procedure decreases the error variance.

Example 2. The data of Ex. 1 are used for spectral analysis. Fig. 8 shows the autocorrelation function of the input signal and the crosscorelation function of the input and output signal, Fig. 9 the corresponding power spectral density functions, Fig 10 the nonsmoothed amplitude response and Fig. 11 the smoothed amplitude response (dotted lines). In Figs. 10 and 11 also the actual amplitude response (solid line) and the band of possible amplitude responses (surrounded by dashed lines) are shown.

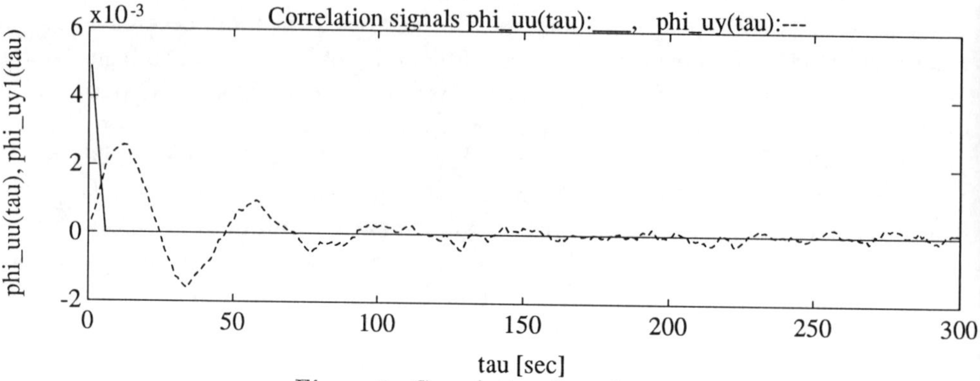

Figure 8: Correlation functions

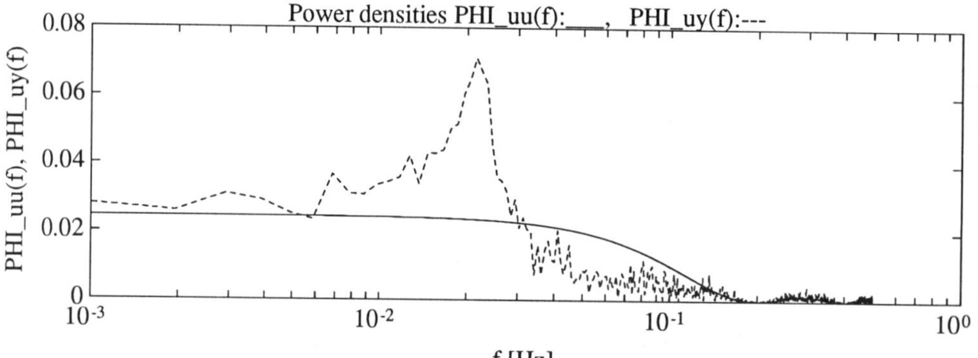

Figure 9: Power spectral density functions

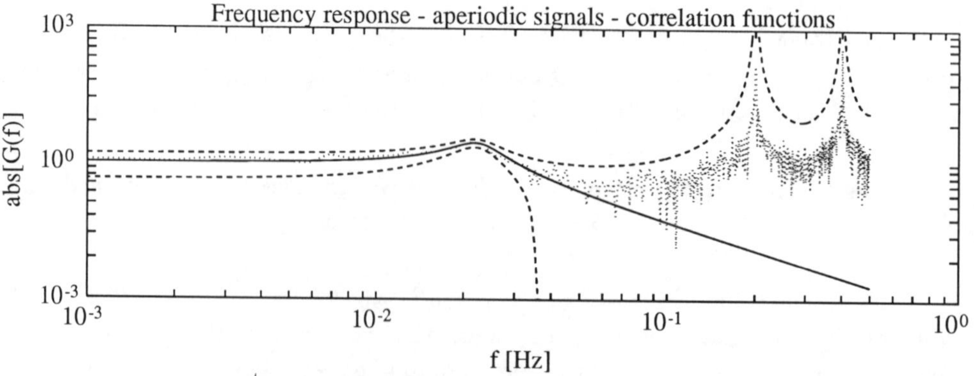

Figure 10: Nonsmoothed amplitude responses

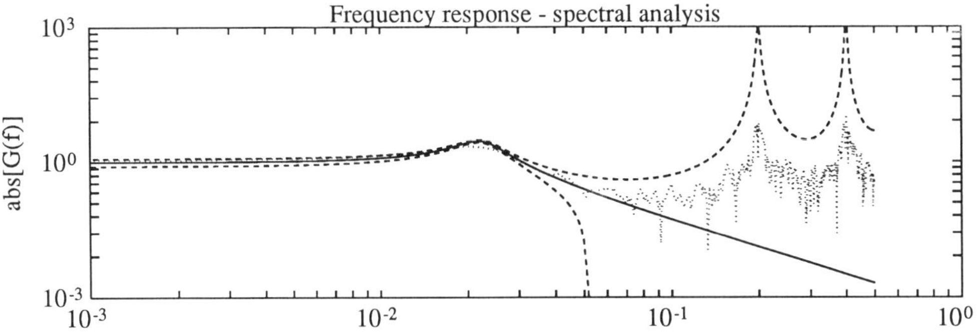

Figure 11: Smoothed amplitude responses

□

3.3. Frequency response analysis

As mentioned in subsection 2.1. the frequency response consists of the amplitude response (which is the ratio of the output and input sinusoidal/cosinusoidal signal's amplitudes) and the phase response (which is the difference of both signal's phases). Frequency response can be measured by its definition, i.e. a sinusoidal/cosinusoidal signal of a particular frequency ω_0 is applied to the input of the system and after the transient response has vanished, the ratio of input and output signal's amplitudes and the difference of their phases is evaluated either by direct measurements in the time domain or by compensating principle or by sampling the output signal at appropriate instants. If the frequency is swept over all interesting frequencies, the frequency response is measured (analysed) per its definition.

In Fig. 1 this identification method corresponds to the path through Eqns. (11) and (13) using sinusoidal input signals. The Fourier analysis (Eq. 11) is not performed since a single frequency signal is applied.

As signals are corrupted by noise, correlation methods are preferred. The orthogonal correlation method, Isermann (1988), represents in Fig. 1 the path through Eqns. (14), (16) and (17) using sinusoidal/cosinusoidal signals. Without loss of generality the sinusoidal signal

$$u(t) = U_0 \sin(\omega_0 t) \tag{44}$$

is applied to the input of the system and in this case its output signal in the steady state will be

$$y(t) = U_0|G(\imath\omega_0)|\sin[\omega_0 t + \varphi(\imath\omega_0)]\ . \tag{45}$$

The corresponding auto and crosscorrelation functions are

$$\phi_{uu}(\tau) = \frac{U_0^2}{2}\cos(\omega_0\tau) \tag{46}$$

$$\phi_{uy}(\tau) = \frac{U_0^2}{2}|G(\imath\omega_0)|\cos[\omega_0\tau + \varphi(\imath\omega_0)]\ . \tag{47}$$

The auto and crosscorrelation functions in the above equations were evaluated over n periods using the following relation

$$\begin{aligned}\int_0^{nT}\sin(\omega_0 t)\sin[\omega_0(t+\tau)+\varphi]dt &= \frac{1}{2}\int_0^{nT}\{\cos(-\omega_0\tau-\varphi)- \\ &\quad -\cos[2\omega_0(t+\tau)+\varphi]\}dt = \\ &= \frac{1}{2}\cos(\omega_0\tau+\varphi)\ .\end{aligned} \tag{48}$$

The crosscorrelation function (47) is evaluated at two points, i.e. for $\tau = 0$ and $\tau = -\frac{T}{4} = -\frac{\pi}{2\omega_0}$ yielding

$$\phi_{uy}(0) = \frac{U_0^2}{2}|G(\imath\omega_0)|\cos[\varphi(\omega_0)] = \frac{U_0^2}{2}\Re[G(\imath\omega_0)] \tag{49}$$

and

$$\phi_{uy}(\frac{-\pi}{2\omega_0}) = \frac{U_0^2}{2}|G(\imath\omega_0)|\cos[\frac{-\pi}{2}+\varphi(\omega_0)] = \frac{U_0^2}{2}|G(\imath\omega_0)|\sin[\varphi(\omega_0)] = \frac{U_0^2}{2}\Im[G(\imath\omega_0)]. \tag{50}$$

The real and imaginary part of the frequency response are obtained using Eqns. (49), (50), (44) and the definition of the crosscorrelation function (Eq. 14) as follows

$$\Re[G(\imath\omega_0)] = \frac{2}{U_0^2 nT_p}\int_0^{nTp} U_0\sin(\omega_0 t)y(t)dt = \frac{2}{U_0 nT_p}\int_0^{nTp} y(t)\sin(\omega_0 t)dt \tag{51}$$

and

$$\Im[G(\imath\omega_0)] = \frac{2}{U_0^2 nT_p}\int_0^{nT_p} U_0\sin(\omega_0 t)y(t-\frac{\pi}{2\omega_0})dt = \frac{2}{U_0 nT_p}\int_{-\frac{\pi}{2\omega_0}}^{nT_p-\frac{\pi}{2\omega_0}} y(t)\cos(\omega_0 t)\,dt. \tag{52}$$

Due to the periodicity of all functions in Eq. (52) the integral limits can be replaced by 0 and nT_p respectively.

Convergence analysis

If the output signal $y_0(t)$ is corrupted by a disturbance signal $n(t)$, see Eq. (19)the frequency response error becomes according to Eqns. (51) and (52)

$$\Delta\Re[G(\imath\omega_0)] = \frac{2}{U_0 n T_p}\int_0^{nT_p} n(t)\sin(\omega_0 t)dt \tag{53}$$

and

$$\Delta\Im[G(\imath\omega_0)] = \frac{2}{U_0 n T_p}\int_0^{nT_p} n(t)\cos(\omega_0 t)dt \ . \tag{54}$$

If the noise $n(t)$ is zero mean and not correlated with the input signal, the frequency response is not biased.

The variance of the frequency response error is

$$E\{|\Delta G(\imath\omega_0|^2\} = E\{\Delta\Re^2[G(\imath\omega_0)] + \Delta\Im^2[G(\imath\omega_0)]\} \tag{55}$$

and becomes for different kinds of disturbances as follows:

1. High frequency quasistationary noise, Eykhoff (1974)

$$\sigma^2_{G_n} = \frac{4}{U_0^2 n T_p}\int_{-nT_p}^{nT_p} \phi_{nn}(\tau)[1 - \frac{|\tau|}{nT_p}]e^{-\imath\omega_0\tau}d\tau. \tag{56}$$

For white noise disturbance $(n(t) = v(t))$ with power spectral density Φ_0 this equation becomes

$$\sigma^2_{G_v} = \frac{4\Phi_0}{U_0^2 n T_p} \ . \tag{57}$$

2. Sinusoidal/cosinusoidal disturbance $n(t) = N_0\cos(\omega t)$, Isermann (1988)

$$\sigma^2_{G_s} = \frac{N_0\sqrt{2}}{U_0} \ \frac{(\frac{\omega}{\omega_0})\sqrt{1+(\frac{\omega}{\omega_0})^2}}{|1-(\frac{\omega}{\omega_0})^2|} \ \frac{|\sin[\pi(\frac{n\omega}{\omega_0})]|}{\pi(\frac{n\omega}{\omega_0})} \ . \tag{58}$$

3. Drift signals $n(t) = d(t)$, Isermann (1988)

$$\sigma^2_{G_d} = \frac{4}{U_0^2 n^2 T_p^2}|D(\omega)|^2 \tag{59}$$

where $D(\omega)$ is the Fourier transform of $d(t)$. For a ramp signal $d(t) = D_0 t$ the frequency response error variance becomes

$$\sigma^2_{G_d} = \frac{4D_0^2}{U_0^2\omega_0^2} \ . \tag{60}$$

The frequency response error variance tends to zero as observation interval nT_p increases toward infinity for high frequency quasistationary noise and sinusoidal disturbances (except sinusoidal disturbances with the same frequency as measuring frequency). For drift disturbances however it does not. The only way to eliminate the influence of drift signals is to filter the measured signals by the corresponding filters.

Example 3. A sinusoidal input signal with the amplitude 1 was applied to the same system as in Examples 1 and 2. The output was corrupted by a broadband noise with the standard deviation 0.1. Fig. 12 shows the input and the output signals for four different frequencies. The frequency response was evaluated with the orthogonal correlation method at five frequencies and the obtained result is shown in Fig. 13 (the amplitude response only - denoted by asterisks). On the same figure also the true amplitude response (denoted by solid line) and the band of possible amplitude responses (surrounded by dashed lines) are shown.

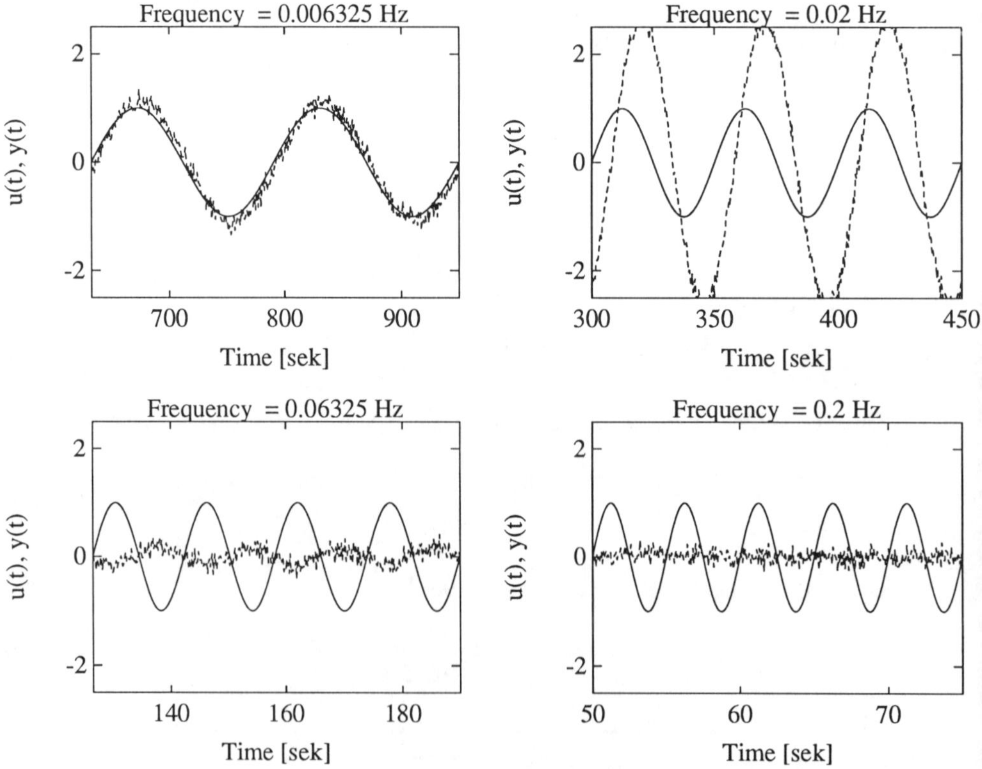

Figure 12: Input-output signals

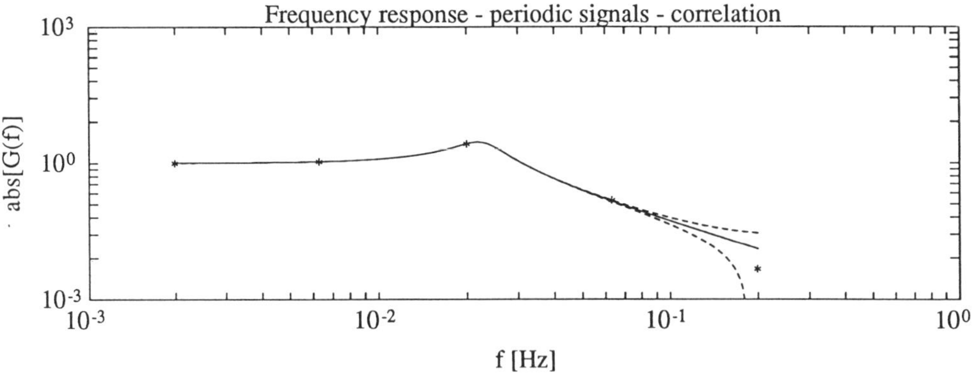

Figure 13: Amplitude response

□

3.4. Correlation analysis

There are only two paths from time signals to nonparametric models in Fig. 1 which we haven't discussed yet. The first one is the direct deconvolution (Eq. 4) which is not used in practice because of very high sensitivity to noise. The second one is the correlation analysis, which is based on the evaluation of the correlation functions (autocorrelation function of the input signal and crosscorrelation function of the input and the output signal) according to Eq. (14) and the deconvolution of them yielding the impulse response of the system (Eq. 15). The frequency response can be then obtained by the Fourier analysis of the impulse response (Eq. 12).

The deconvolution of continuous time signals is a complex mathematical problem, which can be in practice solved only by approximations, Isermann (1988). Another possibility is the use of discrete time signals theory, discussed in Section 4. However, if the input signal is white noise with the power spectral density Φ_0 (broadband noise in practice), no deconvolution is needed since the output signal in this case is just the impulse response

$$g(\tau) = \frac{1}{\Phi_0}\phi_{uy}(\tau) \ . \tag{61}$$

Convergence Analysis

Concerning the convergence issues all results of the convergence of the correlation functions are appliable. The impulse response estimate is unbiased (even for finite observation intervals), if measurement noise n_u is uncorrelated with output disturbance n. The variance of the impulse response error becomes for systems which have no "jumping" ability ($g(0) = 0$ and consequently $\phi_{u_0y_0}(\tau+\xi)\phi_{u_0y_0}(\tau-\xi) = 0$) and without input measurements noise ($n_u = 0$) due to Eqns. (28) and (61) as follows

$$\begin{aligned}\sigma_g^2(\tau) &= E\{\Delta g^2(\tau)\} = \frac{1}{\Phi_0^2 T}\int_{-T}^{T}[\Phi_0\delta(\xi)\phi_{y_0y_0}(\tau) + \Phi_0\delta(\xi)\phi_{nn}(\xi)]d\xi = \\ &= \frac{1}{\Phi_0 T}[\phi_{y_0y_0}(0) + \phi_{nn}(0)] \ .\end{aligned} \tag{62}$$

The autocorrelation function of the output $\phi_{y_0y_0}(\tau)$ can be for white noise input expressed using Eqns. (15) and (61) as follows:

$$\phi_{y_0y_0}(\tau) = \int_0^\infty g(t)\phi_{yu}(\tau - t)dt = \int_0^\infty g(t)\Phi_0 g(t-\tau)dt \ . \tag{63}$$

So finally the variance of impulse response error (Eq. 62) can be expressed as

$$\sigma_g^2(\tau) = \frac{1}{T}[\int_0^\infty g^2(t)dt + \frac{\sigma_n^2}{\Phi_0}] \tag{64}$$

where $\sigma_n^2 = \phi_{nn}(0)$ is the variance of the output corrupting noise.

Example 4. The system of Examples 1, 2 and 3 is excited by approximately white noise (discrete random binary signal) and the output was corrupted in the same way as in Examples 1, 2 and 3. Figs. 14 and 15 show the input and output signals of the system respectively. The impulse response was evaluated as the cross-

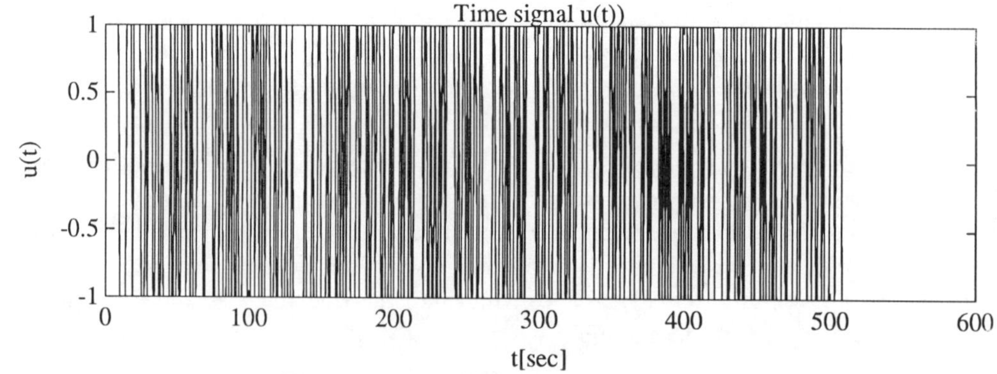

Figure 14: Input signal of the system

correlation function of the output and input signal and is depicted in Fig. 16. Fig. 17 shows the Fourier transform of the impulse response, i.e. the frequency response (the amplitude response only). In Figs. 16 and 17 the actual impulse and amplitude responses respectively are depicted by the solid line as well. In Fig. 16 also the band of possible impulse responses (surrounded by dashed lines) is shown.

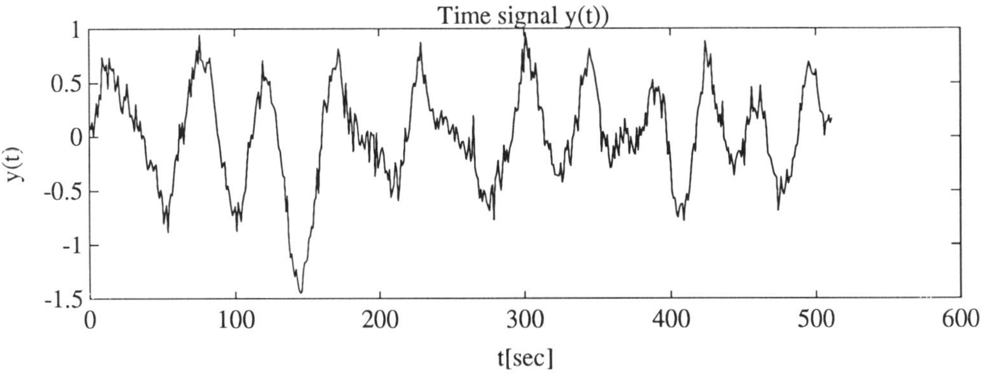

Figure 15: Output signal of the system

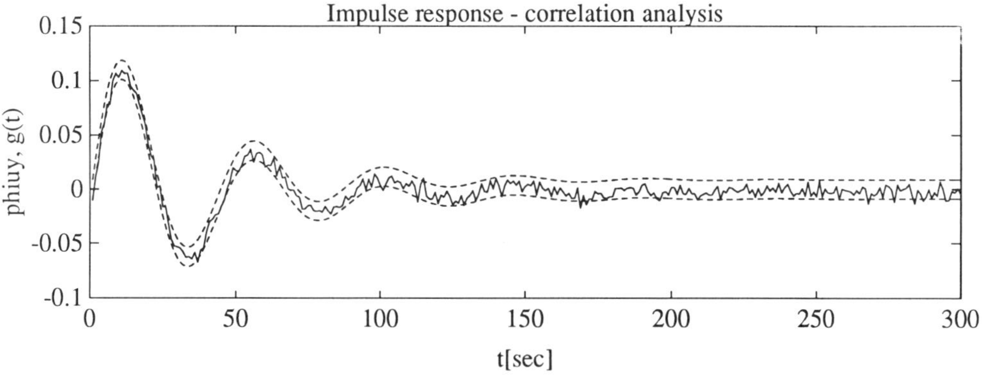

Figure 16: Impulse response

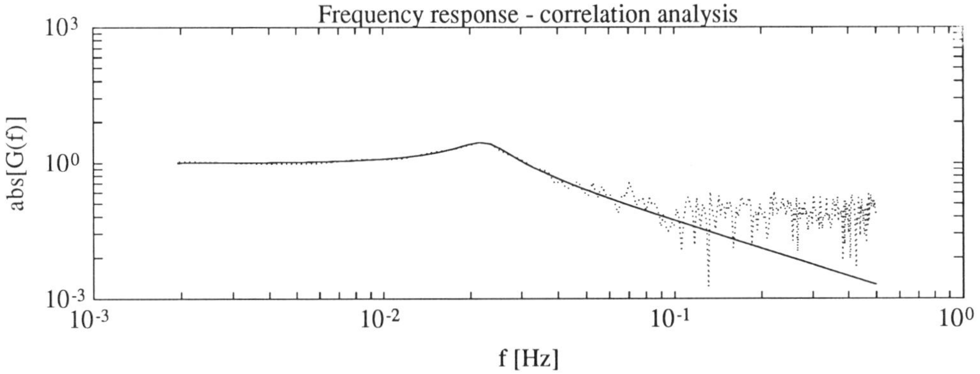

Figure 17: Amplitude response

□

4. Realization of the methods with digital computers

The methods for the estimation of continuous time nonparametric models were discussed in this contribution. Nowadays only digital computers are used in practice and there are two possibilities to cope continuous time systems with them:

- **Discrete time systems:**
  This possibility could be explored by drawing a parallel to Sections 2. and 3. of this contribution for discrete time systems where the z transformation would replace the Laplace transformation, the discrete transfer function $H(z)$ the continuous transfer function $G(s)$, the discrete system's frequency response $H(e^{\imath\omega T_s})$, T_s being sampling interval, the continuous frequency response $G(\imath\omega)$, the discrete impulse response $h(k)$ the continuous impulse response $g(t)$, the sum Σ the integral $\int$ etc.

  One of discrete model nonparametric identification method's advantages is easy deconvolution, which can be solved by standard least squares technique.

  The scope of this chapter doesn't allow us to go into details of discrete nonparametric model identification methods. The reader can find a detailed description of these methods in e.g. Ljung (1987) or Söderström and Stoica (1989).

- **Sampled signals:**
  It is well known that the spectra of sampled signals are periodic and that the spectra of periodic signals are sampled (discrete). In practice the time signals are observed in a finite period of time and they are of course sampled due to the processing by digital computers. Even if these signals are not periodic, they can be supposed to be periodic with the period of at least the observation interval. As such (sampled and periodic) they can be treated by the Discrete Fourier transformation (or its faster version for the number of samples equal to a power of 2 - the Fast Fourier transformation). The resulting frequency spectrum is of course periodic and sampled (discrete). If the original signals are continuous and nonperiodic, the first (base) period of their spectra can be used as the sampled pattern of the true spectrum divided by the amount of the sampling time T_s. If the Shannon sampling theorem is fulfilled, i.e. if the highest frequencies of all signals are less than the half of the sampling frequency, no errors occur. The time period, i.e. the length of the observation interval is the inverse of the spectrum's frequency resolution and the inverse of the sampling time (time signal's resolution) is the wideness of the frequency spectrum (where of course only the first half of points are independant since the spectrum is symmetric for its real and inverse symmetric for its imaginary

part respectively).

If the excitation signals are generated by digital computers (digital to analog converters), the fulfilment of the Shannon's sampling theorem may due to the sharp edges cause some problems, which can be in practice avoided by fast sampling and appropriate filtering.

In practice the continuous time broadband noise is most frequently generated by digital computers. The discrete white noise can namely be generated on digital computers and then converted to the broadband continuous noise on digital to analog converters. In the theory of discrete systems digital to analog converters represent the zero order hold devices, which hold the value of the input sampled (digital) signal up to the next sample instant, where the whole procedure repeats. The frequency response of the zero order hold device is obtained by the inverse Fourier transform of its impulse response and is as follows

$$G_{ZOH}(\imath\omega) = T_s \frac{\sin \frac{\omega T_s}{2}}{\frac{\omega T_s}{2}} e^{-\frac{\omega T_s}{2}} \tag{65}$$

There are two possibilities for the generation of white noise on digital computers:

- White noise with uniform or Gaussian distribution can be generated by corresponding random number generators. Usually evenly distributed random numbers are generated by the multiplicative congruential formula and then converted to a set of Gaussian distributed random numbers, Borrie (1986).
- White noise with binary distribution can be generated by the use of shift registers. A function (usually the equivalence or modulo 2 sum) of two carefully selected stages is fed back to the input of the shift register. The output of any stage jumps randomly from 0 to 1 and viceversa. The resulting signal behaves like a noise and is called Pseudo Random Binary Signal (PRBS). It is periodic with the period of $2^N - 1$ where N is the number of the stages in the shift register. The term "Pseudo Random" is due to the fact that the noise is not random since its sequence can be repeated again and again by setting the same initial conditions of the shift register. A review of the shift registers can be found in Davies (1970).

The power spectral density function of the discrete white noise is constant and is reshaped on the zero order hold device according to the square of its amplitude response. The shape of continuous broadband noise power spectral density function on the output of the zero order hold device has the form $[\frac{\sin\frac{\omega T_s}{2}}{\frac{\omega T_s}{2}}]^2$ and the sampling time of the zero order hold device must be chosen in such a way that the power spectral density function remains approximately constant in the frequecy range we are interested in.

References

Åström, K.J. (1970), *"Introduction to stochastic control theory"*, Academic Press, New York.

Åström, K.J. and P. Eykhoff (1971), "System identification - a survey", *Automatica*, vol.7, pp. 123-162.

Bendat, J.S. and A.G. Piersol (1971), *"Random data: analysis and measurement procedures"*, Wiley-Interscience, New York.

Borrie, J.A. (1986), *"Modern control systems"*, Prentice Hall, Englewoods Cliffs N.J.

Davies, W.D.T. (1970), *"System identification for self-adaptive control"*, Wiley-Interscience, London.

Eykhoff, P. (1974), *"System Identification, Parameter and State Estimation"*, John Wiley, London.

Isermann, R. (1971), *"Experimentelle Analyse der Dynamik von Regelsystemen"*, Bibliographisches Institut, Nr. 515/515a, Mannheim.

Isermann, R. (1988), *"Identifikation dynamischer Systeme I, II"*, Springer, Berlin.

Ljung, L. (1987), *"System identification: Theory for the user"*, Prentice Hall, Englewoods Cliffs N.J.

Norton, J.P. (1986), *"An introduction to identification"*, Academic Press, London.

Sage, A.P., and J.L. Melsa (1971), *"System identification"*, Academic Press, New York.

Sinha, N. K. and B. Kuszta (1983), *"Modeling and Identification of Dynamic Systems"*, Van Nostrand Reinhold Company, New York.

Söderström, T. and P. Stoica (1989), *"System identification"*, Prentice Hall, New York.

From fine asymptotics to model selection

L. Gerencsér*
Department of Electrical Engineering
McGill University
Montréal, Québec, Canada H3A 2A7

and

Zs. Vágó
Department of Mathematics
Faculty of Mechanical Engineering
Technical University of Budapest,
Budapest, Hungary H-1521.

Abstract

A short survey of our recent results in the theory of identification of continuous time linear stochastic systems will be presented. The purpose of these investigations is to obtain fine asymptotic results for the parameter estimator process. A culmination of these results is a theorem on the almost sure asymptotics of a continuous-time stochastic complexity.

1. Introduction and first results.

Let us consider a linear time-invariant stochastic system given by the state space equations

$$dx_t(\theta^*) = A(\theta^*)x_t(\theta^*)dt + K(\theta^*)dw_t \tag{1.1}$$

$$dy_t(\theta^*) = C(\theta^*)x_t(\theta^*)dt + dw_t \tag{1.2}$$

where dw_t is a standard Gaussian white noise process in $\mathbf{R}^m$. The process $(y_t(\theta^*))$ is called the output process, while $(x_t(\theta^*))$ is the state-vector. We need the following conditions:

Condition 1.1. The matrices $A(\theta^*), K(\theta^*), C(\theta^*)$ are defined for $\theta^* \in D \subset \mathbb{R}^p$ where D is an open domain and they are C^∞ function of θ^*. Moreover $A(\theta^*)$ and $A(\theta^*) - K(\theta^*)C(\theta^*)$ are stable for $\theta^* \in D$. The last part of the condition means that $A(\theta^*)$ and $A(\theta^*) - K(\theta^*)C(\theta^*)$ have all their eigenvalues in the left half-plane. It follows that dy_t is stationary process the innovation process of which dw_t.

---

* On leave from the Computer and Automation Institute of the Hungarian Academy of Sciences, Budapest.

N. K. Sinha and G. P. Rao (eds.), Identification of Continuous-Time Systems, 575–587.

To estimate θ^* on the basis of the observation process y_t we proceed as follows: fix a $\theta \in D$ and invert the system (1.1), (1.2) assuming that $\theta = \theta^*$ to get dw_t. The inverse system is given by the equations:

$$dx_t(\theta^*) = (A(\theta^*) - K(\theta^*)C(\theta^*))x_t(\theta^*)dt + K(\theta^*)dy_t(\theta^*) \tag{1.5}$$

$$dw_t = dy_t(\theta^*) - C(\theta^*)dt \tag{1.6}$$

Now if θ is chosen arbitrarily we still can use the equation above to generate a process $d\epsilon_t$ which is an estimation of dw_t. The governing equations are

$$dx_t(\theta, \theta^*) = (A(\theta) - K(\theta)C(\theta))x_t(\theta, \theta^*)dt + K(\theta)dy_t(\theta^*). \tag{1.7}$$

$$d\varepsilon_t(\theta, \theta^*) = dy_t(\theta^*) - C(\theta)x_t(\theta, \theta^*)dt. \tag{1.8}$$

In practice we set zero initial state, i.e. $x_0(\theta, \theta^*) = 0$. However for analysis purposes we assume that $(x_t(\theta, \theta^*))$ is the stationary solution of (1.7) (1.8). Since the difference between the zero-initial state and the stationary solution is known to be $O_M(e^{-\lambda t})$ with some $\lambda > 0$ it is easy to check that all the results we derive using the latter process will hold if we use the former one.

Here we used the following convention: if ξ_n is a sequence of random variables and c_n is a sequence of positive numbers then we write $\xi_n = O_M(c_n)$ if $\xi_N / c_n = O_M(1)$, which in turn means that the sequence ξ_n / c_n is M-bounded, as defined in the Appendix.

It is well known that the negative log-likelihood function can be written as

$$V_T(\theta, \theta^*) = \frac{1}{2} \int_0^T |C(\theta)x_t(\theta, \theta^*)|^2 dt - \int_0^T x_t^T(\theta, \theta^*)C^T(\theta)dy_t \tag{1.10}$$

(c.f e.g. Arató 1984), and from this we get

$$\frac{\partial}{\partial \theta} V_T(\theta, \theta^*) = \int_0^T \frac{d}{dt} \frac{\partial}{\partial \theta} \varepsilon_t(\theta, \theta^*) . d\varepsilon_t(\theta, \theta^*) \tag{1.11}$$

(c.f. e.g. Gerencsér, Gyöngy, Michaletzky 1984). Here $\frac{\partial}{\partial \theta}$ means differentiation both in the M-sense and almost surely (c.f. the Appendix). The maximum likelihood method presented here is analogous to discrete time methods (c.f. Caines 1988.)

It is important to note that (1.10) can be written in a form which resembles the discrete time cost-function to a greater degree. For this purpose we note that if $dz_t = f_t dt + \sigma_t dw_t$ is a scalar-valued stochastic differential then its square is a two-form of the infinitesimal generators dt and dw_t defined as

$$(dz_t)^2 = f_t^2 dt^2 + 2 f_t \sigma_t dt dw_t + \sigma_t^2 (dw_t)^2.$$

The only rule we impose for the generators is $(dw_t)^2 = dt$. Similarly we can define the product of stochastic differentials. In the multivariable case we proceed as in the scalar case but we impose an additional rule on the components dw_{it} of dw_t which says: $dw_{it}dw_{jt} = \delta_{ij}dt$. We shall also use the obvious notation $|dz_t|^2 = (dz_t)^T dz_t$.

With this notation the cost function $V_T(\theta,\theta^*)$ can be written as

$$V_T(\theta,\theta^*) = \int_1^T \frac{1}{2}\frac{1}{dt}(|d\varepsilon_t(\theta,\theta^*)|^2 - |dy_t|^2). \tag{1.12}$$

If we work out the expression following the integral sign it is easily seen that it is actually a one-form of the infinitesimal generators dt and dw_t. The equivalence of (1.10) and (1.12) can be obtained directly by simple verification. From (1.12) the identity (1.11) is easily obtained by differentiation with respect to θ. For this purpose we define the derivative of a parameter-dependent one-form $dz_t(\theta) = f_t(\theta)dt + \sigma_t(\theta)dw_t$ as $dz_{\theta t}(\theta) = f_{\theta t}(\theta)dt + \sigma_{\theta t}(\theta)dw_t$. The derivative of two-forms or higher-order derivatives are defined analogously. Using this formalism and the established fact that differentiation and integration can be interchanged when differentiating (1.10) we get (1.11) without difficulty.

Let $D^o \subset D$ be a compact domain such that $\theta^* \in \text{int}D^o$. Then the maximum-likelihood estimator $\widehat{\theta}_T$ of θ^* is defined as the solution of the equation

$$\frac{\partial}{\partial\theta}V_T(\theta,\theta^*) = 0 \tag{1.13}$$

if the solution is unique in D^0. Otherwise we define $\widehat{\theta}_T$ arbitrarily subject to the condition that $\widehat{\theta}_T \in D^o$ a.s. and $\widehat{\theta}_T$ must be a random variable, i.e. it must be $\mathcal{F}$-measurable.

Let us define the asymptotic cost function $W(\theta,\theta^*)$ by

$$W(\theta,\theta^*) = \lim_{T\to\infty}\frac{1}{T}EV_T(\theta,\theta^*). \tag{1.14}$$

It is known and easily proved that

$$\frac{\partial}{\partial\theta}W(\theta,\theta^*)|_{\theta=\theta^*} = 0 \quad \text{and} \quad \frac{\partial^2}{\partial\theta^2}W(\theta,\theta^*) \geq 0. \tag{1.15}$$

where the latter inequality means that the Hessian of W is positive semi-definite at $\theta = \theta^*$.

Condition 1.2. We assume that the asymptotic log-likelihood equation (1.15) has a unique solution $\theta = \theta^*$ in D and we assume that $R^* = (\partial^2/\partial\theta^2)W(\theta,\theta^*)\Big|_{\theta=\theta^*}$ is positive definite.

This may seem a slightly restrictive condition, however without this we can not hope to get anything close to a consistent estimator. For discrete time ARMA-systems this condition was verified by Åström and Söderström 1974 . For multivariate MA-processes Condition 1.2 was verified in Söderström and Stoica 1982. For both discrete and continuous-time systems a partial result was given in G. Vágó and Gerencsér 1985.

To make the notations simpler we shall use θ subscript to denote derivative w.r.t. θ. A fundamental result which is needed in the proof of all subsequent results is the following 'martingale approximation' of the estimation error. (c.f. Gerencsér, Vágó 1990).

Theorem 1.1. Under Conditions 1.1-1.2 we have

$$\widehat{\theta}_T - \theta^* \;=\; -(R^*)^{-1}\frac{1}{T}\int_0^T \frac{\partial}{\partial\theta}\dot{\varepsilon}_t(\theta^*,\theta^*)dw_t \;+\; O_M(T^{-1}) \qquad (1.16).$$

A discrete time version of this theorem has been proved in Gerencsér 1990a and applied in Gerencsér 1988. The theorem allows generalization for controlled system (c.f. e.g. Duncan and Pasic-Duncan 1989.)

Sketch of the proof of Theorem 1.1: Let us now consider equation (1.12) and write it as

$$0 = V_{\theta T}(\widehat{\theta}_T,\theta^*) = V_{\theta T}(\theta^*,\theta^*) + \overline{V}_{\theta\theta T}(\widehat{\theta}_T - \theta^*) \qquad (1.17)$$

where

$$\overline{V}_{\theta\theta T} = \int_0^1 V_{\theta\theta T}((1-\lambda)\theta^* + \lambda\widehat{\theta}_T,\theta^*)d\lambda$$

After a simple rearrangement we get from (1.17):

$$R^*(\widehat{\theta}_T - \theta^*) = -\frac{1}{T}V_{\theta T}(\theta^*,\theta^*) + (R^* - \frac{1}{T}\overline{V}_{\theta\theta T})(\widehat{\theta}_T - \theta^*). \qquad (1.18)$$

Thus it is enough to prove that the second term on the right hand side is $O_M(T^{-1})$.

First we prove that $R^* - \overline{V}_{\theta\theta T}/T = O(T^{-1/2})$. The proof is based on the following uniform law of large numbers with a rate on the moments:

Theorem 1.2. We have

$$\sup_{\theta\epsilon D^o} |\frac{1}{T}V_T(\theta,\theta^*) - W(\theta,\theta^*)| = O_M(T^{-1/2}) \qquad (1.19)$$

and similar estimates hold for all derivatives of $V_T(\theta,\theta^*)$.

Proof: In the proofs we shall use the properties of L- mixing processes, derived in Gerencsér 1989a. We shall summarize some of the more important properties in the Appendix.

Since $y_t(\theta^*)$ is the output of a stable linear state-space system, it is L-mixing w.r.t. $(\mathcal{F}_t, \mathcal{F}_t^+)$ by Theorem 3.2 where

$$\mathcal{F}_t = \sigma\{w_s : s \leq t\}, \qquad \mathcal{F}_t^+ = \sigma\{w_s - w_{s'} : s, s' > t\}. \tag{1.20}$$

I.e. the σ-algebras $(\mathcal{F}_t, \mathcal{F}_t^+)$ represent the past and the future of the Gaussian white noise process dw_t, respectively. Also since the system-matrices are smooth in θ^* it follows that all derivatives of $y_t(\theta^*)$ w.r.t. θ^* are L-mixing. Furthermore it is easy to see that $y_t(\theta^*)$ and its derivatives are uniformly L-mixing w.r.t. θ^* when $\theta^* \in D$. Similarly $x_t(\theta, \theta^*)$ and its derivatives w.r.t. θ and θ^* are uniformly L-mixing. Let us now introduce the notation

$$\delta V_T(\theta, \theta^*) = \frac{1}{T} V_T(\theta, \theta^*) - W(\theta, \theta^*)$$

which can also be written as

$$\begin{aligned}\delta V_T(\theta, \theta^*) &= \frac{1}{2T} \int_0^T (|C(\theta) x_t(\theta, \theta^*)|^2 - \mathrm{E}|C(\theta) x_t(\theta, \theta^*)|^2) dt \\ &- \frac{1}{T} \int_0^T x_t^T(\theta, \theta^*) C^T(\theta) dw_t \\ &\triangleq \frac{1}{2T} \int_0^T u_t(\theta, \theta^*) dt - \frac{1}{T} \int_0^T v_t(\theta, \theta^*) dw_t.\end{aligned}$$

Here $u_t(\theta, \theta^*), v_t(\theta, \theta^*)$ are L-mixing processes uniformly in θ, θ^*, and the same holds true for all their derivatives w.r.t. θ and θ^*. Therefore we can apply Theorems 3.1.-3.3. to get that

$$\sup_{\theta \in D^o} \frac{1}{\sqrt{T}} \int_0^T u_t(\theta, \theta^*) dt = O_M(1), \qquad \text{and} \qquad \sup_{\theta \in D^o} \frac{1}{\sqrt{T}} \int_0^T v_t(\theta, \theta^*) dw_t = O_M(1)$$

which implies that

$$\sup_{\theta \in D^o} \delta V_T(\theta, \theta^*) = O_M(T^{-1/2}).$$

For the derivatives of $\delta V_T(\theta, \theta^*)$ we can use similar arguments, thus the proof of Theorem 1.2 is complete.

Now a careful analysis of (2.3) will yield the following lemma:

Lemma 1.3. We have

$$\widehat{\theta}_T - \theta^* = O_M(T^{-1/2}). \tag{1.21}$$

Combining Theorem 1.2 and Lemma 1.3. we get the desired property of (1.18), namely that the second term on the right hand side is $O_M(T^{-1})$, from which Theorem 1.1. follows.

2. Characterization of the process $\widehat{\theta}_T - \theta^*$

From Theorem 1.1 a central limit theorem with a rate of convergence can be derived (c.f. Theorem 2.3 below) and from this a well-known construction of Berkes and Philipp 1979 gives the following theorem:

Theorem 2.1. Assume that the underlying probability space is sufficiently rich. Then we have for every $\varepsilon > 0$

$$\widehat{\theta}_T - \theta^* = \frac{1}{T}(R^*)^{-1/2}\overline{w}_T + O_M(T^{-3/5+\varepsilon})$$

where $(\overline{w}_T)$ is a standard Wiener-process in $\mathbb{R}^p$.

Using this theorem a potentially useful statistics can be designed and analyzed. For this purpose let us define an auxiliary process $\overline{\overline{\theta}}_T$ by

$$\overline{\overline{\theta}}_T = \frac{1}{T}\int_0^T \widehat{\theta}_t \, dt.$$

Theorem 2.2. Under the conditions on Theorem 2.1 we have for any $\varepsilon > 0$

$$\widehat{\theta}_T - \overline{\overline{\theta}}_T = \frac{1}{T}(R^*)^{-1/2}\overline{\overline{w}}_T + O_M(T^{-3/5+\varepsilon})$$

where $(\overline{\overline{w}}_T)$ is a standard Wiener-process in $\mathbb{R}^p$.

The proof of Theorem 2.1 relies on the following theorem:

Theorem 2.3. Let (f_t) $t \geq 0$ be a $p \times m$ matrix-valued process which is L-mixing with respect to $(\mathcal{F}_t, \mathcal{F}_t^+)$ and assume that for all $t \geq 0$

$$\mathrm{E} f_t f_t^T = S \tag{2.1}$$

with some constant matrix S. Then for any $\varepsilon > 0$ we have

$$I_T = \int_0^t f_t \, dw_t = \xi_T + O_M(T^{1/4+\varepsilon})$$

where ξ_T is an $\mathbb{R}^p$-valued random variable with normal distribution $\mathcal{N}(0, TS)$.

The proof of this theorem was given in Gerencsér 1990b. The idea of the proof is that condition (2.1) and the fact that (f_t) is L-mixing implies that for any $\varepsilon > 0$

$$\sup_{0 \leq t \leq T - T^{1/2+\varepsilon}} \int_0^t (f_s f_s^T - S) ds = O_M(T^{1/2}).$$

(c.f. Theorem 4.1 of the Appendix.) Using this estimate we can define a stopping time $\tau \leq T - T^{1/2+\varepsilon}$ such that $\tau = T - T^{1/2+\varepsilon}$ with probability at least $1 - O(T^{-m})$ for any $m \geq 1$ and that we can modify the f_t on $[\tau, T]$ to get $\overline{f}_t$ which satisfies

$$\int_0^T \overline{f}_t \overline{f}_t^T \, dt = TS.$$

But then a well-known theorem implies

$$\int_0^T \overline{f}_t dw_t = \xi_T$$

with ξ_T is given in the theorem. Estimating the difference between f_t and $\overline{f}_t$ gives Theorem 2.3.

This technique of proving asymptotic normality for stochastic integrals is quite special but it gives good error terms. In the scalar case it was apparently discovered by many authors independently, Kutoyants being certainly one of them. (c.f. his 1980 book.) However the above result of the vector case is new.

Now following the well-known construction of Berkes and Phillips 1979 we can approximate I_T by a Wiener-process:

Theorem 2.4. Let (f_t) be as in Theorem 2.3 and assume that for any $m \geq 1, q \geq 1$ we have $\gamma_q(\tau) = O(\tau^{-m})$ and assume that the underlying probability space is sufficiently rich. Then for any $\varepsilon > 0$ we have

$$I_T = S^{1/2}\overline{w}_T + O_M(N^{2/5+\varepsilon})$$

where $(\overline{w}_T)$ is a standard Wiener-process.

Remark. The interesting point of this theorem that even if (w_t) is a scalar-valued Wiener-process stochastic integration yields a multivariable Wiener-process, at least approximately.

The proof of Theorem 2.2 is based on the following lemma:

Lemma 2.5. Let $(\overline{w}_t)$ $t \geq 1$ be a standard Wiener-process in $\mathbb{R}^p$ with $w_0 = 0$ and let us define the processes

$$r_t = \frac{1}{t}\overline{w}_t, \qquad \overline{r}_t = \frac{1}{t}\int_0^t r_s ds.$$

Then $\overline{r}_t$ is well-defined and

$$r_t - \overline{r}_t = \frac{1}{t}\overline{\overline{w}}_t$$

where $(\overline{\overline{w}}_t)$ is another standard Wiener-process in $\mathbb{R}^p$.

A proof based on the spectral theory of second order stationary process was given in Gerencsér 1989b. An alternative proof can be obtained using Levy's theorem.

3. Stochastic complexity for continuous-time processes

In this section we present a basic result which provides us with a rigorous tool for model order selection for a continuous-time version of predictive stochastic complexity, in analogy with the discrete time case (c.f. Rissanen 1986,1989, Gerencsér 1988, Gerencsér and Rissanen 1990).

From (1.12) we have

$$dV_t(\theta, \theta^*) = \frac{1}{2}\frac{1}{dt}(|d\varepsilon_t(\theta, \theta^*)|^2 - |dy_t|^2) \tag{3.1}$$

Here $d\varepsilon_t(\theta, \theta^*)$ can be interpreted as an infinitesimal prediction error when the assumed value of θ^* is θ. If we choose $\theta = \widehat{\theta}_t$ at time t then we get an 'honest' prediction error (c.f. Rissanen 1989.) and we can define the cumulative prediction error by

$$\int_0^T |d\varepsilon_t(\widehat{\theta}_t, \theta^*)|^2.$$

Note that the optimal cumulative squared prediction error would be

$$\int_0^T |dw_t|^2 = mT$$

where $m = \dim w_t$. A basic problem in the theory of stochastic complexity is to determine the effect of parameter uncertainity prediction error. The following theorem gives an answer to this problem.

Theorem 3.1. Under Conditions 1.1-1.2 we have with $p = \dim\theta$:

$$\lim_{T\to\infty} \int_0^T \frac{1}{dt}(|d\varepsilon_t(\widehat{\theta}_t, \theta^*)|^2 - |dw_t|^2)/\log T = p$$

almost surely.

Remark It is interesting to note that this continuous time result is simpler than the analogous discrete-time result given in Gerencsér, Rissanen 1990 as Theorem 6.1. The explanation for this is that in continuous time the covariance matrix of the innovation process can be and was assumed to be known, therefore the cumulative squared prediction error obtained as the integral of $|d\varepsilon_t(\theta, \theta^*)|^2$ is equal to the logarithm of the 'density function' of (dy_t) under the assumption that the true parameter is θ, except for a simple linear transformation.

The sketch of the proof : Although the proofs are technical the main steps can be explained relatively simply. For the sake of convenience assume that y_t and w_t are scalar-valued. To prove Theorem 3.1 the starting point is the following Taylor expansion, which is obtained as the result of applying an exact first order Taylor expansion twice:

$$d\bar{\varepsilon}_t^2(\widehat{\theta}_t,\theta^*) = d\bar{\varepsilon}_t^2(\theta^*,\theta^*) + 2\frac{\partial}{\partial\theta}d\bar{\varepsilon}_t(\theta^*,\theta^*)\cdot d\bar{\varepsilon}_t(\theta^*,\theta^*)(\widehat{\theta}_t-\theta^*)+$$
$$2\int_0^1\int_0^1 (\widehat{\theta}_t-\theta^*)^T\cdot(\frac{\partial}{\partial\theta}d\bar{\varepsilon}_t(\theta(\lambda,\mu),\theta^*)^T(\frac{\partial}{\partial\theta}d\bar{\varepsilon}_t(\theta(\lambda,\mu),\theta^*)+$$
$$\frac{\partial^2}{\partial\theta^2}d\bar{\varepsilon}_t(\theta(\lambda,\mu),\theta^*)\cdot d\bar{\varepsilon}_t(\theta(\lambda,\mu),\theta^*)\cdot\lambda(\widehat{\theta}_t-\theta^*)d\mu d\lambda, \tag{3.1}$$

where

$$\theta(\lambda,\mu) = \theta^* + \lambda\mu(\widehat{\theta}_t-\theta^*).$$

with $0<\lambda<1$ and $0<\mu<1$. We replace $\theta(\lambda,\mu)$ by θ^* in (3.1), and estimate the error of this approximation and thus we obtain

$$S_T \triangleq \int_0^T \frac{1}{dt}(d\bar{\varepsilon}_t^2(\widehat{\theta}_t,\theta^*) - dw_t^2) = \int_0^T 2\frac{1}{dt}\frac{\partial}{\partial\theta}d\bar{\varepsilon}_t(\theta^*,\theta^*)dw_t(\widehat{\theta}_t-\theta^*)+$$
$$\int_0^T \frac{1}{dt}(\widehat{\theta}_t-\theta^*)^T\cdot\ \left(\frac{\partial}{\partial\theta}d\bar{\varepsilon}_t(\theta^*,\theta^*)\right)^T\cdot\frac{\partial}{\partial\theta}d\bar{\varepsilon}_t(\theta^*,\theta^*)\cdot(\widehat{\theta}_t-\theta^*)+\zeta_T \tag{3.2}$$

where ζ_T is convergent in the M-sense and almost surely. Setting

$$du_t = \frac{\partial}{\partial\theta}d\bar{\varepsilon}_t(\theta^*,\theta^*)\cdot(\widehat{\theta}_t-\theta^*).$$

S_T can be written as

$$S_T = \int_0^T 2\frac{1}{dt}du_t dw_t + \int_0^T \frac{1}{dt}(du_t)^2 + \zeta_T.$$

Let

$$Q_T = \int_0^T \frac{1}{dt}(du_t)^2.$$

A main step in the proof is to show that

$$Q_T/(\sigma^2 k^* \log T) \to 1 \tag{3.3}$$

in the M-sense and a.s. In the proof of (3.3) we make use of Theorem 1.1 to represent $\widehat{\theta}_t-\theta^*$. Performing an exponential change of time-scale $t=e^s$ we can realize that (3.3) is a strong -law of large numbers in the new time scale (c.f. Gerencsér 1988).

Let us now assume that the same system (1.1) and (1.2) are described in two differemnt ways both descriptions satisfying the conditions of Theorem 3.1, but having different dimensionalities say $p_1<p_2$. E.g. we may think of an AR(p_1) process which

can be considered also as an AR(p_2) process with additional zero parameters. How can we recognize that the dimensionality p_2 can be reduced? Let us define

$$S_T^p = \int_0^T \frac{1}{dt}(d\bar{\varepsilon}_t^p(\widehat{\theta}_t^p, \theta^*) - dw_t^2)$$

when a p-dimensional model is fitted to our data. Then Theorem 3.1 implies that

$$\lim_{T\to\infty} (S_T^{p_2} - S_T^{p_1})/\log T = p_2 - p_1$$

almost surely.

Let (M^p) be a family of increasing model classes of the type (1.1),(1.2) the p-th member of which is parametrized by p parameter, and let p^* be the smallest p such that M^p contains a system whose output we observe. Then the above argument implies that S_T^p is "ultimately minimal" at $p = p^*$ in the region $p \geq p^*$. Since S_T^p is expected to grow linearly for $p < p^*$ we conclude that S_T^p is "ultimately minimal" in p for $p \in [1, \infty)$. This observation may be developed into a full-blown model-selection algorithm, but for this additional work is needed. Among others a computationaly efficient approximation of S_T needs to be developed.

4. Appendix

In this section we summarize a few definitions and theorems we need for this paper. A detailed exposition is given in Gerencsér 1989a. Let $D \subset \mathbb{R}^p$ be a compact domain and let the stochastic process $(x_t(\theta))$ be defined on the parameter set $\mathbb{R}^+ \times D$.

Definition 4.1. We say that $(x_t(\theta))$ is M-bounded if for all $1 \leq q < \infty$

$$M_q(x) = \sup_{\substack{t \leq 0 \\ \theta \in D}} \mathrm{E}^{1/q} |x_t(\theta)|^q < \infty.$$

We shall use the same terminology if θ degenerates into a single point.

Let $(\mathcal{F}_s), s \geq 0$ be a family of monotone increasing σ-algebras, and $(\mathcal{F}_s^+)s \geq 0$ be a monotone decreasing family of σ-algebras. We assume that $(\mathcal{F}_s^+)$ is continuous from the right, i.e. $\mathcal{F}_s^+ = \sigma\{\cup_{0<\varepsilon} \mathcal{F}_{s+\varepsilon}^+\}$. Furthermore assume that for all $s \geq 0, \mathcal{F}_s$ and $\mathcal{F}_s^+$ are independent. For $s < 0$ $\mathcal{F}_s^+ = \mathcal{F}_0^+$. A typical example is provided by the σ-algebras

$$\mathcal{F}_s = \sigma\{w_t : t \leq s\} \qquad \mathcal{F}_s^+ = \sigma\{w_t - w_{t'} : t, t' > s\}$$

where $(w_t, t \geq 0)$ is a standard Wiener-process.

Definition 4.2. A stochastic process $(x_t), t \geq 0$ is L-mixing with respect to $(\mathcal{F}_t, \mathcal{F}_t^+)$ if it is $\mathcal{F}_t$ -progressively measurable, M-bounded and with any q such that $1 \leq q < \infty$ and with

$$\gamma_q(\tau, x) = \gamma_q(\tau) = \sup_{t \geq \tau} \mathrm{E}^{1/q} |x_t - \mathrm{E}(x_t | \mathcal{F}_{t-\tau}^+)|^q \qquad \tau \geq 0$$

we have

$$\Gamma_q = \Gamma_q(x) = \int_0^\infty \gamma_q(\tau)d\tau < \infty. \tag{4.1}$$

It can be shown that $\gamma_q(\tau)$ is measurable and thus the integral (4.1) makes sense.

Example The process (x_t) given by

$$dx_t = Ax_t dt + Bdw_t \qquad x_0 = 0$$

where w_t is an m-dimensional Wiener-process, A is $n \times n$, B is $n \times m$ matrix and A is stable is L -mixing with respect to $(\mathcal{F}_t, \mathcal{F}_t^+)$ given in (1.20).

An important property of L-mixing processes is that if $(x_t), (y_t)$ are L-mixing then (z_t) with $z_t = x_t y_t$ is also L-mixing. This is seen by direct calculations. We have the following moment inequality for L-mixing processes which is similar to Burkholder's-inequality.

Theorem 4.1. Let $(u_t), t \geq 0$ be an L-mixing process with $\mathrm{E}u_t = 0$ for all t. Let (f_t) be a function in $L_2[0.T]$. Then we have for all $1 \leq m < \infty$

$$\mathrm{E}^{1/2m}|\sup_{0 \leq T' \leq T} \int_0^{T'} f_t u_t dt|^{2m} \leq C_m \left(\int_0^T f_t^2 dt\right)^{1/2} M_{2m}^{1/2}(u)\Gamma_{2m}^{1/2}(u)$$

where C_m depends only on m.

References

Arató, M. (1984), *"Linear Stochastic Systems with Constant Coefficients"*, Lecture Notes in Control and Information Science, Springer.

Åström, K.J. and Söderström,T. (1974)," Uniqueness of the Maximum-Likelihood Estimates of the Parameters of an ARMA 1974 Model", *IEEE Transactions on Automatic Control*, Vol.19, pp.769-773.

Berkes, I. and Phillip, W. (1979)," Approximation theorems for independent and weakly dependent random processes", *The Annals of Probability*, Vol.7, pp.29-54.

Caines, P.E. (1988), *"Linear Stochastic Systems"*, Wiley, New York.

Duncan, T.E. and Pasik-Duncan, B. (1989), "Adaptive control of continuous-time linear stochastic systems", *Mathematics of Control, Signals and Systems*, to appear.

Gerencsér, L. (1988), "On Rissanen's predictive stochastic complexity for stationary ARMA processes",submitted to the *Annals of Statistics.*

Gerencsér, L. (1989a), " On a class of mixing processes," *Stochastics*, Vol.26, pp.165-191.

Gerencsér, L . (1989b), "Strong approximation theorems for estimator processes in continuous time", Proc. of Limit Theorems in Probability and Statistics (Pécs, Hungary, July 1989) , North Holland, to appear.

Gerencsér, L. (1990a), "On the martingale approximation of the estimation error of ARMA parameters", accepted for publication in *Systems and Control Letters.*

Gerencsér, L. (1990b), "Strong approximation of vector-valued stochastic integrals", Submitted to *Statistics and Probability Letters.*

Gerencsér, L., Baikovicius, J. (1990),"Change point detection in a stochastic complexity framework", Proc. of the 29th IEEE Conference on Decision and Control, (Honolulu, Hawaii, December 1990).

Gerencsér, L., Gyöngy, I. and Michaletzky, Gy. (1984),"Continuous-Time Recursive Maximum Likelihood Method. A New Approach to Ljung's Scheme", A Bridge Between Control Science and Technology, Proc. of the Ninth Triennal World Congress of IFAC, 2, Identification, Adaptive and Stochastic Control (Budapest, Hungary, July 1984) Pergamon Press, Oxford, pp. 683-683.

Gerencsér, L. and Rissanen, J. (1991),"Asymptotic theory of stochastic complexity", New Trends in Time Series Analysis , (Institute of Mathematics and its Applications, Minneapolis, July 1990), Springer, to appear.

Gerencsér, L. and Vágó, Zs. (1990),"A strong approximation theorem for parameter estimators in continuous time", submitted to the *J. of Mathematical Systems, Estimation and Control*

Kutoyants, Yu. A. (1980), (1984), *"Estimation of Stochastic Processes"* . Publishing House of the Armenian Academy of Sciences, Yerevan and Heldermann, Berlin .

Rissanen, J. (1986), "Stochastic complexity and modeling", *Ann. of Statistics*, Vol. 14, pp. 1080-1100.

Rissanen, J. (1989), *"Stochastic complexity in Statistical Inquiry"*, World Scientific Publisher, Singapore, New York.

Söderström, T. and Stoica,P. (1982), "Uniqueness of Prediction Error Estimates of Multivariable Moving Average Models", *Automatica*, Vol. 18, pp. 617-620.

Vágó, Zs. and Gerencsér,L. (1985), "Uniqueness of the Maximum-Likelihood Estimates of the Kalman-gain Matrix of a State Space Model", Proc. of the IFAC/IFORS Conference on Dynamic Modelling of National Economics, (Budapest, Hungary, June 1985).

Real-time issues in continuous system identification

G.P. Rao
A. Patra
and
S. Mukhopadhyay

Department of Electrical Engineering
Indian Institute of Technology
Kharagpur, India 721 302

Abstract

In this chapter, the issues in real-time identification of continuous systems are considered. Starting with some aspects of plant model forms, measurement systems and preprocessing schemes, continuous model estimation algorithms are discussed. Further postprocessing techniques for the parameter estimates and residuals which may be necessary to satisfy the requirements of specific applications such as adaptive control, fault detection, condition monitoring etc. are indicated. Salient features of hardware and software important for practical implementations are also discussed briefly. A practical example of real-time parameter estimation is presented. The chapter concludes with a summarizing view and a look into the future in the light of emerging technology.

1. Introduction

Over the recent years resurgence of continuous-time approaches to identification, estimation and control of dynamical systems has become noticeable in the scene traditionally dominated by discrete-time methods. Several techniques of identification based on continuous models have been established (Rao 1983, Saha and Rao 1983, Unbehauen and Rao 1987). Several other alternatives have been discussed elsewhere in this book. In order to avert problems of numerical sensitivity, lack of physical significance of parameters, etc., associated with the conventional discrete-time approaches using the shift operator z, certain transformation of the z-plane has been recommended (Middleton and Goodwin 1990). The resulting operator in this case becomes closer to the continuous-time s-operator. Despite the above theoretical developments, there is hardly any report available on the issues in the use of continuous-time model-based techniques for real-time system identification. This chapter attempts to discuss these issues with special emphasis on those aspects which are relevant to continuous model identification.

N. K. Sinha and G. P. Rao (eds.), Identification of Continuous-Time Systems, 589–637.

2. The real-time identification environment

Fig. 1 shows the typical real-time identification environment for a plant which is subjected to the nominal inputs along with disturbances, some of which may be measurable. The analog measurement hardware consists of sensors, samplers, signal processing circuitry, A/D converters, etc. Alternatively or in addition, the measurement system may include digital hardware or software modules for linearization, amplification, filtering etc. The output of the measurement system is available as raw data for identification. Some further processing of this raw data is recommended before its use in the recursive algorithm. This is in view of possible low-frequency disturbances, biases, drifts, etc. introduced by the measurement system. It would also be necessary to remove d.c. values in the plant signals corresponding to steady state operating points, filter out high frequency noise and to compensate for measurable disturbances. In order to improve the numerical conditioning of the computations to follow, some scaling of the data may also be required. Other tasks may include formation of the *averaged samples* from the instantaneous ones or removal of outliers (Dai and Sinha 1990). These tasks are performed by the preprocessing module in Fig. 1.

The preprocessing module delivers data in a form suitable for the data vector formation module. This module is of particular significance to continuous-time model-based algorithms wherein the data vectors are formed with samples of certain *measures* of the signals in the sense of a linear dynamic operation (Unbehauen and Rao 1987) rather than those of the signals themselves as is the case with discrete-time methods. Information concerning the chosen model structure (e.g., denominator polynomial order) and the form of the linear dynamic operator are required at this stage. The data vector structure may be modified on-line based on post-processing indications. Elements of the data vector may have to be reset to arrest undue growths in some schemes (for example those employing data integration).

The next module is a recursive algorithm (e.g., least squares, instrumental variables) which is suitably initiated. It accepts the data vector as its input and delivers the updated estimates of the plant parameters along with the prediction error. The modules for data vector formation and recursive algorithm need not work in synchronism with the data acquisition system. In fact, execution time constraints may force these to be slower. Some parameters of the recursive algorithm may be modified on-line based on post-processing indications. For example, the gain of the algorithm may be monitored and modified to improve the convergence rate or to track time-varying parameters. Similarly, when the input signal to the plant is not sufficiently exciting, parameter updating may be halted.

The post-processing module acts as an interface between the recursive algo-

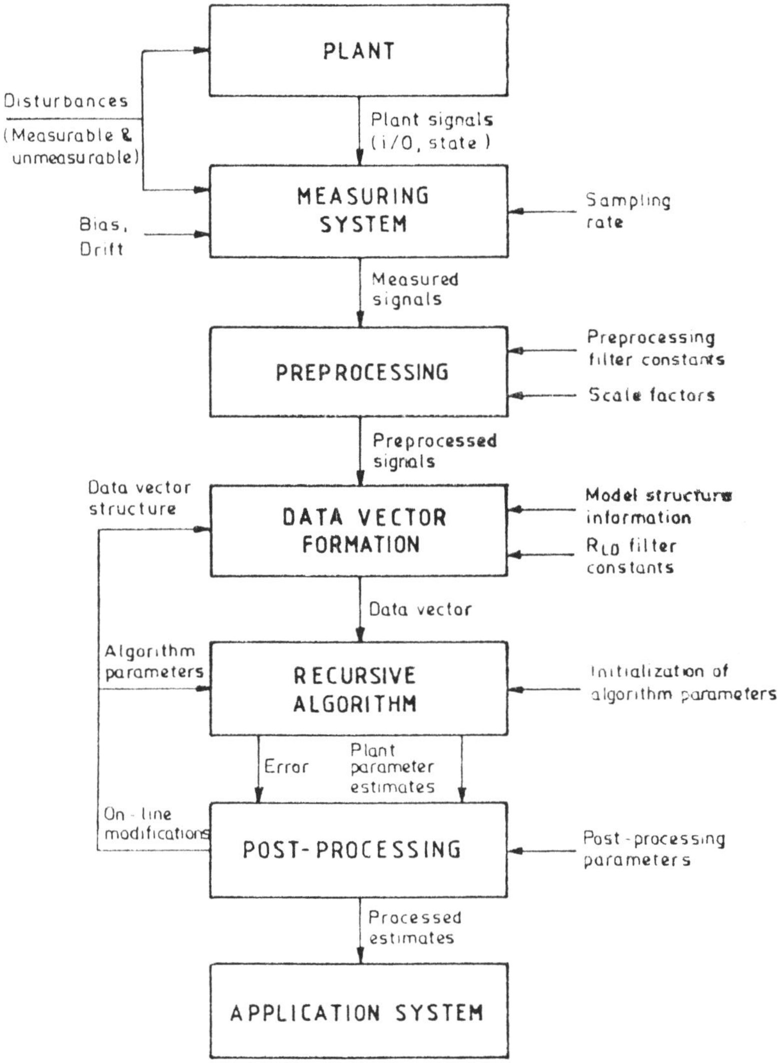

Figure 1: The real-time identification environment

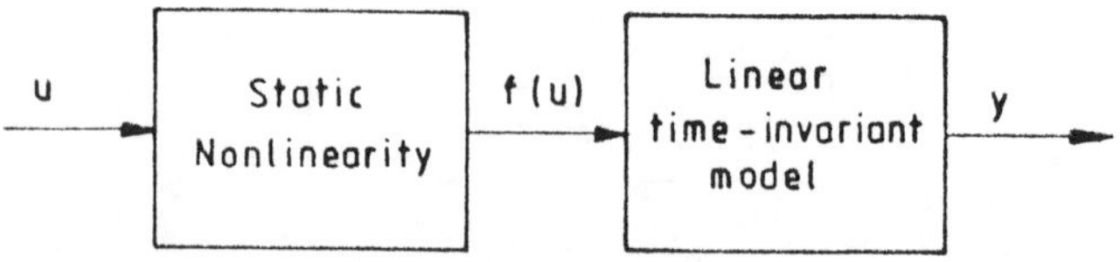

Figure 2: Modeling plants with nonlinear actuators

rithm which provides raw estimates and an application which uses these estimates. Thus it prevents invalid or transient estimates from entering the application. For example, when there is a spike in the estimate, probably due to some data outliers or a sudden increase in the estimator gain, this is suppressed by the post processor. This module also serves to assess and improve the quality of estimates by altering the parameters of the previous two modules on-line.

Examples of application systems which require real-time parameter estimation are adaptive control, fault detection, condition monitoring, prediction, forecasting etc. We now consider in detail, the issues involved in the various stages mentioned here in the following sections.

3. The plant and its model

Our main concern is parameter estimation in continuous-time linear models of dynamical systems. However, since a real-world system exhibits nonlinear behaviour more often than not, use of a linearization procedure is implied. Incremental models are therefore derived around some operating point and whenever the operating point changes, there is a need for estimation.

In some situations, static nonlinearities (e.g., saturation in control valves) can be handled in a straightforward way. As shown in Fig. 2, the nonlinearity denoted by $f(u)$ can be parameterized as

$$f(u) = \boldsymbol{\alpha}^T \mathbf{g}(u) \tag{1}$$

where $\boldsymbol{\alpha}$ is a vector of coefficients and $\mathbf{g}$ is a vector of basis functions. In particular, if the basis functions are $u^k, k = 0, 1, \ldots$, we have the so-called Hammerstein models (Narendra and Gallman 1966). Such models of nonlinear systems which are linear-in-parameters lend themselves to linear regression of the form

$$y = \mathbf{m}^T \mathbf{p} \tag{2}$$

where y is a scalar signal (most often the output), $\mathbf{m}$ the data vector and $\mathbf{p}$ a vector of parameters. If the function $f(u)$ is also unknown, $\boldsymbol{\alpha}$ becomes a subvector of $\mathbf{p}$. For example, when the system is,

$$\dot{y}(t) + a_0 y(t) = f(u) = \alpha_0 u + \alpha_1 u^2$$

we have,

$$\mathbf{p}^T = [\ a_0 \quad \alpha_0 \quad \alpha_1\ ] \ ; \qquad \mathbf{m}^T = [\ -y(t) \quad u(t) \quad u^2(t)\ ]$$

Real-world systems may also exhibit time-varying dynamics. Tracking of slow variations of parameters can normally be achieved by embedding simple *"forgetting the past"* strategies to algorithms for linear time-invariant (LTIV) model identification, as will be discussed later. Fast variations, however, may have to be explicitly modeled by expanding the time-varying coefficients in terms of a basis (Saha and Rao 1983). To illustrate, let us consider the model

$$\dot{y}(t) + a(t)y(t) = bu(t) \tag{3}$$

where $a(t)$ is the time-varying parameter to be estimated. This can be parameterized as

$$a(t) = \mathbf{a}^T \mathbf{h}(t) \tag{4}$$

where $\mathbf{a}$ is a constant parameter vector and $\mathbf{h}(t)$ denotes the vector of known basis functions. For example, if

$$\mathbf{a}^T = [\ a_0 \quad a_1\ ]\,, \quad \mathbf{h}^T(t) = [\ h_0(t) \quad h_1(t)\ ]$$

then

$$\mathbf{p}^T = [\ a_0\ \ a_1\ \ b\ ]\ ,\ \ \mathbf{m}^T = [\ -h_0(t)y(t)\ \ \ -h_1(t)y(t)\ \ u(t)\ ]$$

As the set of basis functions for the representation of time-varying parameters is often to be defined on a definite interval, it may be necessary to have recursions over intervals rather than samples. However this may not always be necessary.

The above discussions show that many aberrations of plant behavior from the ideal LTIV model can be handled by marginal modifications to the original formulation for the ideal case. Thus, we shall restrict our attention to LTIV models for single-input single-output (SISO) systems for more detailed discussions in the sections to follow. We consider a delay-free system of order n given by

$$\sum_{i=0}^{n} a_i \frac{d^i y(t)}{dt^i} = \sum_{i=0}^{m} b_i \frac{d^i u(t)}{dt^i} \tag{5}$$

with $a_n = 1$ and $m \leq n$. A real plant, such as a chemical process is likely to have a delay which may or may not be known. A known delay can be easily handled by delaying the elements of the measurement vector corresponding to the input signal. On the other hand, if the delay is unknown, it can be replaced by its Pade approximant with attendant overdimensioning of the order of the model (Agarwal and Canudas 1987). The transfer function corresponding to (5) is,

$$G(s) = \frac{B(s)}{A(s)} \tag{6}$$

with

$$A(s) = s^n + \sum_{i=0}^{n-1} a_i s^i\ ,\ \ B(s) = \sum_{i=0}^{m} b_i s^i$$

The scheme for parameter estimation using this model is shown in Fig. 3. The linear dynamic operator R_{LD} gives rise to certain measures of the signals distinguished by a superscript '*'. With reference to Fig. 3,

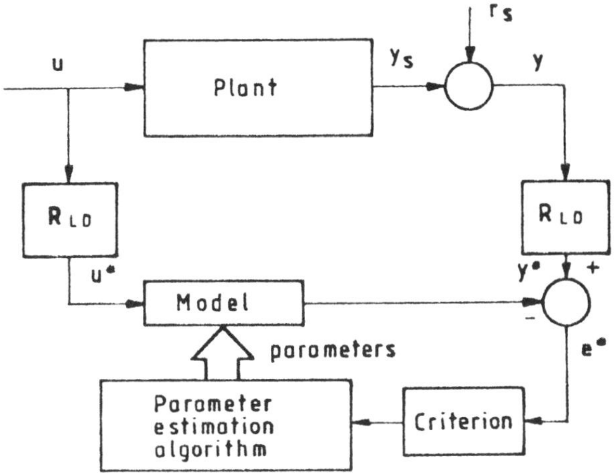

Figure 3: The continuous-time parameter estimator

$$e^*(t) = y^*(t) - y^*_M(t) \tag{7}$$

$$y^*(t) = \mathrm{R}_{LD}\ [y_s(t) + r_s(t)] \tag{8}$$

Depending on the nature of the noise $r_s(t)$ entering the output measurements, one may however minimize a filtered version of the error e^* to obtain unbiased estimates. For example, when $r_s(t)$ is generated by passing white noise $w(t)$ through a filter transfer function

$$G_n(s) = \frac{G_n^+(s)}{A(s)} \tag{9}$$

one should use the filter $\frac{A(s)}{R(s)G_n^+(s)}$ to whiten the error $e^*(t)$. In general one assumes a noise model form

$$\frac{W^*(s)}{V^*(s)} = \frac{C(s)}{D(s)} \tag{10}$$

$$, \quad D(s) = \sum_{i=0}^{n_d} d_i s^i$$

.ere $v^*(t)$ is the equation error given by

$$V^*(s) = R(s)\ [A(s)Y(s) - B(s)U(s)] \ = \ A(s)E^*(s) \tag{11}$$

and $R(s)$ is the Laplace transform of the R_{LD} operator. The assumed structure of the noise model determines the nature of the algorithm (least squares, instrumental variables, generalized least squares etc.). With respect to the general model we define the data vector as

$$\mathbf{m}^T = [\ -\frac{dy^*}{dt} \cdots - \frac{d^n y^*}{dt^n} \mid u^* \cdots \frac{d^m u^*}{dt^m} \mid -\frac{dv^*}{dt} \cdots - \frac{d^{n_c} v^*}{dt^{n_c}} \mid \frac{dw^*}{dt} \cdots \frac{d^{n_d} w^*}{dt^{n_d}}\ ] \tag{12}$$

with $c_0 = 1$ and $d_0 = 1$. The corresponding parameter vector is given by

$$\mathbf{p}^T = [\ a_1 \cdots a_n \mid b_0 \cdots\ b_m \mid c_1\ \cdots\ c_{n_c} \mid d_1\ \cdots\ d_{n_d}\ ] \tag{13}$$

In certain operations R_{LD}, such as repeated integration over $(0, t)$, some initial condition terms additionally appear in $\mathbf{p}$ and measures of known time-polynomial functions appear additionally in the data vector.

It is important to consider whether the plant to be identified is operating in closed loop. In such a case the input signal and the noise entering the system get correlated. There are two ways to handle this situation (Söderström and Stoica 1989). In an indirect scheme one identifies the closed loop system and separates out the known part due to the controller. In another scheme, an additional instrumental variable (IV), uncorrelated with the system signals and noise is injected into the controller output side. The instrumental variable must be chosen appropriately to ensure convergence. Use of an instrumental variable is also suggested when the LS estimation yields biased estimates because of wrong choice of the noise model (Young 1984).

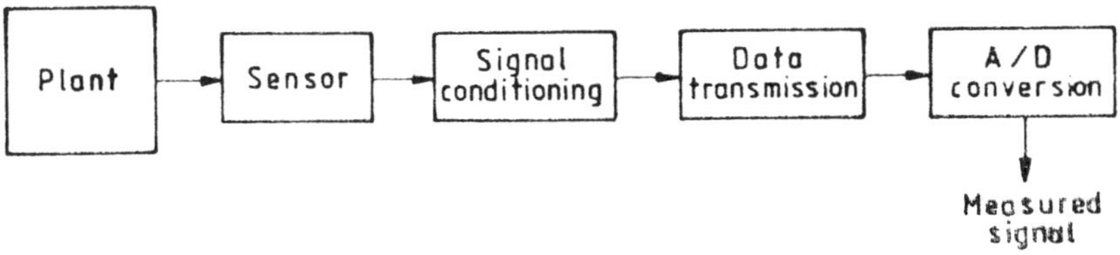

Figure 4: The measurement channel

4. Measurement System

Measurement dynamics

Fig. 4 shows a schematic of a typical measurement channel in a measurement system. Since an identification algorithm makes use of these measured data, influence of sensor nonlinearities, noise in data transmission, loading effects etc. have to be considered. We also have to take corrective action if the input and output signals are passed through dissimilar measurement dynamics. Some of the effects that should be compensated for are

i) sensor nonlinearities, bias, drift and variation in sensitivity;

ii) time constants introduced by the sensors; for example temperature transducers have time constants of the order of a few seconds (McMillan 1990);

iii) delays introduced by the signal transmission block and the sampling and ZOH operation;

iv) time constants introduced by the signal conditioning circuitry which is normally low pass in nature;

v) noise and spikes introduced by long transmission lines; their removal by conventional filters introduces additional dynamics. Therefore a common industrial practice is to take the middle value of the measurements from a set of three parallel sensors.

To compensate for the above effects one has to know the sensor characteristics, the plant environment and its layout. Some times it may be possible to measure some additional variables (e.g., temperature) and use them to correct for measurement errors (such as due to thermal effects).

Presampling filters

In order to avoid aliasing effects due to sampling, normally a low pass analog presampling filter with a sharp cutoff at the Nyquist frequency is used. This also results in significant reduction of high frequency noise, without losing the low frequency information of our vital interest. This signal is then sampled and converted to digital form by means of an A/D converter. If only the output signal is sampled, while the input signal (computed within the computer to be piecewise constant) is fed directly to the estimator, the presampling filter dynamics augments the plant dynamics and may have to be separated after identifying an inflated model. An alternative is to pass the input signal also through an equivalent digital filter.

In some situations, sampling may be done at a high rate (e.g., for control) while the identification rate is lower (e.g., to reduce computational burden). So, records with lower rate of sampling may have to be created either by decimation or by averaging with additional filters to avoid aliasing as suggested in discrete-time literature (Ljung 1987, Söderström and Stoica 1989). In CTM estimation however, the low rate samples of the relevant signal measures can be computed from signal samples obtained at a higher rate, albeit with increased computation as explained below.

Sampling time

The choice of sampling time is crucial in discrete-time model estimation (Ljung 1987). In continuous-time model identification, the data vector must accurately represent the sampled *measures* of the corresponding analog signals. Unless this is done, the estimated parameters will not correspond to the true CT model. In fact, they will belong to some closely related form, as will be shown later. The elements of the data vector should therefore be derived using samples as closely spaced as possible. However, this sampling frequency is not as crucial for estimation accuracy as in the DT case.

Wordlength of ADC

As usual, the wordlength of the ADC should be chosen keeping in view the dynamic range, accuracy requirement and the span of the instrument. If the sensor accuracy itself is low, a shorter wordlength may be adequate even if the dynamic range permits larger wordlength. For signals whose spans are much shorter with respect to the ADC range, programmable gain amplifiers should be used for amplification to increase the resolution. These are now provided on most modern data acquisition systems.

Skewed sampling

Skewed sampling, which occurs if a single ADC is used to convert analog signals in n channels through a multiplexer, introduces a fractional delay in plant characterization (Wellstead and Zanker 1982). In the discrete-time case, this gives rise to non-minimum phase behaviour as illustrated by Rao and Sinha in one of the earlier chapters of the book. However, in the CT approach, the signal measures (e.g., integrals, block pulse components, PMFs) are little affected by skewing. If the delay is a small fraction of the sampling time, it can be compensated for by interpolation (Saha and Rao 1983). However, with the simultaneous sample and hold modules available with modern data acquisition systems, skewing can be drastically reduced (Burr Brown 1987).

5. Preprocessing of data

The measurement system may not have all the capabilities to filter out noise or to remove outliers and on its own may introduce nonlinearities, low frequency disturbances, drift, bias etc. Therefore, the measured data has to be further processed in the preprocessing block. Also, the model structure may necessitate computation of incremental or other types of variables from the measured versions. We consider some aspects of such preprocessing in the following.

Removal of d.c. value

When a linearized incremental model is being identified, the nominal d.c. values related to the operating point have to be subtracted from the measured data. This may be accomplished by the following procedure:

i) Compute the mean over a time window $(0, NT)$. For the input signal,

$$\bar{u} = \frac{1}{N} \sum_{k=0}^{N} u(t - kT) \tag{14}$$

ii) Using *a priori* knowledge of the operating characteristic, compute the corresponding steady state output $\bar{y}$.

iii) Compute $\tilde{u} = u - \bar{u}$ and $\tilde{y} = y - \bar{y}$.

iv) Use $\tilde{u}$ and $\tilde{y}$ for estimation.

A finite data window allows detection of changes in the operating point. Another approach would be to compute an exponentially weighted mean. The window length N or the values of the weights may be chosen depending on the expected rates of variations in the plant dynamics.

Data differencing

If the input/output signals contain unknown d.c. values or drift effects caused by measurement and not due to the operating point, there are two ways to handle this. One approach is to explicitly consider an unknown d.c. term in the model, which increases the size of the parameter vector. Another alternative is to compute differences between successive samples and use these for estimation. At a fixed operating point, the equilibrium values u and y also automatically get removed. Data differencing, however, changes the noise model structure (Wahlberg 1988) although the system model remains unchanged.

The initial conditions of the differenced data model differ from those of the original model. Consequently, in situations where the initial condition terms are also estimated and used in further computations, say, in simultaneous state and parameter estimation (Mukhopadhyay and Rao 1991), appropriate modifications to the state space model have to be made.

However, it should be generally remembered that differencing, particularly on rapidly sampled and noisy data, would significantly accentuate the noise. If this has to be avoided, one has to use the first approach with additional computational burden.

Removal of outliers

The occasional large errors that may appear in a few measurements are commonly called *outliers*. These can considerably disturb the estimates, particularly if the estimator gain is high during their occurrence. Detection of such data can be done off-line by careful visual observation. Doing this on-line is quite difficult. One way to remove the effect of outliers on data is to use spectral approximation schemes in which the coefficients can be computed in some robust way (Dai and Sinha 1990). This technique is well-suited to CT model identification (Unbehauen and Rao 1987). However, most of the spectral approximation methods are inapplicable to real-time situations since the coefficients can be computed only after a certain time-interval elapses. In on-line situations, merely an unusually high/low value in a signal does not guarantee it to be an outlier. Instead, careful reasoning based on the input and output records (including the past), as well as some knowledge of the system time constant, actuator, sensor and plant constraints will be helpful in the detection and elimination of outliers. Some guidelines are:

i) If any variable exceeds a physical plant constraint such as actuator limit, operating speed, voltage, etc. then it is an outlier.

ii) If the ratio of the output to the maximum input amplitude of an overdamped lowpass system is greater than the maximum expected steady state gain (when all initial transients have died down), then there probably is an outlier.

iii) If the rate of rise of the output is greater than the maximum permissible limit (which is computed based on apriori knowledge about plant time constants and steady state gain), there may be an outlier.

In none of the above cases, one can be absolutely sure of the occurrence of an outlier, since faults developed in the plant or the measurement system may also manifest themselves in the same way.

An approach to handle outliers is to reduce the estimator gain for such data points. This may be done by modeling the disturbances to be made up of random variables from two different distributions. If these are gaussian, the noise distribution N may be written as

$$\mathrm{N} = \mathrm{N}(\cdot \mid 0, \sigma_1^2, \sigma_2^2) = (1-\epsilon)\mathrm{N}_1(\cdot \mid 0, \sigma_1^2) + \epsilon \mathrm{N}_2(\cdot \mid 0, \sigma_2^2), 0 \leq \epsilon \leq 1 \tag{15}$$

with zero-means and standard deviations σ_1 and σ_2 implying that the disturbance value occurring from N_1 is with probability $(1-\epsilon)$. The following rule may then be set up to detect the noise variance for use in least squares estimation:

$$\left.\begin{array}{rl} e(kT) = & y(kT) - \mathbf{m}^T(kT)\hat{\mathbf{p}}(kT-T) \\ \text{If} \quad \mid e(kT) \mid < & \mathrm{M} \\ \text{then} & \sigma = \sigma_1 \\ \text{else} & \sigma = \sigma_2 \end{array}\right\} \qquad (16)$$

The bound M has to be known. If the variance of the disturbance ordinarily occurring is known approximately as $\hat{\sigma}$, then we may take M= $3\hat{\sigma}$ (Puthenpura and Sinha 1986).

Care should be taken to avoid outliers in the measurement system as well. The *middle value of three* principle has been found to be quite effective in this case.

Data scaling

Scaling has been extensively discussed in the literature in general as well as with reference to control (Hanselmann 1987). In the context of identification, three kinds of scaling have to be considered.

i) Data acquisition scaling: The data from an ADC generally does not represent the physical variables directly. For example, a temperature in the range $(0°, 1000°)$ may be represented in the voltage range (−10V,+10V) and further brought into the range $(-2^{\alpha-1}, +2^{\alpha-1})$ in the processor arithmetic, α being the ADC wordlength. The system parameters satisfying i/o equations with such values are therefore scaled versions of the actual values. Sometimes the range may be unsymmetrical and should be suitably handled.

ii) Scaling of data vector elements: This is conceptually similar to the state vector scaling problem arising in the control compensators. The data vector elements which are possibly state variable or Poisson filtered versions of the input and output, may be computed as partial state of such filters (Gawthrop 1987).

We first obtain a reasonable bound on the state vector under realistic worst case assumptions based either on simulation or analysis from input-output records and the state variable filters. Consider the case of generating the input measures in the data vector. The state variable or Poisson filter equations are scaled as follows:

$$\left.\begin{aligned} x_{s,k+1} &= \mathbf{S}_x^{-1}\mathbf{A}\mathbf{S}_x x_{s,k} + \mathbf{S}_x^{-1}\mathbf{B}\mathbf{S}_u u_{s,k} \\ &= \mathbf{A}_s x_{s,k} + \mathbf{B}_s u_{s,k} \\ x_{s,k} &= \mathbf{S}_x^{-1} x_k \\ u_{s,k} &= \mathbf{S}_u^{-1} u_k \end{aligned}\right\} \tag{17}$$

where the matrices $(\mathbf{A}, \mathbf{B})$ correspond to the original state variable or Poisson filter.

iii) Scaling for computation: Certain computations such as scalar products require scaling since the partial sums may overflow. This may require an additional scaling by 2^{-p}. The correction for this scaling may be made by shifting the result by p-bits. The covariance update equation in least squares may be split up in such a way that this takes care of overflow. That is,

$$\mathbf{P}_{k+1} = \frac{[\mathbf{P}_k \mathbf{m}_{k+1}][\mathbf{m}_{k+1}^T \mathbf{P}_k]}{1 + [\mathbf{m}_{k+1}^T [\mathbf{P}_k \mathbf{m}_{k+1}]]} \tag{18}$$

can be obtained by computing the bracketed terms as a set of scalar products. It may also be necessary to scale the elements of the data vectors individually to represent each of the terms with sufficient precision. If the output and the input terms differ widely in magnitude, they may be scaled differently. When unbounded signals may occur, such as in the identification of unstable systems, this aspect should be given special attention. It was shown in the context of discrete-time model identification that numerical conditioning can generally be improved by certain linear transformation of the parameterization (Goodwin 1988).

6. Digitally realizable forms for models of continuous-time (CT) systems

The various approaches:

An important issue in real-time identification and control of dynamical systems is the development of recursive algorithms using known and measurable information in some discrete form. Starting with a given continuous-time model (CTM) a structure for the required recursive algorithm has to be developed. In the traditional discrete-time methods, the CTM is discretized into the shift operator form (z or q) and the resulting discrete-time model (DTM) (often a difference equation) becomes the basis

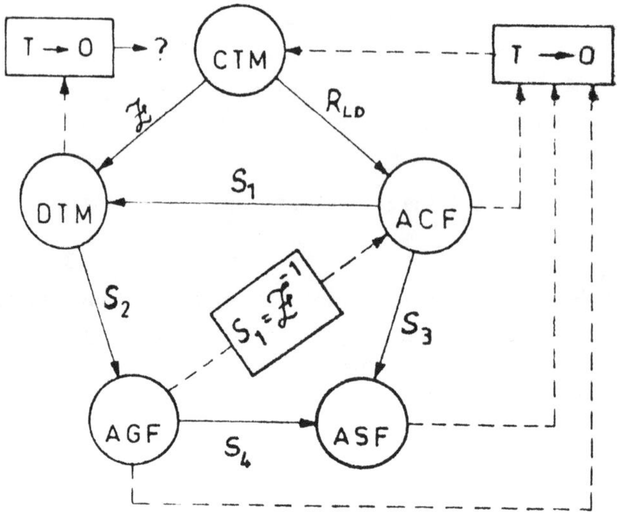

Figure 5: Various model forms for continuous-time systems

for implementation in a digital processor. Such algorithms are now well established and in wide use. Hanselmann (1987) discusses several issues in the implementation of such algorithms for control. In CT approaches, a model form which possesses a close and well conditioned link with the original CTM is adopted. It only requires some marginally additional considerations along with the existing methodology for its implementation. Presently, our immediate aim is to obtain such alternative digitally realizable forms from the original CTM to suit the tasks of identification and control. In this chapter we often refer to such forms as ***algebraic*** as the computations involved therein are to be performed digitally using algebraic operations such as addition, multiplication etc. Fig. 5 (Rao et al. 1991) shows how the various model forms may be obtained from the CTM as well as other such ones and their interrelationships.

The algebraic forms, namely, the algebraic continuous form (ACF), the algebraic gamma form (AGF), and the algebraic simulation form (ASF) are possible. From the given CTM these can be reached by the paths indicated in Fig.5 from which two possibilities are evident. One takes us directly to an ACF through the path denoted by R_{LD} signifying digital approximation of a linear dynamic operation to remove the derivatives occurring in a CTM. In the other, the traditional shift operator form DTM becomes an intermediate stage. The path $\mathcal{Z}$ denotes the traditional sampling and model discretization to reduce the calculus of CT systems to an algebra of a DTM. Several options for the choice of an R_{LD} exist. These can be classified broadly into two categories. One involves prefiltering the model input-output description. The prefilter transfer function should have a relative order at least as high as that of the system model. Poisson filter chains (PFC) (Saha and Rao 1983) and state variable filters (SVF) (Young 1981) used in system identification fall into this category. When a prefilter in the specific form of a chain of integrators is chosen, the model differential equation is converted into an integral equation (IE) used widely in continuous model identification.

In order to reduce the calculus of the IE's into an implementable algebra, we can use the orthogonal functions approach in which the signals are expressed in terms of a basis set of orthogonal functions and the integral operation is approximated by a corresponding operational matrix (Rao 1983, Unbehauen and Rao 1987, Patra 1989). Of the various systems of orthogonal functions, block-pulse functions (BPF) alone give rise to time recursions suitable for real-time implementation. Alternatively the integrals can also be realized by numerical methods using sample based approximations of continuous signals.

Paths $\mathcal{S}_1 - \mathcal{S}_4$ denote some restructuring operations which will be explained later. In all the algebraic forms shown here, a DT variable T is involved. This is the sampling time for DTM and is in a similar spirit in the other forms. As $T \to 0$, the forms AGF, ASF and ACF converge to the CTM but the DTM does not (Middleton and Goodwin 1990, Mukhopadhyay 1990). Among these, we consider now in some detail the IE model forms as derived from a CTM. Our present discussion is limited to SISO systems, and its extension to the multi-input multi-output (MIMO) case is straightforward.

Integral equation (IE) models

Repeated integration happens to be the specific form of R_{LD} leading to IE models. This can be performed in two ways:

i) Stretched window integration

The j-th integral of a signal $x(t)$ is obtained as

$$x_t^j(t) = \int_0^t \int_0^{t_1} \cdots \int_0^{t_{j-1}} x(t_j)\ dt_j\ dt_{j-1} \cdots dt_1 \tag{19}$$

ii) Moving window integration

The j-th integral of a signal $x(t)$ is obtained for a fixed τ as

$$x_\tau^j(t) = \int_{t-\tau}^{t} \int_{t_1-\tau}^{t_1} \cdots \int_{t_{j-1}-\tau}^{t_{j-1}} x(t_j)\, dt_j\, dt_{j-1} \cdots dt_1 \tag{20}$$

If $\tau = t_1 = t_2 = \cdots = t, x_\tau^j(t) = x_t^j(t)$.

Integrating eqn.(5), we get the following forms:

$$\begin{aligned} \mathcal{E}_t(t) \quad : \quad & \sum_{i=0}^{n} a_i \left\{ y_t^{n-i}(t) - \sum_{j=0}^{i-1} \frac{t^{j+n-i}}{(j+n-i)!}\, y^{(j)}(0) \right\} \\ = \quad & \sum_{i=0}^{m} b_i \left\{ u_t^{n-i}(t) - \sum_{j=0}^{i-1} \frac{t^{j+n-i}}{(j+n-i)!}\, u^{(j)}(0) \right\} \end{aligned} \tag{21}$$

$$\begin{aligned} \mathcal{E}_\tau(t) \quad : \quad & \sum_{i=0}^{n} a_i \left\{ \sum_{j=0}^{i} (-1)^j\, {}^i\mathrm{C}_j\, y_\tau^{n-i}(t - j\tau) \right\} \\ = \quad & \sum_{i=0}^{m} b_i \left\{ \sum_{j=0}^{i} (-1)^j\, {}^i\mathrm{C}_j\, u_\tau^{n-i}(t - j\tau) \right\} \end{aligned} \tag{22}$$

where the superscripts ^{j} and $^{(j)}$ denote respectively j-times repeated integration and differentiation of the related signal with respect to time. Notice that the initial conditions are absent in the moving window IE. It can be shown that

$$x_\tau^i(t) = \sum_{j=0}^{i} (-1)^j\, {}^i\mathrm{C}_j\, x_t^i(t - j\tau) \tag{23}$$

and

$$\mathcal{E}_\tau(t) \equiv \sum_{k=0}^{n} (-1)^k\, {}^n\mathrm{C}_k\, \mathcal{E}_t(t - k\tau) \tag{24}$$

where '≡' implies 'is equivalent to'.

The macrointerval $(t-\tau, t)$, may be independent of the sampling or averaging subinterval T. It will be convenient if the length of the macrointerval is chosen as an integral multiple of T.

Approximation of integrals

The integrals in equations (19) and (20) should now be approximated in ways which are suitable for on-line realization. The approximations are to be in terms of approximations of continuous signals based on measures of the related signals such as samples or spectral coefficients on an appropriate orthogonal basis. The following possibilities exist:

i) Block-pulse function approximation (BPFA)

The k-th BPF coefficient x_k of a signal $x(t)$ is its average value in the k-th segment of time of width T.

ii) Two point step approximation (TPSA)

The TPSA differs from BPFA only in the definition of x_k, the step amplitude, which is given by

$$x_k = \alpha x(kT - T) + (1-\alpha)x(kT), 0 \leq \alpha \leq 1 \tag{25}$$

$\alpha = 0$, 1, 0.5 respectively correspond to the forward and backward rectangular and trapezoidal rules of integration. Repeated integrals can be approximated by one shot operational matrices in the case of BPFA (Rao 1983). A similar approach is also possible with TPSA (Mukhopadhyay 1990). The orthogonal components of the repeated stretched window integrals over N segments are given by

$$\mathbf{x}_t^j \approx \mathbf{E}_{tj}^T \mathbf{x}, \tag{26}$$

where

$$\mathbf{E}_{tj}^T = \frac{T^j}{(j+1)!} \begin{bmatrix} f_{j,1}^* & 0 & \cdots & 0 \\ f_{j,2}^* & f_{j,1}^* & \cdots & 0 \\ \vdots & \vdots & \vdots & \vdots \\ f_{j,N}^* & f_{j,N-1}^* & \cdots & f_{j,1}^* \end{bmatrix}$$

in which

$$\begin{aligned} f^*_{j,1} &= 1 \\ f^*_{j,i} &= i^{j+1} - 2(i-1)^{j+1} + (i-2)^{j+1}, i = 2, 3, \ldots N \\ \mathbf{x} &= [x_1\ x_2 \cdots\ x_N]^T \end{aligned}$$

and

$$\mathbf{x}^j_t = [x^j_{t,1}\ \cdots\ x^j_{t,N}]^T$$

For step approximations such as BPFA or TPSA

$$x^j_t(kT) \approx \sum_{i=1}^{k} x_i \left[\frac{T^j}{j!} + \sum_{l=1}^{j-1} \frac{T^l}{l!} \frac{(kT - iT)^{j-l}}{(j-l)!} \right] \tag{27}$$

where, for instance with TPSA

$$x^j_{t,k+1} = \alpha x^j_t(kT) + (1-\alpha) x^j_t(kT + T) \tag{28}$$

The recursive formulae with BPFA and TPSA are respectively as follows:

$$x^j_{t,k} = \frac{T^j}{(j+1)!}[x_k + j x_{k-1}] + \sum_{i=0}^{j-1} \frac{T^i}{i!} x^{j-i}_{t,k-1} \tag{29}$$

$$x^j_{t,k} = \frac{T^j}{j!}[(1-\alpha)x_k + \alpha x_{k-1}] + \sum_{i=0}^{j-1} \frac{T^i}{i!} x^{j-i}_{t,k-1} \tag{30}$$

where $x^j_{t,k}$ denotes the approximate value of $x^j_t(t)$ over the k-th interval (say BPFA)

In τ type integration if we let $\tau = \lambda T$,

$$x^i_{\tau,k} = \frac{T^i}{(i+1)!} \sum_{j=0}^{i} (-1)^j\, {}^i\mathrm{C}_j\, [x_{k-j\lambda} + i x_{k-j\lambda-1}] + \sum_{l=0}^{i-1} \frac{T^l}{l!} \sum_{j=0}^{l} (-1)^j\, {}^l\mathrm{C}_j x^{i-l}_{\tau,k-j\lambda-1}. \tag{31}$$

Repeated use of single integration formula gives the less accurate version

$$x^{i}_{T,k} = \frac{T}{2}\{x^{i-1}_{T,k} + x^{i-1}_{T,k-1} - x^{i-1}_{T,k-\lambda} - x^{i-1}_{T,k-\lambda-1}\} + x^{i}_{T,k-1} \tag{32}$$

The one-shot formula with TPSA appears in the recursive form as

$$x^{i}_{T,k} = \frac{T^i}{i!}\sum_{j=0}^{i}(-1)^j\, {}^i\mathrm{C}_j[(1-\alpha)x_{k-j\lambda}+\alpha x_{k-j\lambda-1}]+\sum_{l=0}^{i-1}\frac{T^l}{l!}\sum_{j=0}^{l}(-1)^j\, {}^l\mathrm{C}_j\, x^{i-l}_{T,k-j\lambda-1} \tag{33}$$

and by repeated use of single integration we get

$$x^{i}_{T,k} = T\ [(1-\alpha)\{x^{i-1}_{T,k-\lambda} - x^{i-1}_{T-k\lambda}\} + \alpha\{x^{i-1}_{T,k-1} - x^{i-1}_{T,k-\lambda-1}\}] + x^{i}_{T,k-1}. \tag{34}$$

Realization of $\mathbf{R}_{LD}$ by filters

Zhao (1990) describes the use of a dedicated linear integral filter (LIF) to perform moving window integration. There are several publications of Sagara and Zhao listed in (Zhao 1990) concerning the application of the LIF in problems of system identification. The general form of the LIF is shown in Fig. 6. With $j = 0, 1, \ldots n$ we have $n + 1$ filters of the same kind. Such filters can be easily constructed from standard digital hardware.

Stretched window integration with exponential time weighting gives rise to the so called Poisson moment functionals (PMF) in place of pure integrals. The corresponding filter is known as Poisson filter chain (PFC). Each element of the PFC has transfer function of the form $1/(s+\varsigma), \varsigma \geq 0$. If $\varsigma = 0$ this reduces to stretched window integration. The PFC can be realized on the digital computer by a state space model (Saha and Rao 1983). The state space model has, as input, the signal whose PMF's are required while the state vector contains the PMF's of orders $0, 1, 2 \ldots$. The CT state equation can be realized digitally based on a suitable approximation (say BPFA or TPSA) and the PMF's can be obtained recursively. The PMF characterization of signals is presently known only in identification of CT systems. However, its special case, $\varsigma = 0$, is applicable in identification, simulation and dynamic model implementation in a digital computer. Saha et al. present a detailed account of the various aspects of the PMF method elsewhere in the book. The state variable filter (Young 1981) is a generalisation of the PFC.

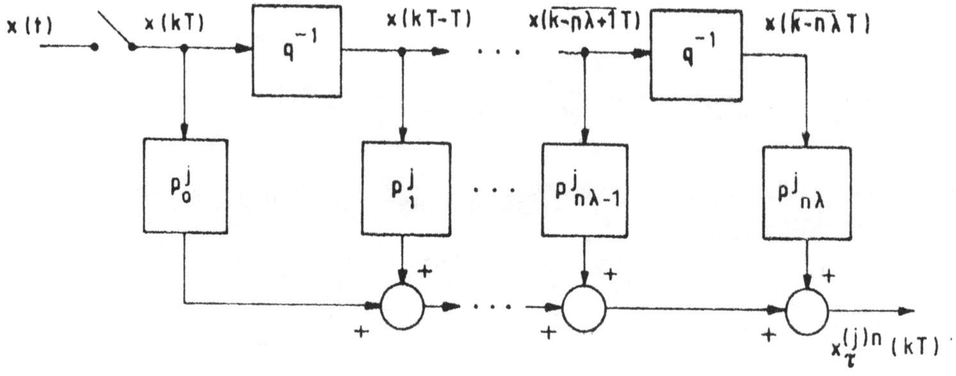

Figure 6: The Linear Integral Filter (LIF)

Restructuring recursive IE forms of CT models for simulation

To show the rearrangement of the recursive IE to get an algebraic simulation form, consider BPFA applied to eqn.(21). Collecting all terms containing y_k on one side and dividing throughout by its coefficient we get

$$y_k = -\sum_{i=0}^{n} \frac{a_i^*}{a^*} y_{t,k-1}^{n-i} + \sum_{i=0}^{m} \frac{b_i}{a^*} u_{t,k}^{n-i} + \sum_{i=0}^{n-1} \frac{g_i^*}{a^*} h_{t,k}^{n-i-1} \tag{35}$$

where

$$\begin{aligned}
a^* &= \sum_{i=0}^{n} a_i \frac{T^{n-i}}{(n-i+1)!} \\
a_i^* &= \sum_{j=0}^{i} a_j \frac{T^{i-j}}{(i-j)!}, i = 0, \ldots (n-1); \\
a_n^* &= \sum_{j=0}^{n-1} a_j \frac{(n-j)T^{n-j}}{(n-j+1)!} \\
h(t) &= \text{unit step function} \\
g_i^* &= \text{initial condition terms}
\end{aligned}$$

As $T \to 0,\ a^* \to a_n,\ a_n^* \to 0,\ a_i^* \to a_i\ \ \forall\, i = 0, 1, \ldots (n-1)$.

Digitally realizable models for CT systems via DTM

Although the DTM is readily available quite often for a CTM, to avoid direct use of DTM for reasons discussed by Rao and Sinha in an earlier chapter of this book and by Unbehauen and Rao (1990), an algebraic gamma form (AGF) or an ASF may be derived from the DTM by transforming the shift variable z as

$$\gamma = \frac{g_1 z + g_2}{g_3 z + g_4}, \tag{36}$$

where $g_i \in \mathbb{R}$ and $g_1 = 1$ without loss of generality. $g_i, i = 2$–4 in eqn.(36) are chosen such that γ approximates s (or $j\omega$) in some sense and $\lim_{T\to 0} \gamma = s$. Some well known choices are

$$\left.\begin{array}{lcll} \gamma & = & \dfrac{z-1}{T} & (a) \\ \gamma & = & \dfrac{1-z^{-1}}{T} & (b) \\ \gamma & = & \dfrac{2}{T}\dfrac{z-1}{z+1} & (c) \end{array}\right\} \tag{37}$$

The first two are derivative approximations and (37c) is the bilinear or Tustin transformation. (37a) corresponds to the well known δ-operator ($\gamma = \delta$) (Middleton and Goodwin 1990). For low frequency matching (36) can be written as

$$\gamma = \frac{z-1}{\alpha T z + (1-\alpha)T}, \qquad 0 \le \alpha \le 1 \tag{38}$$

where α is a free parameter. Eqns.(37a-c) correspond respectively to $\alpha = 0, 1$ and 0.5. The digital realization of the operator γ^{-1} is shown in Fig. 7, and is marginally more expensive than the usual delay element of the DT approaches. The value of α may be chosen on the basis of certain criteria (Mukhopadhyay 1990)

a) Minimal frequency error over the range $(0, \omega_1)$: $\alpha = 0.5$

b) Matching first derivative w.r.t. ω at low frequencies: $\alpha = 0.5$

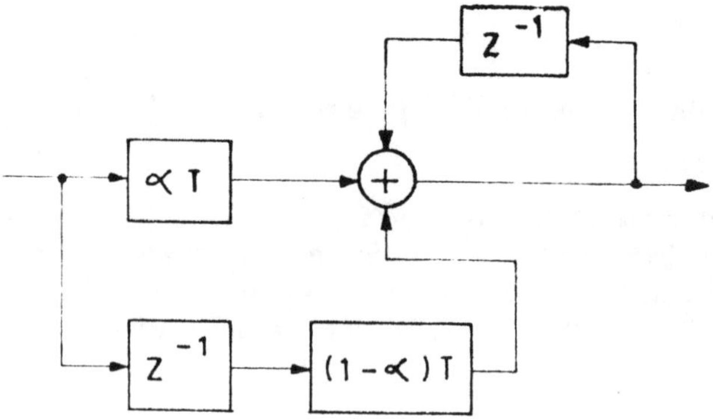

Figure 7: Realization of the γ^{-1} operator

c) Maximal execution speed, $\alpha = 1$ or 0.

If the DTM is given by

$$G_z(z) = \frac{\sum_{i=0}^{n} b_{zi} z^i}{\sum_{i=0}^{n} a_{zi} z^i} \tag{39}$$

the corresponding AGF is in the form

$$G_\gamma(\gamma) = \frac{\sum_{i=0}^{n} b_{\gamma i} \gamma^i}{\sum_{i=0}^{n} a_{\gamma i} \gamma^i} \tag{40}$$

where, for instance,

$$b_{\gamma_k} = T^k \sum_{i=0}^{k} b_{zi} \sum_{\substack{j,l \,|\, j+l=k \\ 0 \le l \le i \\ 0 \le j \le n-i}} (-1)^{n-j}\ {}^{n-i}C_j\ {}^{i}C_l\ \alpha^j (1-\alpha)^l \tag{41}$$

From CTM to AGF

By specifying the method of discretization implicitly, an AGF can be obtained from the CTM directly by substituting either for s or for terms in some form. It has been shown by Mukhopadhyay (1990) that as $T \to 0$ the AGF $\to$ CTM in all respects and that it has superior properties than the DTM in respect of sensitivity of roots, coefficients and transfer function magnitude. The Diophantine equations of pole placement controller design are found to be better conditioned in the case of AGF.

7. Algorithms

For a system modeled by eqn.(5) with $m = n - 1$ and noise filter characterized by eqn.(10), $n_c = n_d = n$, the data vector at the k-th stage may be written as (Unbehauen and Rao 1987)

$$\mathbf{m}_k^T = [-y_{1,k}^* \cdots - y_{n,k}^* | u_{0,k}^* \cdots u_{n-1,k}^* | - v_{1,k}^* \cdots - v_{n,k}^* | w_{1,k}^* \cdots w_{n,k}^*] \tag{42}$$

and the parameter vector is

$$\mathbf{p}^T = [\ a_1 \cdots a_n \mid b_0 \cdots b_n \mid c_1 \cdots c_n \mid d_1 \cdots d_n\ ] \tag{43}$$

In the data vector a term such as $u_{j,k}^*$ represents the chosen measure of the j-th derivative of $u(t)$ at the k-th stage. Of the several possible algorithms (LS, IV, GLS) we consider the LS and IV schemes incorporating a weighting strategy, by reducing the vectors $\mathbf{m}$ and $\mathbf{p}$ appropriately. The data vector $\mathbf{m}$ now contains terms due to input and output only. Likewise, $\mathbf{p}$ consists of a-s and b-s only. The algorithm is as follows:

$$\left.\begin{array}{rcll} \hat{\mathbf{p}}_{k+1} & = & \hat{\mathbf{p}}_k + \mathbf{q}_{k+1}\hat{v}_{k+1}^* & (a) \\ \mathbf{q}_{k+1} & = & \mathbf{P}_k\mathbf{h}_{k+1}(1 + \mathbf{m}_{k+1}^T\mathbf{P}_k\mathbf{h}_{k+1})^{-1} & (b) \\ \mathbf{P}_{k+1} & = & \frac{1}{\rho_{k+1}}(\mathbf{P}_k - \mathbf{q}_{k+1}\mathbf{m}_{k+1}^T\mathbf{P}_k) & (c) \\ v_{k+1}^* & = & y_{k+1}^* - \mathbf{m}_{k+1}^T\hat{\mathbf{p}} & (d) \end{array}\right\} \tag{44}$$

$\rho_k < 1, k = 1, 2, \ldots$ is a forgetting factor which helps in tracking slowly varying parameters. The instrumental variable (IV) vector (Young 1984)

$$\mathbf{h}_k^T = [-y^*_{a_{1,k}} \cdots - y^*_{a_{n,k}} | u^*_{0,k} \cdots u^*_{n-1,k}] \tag{45}$$

The IV signal is derived from an auxiliary model

$$G_a(s) = \frac{B_a(s)}{A_a(s)} \tag{46}$$

whose output measures can be computed as

$$y^*_{a_{0,k}} = \mathbf{h}_k^T \mathbf{p}_{a_k} \tag{47}$$

The parameters of the auxiliary model are updated as

$$\mathbf{p}_{a_k} = (1-\sigma)\mathbf{p}_{a_{k-1}} + \sigma \hat{\mathbf{p}}_{a_{k-d}}, \qquad 0 < \sigma < 1, \quad d > 1 \tag{48}$$

The structure of $\mathbf{p}_a$ is similar to that of $\mathbf{p}$.

The ordinary LS scheme is run for some initial length of time with $\mathbf{h}_k = \mathbf{m}_k$, $\hat{\mathbf{p}}_0 = \mathbf{0}$, $\mathbf{P}_0 = \beta\, \mathbf{I}$, $\beta \gg 1$, after which the IV scheme is invoked.

The outputs of the LIF directly correspond to the entries of $\mathbf{h}$ and $\mathbf{m}$. In the other schemes, the entries are derived from formulae (29–34). The vectors in (42) and (43) will additionally contain initial condition terms such as those indicated in eqn. (35) in the case of stretched window integration. In the moving window case, these terms do not appear. Eqn. (35) and a similar ASF such as eqn. (46) can be used to estimate the measures of $w(t)$ and $v(t)$ for a GLS scheme. The PMF version of this algorithm with all its variants is discussed by Saha et al. elsewhere in this book.

Simpler alternatives to the elaborate algorithm of eqn. (44) are also possible. For instance, a stochastic approximation scheme generates a gain sequence (Åström and Eykhoff 1971).

$$\mathbf{q}_k = \frac{1}{k}\, \mathbf{M}\, \mathbf{m}_{k+1} \tag{49}$$

in which $\mathbf{M}$ is positive definite. Choosing $\mathbf{M} = \mathbf{I}$ gives a simple algorithm of interest at the cost of increased parameter error variance.

System reference adaptive model (SRAM) techniques which are based on output error methods of model reference adaptive control can also be used. An extensive presentation of these techniques by Unbehauen may be found elsewhere in this book. To show a simple possibility in a parallel model updating scheme, we have (Ljung and Söderström 1987)

$$\hat{\mathbf{p}}_{k+1} = \hat{\mathbf{p}}_k + \mathbf{P}_{k+1}\mathbf{m}_{k+1}(y^*_{k+1} - \hat{y}^*_{k+1}) \tag{50}$$

where the gain matrix $\mathbf{P}_{k+1}$ may be kept constant or updated in the spirit of least squares as

$$\mathbf{P}_k = \mathbf{R}_k^{-1} = \left[\sum_{i=1}^{k} \mathbf{m}_i\ \mathbf{m}_i^T\right]^{-1} \tag{51}$$

so that

$$\mathbf{R}_{k+1} = \mathbf{R}_k + \mathbf{m}_{k+1}\ \mathbf{m}_{k+1}^T \tag{52}$$

in which $\mathbf{m}_{k+1}$ contains measures of the model output $\hat{y}(t)$.

With stretched window integration or a low bandwidth low pass PFC, the initial condition terms appearing as additional unknowns in the estimation equations have also to be estimated with the usual system parameters. In a PFC with large ς, these can be ignored. However, a wideband PFC would be less immune to noise. The estimated initial conditions can be gainfully used for simultaneous state estimation in a special canonical structure of the systems (Mukhopadhyay and Rao 1991).

In the case of MIMO systems, the model can be decomposed into a set of MISO models. The estimation algorithm can be extended to the MISO model by suitably defining the data vector incorporating all input signals. At the same time the corresponding parameters of these input sections are to be included in the parameter vector. The MISO model is one with a least common denominator of all its elements. With increased number of inputs and outputs, the overall computational complexity of the estimation algorithm in real-time should be ascertained. If the system model is too large to be handled enblock, some decomposition and decentralized schemes will be necessary (Rao 1985).

8. Post-processing

The need for post-processing

The sequence of estimates $\hat{\mathbf{p}}_k$ resulting from the estimation algorithm are often processed further before they are used in the application system. This may be necessary:

i) To reduce the noise sensitivity of the estimates, especially in tracking applications where the estimator gain is kept reasonably high to achieve fast parameter tracking ability.

ii) To incorporate *a priori* knowledge regarding the plant. For example, the plant may be known to be stable and minimum phase. If the estimated parameters $\hat{\mathbf{p}}_k$ at a given instant k do not represent such a system then it may be necessary to obtain a $\hat{\mathbf{p}}'_k$ nearest, in some sense, to $\hat{\mathbf{p}}_k$ and having the above properties. Such estimates, lacking known plant properties may occur during transient periods of estimation or when the identifiability conditions are not satisfied well.

iii) To assess the *quality* of the estimates. This usually involves checking how well the estimated model can explain the input-output behaviour of the plant or some of its measures.

iv) To monitor the *condition* of the estimation algorithm and, if necessary, perform on-line modifications of the algorithm parameters.

v) To avoid singularities or ill-conditioning that may occur due to the use of the estimates directly in the application system. Thus, in adaptive control, it may be necessary to check whether the parameters satisfy the requirements on the plant model (e.g., irreducibility) of the controller design module (e.g., pole placement).

In this section we consider the techniques of post-processing to meet one or more of the above objectives. Before we proceed to discuss these in detail it may be mentioned that some of these may be embedded in the estimation algorithm itself.

Processing of the parameter estimates

Filtering the estimates

Under noisy conditions, especially when the estimator gain (e.g., norm of the covariance matrix) is substantial, the parameter estimates may exhibit significant perturbations about a mean value even if the plant is time-invariant. Also disturbance peaks and outliers may contribute to sudden and large jumps in the parameters. This may result in unnecessary redesigns in adaptive control and give undesirable high perturbation inputs to the plants. In a fault detection environment a false detection may occur. Therefore to shield the application system from such parameter perturbations, low pass filtering of the estimated parameters is recommended. Some filtering schemes suggested in the literature (Isermann and Lachmann 1987) are

A. Recursive low pass filtering

$$\hat{\mathbf{p}}_{f,k} = \mathbf{F}\ \hat{\mathbf{p}}_{f,k-1} + (\mathbf{I} - \mathbf{F})\hat{\mathbf{p}}_k \tag{53}$$

where $\hat{\mathbf{p}}_f$ denotes the filtered parameter vector and $\mathbf{F}$ is a diagonal matrix whose elements determine the time constants used for filtering each parameter in $\hat{\mathbf{p}}$ depending on its expected rate of variation.

B. Finite memory recursive filtering

$$\hat{\mathbf{p}}_{f,k} = \frac{1}{N_W} \sum_{j=k-N_W+1}^{k} \mathbf{F}_{j+N_W-k}\ \hat{\mathbf{p}}_j \tag{54}$$

where N_W is the number of data points considered at a time and $\mathbf{F}_i$-s contain the weighting coefficients.

C. Recursive switched low pass filtering

$$\left.\begin{array}{rcl} \hat{\mathbf{p}}'_{f,k} & = & \mathbf{F}\ \hat{\mathbf{p}}_{f,k-1} + (\mathbf{I} - \mathbf{F})\hat{\mathbf{p}}_k \\ \mid \hat{\mathbf{p}}_k - \hat{\mathbf{p}}'_{f,k} \mid & \leq & \epsilon_0 \Rightarrow \hat{\mathbf{p}}_{f,k} = \hat{\mathbf{p}}'_{f,k} \\ \mid \hat{\mathbf{p}}_k - \hat{\mathbf{p}}'_{f,k} \mid & > & \epsilon_0 \Rightarrow \hat{\mathbf{p}}_{f,k} = \hat{\mathbf{p}}_k - \epsilon_0\ \mathrm{sgn}[\hat{\mathbf{p}}_k - \hat{\mathbf{p}}'_{f,k}]\ ,\quad \hat{\mathbf{p}}'_{f,k} = \hat{\mathbf{p}}_k \end{array}\right\} \tag{55}$$

where sgn[·] evaluates a vector containing +1's or −1's depending on the sign of the elements of its argument.

Checking the estimates for desired/expected properties

When some *a priori* knowledge regarding the plant dynamics is available (say, that the plant is stable, minimum phase), or when the estimated models are required to possess certain properties (e.g., irreducibility, no transmission zero at origin) checks for these attributes have to be performed regularly or whenever necessary. While standard algorithms exist for such purposes the computational constraints are the deciding factors for determining the suitability of adopting a particular algorithm in the real-time context. From this point of view the following are recommended for the various cases mentioned above

i) Stability: Routh-Hurwitz tests on the denominator polynomial.

ii) Phase minimality: Routh-Hurwitz test on the numerator polynomial.

iii) Coprimeness: Euclid's algorithm or tests on Sylvester's, MacDuffee's or Bezout's resultant (Kailath 1980).

iv) Zero at the origin: Coefficient of $s^0 = 0$ for the numerator polynomial and not for the denominator polynomial.

Projection of the parameter estimates

Very often, based on physical considerations or otherwise, one can construct bounds on the plant parameter values. Consequently, it is possible to determine a subspace of the parameter space which contains the true or 'best' estimate of the plant. To gain advantage from the knowledge of the subspace it is then necessary to find, for each update on the parameter vector a corresponding vector lying within the subspace. This is usually achieved via a projection strategy. The new parameter vector can be used for further processing in the application system. However, if the parameter vector obtained from projection is also to be used as the past estimate in the recursive algorithm, then it must also be ensured that the use of this projected vector instead of the one estimated by the algorithm does not adversely affect the estimator properties (e.g., convergence rate). In fact, use of suitable projection schemes can improve the performance of the estimation algorithm in the form of improved convergence rate, robustness to noise and data outliers (Middleton and Goodwin, 1990). The following algorithm may be used to accomplish this objective (Goodwin and Sin, 1984).

If $\hat{\mathbf{p}}$ lies outside $\mathbf{C}$, we select $\hat{\mathbf{p}}'$ as

$$\hat{\mathbf{p}}' = \arg\min_{\hat{\mathbf{p}} \in \mathbf{C}} \left\{ (\hat{\mathbf{p}} - \hat{\mathbf{p}}')^T \mathbf{P}^{-1} (\hat{\mathbf{p}} - \hat{\mathbf{p}}') \right\} \tag{56}$$

The steps to achieve this are

1. Transform the coordinate basis for the parameter space by defining

$$\mathbf{p}^O = \mathbf{P}_k^{-\frac{1}{2}}\hat{\mathbf{p}}_k \tag{57}$$

where

$$\mathbf{P}_k^{-1} \stackrel{\text{def}}{=} \mathbf{P}_k^{-\frac{T}{2}}\ \mathbf{P}_k^{-\frac{1}{2}}$$

and denote by $\bar{\mathbf{C}}$ the image of $\mathbf{C}$ under the linear transformation $\mathbf{P}_k^{-\frac{1}{2}}$.

2. Orthogonally project the image $\mathbf{p}_k^O$ of $\hat{\mathbf{p}}_k$ under $\mathbf{P}_k^{-\frac{1}{2}}$ onto the boundary of $\bar{\mathbf{C}}$ to yield $\mathbf{p}_k^+$, where

$$\hat{\mathbf{p}}_k^O = \mathbf{P}_k^{-\frac{1}{2}}\hat{\mathbf{p}}_k \tag{58}$$

3. Put

$$\hat{\mathbf{p}}_k = \hat{\mathbf{p}}'_k \stackrel{\text{def}}{=} \mathbf{P}_k^{\frac{1}{2}}\hat{\mathbf{p}}_k^+ \tag{59}$$

Note that the close link of the CTM parameters with the physical plant dynamics enables one to construct the subspace $\mathbf{C}$ in a more straightforward manner.

Assessment of the estimated model

The need of assessing the estimated model

Parameter estimation is usually carried out under several assumptions regarding

i) Model order: An over parameterized model may have redundant pole zero pairs which are closely spaced.

ii) Plant delay: Often the plant delay is assumed to be known. However, if it is not known accurately enough, phase mismatch may occur especially at high frequencies.

iii) No correlation between noise and plant input: The assumption is invalidated when the plant is operating in closed loop. This, in turn, may bias the parameter estimates.

iv) Noise properties: In various schemes, for achieving accurate estimates, spectral properties of the noise, knowledge of its variance etc. are assumed. Errors in these will affect the estimates.

v) Signal properties: Persistency of excitation is often required to ensure convergence. In control applications however a persistently exciting input is unlikely. These may cause slow convergence.

From the above, it becomes clear that many of the assumptions made for rapidly obtaining good parameter accuracy may be violated in practice. It is therefore important to assess the quality of the estimated model before it is used in the application system. If the identification scheme is off-line, several possibilities exist for decision making on the basis of visual observation of the parameter trajectories for convergence, matching of predicted and observed output, testing properties of residuals etc. In a real-time setting, however, these choices become restricted due to computational constraints. In the following we discuss some methods for validation of the estimated model in real-time.

Tests on residuals

In the case where the system characteristics remain invariant over substantial periods of time, it is possible to experience parameter "convergence". It may then be required to ascertain whether this converged set of parameters is "good" in some sense. This is typically tested by computing the deviation of the model output from the system output measurements. These residuals may be defined in various ways such as,

$$\left.\begin{aligned} \epsilon_{1,k} &= y_k - \mathbf{m}_k^T \hat{\mathbf{p}}_{k-1} \\ \epsilon_{2,k} &= y_k - y_M(k, u, \hat{\mathbf{p}}^*, \hat{\mathbf{x}}_i) \end{aligned}\right\} \tag{60}$$

where $\hat{\mathbf{p}}^*$ is some measure (e.g., filtered versions) of the parameter estimates, $\hat{\mathbf{x}}_i$ is the initial state or its estimate and y_M is a "simulated" output using a suitable method. Alternatively one may use

$$\epsilon_{3,k} = \mathcal{F} * \epsilon_{i,k} \; ; \quad i = 1, 2 \tag{61}$$

where $\mathcal{F}$ stands for a filter impulse response and '$*$' denotes convolution.

An approach is to test the whiteness of these residuals. Note that the spectral properties of ϵ_1, ϵ_2 and ϵ_3 are not the same. For additive white noise at the output, and true parameters ϵ_2 is likely to be white and may be tested directly. ϵ_1 is however not white. In such cases the test should be performed on its filtered version ϵ_3. The filter should be chosen such that ϵ_3 is expected to be white with true parameters. Note that this may require knowledge of the polynomial $A(s)$ or any noise filter that may be present. It is a common practice to substitute the latest available estimates of these in the computations. The estimates of $A(s)$ are available but to get the estimates of the noise filter one needs to adopt the GLS scheme. Keeping this in mind we consider a sequence ϵ_k which is expected to be white and compute

$$\hat{R}_\epsilon(l) = \frac{1}{N} \sum_{j=k+1}^{N+k-l} \epsilon_{j+k}\, \epsilon_j \tag{62}$$

The whiteness of the residuals are inferred from standard tests such as the one based on 2-SE limits.

The residual is made up of two parts. One is due to the model-plant mismatch which may be due to poor parameter estimates or due to unmodeled dynamics and is caused by the process input. The second part consists of a filtered noise signal which is usually uncorrelated with the input. Thus, assuming the plant has been modeled adequately in the input frequency band, the residuals will be correlated significantly with the input as long as a model-plant mismatch exists. One can therefore compute

$$\hat{R}_{\epsilon u}(l) = \frac{1}{N} \sum_{j=k-N}^{k-l} \epsilon_{j+l}\, u_j \tag{63}$$

As before, uncorrelatedness between ϵ and u may be tested. The following possibilities exist in this context (Söderström and Stoica 1989).

A. $\hat{R}_{\epsilon u} \approx 0$ both for $l > 0$ and $l < 0$ imply that the model is accurate and there is no feedback.

B. $\hat{R}_{\epsilon u} \approx 0$ for $l < 0$ and $\hat{R}_{\epsilon u} \neq 0$ for $l > 0$ implies that the model is accurate but the plant is operating in close loop.

However, if u_k is white - not a likely situation, then $R_{\epsilon u} = 0$ even if the model is inaccurate.

Monitoring and control of estimator performance

The need for monitoring

The performance of the estimation algorithm needs to be monitored to avoid problems due to ill-conditioning as well as to extract desired performance out of the estimator. For example, the norm of the covariance matrix in ordinary LS estimation decreases continually and for very small gain of the estimator rapid tracking of variations in the plant dynamics cannot be achieved. On the other hand, keeping a high value of gain as in constant trace algorithms cause severe sensitivity of the estimates to noise. A mechanism is therefore necessary to judiciously boost the covariance gain whenever parameter tracking is called for. In another situation if the input signal is not persistently exciting the estimator has to be switched off to avoid bursts in the estimates. In this section we discuss how to detect the need for taking corrective action in such cases as well as the strategies for correction.

Monitoring/controlling the estimator gain

This is necessary when the system parameters are time varying. The time variation is usually considered slow enough for the estimator to catch up with the changing plant dynamics. There exist techniques in the literature to track slowly varying parameters. Fixed gain estimators in SRAM approaches posses good tracking capability, particularly if the norm of the gain matrices are high. However, as mentioned before, they are highly sensitive to noise. Therefore, the gain sequence modification strategy should strike a compromise between tracking and noise sensitivity. Estimators (e.g., LS, Stochastic Approximation) having an inherently decreasing gain sequence, require some modifications to achieve a parameter tracking capability. Some times the basic algorithm itself is modified to give less importance to old data (e.g., exponential or sliding window data forgetting). The weighting factors, window width etc. can be monitored on line. Several strategies are possible.

A. Weighted exponential forgetting: The covariance matrix $\mathbf{P}$ may be updated as

$$\mathbf{P}_k^{-1} = \lambda_k \mathbf{P}_{k-1}^{-1} + \mathbf{m}_k\, \mathbf{m}_k^T \tag{64}$$

λ_k defines the weighting profile. Its choice can also be made adaptive.

B. Recursive prediction error scheme:

$$\mathbf{R}_k = [1-\gamma_k]\mathbf{R}_{k-1} + \gamma_k \mathbf{m}_k\, \mathbf{m}_k^T \tag{65}$$

Here the gain direction is fixed by the matrix $\mathbf{R}$ and the gain step by γ_k (Ljung, 1987).

C. Switched forgetting factor algorithm: In a weighted LS scheme we may have

$$\lambda = \begin{cases} \lambda_1 & \text{if } \|\mathbf{P}_k\| < \mathbf{P}_{\max} \\ \lambda_2 & \text{if } \|\mathbf{P}_k\| \geq \mathbf{P}_{\max} \end{cases} \quad 1 > \lambda_2 > \lambda_1 > 0 \tag{66}$$

where $\|\cdot\|$ stands for a suitably defined norm of the covariance such as trace.

D. Kalman filter framework: In a Kalman filter structure for the estimation of time varying parameters one has

$$\mathbf{P}_{k+1} = \mathbf{P}_k - \frac{\mathbf{P}_k\ \mathbf{m}_{k+1}\ \mathbf{m}_{k+1}^T\ \mathbf{P}_k}{r_{2,k+1} + \mathbf{m}_{k+1}^T\ \mathbf{P}_{k+1}\ \mathbf{m}_{k+1}} + \mathbf{R}_{1,k+1} \tag{67}$$

where $\mathbf{R}_1$ and $r_{2,k}$ are parameter and measurement covariances which also need to be estimated (Ljung and Gunnarsson 1990).

E. Covariance resetting: A degenerate version of the scheme is the covariance resetting scheme (Goodwin and Mayne, 1987) where we have

$$\mathbf{P}_{W\times j} = \mathbf{P}_j^0\ , \quad j = 1, 2, \ldots \tag{68}$$

where W is the length of the time-window after which resetting is performed. $\mathbf{P}_j^0$ is an initialization matrix whose norm reduces with increasing j.

It has been shown that even ad hoc modifications of the covariance are applicable provided that i) whenever the covariance is modified, it is increased and ii) there is an upper bound to the trace of the matrix.

The above algorithms are all applicable with advantage in CT parameter estimation. However it is pointed out for caution that CT estimators are more sensitive to covariance modification and produce large parameter transients. These therefore must be used with care. For example in adaptive control using weighted LS estimation of an IE model, it has been observed by the authors, that the forgetting factor should be chosen between 1.0 and 0.99. The above strategies may also be useful in improving the convergence rate in estimation of time-invariant systems.

Switch on/off the estimator

A typical problem of adaptive control is that the control and estimation objectives impose conflicting demands on the system operations. That is, while a steady state operation is desirable from control point of view, the signals in steady state provide poor plant excitation. As a result, the parameter estimator may exhibit burst phenomena. Injection of an additional perturbation signal, as has been suggested in the adaptive control literature (Patra and Rao 1989), may not be admissible due to stringent output regulation requirement. Then, an alternative is to 'switch off' the estimation algorithm whenever insufficiency of excitation is sensed. Since persistency of excitation manifests itself in the positive definiteness of a certain gain matrix ($\mathbf{P}_k$ in LS estimation), the condition number of such a matrix may be used to formulate a stop/start rule for the estimator. However, before checking the condition number of the covariance matrix, the same should be scaled appropriately as widely varying magnitudes of the data vector elements may cause a high condition number of the covariance. Note that in CT parameter estimation methods (e.g., the IE approach) the data vector often has such elements. An alternative approach is to check the norm of the Kalman gain vector $\|\mathbf{P}_k\, \mathbf{m}_k\|$. If these exceed some upper bounds the estimation is stopped. For details the reader is referred to (Sripada and Fisher 1987).

Detection of abrupt parameter changes

In a scheme where the covariance is modified from time to time, it is necessary to determine the appropriate times for modification. Normally the gain is reset when the plant dynamics (say due to change in the operating point) changes and the estimator is required to track this variation and obtain the new set of parameters describing the plant's changed dynamics. Here we describe a scheme originally devised for DTMs by Hägglund (1984) and successfully applied by Mukhopadhyay (1990) in CTM estimation for adaptive control.

Under the assumptions that i) system parameters are constant ii) the estimated parameters are close to the true values iii) the output noise is zero mean random and iv) the forgetting factor is close to unity, we have

$$\mathcal{P}(\triangle\hat{\mathbf{p}}_k^T\, \triangle\hat{\mathbf{p}}_{k-1} > 0) \approx \mathcal{P}(\triangle\hat{\mathbf{p}}_k^T\, \triangle\hat{\mathbf{p}}_{k-1} < 0) \tag{69}$$

where $\mathcal{P}$ denotes a probability measure and

$$\triangle\hat{\mathbf{p}}_k = \hat{\mathbf{p}}_k - \hat{\mathbf{p}}_{k-1}$$

On the other hand when there is an abrupt change of parameters

$$\mathcal{P}(\triangle\hat{\mathbf{p}}_k^T\, \triangle\hat{\mathbf{p}}_{k-1} > 0) > \mathcal{P}(\triangle\hat{\mathbf{p}}_k^T\, \triangle\hat{\mathbf{p}}_{k-1} < 0) \tag{70}$$

Then, the following steps can be used to detect abrupt parameter changes

Step 1: $\triangle \mathbf{p}_{f,k} = f_p \, \triangle \mathbf{p}_{f,k-1} + (1 - f_p)\triangle \hat{\mathbf{p}}_k, 0 \leq f_p \leq 1$: (parameter filtering)

Step 2: $S_k = \text{sgn}(\triangle \hat{\mathbf{p}}_k^T \, \triangle \mathbf{p}_{f,k})$: (extract sign)

Step 3: $R_k = f_s R_{k-1} + (1 - f_s) S_k, 0 \leq f_s \leq 1$: (weighted sum)

Step 4: If $R_k > R_0$, boost the gain of the covariance matrix: (parameter change detected).

The 'sgn' function in step 2 makes the sequence $\{S_k\}$ insensitive to noise variances. Parameter filtering in step 1 should not suppress the variations in $\mathbf{p}_k$ totally. The values of R_0 and f_s are chosen based on a trade off between detection speed and reliability.

Resetting of integrals

Computational schemes with integration over $(0, t)$ are usually faced with the possibility of "integral wind up". Although overflow is unlikely in floating point operations, ill-conditioning is inevitable. With fixed point arithmetic, overflow also becomes a potential problem. To avert integrator wind up and to improve the numerical conditioning of the problem, scaling may be adopted as discussed previously. An alternative strategy is to 'reset' the integrals, thereby shifting the time origin, whenever the integrals exceed some predefined bound.

Another need for resetting the integrals arises if a parameter change in detected. This is because the integrals contain data generated by the old plant dynamics which is no longer representative of the plant. In such a situations a simultaneous covariance reset is also likely. The magnitude of the covariance matrix should not be too large since it causes violent perturbations to the parameters especially in the CT case.

9. Applications and their specific demands

Real-time identification is usually carried out to achieve specific objectives related to control, fault detection and isolation (FDI) or monitoring in a situation where the plant characteristics are not properly known or changing or both. It thus forms part of a bigger task such as adaptive control, auto tuning, adaptive state estimation, FDI etc. A detailed discussion on these applications is certainly outside the scope of this chapter. But it is necessary to outline briefly the specific demands made on the estimator by the various application and vice versa.

Adaptive control

The field of adaptive control has developed mostly in the realm of DT models. However, there have been many attempts in the recent times, in view of the considerable advantages to be gained with CTM estimation over the DT (Durgaprasad et al. 1991, Mukhopadhyay 1990) to use CTM based schemes (Gawthrop 1987, Patra and Rao 1989, Mukhopadhyay 1990). The proposed CT schemes are mostly indirect, i.e., where the plant parameters themselves are estimated and used in a later stage of control design.

In adaptive control the following points related to the estimator deserve proper attention.

A. For best performance the available *a priori* knowledge should be used to the fullest extent to aid the estimator in searching out the so called 'true' parameter. This may be utilized in the form of good initial estimates, projections through bounds on certain parameters, stability checks, signal bounds to remove outliers, choosing proper system order and time-delay, incorporating proper noise models etc.

B. Fast tracking and convergence of parameters is necessary for fast adaptation. However, the speed of adaptation is always to be decided based on application speed requirements, signal to noise ratio, reliability etc. On the other hand, the speed of adaptation decides factors such as computing speed, choice of estimation algorithms etc.

C. Robustness of an adaptive controller depends crucially on the success of the estimator. Proper care should be taken to guard against ill-conditioning of the estimator by data scaling, maintaining persistency of excitation by injecting external signals whenever permissible etc.

D. Low sensitivity to noise is a desirable feature in almost all applications as in adaptive control. CTM estimation is advantageous from this point view. However, parameter filtering goes further in reducing this sensitivity.

E. During the transient phase of estimation the estimates should not be used for control. The commonly adopted approach is to replace the adaptive controller by a nominal controller during such phases. A more complex approach is to shape the transients such that at all points on the parameter trajectory from one steady state to another, the designed control is well defined and meaningful. This is possible through parameter projection.

Fault detection

In such an application one is interested to sense changes in parameters of a system – their nature and extent in order to decide, based on other observations, on the cause of such a change to take quick corrective action. While there may be a change of parameters due to a change of the operating point, it is in the course of normal operation but a change of parameter due to, say, malfunctioning of a control valve, is abnormal. It is therefore desirable that the parameters being estimated in such an application have close relationships with the physical variables and characteristics of the process. From this point of view CT model estimation may be preferred in such applications in comparison with DTMs (Isermann 1984). The following points deserve attention in the context of parameter estimation.

A. Accurate estimation of the parameters of interest.

B. Low noise sensitivity to minimize false alarms.

C. Parameter tracking ability both in respect of slow (soft faults due to aging, environmental conditions etc.) and abrupt (hard faults) changes in parameters.

D. Identifiability of the fault from estimated parameters. This relates to model parameterization.

E. Fault detection normally requires significant *a priori* information about the plant. This should be utilized to improve estimation.

Other applications

Among these is condition monitoring of a process. This is often a part of an FDI system. Among the several process parameters considerable care is to be taken to choose the proper ones that reflect the health of the process properly. Interpretation of the trends also requires considerable engineering experience. With modern computing systems it is possible to provide the operator with a lot of graphic and expert support. State estimation is another application which has been extensively used in areas such as power systems, adaptive filtering and control etc. Problems in power systems are typically large scale. Suitable decomposition of the problem into subproblems is therefore necessary to meet the computational and the application timing requirements. Often in adaptive filtering fast computation is necessary. Special purpose hardware and algorithms that exploit possible parallelisms of computation (e.g., systolic matrix computations) have been evolved (Chisci 1988). However, this area is largely dominated by discrete-time techniques.

10. Hardware and software support

We discuss the requirements of hardware and software for real-time estimation keeping in mind two facts. Firstly, any discussion on hardware and programming aspects is going to be obsolete and incomplete in the face of the tremendous pace of development and the vast array of products. Secondly, there is little experience available on true real-time implementation of estimators. Adaptive controllers have already hit the market for some years now (Åström and Wittenmark 1990) though all of them do not employ an estimator. There are also some issues common to real-time control and real-time estimation. A comprehensive discussion on the control aspects is available in (Hanselmann 1987).

Hardware

Standard personal computers or microprocessor based board level systems with process interfaces and numeric coprocessor are available for implementation of estimators. Such systems often form an intermediate supervisory layer in a hierarchical control structure receiving data from a host of sensors in a number of control loops and also sending processed information to a central supervisory computer via data links. The instrumentation front end of such a system typically includes programmable gain analog input channels for voltage (e.g., ± 10V) or current signals (e.g., 4–20mA), digital inputs, analog voltage and current outputs. Programmable counters, communication channels via serial/parallel data ports and numeric coprocessor are other hardware requirements in addition to the system firmware. Manufacturers provide facility of customizing the hardware with the application to a limited extent (Burr Brown 1987). More rudimentary and dedicated processors called microcontrollers are available at present for control purposes (Intel 8096). For most applications the ordinary processors that are provided with the microcomputer/ microprocessor systems suffice, but for very high speed applications VLSI signal processing chips (e.g., TMS 32020 from TI) offering great escalation of computing speed, albeit at a significantly higher cost, may be used. They however normally do not provide easy high level programming support or floating point computations. Particular requirements on the hardware for implementing estimators are the following:

A. Larger memory than for controllers.

B. A reduced instruction set architecture to enhance speed at nearly no loss of flexibility.

C. Architectures with special array processing capabilities to enhance speed.

D. Separate instruction and data buses, as in Harvard architecture, for maximum throughput.

E. A parallel hardware multiplier.

In addition, one may require floating point computational facility. However, this would incur more cost especially if fast computation is simultaneously required. In an industrial environment sophistication and complexity of hardware is often traded off with reliability and maintainability and above all, the cost.

Software

Real-time implementation of a software for identification requires the following features of the supporting software environment.

A. The operating system should be capable of real-time tasks scheduling.

B. Provision for simultaneous (transparent to the user) execution of interactive jobs and non-interactive background jobs.

C. Software tools for developing operator interfaces.

D. An efficient interface between the high level/assembly language routines and the instrumentation front end which should be fully software programmable.

E. General tools for modular development of software.

The above requirements refer to an implementation on a microcomputer system. For implementations on more rudimentary board level microprocessor hardware, assembly language programming is necessary for stringent optimization of memory and computational time. For mathematical procedures (matrix inversion, computation of condition number etc.) numerically robust algorithms should be employed. Such procedures are available as parts of commercial software packages like MATLAB (Ljung 1987) and other mathematical libraries. Development of a good operator interface can be a crucial factor in the success of such automatic systems and their acceptance among the process operators. Data validation, operator guidance, proper graphic presentation, selective data logging, presettable alarm levels, operator invoked plant diagnosis and expert suggestion features, may be provided in a practical software. Among the general guidelines are software maintainability, upgradability and portability. Packages for identification of CT models are rare. Patra et al. (1988) have reported on a package development.

11. An example of implementation

Real-time parameter estimation using block-pulse functions

A two-input one output plant was deliberately simulated with the nominal transfer function matrix

$$G(s) = \left[\frac{1}{s+2} \quad \frac{1}{s+1}\right]$$

on an aged analog computer known for its drift and lack of accuracy. The parameter and data vectors are defined for an r-input system as

$$\mathbf{p}^T = [a_1 \cdots a_n | b_{1,0} \cdots b_{1,m_1} | \cdots | b_{r,0} \cdots b_{r,m_r} | g_0 \cdots g_{n-1}]$$

$$\mathbf{m}^T = [-y^1 \cdots - y^n | u_1^{n-m_1} \cdots u_1^n | \cdots | u_r^{n-m_r} \cdots u_r^n | h \cdots h^{n-1}]$$

where g_i and h respectively denote initial condition terms and unit step function. The superscript indicates the order of stretched window integration on the signal. The parameter and covariance update equations are set up with $r = 2$. An algorithm with eight major steps based on LS estimation was created. The following are some salient features in the actual implementation (Chakravarty et al. 1989).

i) Synchronously sampled i/o signals by external triggering through timing pulses from a microprocessor.

ii) 12 - bit A/D conversion.

iii) Suitable choice of the programmable gains of i/o channel amplifiers to avoid under/overflows.

iv) HP1000, A-700 minicomputer system with its analog i/o subsystem.

v) The d.c values in signals were estimated by the relation

$$x_{dc}(l+1) = \alpha x_{dc}(l) + (1-\alpha)x(l+1)$$

and removed from $x(l)$.

vi) Block-pulse components are considered as the signal measures.

vii) The resetting scheme is characterized by

θ: resetting interval

τ: running time index

λ: time index in a frame between two consecutive resettings.

ω: Frame or window index.

$\lambda = \{(\tau - 1)\text{mod}\theta + 1\}$

$\omega = \{(\tau - 1)\text{div}\theta + 1\}$

Consequently, the entries in the data vector are defined in the respective intervals. For instance, $x^i(\omega, \lambda)$ denotes the BPF approximation of the i-times repeated integral from t_1 to t_2 where $t_1 = (\omega - 1)\ \theta t_b,\ t_2 = \{(\omega - 1)\ \theta + \lambda\}t_b,\ t_b =$ width of each BPF. A GLS scheme was implemented. In the light of the resetting scheme, the various replacements of variable labels are as follows:

$$\mathbf{p}(\tau) \rightarrow \mathbf{p}(\omega, \lambda), \mathbf{m}(\tau) \rightarrow \mathbf{m}(\omega, \lambda), \mathbf{P}(\tau) \rightarrow \mathbf{P}(\omega, \lambda),$$

$$\mathbf{p}(\tau - 1) \rightarrow \mathbf{p}(\omega, \lambda - 1), \text{and} \mathbf{P}(\tau - 1) \rightarrow \mathbf{P}(\omega, \lambda - 1)$$

The initiation is as follows:

$$\mathbf{P}(\omega, 0) = \sigma\rho^{(\omega-1)}\mathbf{I},\ \sigma \gg 1,\ 0 < \rho \leq 1 \text{ depending on noise}$$

Table 1 presents the results with various kinds of inputs including Legendre polynomials.

$P(x, y, z)$: PRBS sequence with amplitude $+x$ volts and periodic length $2^y - 1$, each value of the sequences was stretched for z units of time.

$L(x, y)$: Legendre polynomial of order y with maximum amplitude of x volts.

$S(x, y)$: Square wave of amplitude xV and period y units.

n_n: The order of the noise filter with zero relative order.

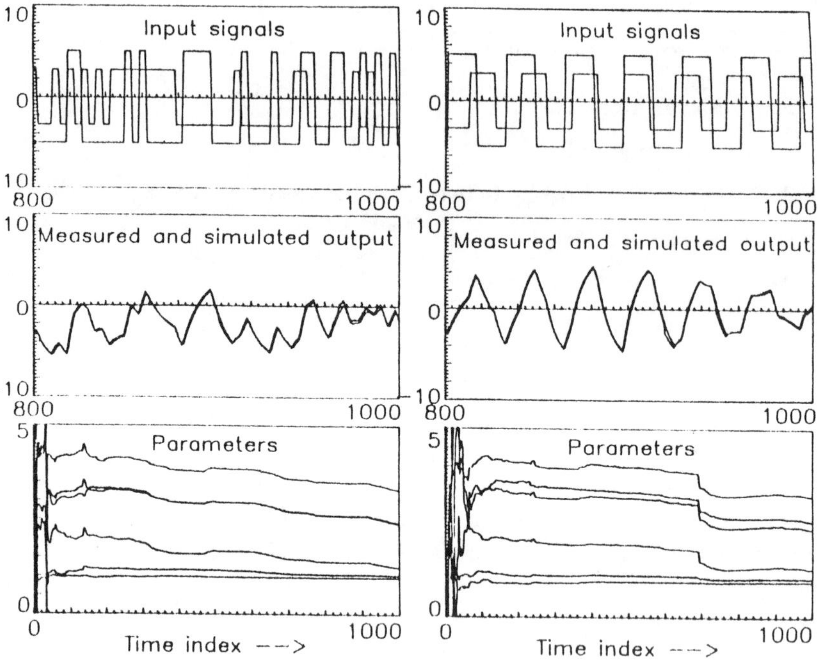

Figure 8: Real-time parameter estimation using block pulse functions

Table 1: Real-time parameter estimation in a two-input single-output system: Simulation details and results

Case	t_b	n_n	u_1	u_2	θ	ρ	α	$\hat{a}_2$	$\hat{a}_1$	$\hat{b}_{1,1}$	$\hat{b}_{1,0}$	$\hat{b}_{2,1}$	$\hat{b}_{2,0}$
1.	0.1	1	$P(3,11,1)$	$P(5,11,1)$	1000	–	1.00	1.852	2.504	1.106	0.904	1.663	0.930
2.	0.1	1	$P(3,9,4)$	$P(5,9,4)$	1000	–	1.00	2.917	3.842	1.662	1.051	2.935	0.975
3.	0.1	1	$S(3,10)$	$S(5,100)$	1000	–	1.00	2.597	3.602	1.601	1.008	2.592	1.016
4.	0.1	1	$S(3,28)$	$S(5,32)$	1000	–	1.00	2.479	2.956	1.145	0.961	2.112	0.896
5.	0.1	1	$L(3,11)$	$L(5,6)$	1000	–	1.00	2.365	2.809	1.169	0.760	2.353	0.471
6.	0.1	1	$P(3,11,1)$	$P(5,11,1)$	1000	–	0.99	3.548	3.914	2.199	0.972	3.323	0.900
7.	0.1	1	$P(3,9,4)$	$P(5,9,4)$	1000	–	0.99	2.480	3.389	1.361	0.991	2.493	0.970
8.	0.1	1	$S(3,10)$	$S(5,100)$	1000	–	0.99	3.488	4.382	2.605	1.007	3.498	0.959
9.	0.1	1	$S(3,28)$	$S(5,32)$	1000	–	0.99	2.182	2.552	0.877	0.914	1.712	0.873
10.	0.1	1	$L(3,11)$	$L(5,6)$	1000	–	0.99	1.986	4.339	1.132	1.177	1.946	2.329
11.	0.1	2	$P(3,11,1)$	$P(5,11,1)$	1000	–	1.00	1.935	2.291	0.907	0.913	1.599	0.904
12.	0.1	2	$P(3,9,4)$	$P(5,9,4)$	1000	–	1.00	2.408	3.302	1.233	1.024	2.386	0.961
13.	0.1	2	$S(3,10)$	$S(5,100)$	1000	–	1.00	2.724	3.692	1.696	1.006	2.700	0.997
14.	0.1	2	$S(3,28)$	$S(5,32)$	1000	–	1.00	2.518	3.192	1.260	1.017	2.309	0.935
15.	0.1	2	$L(3,11)$	$L(5,6)$	1000	–	1.00	4.856	5.400	2.524	1.498	4.789	0.581
16.	0.2	2	$P(3,11,2)$	$P(5,11,2)$	500	0.01	1.00	1.643	2.685	0.836	0.959	1.679	0.980
17.	0.3	2	$P(3,11,3)$	$P(5,11,3)$	500	0.01	1.00	1.666	2.663	0.826	0.914	1.651	0.963
18.	0.5	2	$P(3,11,5)$	$P(5,11,5)$	500	0.01	1.00	1.341	2.342	0.668	0.817	1.327	0.956

The other constants chosen were $\tau_{\max} = 1000$ and $\alpha = 0.75$, $\mathbf{p}_0 = \mathbf{0}$ and $\sigma = 5 \times 10^4$.

Fig. 8 shows the signals and parameters convergence patterns for cases 12 and 14 of Table 1. In the case of large t_b, covariances and integrals need resetting.

12. Conclusions

In this chapter we have attempted to draw the attention of the reader to the relevant issues in real-time identification of continuous systems. Issues relating to measurement and pre-processing of data required due to the application configuration, model structures etc. have been discussed. These factors often do not receive adequate treatment in the literature. Moreover, majority of the existing guidelines on these issues pertain to DTM estimation. However, needless to emphasize, only good and representative data can lead to accurate estimates.

CTM estimation contrasts itself with its DT counterpart in the formation of the data vector. In this case, we have various filtered measures of the i/o signals instead of their delayed samples. These are discussed with particular reference to integral filtering and its digital realizations. Many estimation algorithms can possibly be used and there exists much literature on them. Here we have briefly discussed LS, GLS, IV and some simple gradient schemes.

Further processing of the estimates is necessary to ensure estimation accuracy, good tracking of time-varying dynamics, low noise sensitivity complying with the known plant properties of the estimated model. It is also necessary to ensure well-conditioning of the estimator itself. These have been discussed in some detail.

An example of real-time identification of a multi-input system has been presented for various cases of sampling time, inputs, forgetting factors and noise model order etc. Necessary and desirable features of the hardware and software support for implementation of real-time estimators have also been discussed in brief.

New techniques are emerging in computer science that may have significant impact on real-time estimation in general. One of the significant developments appears to be the expert system. The value of practical engineering experience cannot be underestimated and such qualitative information based judgments built into an expert software can provide significant aid to the operating personnel. While a real-time estimator can form part of an expert system, an expert system can also be used as on-line adviser for plant identification (Betta and Linkens 1990) say, for retuning of control loops. Another emerging area is that of neural computing. These systems have a so called 'learning' ability and therefore can be applied to situations where nice and complete structural information about the process is not available. For an application of these in system identification see (Narendra and Parthasarathy, 1990). Distributed and parallel computing techniques are also significant for applications of growing size and speed. However, very little has been reported on their application in system identification.

Recent research has shown that in a variety of adverse situations, use of continuous-time models gives vastly improved estimates. *Smart* controllers with adaptive features, which presently employ discrete-time model estimation techniques only, can therefore incorporate continuous-time modeling — promising several advantages.

References

Agarwal, M. and C.A. Canudas (1987), "On-line estimation of time-dalay and continuous-time process parameters", *Int. J. Control*, Vol.46, No.1, pp. 295–311.

Åström, K.J. and P. Eykhoff (1971), "System identification – A survey", *Automatica*, Vol.7, pp. 123–162.

Åström, K.J. and B. Wittenmark (1989), "*Adaptive Control*", Addison Wesley.

Betta, A. and D.A. Linkens (1990), "Intelligent knowledge-based system for dynamic system identification", *Proc. IEE Pt. D*, Vol. 137, No. 1, pp.1–12.

Burr Brown Corporation (1987), "*The Handbook of Personal Computer Instrumentation*".

Chakravarty, B.B., N. Mandayam, S.Mukhopadhyay, A. Patra and G.P. Rao (1989), "Real-time parameter estimation via block pulse functions", *Proc. SICE-89*, Matsuyama, Japan, pp. 1095–1098.

Chisci, L. (1988), "High speed RLS parameter estimation by systolic like arrays", in *Advanced Computing Concepts and Techniques in Control Engineering* (Ed. M.J. Denham and A.J. Laub), Springer.

Dai, H. and N.K. Sinha (1990), "Robust coefficient estimation of Walsh Functions", *Proc. IEE, Pt. D*, Vol.137, No.6, pp. 357–363.

Durgaprasad, G., G.P. Rao, A. Patra and S. Mukhopadhyay (1991), "Indirect methods of parameter estimation of discrete time models", *Proc. 9th IFAC/ IFORS Symp. on Identification and System Parameter Estimation*, 8–12 July, 1991, Budapest, Hungary.

Gawthrop, P.J. (1987), "*Continuous-time Self Tuning Control, Vol. 1 - Design*", Research Studies Press, Lechworth, England.

Goodwin, G.C. (1988), "Some observations on robust stochastic estimation and control", *Preprints 8th IFAC/ IFORS Symp. on Identification and System Parameter Estimation*, Beijing, China.

Goodwin, G.C. and D.Q. Mayne (1987), "A parameter estimation perspective of continuous-time adaptive control", *Automatica*, Vol.23, No.1, pp 57–70.

Goodwin, G.C. and K.S. Sin (1984), "*Adaptive Filtering Prediction and Control*", Prentice Hall.

Hägglund, T. (1984), "Adaptive Control of systems subject to large parameter changes", *Proc. 9th IFAC World Congress*, Budapest, pp. 993–998.

Hanselmann, H. (1987), "Implementation of digital controllers – A survey", *Automatica*, Vol.23, No.1, pp. 7–32.

Isermann, R. (1984), "Process fault detection based on modelling and estimation methods – A survey", *Automatica*, Vol.20, No.4, pp. 387–404.

Isermann, R. and K.-H. Lachmann (1987), "Adaptive controllers – Supervision level", in *Systems and Control Encyclopedia*, M.G.Singh (Ed.), Pergamon Press.

Kailath, T. (1980), "*Linear Systems*", Prentice Hall.

Ljung, L (1987), "*System Identification - Theory for the User*", Prentice Hall.

Ljung, L and S. Gunnarsson (1990), "Adaptation and tracking in system identification - A survey", *Automatica*, Vol.26, No.1, pp. 7–21.

Ljung, L and T. Söderström (1983), "*Theory and Practice of Recursive Identification*", MIT Press.

McMillan, G.K. (1990), "*Tuning and Control Loop Performance*", Instrument Society of America.

Middleton, R.H. and G.C. Goodwin (1990), "*Digital Control and Estimation - A Unified Approach*", Prentice Hall.

Mukhopadhyay, S. (1990), "*Continuous-time Models and Approches for Estimation and Control of Linear systems*", Ph.D. Thesis, Department of Electrical Engineering, IIT Kharagpur, India.

Mukhopadhyay, S. and G.P. Rao (1991), "Integral equation approach to joint state and parameters estimation for MIMO systems", *Proc. IEE Pt. D*, (To appear).

Narendra, K.S. and P.G. Gallman (1966), "An iterative method for identification of nonlinear systems using Hammerstein models", *IEEE Trans. Automatic Control*, Vol.AC–11, p. 546.

Narendra,K.S. and K. Parthasarathy (1990), "Identification and control of dynamical systems using neural networks", *IEEE Trans. on Neural Networks*, Vol.1, No.1, pp. 4–27.

Patra, A. (1989), "*General Hybrid Orthogonal Functions and Some Applications in Systems and Control*", Ph.D. Thesis, Department of Electrical Engineering, IIT Kharagpur, India.

Patra, A., P.V. Bhaskar and G.P. Rao (1988), "A package for simulation and parameter estimation of continuous-time dynamical systems", *Proc. 8th IFAC/ IFORS Symp. on Identification and System Parameter Estimation*, Beijing, pp. 1959–1963.

Patra, A. and G.P. Rao (1989), "Continuous-time approach to self-tuning control - Algorithms, implementation and assessment", *Proc. IEE Pt. D*, Vol.136, No.6, pp. 333–340.

Puthenpura, S. and N.K. Sinha (1986), "Robust bootstrap method for joint estimation of states and parameters of a linear system", *J. Dynamic Systems Measurement and Control*, Vol.108, pp. 255–263.

Rao, G.P. (1983), "*Piecewise Constant Orthogonal Functions and Their Application to Systems and Control*", Springer Verlag, Berlin Heidelberg New York Tokyo.

Rao, G.P. (1985), "Decomposition, decentralization and coordination of identification algorithms for large-scale systems", *Proc. 7th IFAC Symp. on Identification and System Parameter Estimation*, York, U.K., pp. 297–301.

Rao, G.P., H. Unbehauen, S. Mukhopadhyay and A. Patra (1991), "From calculus to algebra in models of continuous-time systems", *Proc. 9th IFAC/IFORS Symp. on Identification and System Parameter Estimation*, 8-12 July 1991, Budapest, Hungary.

Saha, D.C and G.P.Rao (1983), "*Identification of Continuous Dynamical Systems - the Poisson Moment Functional (PMF) Approach*", Springer Verlag, Berlin Heidelberg New York Tokyo.

Söderström, T. and P. Stoica (1989), "*System Identification*", Prentice Hall.

Sripada, N.R. and D.G. Fisher (1987), "Improved least squares identification", *Int. J. Control*, Vol.46, No.6, pp. 1889–1913.

Unbehauen, H. and G.P. Rao (1987), "*Identification of Continuous Systems*", North Holland, Amsterdam.

Unbehauen, H. and G.P. Rao (1990), "Continuous-time approaches to system identification - A survey", *Automatica*, Vol.26, No.1, pp. 23–35.

Wahlberg, B. (1988), "On continuous-time system identification", *Preprints 8th IFAC/IFORS Symp. on Identification and System Parameter Estimation*, Beijing.

Wellstead, P.E. and P. Zanker (1982), "Techniques of self-tuning", *OCAM, Special Issue on Selftuning Control*, Vol.3, No.4, pp. 305–322.

Young, P.C. (1981), "Parameter estimation for continuous-time models - A survey", *Automatica*, Vol.7, No.1, pp. 23–29.

Young, P.C. (1984), "*Recursive Estimation and Time Series Analysis*", Springer Verlag, Berlin.

Zhao, Z.-Y. (1990), "*Linear Integral Filter Approach to Identification of Continuous-time Systems*", Ph.D. Thesis, Department of Electrical Engineering, Kyushu University, Fukuoka, Japan.

International Series on

MICROPROCESSOR-BASED SYSTEMS ENGINEERING

Editor: Professor S. G. Tzafestas, *National Technical University, Athens, Greece*

1. S.G. Tzafestas (ed.): *Microprocessors in Signal Processing, Measurement and Control*. 1983 ISBN 90-277-1497-5
2. G. Conte and D. Del Corso (eds.): *Multi-Microprocessor Systems for Real-Time Applications*. 1985 ISBN 90-277-2054-1
3. C.J. Georgopoulos: *Interface Fundamentals in Microprocessor-Controlled Systems*. 1985 ISBN 90-277-2127-0
4. N.K. Sinha (ed.): *Microprocessor-Based Control Systems*. 1986 ISBN 90-277-2287-0
5. S.G. Tzafestas and J.K. Pal (eds.): *Real Time Microcomputer Control of Industrial Processes*. 1990 ISBN 0-7923-0779-8
6. S.G. Tzafestas (ed.): *Microprocessors in Robotic and Manufacturing Systems*. (forthcoming) ISBN 0-7923-0780-1
7. N.K. Sinha and G.P. Rao (eds.): *Identification of Continuous-Time Systems*. Methodology and Computer Implementation. 1991 ISBN 0-7923-1336-4

KLUWER ACADEMIC PUBLISHERS – DORDRECHT / BOSTON / LONDON